INTRODUCTION
TO
ERGONOMICS

INTRODUCTION TO ERGONOMICS

R. S. Bridger, Ph.D.

McGRAW-HILL, INC.

New York   St. Louis   San Francisco   Auckland   Bogotá   Caracas   Lisbon
London   Madrid   Mexico City   Milan   Montreal   New Delhi
San Juan   Singapore   Sydney   Tokyo   Toronto

This book was set in Times Roman by Ruttle, Shaw & Wetherill, Inc.
The editors were Eric M. Munson and James W. Bradley;
the production supervisor was Denise L. Puryear.
The cover was designed by Fai Yang, DFA Communications, N.Y.;
cover concept by R. S. Bridger and R. Araya-Castillo.
Printed and bound by Quebecor-Book Press

INTRODUCTION TO ERGONOMICS

This book is printed on acid-free paper.

2 3 4 5 6 7 8 9 0   QBP/QBP   9 0 9 8 7 6 5

ISBN 0-07-007741-X

Library of Congress Cataloging-in-Publication Data

Bridger, R. S.
    Introduction to ergonomics / R. S. Bridger.
       p.   cm.
    Includes bibliographical references and index.
    ISBN 0-07-007741-X
    1. Human engineering.   I. Title.
TA166.B72   1995
620.8'2—dc20   94-24688

ABOUT
THE AUTHOR

R. S. (BOB) BRIDGER is Senior Specialist Scientist at the University of Cape Town Medical School and Groote Schuur Hospital, Cape Town. He studied psychology at the University of London (Goldsmith's College) and ergonomics at University College London where he was awarded the M.Sc. degree. He has a Ph.D. in functional anthropometry from UCT.

He is responsible for all ergonomics teaching at UCT and has run courses for a wide range of students, including undergraduate engineers, psychologists, and physical therapists. At the postgraduate level, he teaches engineers, computer scientists, and medical/paramedical students.

He has published numerous articles in journals such as *Human Factors,* the *International Journal of Industrial Ergonomics,* and *Industrial Engineering.* He is interested in most areas of ergonomics and their application in developing countries.

CONTENTS

PREFACE

This book has been written to provide a general introduction to ergonomics. It may be used as precourse reading and as a first-year text for ergonomics students. It will be suitable for students of other disciplines whose program of study includes a course in ergonomics. The book will also interest professionals who wish to obtain a broad introduction to the principles of ergonomics and the design issues which ergonomics addresses. The book is not suitable for ergonomists in need of an advanced text or for those whose interest in ergonomics is limited only to research or to specialized areas of application.

An attempt has been made to describe the basic principles of ergonomics at a common level of detail across disciplines in order to avoid any suggestion that one area is more important or relevant than the others. A truly ergonomic approach to design accommodates people's requirements at all levels.

No attempt has been made to convert the novice reader into a full-fledged practitioner, an undertaking that requires a considerable period of formal training and several years of practical experience. The goal is to assist students with their first steps on this journey and to provide the basic concepts needed to study a wide range of more advanced literature. A bibliography containing suggestions for further reading has been included.

Any book written for a wide audience on a subject as vast as ergonomics will have certain limitations. If there is any general bias in the presentation, it stems from my belief that the best way to begin the study of ergonomics is by acquiring an understanding of its fundamental principles and scope—in preference to the rote learning of data and guidelines about particular topics. Many good ergonomics "cookbooks" are available, as well as standards and guidelines published by national and international bodies. Some of these publications, which can be used in conjunction with this book, are referenced in the text and in the bibliography. This book will provide readers with a deeper understanding of the thinking behind legislation and published recommendations and guidelines.

The approach chosen for the text has been to introduce ergonomics from the person outward rather than from some other perspective. The student requires fundamental knowledge about human structure, function, and behavior before being able to participate as a member of a design team. Introductory chapters on anatomy, physiology, and psychology have been included, which precede the corresponding chapters describing the practical application of the basic principles. With this approach in mind, I have left discussion of the wider problems relating to systems design and the incorporation of ergonomics in the design process to the final chapter, and the reader is referred to more advanced literature for further discussion of these matters.

In addition to basic principles, I have attempted to convey something of the breadth of ergonomics, how it is relevant to both high-technology issues in complex systems and problems of daily living in the third world. A particular concern has been to stress that ergonomics is an independent scientific subject and because of this, it does not depend on any particular area of application. It is not a subdiscipline of industrial engineering, architecture, or occupational medicine, although it interfaces with all these areas. To emphasize this point, I have tried, wherever possible, to use examples of generic rather than application-specific problems and to show how basic knowledge and research findings can be applied to the improvement of everyday working life.

A small number of essays and exercises are included at the end of each chapter. Their purpose is to enable students to process the chapter information more actively and to help bring the concepts and ideas out of the book and into the real world. One of the best ways to learn about ergonomics is through the encounter with real problems, however simple they may appear to be.

Ergonomics is a highly interconnected discipline, part of a wider family of sciences whose common goal is to solve the emerging human problems of a technologically advancing world. At the end of each chapter, I have added a special footnote in which a step beyond the usual boundaries of ergonomics is taken. I hope these short forays outside the discipline will convey to the reader some of the interconnectedness of ergonomics.

Thanks are due to the management of UCT Medical School and Groote Schuur Hospital as well as to Professor George Jaros for providing a supportive environment. Also to Dr. Wayne Capper, Dr. Mike Lambert, Professor Tim Noakes, Dr. A. Jongens, and Mr. Reg Sell for commenting on various chapters, and to all at McGraw-Hill and the following reviewers for support, encouragement, and constructive criticism: David Cochran, University of Nebraska; Steven Johnson, University of Arkansas; William Marras, The Ohio State University; William Moor, Arizona State University; and Christopher Wickens, University of Illinois. Y, a Rina Araya-Castillo, muchísimas gracias por los dibujos.

R. S. Bridger
Department of Biomedical Engineering
UCT Medical School and Groote Schuur Hospital
Cape Town, South Africa
October, 1994

INTRODUCTION TO ERGONOMICS

1

INTRODUCTION

Ergonomics is concerned with the design of systems in which people carry out work. Its name comes from the Greek words *ergon* which means "work" and *nomos* which means "law." All worksystems consist of a human component and a machine component embedded in a local environment (Figure 1-1).

It can readily be seen that even in a simple worksystem consisting of one human and one machine in an environment, six directional interactions are possible (H>M, H>E, M>H, M>E, E>H, E>M) and four of these involve the human component (Table 1-1). When designing any system where humans and machines work together to produce something, we need to know about the characteristics of the people involved and be able to apply this knowledge to the design. This activity is the fundamental function of ergonomics.

Ergonomics aims to ensure that human needs for safe and efficient working are met in the design of worksystems.

The ability of people to do their jobs is influenced by both physical design and job content. A major goal of the present text is to describe both the physical and the psychological aspects of system design which influence human performance and some of the ways a knowledge of human characteristics can be used to optimize the design of systems.

In this chapter, the ergonomic approach to the analysis of worksystems is discussed first using simple models. Next, the history of ergonomics is briefly related, and the present state of the discipline described. Then, some current issues and future trends are introduced.

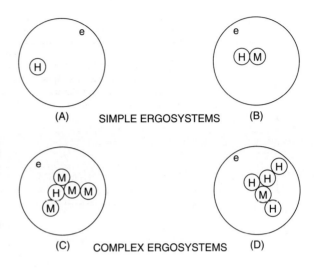

FIGURE 1-1 Examples of ergosystems. (A) and (B) are simple worksystems. (A) represents a human alone in an environment; in (B), one machine is added. (C) and (D) are complex worksystems. In (C), one human interacts with several machines; in (D), several humans use one machine. E = local environment, M = machine component, H = human component. (From Bridger and Jaros, 1986.) This book first examines the human component, then explores the interactions with the rest of the worksystem.

SIMPLE AND COMPLEX WORKSYSTEMS

To exemplify the worksystem concept, we can consider a stone-age human in an environment using a flint knife as a machine to enhance the efficiency of changing the size and shape of objects. A knife extends the capabilities of the hands. The human-machine interactions concern the usability of the knife (e.g., human strength interacting with the mass of the knife). The machine-human interactions concern the fit between the hand and the knife and the avoidance of damage to the hand by the knife. The introduction of the knife may introduce new machine-environment interactions (the knife makes it possible for the human to fell trees, which changes the environment), environment-machine interactions (the knife becomes blunt through continued use), human-environment interactions (the fertile soil is eroded as a result of the loss of trees), and finally, environment-human interactions (the eroded environment may no longer sustain human life).

Worksystems are purposeful, goal-directed systems which produce a clearly identifiable output for a previously defined purpose. By "work" is meant the application of effort in the pursuit of a goal rather than simply the application of force over a distance or an increase in metabolic activity over a basal level. The complexity of a worksystem may be increased by introducing further human and/or machine components or by enlarging the local environment. Each of the components of a worksystem may interact either directly or indirectly with the others. For example, the machine component may change the state of the environment (by emitting noise or heat, for example) and this may affect the human component.

The output of a worksystem may arise directly from the machine (for example, a product produced by the machine under human supervision or control) or directly from the human component (for example, a decision made by a manager on the basis of a computation done by a computer). The output of a worksystem may itself be human, as in hospitals whose output is patients who have received treatment. **The scope of er-**

TABLE 1-1 BASIC INTERACTIONS AND THEIR EVALUATIONS IN A WORKSYSTEM*

Interaction	Evaluation
H>M: The basic control actions performed by the human on the machine. Application of large forces, fine-tuning of controls, provision of raw materials, maintenance, etc.	*Anatomical:* Body and limb posture and movement, size of forces, cycle time and frequency of movement, muscular fatigue *Physiological:* Work rate (oxygen consumption, heart rate), fitness of workforce, physiological fatigue *Psychological:* Skill requirements, mental workload, parallel/sequential processing of information, compatibility of action modalities
H>E: Effects of the human on the local environment. Humans emit heat, noise, carbon dioxide, etc.	*Physical:* Objective measurement of working environment. Implications for compliance with standards
M>H: Feedback and display of information. Machine may exert forces on the human as a result of vibration, acceleration, etc. Machine surfaces may be excessively hot or cold and a threat to the health of the human.	*Anatomical:* Design of controls and tools *Physical:* Objective measurement of vibrations, reaction forces of powered machines, noise and surface temperatures in the workspace *Physiological:* Does sensory feedback exceed physiological thresholds? *Psychological:* Application of grouping principles to design of faceplates, panels, and graphic displays. Information load. Compatibility with user expectations
M>E: Machine may alter working environment by emitting noise, heat, noxious gases.	Mainly by industrial/site engineers and industrial hygienists
E>H: The environment, in turn, may influence the human's ability to interact with the machine or to remain part of the work system (because of smoke, noise, heat, etc.).	*Physical-physiological:* Noise, lighting, and temperature surveys of entire facility
E>M: The environment may affect the functioning of the machine. It may cause overheating or freezing of components, for example. Many machines require oxygen to operate. Oxygen is usually regarded as unlimited and freely available rather than as part of the fuel.	By industrial/site engineers, maintenance personnel, facilities management, etc.

*H = human, M = machine, E = environment, > = causal direction.

gonomics is extremely wide and is not limited to any particular industry or application.

All worksystems have a physical or functional boundary around them which separates them from adjacent systems. Worksystems communicate with each other by means of their respective inputs and outputs. The output of a particular worksystem may be the input to connected worksystems and vice versa. **Systems analysis** is the name of the discipline which studies the structure and function of worksystems and provides the means by which simple systems may be combined to form more complex systems. Systems analysis is an integral part of all advanced work in ergonomics.

This book introduces ergonomics at the level of simple worksystems. It aims to provide the reader with a model of a person working at a machine and to show how the model can be applied to generate design ideas and to evaluate and choose between design options.

Improving System Performance and Reliability

There is a historical tendency to seek technical means of improving system performance and to blame accidents and breakdowns on "human error." Bailey (1982) describes a similar tendency to equate poor system performance with poor human performance. **Detailed analyses of accidents and near accidents reveal that human error is almost never the sole cause of poor system performance.** Bailey cites a number of examples, including the Three Mile Island nuclear incident and aircraft accidents to illustrate this point. In order to understand why accidents, errors, or any unexpected system behaviors occur, one must look beyond human behavior to the rest of the system. Important factors which need to be investigated are:

1 Design of system components, particularly human-machine interfaces
2 State of the system leading up to the incident (e.g., stable/unstable, quiet/busy, on course/off course, etc.)
3 Operator's mental and physical workload
4 Work organization (e.g., shift system and time during shift, supervision, design of work groups)
5 External factors (e.g., weather)

It is frequently the case that one or more of these factors increased the likelihood of "human error." Human error and inefficiency are often inadvertently programmed into systems by design deficiencies which do not take account of the characteristics and limitations of humans. Through its emphasis on the human element, ergonomics aims to design out deficiencies in existing systems and design in reliability and good performance into new systems so that good human performance leads to good system performance and so that system design does not degrade human performance or potential.

The purpose of ergonomics is to enable a worksystem to function better by improving the interactions between the human component and the other components. Better functioning can be defined more closely, for example, as more output from fewer inputs to the system (greater productivity) or increased reliability and efficiency (a lower probability of inappropriate interactions between the system components). The precise definition usually depends on the particular context. Whatever definition is used should, however, be made at the level of the total worksystem and not just one of the components. Improved machine performance which increases the psychological or physical stress on workers or damages the local environment would not constitute improved performance of the total worksystem or better attainment of its goals. Workstation redesign to make workers more comfortable is an incorrect reason for the application of ergonomics if it is done superficially, for its own sake, and not to improve some aspect of the functioning of the total worksystem (such as reduced absenteeism and fewer accidents as a result of better working conditions).

Ergonomics in Practice

The practice of ergonomics requires that knowledge about human anatomy, physiology, and psychology be applied to the design of worksystems. Particular emphasis is placed on the design of the human-machine interface to ensure increased safety and usability of equipment and the removal of harmful stressors.

There are two ways in which ergonomics has an impact upon systems design in practice. First many ergonomists work in research organizations or universities and carry out basic research to discover the characteristics of people that need to be allowed for in design. This research often leads, directly or indirectly, to the drafting of standards, legislation, and design guidelines. Second, many ergonomists work in a consultancy capacity either privately or in an organization. They work as part of a design team and contribute their knowledge to the design of the human-machine interactions in worksystems. This work often involves the application of standards, guidelines, and knowledge to specify particular characteristics of the system.

DESCRIPTION OF HUMAN-MACHINE SYSTEMS

Ergonomics is a multidisciplinary subject and in order for it to be applied in a consistent and coherent way, a model or framework is required which specifies its areas of application, boundaries, and limitations. Such a framework cannot be derived from the study of anatomy, psychology, or physiology alone since these sciences are focused at the level of the human component rather than the human as part of a worksystem. Several attempts have been made to describe the role of the human component in worksystems with varying degrees of success (some of these are presented in later chapters). The most appropriate for the present introductory discussion is derived from the empirical model of Leamon (1980). Leamon's human-machine model is compatible with the worksystems framework presented here and has the advantage of providing a closer analysis of the interactions between the human component and the rest of the worksystem. The model is presented in Figure 1-2.

As can be seen, the model divides a simple human-machine worksystem into nine components—three from the human, three from the machine, and three in the local environment. The model was originally developed in the context of industrial ergonomics, but is used more widely in the present discussion.

The Human Components of a Worksystem

The human body is part of the physical world and obeys the same physical laws as other animate and inanimate objects. The goal of ergonomics at this level is to optimize the interaction between the body and its physical surrounds, that is, to ensure that physical space requirements are met (using data on human anthropometry) and that internal and

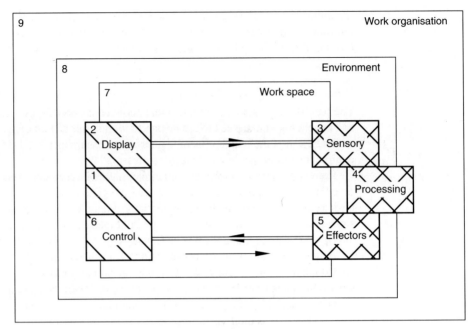

FIGURE 1-2 A human-machine model. (From Leamon, 1980.)

external forces acting on the body are not harmful. This requires an understanding of posture—how the average alignment of the body parts affects the loading of joints, bones, and muscles as, for example, when standing or sitting. Additionally, the body's response to external loading, brought about by task requirements such as lifting or environmental influences such as acceleration, needs to be considered.

The human body consists of a jointed skeleton, muscles, connective tissues, sense organs, and an information-processing center—the brain. It is important to distinguish between the effectors and the senses themselves and the physiological and psychological processes which **support and sustain** working behavior. Many ergonomic problems arise not because the operator is unable to carry out the task, but because the effort required overloads the sustaining and supportive processes and causes fatigue, injury, or errors.

The Effectors The three primary effectors are the hands, the feet, and the voice. More generally, the musculoskeletal system and body mass can be regarded as effectors because no purposeful physical activity of the limbs can be carried out without maintenance of the posture of the body and stabilization of the joints. Most skilled motor behaviors involve the generation of forces indirectly through shifts in body weight rather than directly through the action of isolated body segments. The primary effectors are the mechanisms by which information is entered into a machine or passed from one human to the other. The hands and the voice are particularly interesting—the human hand with its opposable thumb is suited to grasping objects and using tools. A knowledge of the anatomy of the hand is important for evaluating the possibilities for human manipulation of objects. The power grip, in which objects are held at the root of the fingers and palm, can be contrasted with the precision grip (used to hold a pen, for example), in which objects are held in the tips of the fingers.

Voice-recognition systems make possible the use of voice as an effector and can provide an extra channel of communication between the human and machine components. All three effectors are supported by musculoskeletal, physiological, and information-processing processes.

The Senses The senses are the means by which we are made aware of our surroundings. Human beings are often said to have five senses—sight, hearing, touch, taste, and smell. The existence and nature of the "sixth sense" remains debatable, although a sense of balance and of the body's position in space and a sense of the passage of time are some of the less controversial candidates.

From an evolutionary point of view, animals must be aware of those aspects of their environment which are important to their survival. Light is electromagnetic radiation to which the human eye is sensitive. For other animals, including certain insects, nonvisible electromagnetic radiation, including some infrared wavelengths, constitutes "visible light" inasmuch as their visual systems can respond when exposed to radiation at these wavelengths. Similarly, the sound frequencies which the human ear can detect range from a few hertz (Hz) to 20,000 Hz. Other animals (dogs are a well-known example) are sensitive to much higher frequencies.

Vision is the dominant sense in humans, followed by hearing. For this reason, the present discussion is largely restricted to these two senses. However, the tactile sense is important in the design of controls, and kinesthetic feedback is essential for all manual activities. Olfaction, the sense of smell, is less commonly thought of as an information channel in human-machine interaction. In many species, from moths to bloodhounds, an acute sense of smell is crucial. In humans, the olfactory sense is little used except in relation to food. Olfactory displays such as perfumes and aftershaves are notable exceptions. Smoke can be thought of as an olfactory warning signal for the presence of fire. Chemicals with unpleasant odors are often added to otherwise odorless but noxious substances to warn people to avoid them.

Although vision and hearing may be dealt with separately at an introductory level (as they are here), people typically utilize a combination of senses to carry out an activity. Vision is often oriented by hearing—a young baby will move its head to fixate a rattle placed at one side. Vision is often complemented by touch. When encountering a novel object, we may hear someone say, "Let me see that," while at the same time reaching for it. In this example, the word "see" really means "see and touch." Babies often place objects in their mouths in order to "see" them. Professor Eric Laithwaite once described the mouth as a tactile magnifying glass—a hole in the tooth appears much larger when felt with the tongue than with the finger.

Supportive Processes For present purposes, the term "processing" is used more widely than in the sense used originally by Leamon. In order to carry out work activities, the human component requires energy and information. Physiological processes provide energy to the working muscles and dissipate waste products. The brain can be regarded as an information-processing center which contains low-level programs to control the basic sensorimotor work activities and higher-level cognitive processes which make possible the planning, decision-making, and problem-solving activities of work. The human operator can be thought of as both a user and a source of energy. Energy, obtained from the action of metabolic processes on ingested food and oxygen, is essential to maintain life. An adequate intake of energy makes physical and mental work possible. A worker can also be seen as a source of energy—as an engine which converts the energy stored in food into purposeful physical activity by the action of the effectors. An understanding of these basic processes is essential in work design to determine workers' capacity for physical work and to investigate the factors (such as climate and individual differences) which influence work capacity.

Information is obtained via the senses, as feedback from the joints and muscles and from memory. Modern approaches model the brain as an information-processing system like a computer—an analogy which has some value in directing attention to the types of programs that underlie human information processing, the limitations of the system, and the circumstances under which it can break down. Of particular interest are the implications of this approach for information design. This approach raises the question of how best to design the information content of jobs to be compatible with the information acquisition and storage characteristics of the human information-processing system.

Although the computer analogy has some value, it is also clear that, in many ways, humans process information quite differently from computers. Apart from the conative aspect, humans excel at activities such as recognizing a face or classifying an event on the basis of incomplete data whereas computers excel at numerical computation and logical problem solving. From a structural viewpoint, computer and human information-processing systems also differ. Memory and processing (the CPU) are separate in the computer system whereas in humans, processing can almost be thought of as that part of memory which is active. An ergonomic approach would suggest that the strengths of each of these information-processing systems should be used to support the weaknesses of the other in a complementary way.

Finally, energy and information can lead to purposeful work activity only if the human is sufficiently **motivated**. Motivation is the force which directs behavior and is here regarded as a supportive process. Some would argue that motivation lies in the realm of industrial psychology rather than ergonomics. Although this belief may be true, the ergonomist cannot ignore such a fundamental determinant of human behavior—no worksystem can operate in a goal-directed and purposeful way if the human component itself is not goal-directed and purposeful.

The Machine Components of a Worksystem

The word "machine" is used here in a general sense to include any manufactured device which augments some aspect of human behavior. A subsistence farmer using simple tools to produce food in the local environment can be regarded as a human-machine system, with the tools constituting the machine component. The prototypical human-machine system of ergonomics is better illustrated by the car driver or machine operator, where the machine component and the links between the human and the machine (the displays and controls) are much more tangible. Recent developments in information technology have shifted much of the attention of ergonomics to information systems in which the worksystem is abstract and has no unique spatial location. A piece of software on a computer network is an example of such a machine. The local environment may be the network itself and many users, in different locations, may interact with the machine at the same time. Only an abstract, context-independent approach to the definition of ergonomics can handle worksystems of such physical and technological variety.

The Controlled Process   This is the basic operation of the machine on its local environment as controlled by the human. Digging a vegetable garden with a spade is an example of such a process—one of the functions that is required in order for the system to produce output. Nuclear fission is another example of a controlled process which is used to produce electricity in a nuclear power station. In information systems, the controlled process is often more difficult to categorize because of the abstract nature of the concepts used in these systems. Automatically sorting files, sending electronic mail, and searching a database for an item of information or a directory for a particular entry are examples of controlled processes in information systems.

Displays   In simple worksystems, the display is often just the action of the machine on its local environment. The process is its own display (as in chopping wood with an ax). With increasing technological sophistication, the distance between the controlled process and the human component is increased and artificial displays have to be designed. Driving a car or operating a lathe are intermediate in the sense that the display comes both directly from the controlled process (the view of the road or the action of a machine tool on a part) and indirectly from gauges, dials, etc. The complexity and hazardous nature of the controlled processes in nuclear power generation or chemical process industries necessitate the use of artificial displays. In these systems, the human operator has no direct access to the controlled process and interacts with the machine entirely by means of artificial displays. The way these displays are designed need have no one-to-one correspondence with the actual controlled processes they represent. In information systems, the controlled process may be so abstract that it cannot be faithfully represented at all using physical displays (for example, compiling of code written in a high-level language). The designer's challenge is then to construct an interface which will provide the human with an appropriate metaphor or way of thinking about the process.

Controls   Human interaction with machines depends on the provision of suitable controls which can be acted on by the effectors. With simple technology, the machine component itself is often the control. The handle of an ax is a lever which enables the cutting edge to be accelerated toward the target via the pivoting action at the hand-handle interface (an improvement on the Paleolithic handheld ax where the ax head itself is the control and the acceleration of the cutting edge depends on the pivoting action of the arm at the elbow and shoulder joints). At this level, the design requirements center on the interaction of the effectors and the control and on the mechanical advantage provided by the design of the control.

Controls are also an important source of feedback during the execution of control actions. For example, the resistance to cutting of a saw tells us something about the sharpness of the blade or the hardness of the wood; the resistance to turning of a car steering wheel provides feedback about the road surface or the tire pressures.

The Local Environment

The term "local environment" is used here in a general way. It is the place and the circumstances in which work is carried out and consists of the physical workspace, the physical environment, and the social and technical constraints under which the work is done.

Workspace   The workspace is the three-dimensional space in which work is carried out. In simple worksystems, the workspace may be just the place in which work is being carried out at any point in time as the human moves from one location to another. In more complex systems, workspaces usually become fixed, which introduces design issues such as the need to determine the workspace dimensions. This requires that the dimensions of the machine, the anthropometry of the human, and the activities required of both machine and human to carry out the work are considered. There are many other considerations in the design of the workspace, such as the choice and layout of furniture, and these are discussed in a later chapter.

The Physical Environment   Many aspects of the physical environment can affect humans. The worksystems approach points to those which have an influence on the way the human and machine components interact. Noise, vibration, lighting, and climate are of most concern to the ergonomist. Leamon (1980) states that contamination and pollution of the environment are matters best dealt with by industrial hygienists, presumably because they have direct effects on the health of the human component irrespective of any other worksystem factors. However, an awareness of these aspects is also important from an ergonomic perspective because they may have effects on human abilities and motivation as well as on health.

Work Organization   Work organization at its most basic level refers to the immediate organization of human-machine interaction—the rate of work, whether the human works at his or her own pace or whether the machine sets the pace, and whether people work alone or are dependent on others. More broadly, it refers to the organizational structure in which the work activity is embedded, the technical system, and the social system which supports it.

ERGONOMICS AND ITS AREAS OF APPLICATION IN THE WORKSYSTEM

One of the problems facing the ergonomist both in the design of new worksystems and in the evaluation of existing ones is to ensure that all aspects are considered in a systematic way. The human-machine approach enables key areas to be identified irrespective of the particular system so that ergonomics can be applied consistently in different systems.

The first step is to describe the worksystem and its boundaries. This step enables the content and scope of the application of ergonomics to be specified. Next, the human and machine components and the local environment are defined and described in terms of their main components, as depicted in Figure 1-2. Following this step, the interactions between the various components can be analyzed to identify the points of application of basic knowledge to the design-evaluation process. One example of such an interaction is that between the displays and the workspace—this interaction directs attention to the positioning of the displays in the workspace so that the operator can see them when working. Another example is the interaction between the effectors and the workspace, which introduces considerations about the space requirements for body movements required by the task. Table 1-2 summarizes the basic ergonomic interactions between the components of a worksystem.

This book has been written from the human outward to provide a general introduction to ergonomics. It contains several chapters which describe the basic processes which support behavior at the physical-musculoskeletal, physiological, and psychological levels. The remaining chapters discuss various aspects of human-machine systems in terms of basic human characteristics and their interactions with other parts of the worksystem. For example, vision and lighting are discussed together in relation to the workspace and the requirements for seeing the displays. In addition, light is part of the local environment and has psychological and biological effects on workers which should also be taken into account in any design of a worksystem. Sound and hearing are discussed together both fundamentally and in terms of the effects of noise on task performance and people's satisfaction with the environment. The design of symbolic infor-

TABLE 1-2   COMMON INTERACTIONS BETWEEN THE COMPONENTS OF A WORKSYSTEM*

Interaction	Design issue
Display-workspace	Location of displays
Display-environment	Effects of lighting, vibration, noise on legibility
Senses-workspace	Sensory access to task
Senses-environment	Environmental requirements for operation of the senses
Processing-environment	Effects on perception and cognition
Processing-organization	Skill levels, training, fatigue, motivation
Effectors-workspace	Determination of workspace envelope
Effectors-environment	Effects on vibration and climate on effectors
Controls-workspace	Task description needed to optimize control layout
Controls-environment	Effects of environment on usability of controls

*Adapted from Leamon (1980).

mation is discussed in relation to the characteristics of the human memory system and the processing of language. The basic scientist might find this approach unsatisfactory or even confusing because issues from different areas of science are treated under the same heading. From the perspective of ergonomics the approach can be defended: In real worksystems, conceptually different issues do overlap and interact, sometimes in complex ways, and investigations which are restricted to one core science and one level of analysis are almost always unsatisfactory. An analysis of noise or lighting, for example, that took only numerical measurements and ignored questions of environmental quality and people's attitudes would clearly be incomplete. Similarly, an investigation of workspace design and musculoskeletal pain that took only physical factors into account and ignored work organization and mental workload would also be lacking. In practice, the ergonomist must take a holistic view of worksystems and be able to apply knowledge from different areas of science to the analysis of problems and the specification of designs.

A BRIEF HISTORY OF ERGONOMICS

Ergonomics came about as a consequence of the design and operational problems presented by new worksystems which had evolved with the advance of technology. It owes its development to the same historical processes which gave rise to other worksystem disciplines such as industrial engineering and occupational medicine.

In worksystems based on simple technology, technological advances make possible direct improvements in the functioning of the system. The replacement of stone tools with copper and then iron implements is a well-known example from history. Another well-known example is the development of the steam engine which gave rise to the industrial revolution. The steam engine enabled great improvements to be made in production because of the vastly increased power which it made available. It should be noted that both railways and underground mines existed before the development of the steam engine. The carriages pulled along the rails and the pumps used to pump water out of the mines were powered, however, by horses or by humans rather than machines. With the replacement of horse power with steam power, the scale at which production could take place was increased (note that the material advances of the industrial revolution resulted from the new source of power, not from increased efficiency—the steam engine is an extremely inefficient way of converting chemical energy into mechanical energy). Singleton (1972) has called this period in history the **era of pure engineering** since improvements in the performance and reliability of the machine were all that was needed to bring about improvements in the performance of the worksystem.

Scientific Management and Work Study

As industrial systems became more complex and the scale of industrial society grew, it became clear that improvements in the reliability of technology, beyond a basic level, did not guarantee corresponding improvements in system performance. Scientific management, developed by F. W. Taylor at the turn of the century, was one of the first attempts at maximizing productivity by improving the design of tasks rather than the design of machines. It was based on the idea of "rational economic man"—the view that given the incentive and the opportunity to maximize output, an individual worker would do so—and the assumption that there was one best way of doing the job. It followed from these notions that productivity would be maximized if rational economic men were given properly designed jobs. Emphasis was placed on the design of rest periods and bonus rates as a means of motivating workers. Taylor's contribution was remarkable in many ways—in particular because it regarded manual work as something worthy of analysis and study rather than disdain.

A complementary approach was developed by the Gilbreths at around the same time. They analyzed the methods used to carry out a task. A task would be broken down into "elements"—the basic movements and procedures required to perform a task. Redesigning and reconfiguring these elements, it was hoped, would eliminate unproductive movements and procedures and enhance productivity.

These developments were the forerunners of time and motion study and human engineering. The historical significance of this work is that for the first time, companies employed specialists to investigate human-machine interaction and to design tasks. Tasks were no longer seen as fixed by the design of the machine and workplace or by tradition but instead were seen as something to be optimized. This view facilitated the use of assembly- and production-line techniques in which a complex or skilled task was broken down into a series of easily learned subtasks, each performed by a different person.

Time and motion study (methods engineering) can be criticized on many grounds—for example, that it looks only at the superficial features of task performance, that it makes unwarranted assumptions about people, and that it is little more than common sense. Moores (1972) attributed the success of the methods engineer to being the first such specialist to arrive on the industrial scene. However, methods engineering rapidly penetrated a wide range of industries and many of its basic techniques are still used today.

As technology developed, more and more aspects of system behavior were quantified in an attempt to increase productivity and efficiency, and more scientific principles were sought to interpret the data—from basic mathematics and physics to biological, psychological, and physiological principles. New fields of investigation and new professions arose to meet the demands of managing and developing industry.

Occupational Medicine

Occupational medicine had its origins in the eighteenth century when Ramazzini (1717) wrote his "Treatise on the Diseases of Tradesmen" but became more formalized at the beginning of the twentieth century. Around the period 1914–1918, a number of government institutions were founded in Britain as interest in working conditions spread to scientists and medical doctors. The Health and Munitions Workers Committee studied conditions in munitions factories and factors influencing productivity, such as the length of the working day. It subsequently became the Industrial Health Research Board and its area of interest was fairly wide, including the effects of ventilation, heat, and shiftwork on health and the effect of training on performance. Recommendations were made at about this time for a variety of aspects of industrial work, including the types of food served in factory cafeterias, taking into account the likely nutritional deficiencies of the workforce and the demands of work. This work was appropriate in Britain in the early years of the twentieth century and is still appropriate in developing countries today. Attention was also directed at fatigue and the problems of repetitive work. Vernon (1924) investigated postural and workspace factors related to fatigue and concluded:

Any form of physical activity will lead to fatigue if it is unvarying and constant.

It is from foundations such as these that industrial physiology and occupational health have arisen (it is surprising, given the current high incidence of cumulative trauma disorders, how few industries have acted upon Vernon's valid conclusion).

Occupational Psychology

Occupational psychology, as distinct from personnel management and psychometrics, developed in the 1920s and 1930s. The approach of Taylorism had been to regard the worker as an isolated individual whose output was affected by physical factors such as fatigue, poor job design, etc. The social context in which work took place was ignored. The human relations school drew attention to the importance of group norms and the social rewards of worker interaction, and shifted the emphasis away from the individual worker to the work group. This emphasis on the importance of the social structure to support the organization of technology has persisted. After World War II it reemerged in the form of sociotechnical systems theory. Trist and Bamforth (1951), for example, investigated the social and psychological consequences of mechanized coal mining in the context of a reportedly higher incidence of psychosomatic disorders among miners working under mechanized conditions. They pointed out that the mechanized method owed its origins and scale to the factory floor rather than to coal mines. In coal mines, a different form of social organization was necessary because of the intrinsically unpredictable nature of the working environment in a mine compared with a factory. The organization of technology, the social organization, and the local environment had to be seen as interconnected and designed to be mutually compatible if low productivity and pathological psychological stresses were to be avoided. Sociotechnical systems theory is discussed in the final chapter of this book; its emphasis on the importance of the social organization of work for effective functioning of a worksystem is of clear relevance to ergonomics. That the theory has survived to the 1990s (e.g., Hitchcock, 1992) attests to the value of its fundamental tenets.

Human Performance Psychology

Human performance psychology had its roots in the practical problem of how to reduce the time taken to train a worker to carry out a task. Taylor and the Gilbreths had gone some way to the solution of this problem by simplifying tasks. This approach caused the content of previously individual skills to be embodied at the level of the technical system directly under management rather than worker control. The effect was to facilitate the coordination of workers on a large scale and the use of machine power to boost productive capacity. However, the emergence of new, more complex machines and manufacturing processes created new jobs whose skill requirements could not be met from the existing labor pool. Traditional methods of training (such as apprenticeship to and sitting at the elbow of a skilled craftsperson) were not always appropriate to the industrial system with its fundamental requirement to standardize all aspects of the production process. Psychologists began investigating the variables influencing the learning of work skills—for example, whether tasks should be broken down into elements to be learned separately and then brought together, the advantages and disadvantages of "massed" versus "spaced" practice, and the duration of rest periods. After World War II, interest in training continued. The cybernetic approach investigated the use of feedback (knowledge of results) and its effects on learning. The theoretical ideas of B. F. Skinner and the behaviorist school of psychology were implemented in the form of "programmed learning." Behaviorism saw learning as the chaining together of stimulus-response pairs under the control of reinforcing, or rewarding, stimuli from the environment. In programmed learning, the material to be learned was presented in a stepwise fashion and the order of presentation of information to be learned was determined by whether trainees' previous responses were correct or incorrect.

The pressure for the productive and efficient use of machines was amplified by the demands of the Second World War and brought psychologists into direct contact with the problems of human-machine interaction. The famous Cambridge psychologist Sir Frederick Bartlett built a simulator of the Spitfire aircraft and investigated the effects of stress and fatigue on pilot behavior. This investigation led to an increased understanding of individual differences in response to stress and enabled the breakdown of skilled performance to be described in psychological, rather than machine-based, terms. Perceptual narrowing, which occurs as a result of fatigue or as a maladaptive response to severe stress, is an example. Craik (1943) studied a class of tasks known as "tracking tasks" (which involve following a target, as in gunnery or steering a vehicle). The beginnings of user-interface (display and control) design emerged from this study in the form of recommendations for gear ratios and lever sizes. These investigations developed into an important area for research and development in ergonomics, particularly in post-war military applications but also in civilian vehicle design and in the aerospace industry. Craik also developed some of the key ideas which were to emerge many years later in the field of cognitive ergonomics—particularly the notion that users of equipment carry some **internal representation** of the device in their heads, which enables them to predict its behavior in response to their control actions.

Operations Research

Operations research attempts to build mathematical models of industrial processes. It was also stimulated to grow by the demands for prediction and control brought about to satisfy the requirements of the military during the Second World War. It had become clear that further advances in system performance would depend on how well technology was *used*, not just how well it was *designed*. This shift of attention from the machine to the man-machine system gave birth to the subject of ergonomics.

FMJ vs. FJM

A number of general trends can be identified in the historical review. First, organizations attempted to increase their productivity by introducing new methods and machines. In the era of pure engineering this approach worked because great improvements in machine design were possible (many existing processes had not been mechanized at all until then). Second, attempts to increase productivity tried to optimize the design of tasks to minimize apparently unproductive effort. After the First World War a movement arose, stimulated by the development of psychometric testing, which tried to develop tools to objectively measure various human characteristics such as intelligence or personality.

Attempts to **"fit the man to the job" (FMJ)** were based on the idea that productivity or efficiency could be improved by selecting workers with the right aptitudes for a particular job. This approach, which forms one of the roots of modern occupational psychology, is based on the assumption that important aptitudes for any particular job really do exist and that they can be identified and objectively measured. This assumption is certainly true in the sense of selecting people with formal qualifications or skills to fill particular posts. It is also true for some jobs—even today physically strenuous jobs such as a fire fighter or lifesaver or jobs such as a military aircraft pilot are restricted to individuals with certain specific aptitudes and physical characteristics. However, the assumption is not always true and is still a source of controversy. Trade unions, for example, may object to management attempts to select workers with "strong backs" for a particular job, arguing that the problem lies with the job, not the worker and that the job should be redesigned to be performable by anyone in the workforce.

An alternative approach, which is the guiding philosophy of ergonomics, is known as **"fitting the job to the man" (FJM)**. Much of the early human engineering and workspace design attempted to design job tasks to suit the characteristics of the worker. The underlying assumptions of the FJM approach are that a suitable set of worker characteristics can be specified around which the job can be designed, and that this can be done for any job. **A large part of this book is devoted to describing these characteristics at the anatomical, physiological, and psychological levels and explaining their design implications.**

It has become clear that **FJM is almost always the superior approach to the design of worksystems.** FJM can be carried out at all levels, as is described throughout this book. For example, at the physical level, the National Institute of Occupational Safety and Health (NIOSH) has developed a method for specifying the loads to be handled in lifting tasks which will minimize the risk of injury to healthy workers. Hazardous loads can be identified and redesigned. Acceptable and maximum temperatures for carrying out physical work have been specified by OSHA and enable the climate to be designed appropriately for workers. At the level of human performance psychology, it is known that people have expectations about the relationships between the movements of controls and the corresponding movements of controlled devices. Although operators can learn counterintuitive relationships, this practice is almost always unsatisfactory and increases the risk of error. Ergonomics, on the other hand, stresses that control-display relationships be designed to be compatible with users' expectations. At the level of language, it is known that people can be trained to work with badly worded instuctions, manuals, or software interfaces but become frustrated under normal operation and are more likely to make comprehension errors when under stress. Rules, derived from research in psycholinguistics, exist to design language which is easy to comprehend. Finally, if the designer's conception of a system is different from users' conceptions, then users will have difficulties in learning to use and in using the system. Ergonomics stresses the need for cognitive compatibility in system design.

It can be seen from this that ergonomics "fits the job to the man" at the biomechanical, physiological, behavioral, linguistic, and cognitive levels.

Only under extreme circumstances is the FMJ approach taken, as when acclimatizing workers who have to work in hot conditions which cannot be changed. Even in such a situation there are many FJM options, such as designing a better work-rest schedule, providing protective clothing, or building a "cool spot."

Human Factors and Ergonomics

In Britain, the field of ergonomics was inaugurated after the Second World War (the name was invented by Murrell in 1949 despite objections that people would confuse it with economics). The emphasis was on equipment and workspace design and the relevant subjects were held to be anatomy, physiology, industrial medicine, design, architecture, and illumination engineering. In Europe, ergonomics was even more strongly grounded in biological sciences. In the United States, a similar discipline emerged (known as "human factors"), but its scientific roots were grounded in psychology (applied experimental psychology, engineering psychology, and human engineering).

Human factors and ergonomics have always had much in common but their development has moved along somewhat different lines. Human factors puts much emphasis on the integration of the human considerations into the total system design process. It has achieved remarkable success in the design of large systems in the aerospace industry, notably in the National Aeronautics and Space Administration (NASA) and the U.S. space program. European ergonomics is sometimes more piecemeal and traditionally has been more tied to its basic sciences or to a particular topic or application area.

Despite these differences, the reader should not be concerned about the two terms. In the United States, the Human Factors Society has recently changed its name to the Human Factors and Ergonomics Society. Presumably this change was made to indicate that the two areas now have so much in common that one society can represent the interests of people who see themselves as working in only one or the other areas. In the author's opinion, few objections would be raised today if the title of the present book were *Introduction to Human Factors*.

Both human factors and ergonomics take the FJM approach and state that jobs should be made appropriate for people rather than the other way around. Since this is an introductory book, much attention has been devoted to describing those basic human characteristics which have been investigated by ergonomics in the context of job design. This is not to say that other human characteristics (e.g., personality, intelligence) are not important, only that they have not been investigated in the same degree as other, more classically "ergonomic" characteristics. As is discussed in the final chapter, the boundaries of ergonomics are not clear and there is still debate about where ergonomics finishes and where other worksystem disciplines begin.

Modern Worksystems

Modern worksystems operate under much tighter constraints than their predecessors. Labor is often the major running cost in any organization—greater than raw materials, energy, and equipment. According to Stamper of The Boeing Company (Stamper, 1987), his company paid three times more for employee health care in the 1980s than it did for aluminum to build its airplanes. In a similar vein, Chrysler Corporation found that an insurance company was its single biggest supplier—not a steelmaker, as might otherwise be expected.

The technical subsystem of modern organizations is often designed to operate in a finely tuned manner, which introduces great demands for the coordination of large numbers of people and machines. Modern worksystems employ large numbers of specialists in fields such as industrial engineering, work study, personnel management and selection, operations research, ergonomics (which sometimes falls under one of the other disciplines in an organization), and facilities management. All the specialists are concerned, in one way or another, with issues about people and technology, and the emphasis is on optimizing the functioning of the total system rather than on one or other group of people or machines.

MODERN ERGONOMICS

Modern ergonomics contributes to the design and evaluation of worksystems and products. Unlike in earlier times when an engineer designed a whole machine or product, design is a team effort nowadays. The ergonomist usually has an important role to play both at the conceptual phase and in detailed design as well as in prototyping and the evaluation of existing products and facilities. Modern ergonomics contributes in a number of ways to the design of the worksystem (Table 1-3). These activities should be seen as an integral part of the design and management of systems rather than as "optional extras."

Standard Format for Describing Human-Machine Systems

The human-machine model and the worksystems framework provide a standardized, albeit empirical, way of describing worksystems, irrespective of the application area and independently of any particular core science of ergonomics. They can be used to generate checklists and methodologies for evaluating prototypes or existing systems. It is often the case that some of the design issues in a particular system will stand out more than others and perhaps receive undue attention from the design team. A standard format for describing worksystems forces the ergonomist to consider all the issues, at least in the early stages.

The two most important first steps when using the human-machine model are to de-

TABLE 1-3   CONTRIBUTION OF MODERN ERGONOMICS IN SYSTEMS DESIGN AND MANAGEMENT

1. A standard format for describing human-machine systems
2. Identification, classification, and resolution of design issues involving the human component
3. Task and human-machine interaction analysis
4. Specification of system design and human behavior. Implementation of controls
5. Identification of core trends in human and biological science and their implications for system design and management
6. Generation of new concepts for the design and analysis of human-machine systems
7. Evaluation of the sociotechnical implications of design options

scribe the technology and to describe the user or operator. Machines are normally well described, and plenty of information in the form of manuals and textbooks is available. Designers of human-machine systems normally operate with much less detailed and less formal information about people and make assumptions about people in a way that they would never do about machines.

A major task of the ergonomist is to describe the human at all levels appropriate to the particular system.

This task can entail a physical description in terms of user dimensions and abilities, including physiological factors such as age and fitness and their implications for human function. It also entails a psychological description, which includes details of operator/user skills, knowledge, experience, and motivation, and may include detailed considerations such as preferred ways of working, jargon, etc.

Identification and Classification of Design Issues

A major role of ergonomics is to identify design issues which involve the human component of the worksystem and to classify them in order to render them amenable to further analysis using appropriate knowledge.

All worksystems are specific, yet many published design guidelines are general. For this reason, there is normally no body of literature appropriate to all the needs of a particular application or industry, and a standard method is required to enable system-specific design issues to be identified and interpreted in general terms amenable to analysis using available data. For example, many of the issues in office design are the same whether it is an office in a bank or an insurance company or is the administrative center of a heavy industry, but often the people working on the project will know only about the problems of their own particular industry.

Dray (1988) provides an interesting example of the classification problem in applied ergonomics. She describes how complaints of headaches by visual display terminal (VDT)-workers in a modern office cannot always be dealt with using conventional ergonomics approaches to redesign the workspaces. Rather than being a workspace design problem, the complaints were caused by problems in work organization—particularly, uncertainty caused by the introduction of new systems into the organization. This case illustrates that although physical changes in a worksystem may be accompanied by physical complaints, they may not be the cause of these complaints. Changes in one aspect of the work environment are often accompanied by other changes (in organizational structure of management, for example) which are easily overlooked. Proper classification of design issues and their human implications is a crucial step in the design or redesign process.

There is often a tendency to attribute accidents, breakdowns, and low productivity to the human component of worksystems. **Human behavior at work takes place in the context of a system and is shaped by the way the system is designed.** System malfunctions which involve humans must therefore be analyzed in context, which requires that the focus of the analysis be shifted from the human to the human-machine system. For example, the design of the machine may invite the human to make a particular type of error or the machine itself may not be immune to the types of errors which humans are known to make. The pattern of work organization may cause undue fatigue and increase the likelihood of inappropriate actions by the human. It is the task of the ergonomist to investigate problems of reliability from a systems perspective to determine the relative contributions of inappropriate human behavior and inappropriate system design. Further analysis can then be undertaken to determine which aspects of the design of the worksystem degrade performance and how the system can be redesigned to solve the problem

Task and Human-Machine Interaction Analysis

Tasks can be analyzed by breaking them down into their various components and subcomponents in a structured way to reveal the behavior required of the human and the context in which the behavior takes place in the worksystem. **Task analysis** provides a system-specific context for the application of the fundamental ergonomics principles.

Task analysis is undertaken either informally or formally in the design and evaluation of human-machine interaction. When it is undertaken informally, it depends on the designer's or engineer's intuitive judgments about how operators interact with machines and the appropriate ways of designing this interaction.

Ergonomists have developed more formal ways of analyzing tasks. These include hierarchical representation of task behaviors, observational techniques for obtaining data about the behavior involved in carrying out a task, and methods for representing the dynamic aspects of human-machine interaction. The outcome of a task analysis consists of the following:

1 A description of the behaviors required to carry out the task
2 A description of the system states which occur when the task is carried out
3 A mapping of the task behaviors onto the system states

This information can be used for a variety of purposes:

1 Evaluation or the design of the human-machine interface
2 Identification of the skills needed by an operator of the system
3 Design of training materials and operating instructions

4 Identification of critical elements of the task to predict or evaluate the reliability of the system

Table 1-4 gives an example of the stages involved in the hierarchical decomposition of a job from a job title to specific operations.

This type of analysis can produce enormous quantities of documentation, as can be imagined. However, in complex systems, it is often essential to describe tasks at this level of detail because even small changes in equipment or procedures can affect the mapping between required human behaviors and system states. Table 1-5 gives examples of the types of questions the systems designer must ask when looking at the mapping between individual task behaviors and system states. A basic question is, How are operator task behaviors to be invoked and how will the system respond to their execution?

Task analysis yields data about system operation and human behavior, which provides a context for the application of ergonomics principles and guidelines. There is sometimes a tendency for ergonomists to direct their attention to problems and issues

TABLE 1-4   LEVELS OF DESCRIPTION IN TASK ANALYSIS: AN EXAMPLE*

Job Title: Fuel Pump Station Operator

Assignments

A1	Inspect environment, equipment, and machinery.
A2	Execute start-up procedures for pumping.
A3	Monitor system condition when running.
A4	Execute close-down procedures.
A5	General maintenance.

Segments

(A1)	S1	Conduct general inspection: safety, lighting, housekeeping.
(A2)	S2	Start up filtration of fuel in reservoir tank.
(A3)	S3	Periodically inspect pumps.
(A4)	S4	Close down pump motors.
(A5)	S5	Change the filter elements.

Tasks

(A1S1)	T1	Replace broken lamps in station.
(A2S2)	T2	Start up transfer pump.
(A3S3)	T3	Check torque settings on pump retaining clamps.
(A4S4)	T4	Close down pump motor 1.
(A5S5)	T5	Remove filter element 1.

Operations

(A1S1T1)	O1	Switch off power to lighting unit at wall switch.
(A2S2T2)	O2	Press green button on transfer pump housing.
(A3S3T3)	O3	Using a torque wrench, tighten nuts in clockwise direction until torque setting is displayed.
(A4S4T4)	O4	Press red button on pump motor 1.
(A5S5T5)	O5	With an 8mm spanner (wrench), loosen nuts retaining filter element housing.

*There are many more segments, tasks, and operations. Those that are shown are included to illustrate the detail required at the different hierarchical levels.

TABLE 1-5   MAPPING OF TASK BEHAVIOR AND SYSTEM OPERATION IN SYSTEM DESIGN AND EVALUATION: AN EXAMPLE

Task: To transfer fuel from mobile tanker to on-board storage unit

Mapping Elements

1. Indications and when to do the task
   When requested by captain
   When indicated by fuel gauge
   With engines switched off

2. Control objects and operation
   Transfer valves, transfer hose, hose retaining clamps, transfer pump controls.
   Slide hose into transfer valve orifices on tanker and storage unit (outlet and inlet).
   Use retaining clamps to secure hose to transfer couplings on tanker and storage unit.
   Open transfer valves on tanker and storage unit.
   Press green button on transfer pump.

3. Precautions
   Valves in open position and hose secured
   Fuel level in mobile tanker above red line
   Tanker on level surface and brakes in "on" position

4. Feedback modality and indication of response adequacy
   Visual/kinesthetic
     Valves click into open position
   Auditory/kinesthetic
     Pump vibrates/whines if flow impeded
   Visual
     Transfer pump pressure gauge reads 150–250 kPa
     Absence of leaks
     Fuel level in storage unit gauge rises

5. Fault diagnosis and maintenance
   If pump "whines" or flowmeter indicates blockage, press red button to stop pump.
   Check
     Transfer valves open
     Hose not twisted or crushed
     Fuel level in tanker
     Pressure gauge on transfer pump

which they themselves regard as important or which they know a great deal about. Leamon (1980) described this tendency as the widely held view that physiologists practicing ergonomics propose physiological solutions to problems and psychologists practicing ergonomics propose psychological solutions to the same problems. A task analysis and task description is one of the cures for this idiosyncratic form of tunnel vision—it assists the design team in prioritizing issues according to their importance to **system operation** regardless of the personal preferences and skill profiles of the team's members.

Specification of System Design and Human Behavior

After many years, research in ergonomics and its related disciplines has yielded standards and guidelines for tighter control of working conditions. The most common areas

TABLE 1-6   SOME SOURCES OF PUBLISHED STANDARDS AND GUIDELINES

	United States	Europe
Lighting	IES Handbook, Illumination Engineering Society (IES) of North America	British Standard CP3, British Standards Institution (BSI)
		CIBS Code for Interior Lighting, London, UK
		Commission Internationale de l'Eclairage (CIE) publications
Noise	Industrial Noise Manual, American Industrial Hygiene Association	International Organization for Standardization (ISO) publications, e.g., ISO 1999 (1972)
	Occupational Safety and Health Administration (OSHA) exposure limits	BSI publications
	American National Standards Institute (ANSI) publications	
Climate	American Society of Heating, Refrigerating, and Air Conditioning Engineers (ASHRAE) publications	ISO publications, e.g., ISO 7243 (1982)
	OSHA publications	BSI publications (reactions to surface temperatures)
Manual handling	NIOSH (National Institute of Occupational Safety and Health) Work Practices Guide (1981, weight limit equation revised 1993).	University of Surrey, U.K. Force Limits in Manual Work
	DHHS (Department of Health and Human Services) (NIOSH) Publication no. 94-110	European Community (EC) Directive on Manual Handling (1991) HSC (Health and Safety Commission, U.K.) Proposals for Regulation and Guidance, 1992
Seating/VDTs	ANSI/HFS 100-1988 National standard for Human Factors Engineering of VDT Workstations	EC Directive on Work with Display Screen Equipment, 1990.
		HSC Proposals for Regulation and Guidance, 1992
Anthropometry	NASA Antropometric Source Book	Pheasant, 1986.
CTDs	ANSI Z365	

are lighting design, noise control, the indoor climate, the design of manual handling tasks, and seating. Table 1-6 summarizes readily available sources of this information in the United States and Europe.

Although it is necessary for the modern ergonomist to know about these controls and to be able to apply them to the design of systems, this is not sufficient. The professional ergonomist must fully understand the rationale for a particular control, the scientific principles on which it is based, the consequences of misapplication, and any limitations in the knowledge used to specify the control itself. A major emphasis of the present text is to enable the novice to develop a deeper understanding of this type of information.

Standards change over time and vary between countries, so there is little use in the ergonomist's learning them by rote. More importantly, the ergonomist must know *when* to look for this type of information, *how* to find it, and *how* to apply it.

The ergonomist must also be able to specify **appropriate human behaviors and actions** in the operation of a system. Much of the existing literature allows such specification to be done on a **default** basis. Ergonomists are good at deciding how tasks shouldn't be designed—either because of danger or health hazards or high probability of human error which will reduce the reliability of the system. The more positive approach requires that implementable human behaviors be specified at the design stage, which will result in the system's functioning at a specified level. Empirical techniques such as task analysis and description enable human behavior in the operation of systems to be studied systematically. However, there is a lack of formal procedures for specifying behavior, although attempts are being made to integrate human factors and systems engineering in a formal way (Dowell and Long, 1989).

Identification and Analysis of Core Trends

An important role for ergonomics generally and for the ergonomist working in a large organization is to act as an interface between developments in basic human and biological sciences and organizational needs. The ergonomist is frequently one of the few members, if not the only member, of the design team with formal training in these areas capable of interpreting the latest legislation, findings, and reports. The current confusion and superficiality of the debate surrounding repetition strain injuries exemplifies the need for a more critical application of fundamental principles. Until recently, no theoretical explanation or conceptual model of these injuries had been developed, and even authoritative sources such as Pheasant (1992) were able to debate the very existence of the problem.

Ergonomics has generated many design guidelines and recommendations, which are frequently made available to engineers and designers. This information is usually of a general nature and cannot always be used in a straightforward "cookbook" fashion. It is the function of the ergonomist to use background knowledge of human sciences to interpret general guidelines so that they are appropriate to the particular system. For example, design manuals often give the dimensions of workstations irrespective of the particular industry. Because the ergonomist knows the rationale for a given guideline or specification, he or she can decide whether it needs to be modified in accordance with local requirements.

Generation and Implementation of New Concepts

The design of the human-machine interface is the classical point of departure for the application of ergonomics. A very large body of literature is available to assist designers with the design of the user interface. The ergonomist has an important role to play at the concept generation stage in trying to anticipate what the demands of the new system will be and how they will affect operators. Organizations sometimes exhibit a certain inertia, which manifests itself as a desire to "do things in the way we've always done them." The ergonomist must analyze the reasons for current or proposed designs and suggest improvements and alternative concepts. Technological advances now offer many new concepts for the design of interfaces which are radical departures from traditional methods. Control of machines by voice and the use of synthetic speech as a display are examples. Computer-generated two- and three-dimensional graphic displays offer designers much more flexibility and capacity for information display than their electromechanical counterparts. Such displays have application in the aerospace and process industries and in the design of everyday products such as the automobile dashboard.

It is necessary to develop a sensitivity to the cost-benefit implications and practicalities of new design ideas. This is particularly true when an existing system is being improved. Several schemes exist so that recommendations for ergonomic improvements can be prioritized. The category chosen for a recommendation depends on the need for change, as perceived by the ergonomist, and the implementation costs in terms of money, time, expertise, impact on day-to-day running, etc. One way of prioritizing recommendations (used, for example, by the University of Vermont Rehabilitation Engineering Center) is as follows:

1 Implement recommendation immediately (e.g., there is a serious design flaw threatening employee health or system reliability, or there is contravening legislation).

2 Implement recommendation soon (e.g., the current way of working is unsatisfactory, but there is no immediate danger).

3 Implement when equipment is shut down (e.g., if stoppages are expensive and there is no immediate danger, wait until the system is shut down for regular maintenance or repair and then implement the idea).

4 Implement when cost-benefit ratio is acceptable (e.g., wait until the financial situation improves or implementation costs are lower).

5 Implement when equipment is built or purchased (e.g., phase in new products or items on a replacement basis as old ones are discarded).

Evaluation of Sociotechnical Implications

The design of new systems and the redesign of existing ones can have serious implications for the organizational climate. Technological and organizational changes can have profound effects on the working lives of individuals, and it is part of the ergonomist's function to determine what these effects might be and to anticipate future problems. There is a growing body of literature on the sociotechnical aspects of technological change and the **importance of involving employees as part of a participatory approach to system design** (e.g., Sell, 1986).

FUTURE DIRECTIONS FOR ERGONOMICS

Ergonomics as a discipline is developing in a number of directions. Technological developments, particularly in information and communications technology, have increased the amount of attention paid to problems of human information processing and its integration with machine information processing. The corpus of recommendations and guidelines for interfacing these two information-processing systems is growing. The time that it takes to develop new systems is becoming increasingly shorter. The same cannot be said for the time taken to implement the ergonomic component of system development. Collaboration with computer scientists and systems designers is taking place to better integrate ergonomics with the rest of the system development process.

Some general developments in physical ergonomics can also be identified. Ergonomists have often claimed that their discipline can protect the health of the workforce. As a general, rather than a specific, claim, this is insupportable and the challenge now is to prove it. This challenge has stimulated epidemiological research into the effects of good and bad design on health. Ergonomists are now collaborating more with occupational health practitioners and epidemiologists, and medical schools at some universities are becoming interested in the subject.

Another important development concerns the drafting of standards, guidelines, and controls. Given that unsatisfactory conditions can now be identified, the challenge to ergonomics is to draft implementable, objectively assessable standards which can be taken up by industry or included in legislation for workplace design. The NIOSH approach to the design of lifting tasks described in a later chapter is a good example of such a standard.

A recent trend in ergonomics has been the upsurge in interest in the problems of industrially developing countries and of transferring "first world" technology to them. There is a small but growing body of literature on this subject and groups such as that of Professor H. Shahnavaz at The University of Lulea, Sweden, are actively involved. Many of the techniques developed by ergonomists in industrialized countries can be directly applied to the study of problems in developing countries. Whether standards, guidelines, and design concepts can be transferred in so straightforward a manner is more debatable.

SUMMARY

Ergonomics is a multidisciplinary subject which can be applied to the study and design of the human component of worksystems. Its areas of application are not limited by the particular technology or by the scale of the system. Ergonomics provides a standardized approach to the analysis of worksystems with the emphasis on the interactions between humans and machines. It has some areas of overlap with work study but is more scientific (and sometimes less quantitative) in approach. Its related disciplines are industrial medicine, industrial engineering, industrial psychology, and systems design and analysis.

Historically, ergonomics can be seen to have arisen as a response to the need for rapid design of complex systems. As technology becomes more complex and worksystems operate under increasingly severe constraints, good ergonomic design becomes increasingly important. Thus, "technology push" can be identified as one of the main factors influencing the direction and growth of the subject. The modern ergonomist has an important role to play as a source of scientific information about humans (a scarce commodity in many organizations), as a generator of knowledge about the human component of a worksystem, and as a member of a design team.

ESSAYS AND EXERCISES

1 The Human Factors Society recently changed its name to The Human Factors and Ergonomics Society. Discuss possible reasons for this change. Illustrate your answer with material from the present chapter and from other sources of your choice.
2 Write a task description which describes, in detail, how to make a cheese sandwich. (Hint: Begin with the following segments—preparing the work area, preparing the ingredients, making the sandwich, storing the ingredients, cleaning the equipment.)
3 Describe the present scope and concerns of ergonomics. Obtain back issues for the last 5 years of two or more of the following journals: *Human Factors, Ergonomics, International Journal of Industrial Ergonomics*, and *Applied Ergonomics*. Summarize, using key words of your choice, the topics and research questions being dealt with in each of the papers in these journals.

FOOTNOTE: Short Notes on the History of Human-Machine Systems

The widsom of the Greeks bears only flowers, not fruit.

Abulshason Yehuda Halevi, a poet of medieval Spain*

It will scarcely be believed that an invention so eminently scientific, and which could never have been derived but from the sterling treasury of science, should have been claimed on behalf of an engine-wright of Killingworth of the name of Stephenson—a person not even possessing a knowledge of the rudiments of chemistry.

From the biography of Sir Humphrey Davy (Source: Rolt, 1963)

One of the characteristics of Homo sapiens is the ability to make and use tools; even in the simplest human worksystem, it can be taken as axiomatic that a machine of some sort will be used.

In hunter-gatherer societies, the level of technology is low but the social bonds between people and the knowledge of the environment are well developed. This is appropriate since the survival of these people depends on cooperation and the ability to move freely around large areas of unoccupied land searching for food and water. Possessions are therefore lightweight and made out of readily available materials. In technologically more advanced societies, the emphasis may shift toward machines and more complex material possessions but is often accompanied by a corresponding reduction in the complexity of social interaction and knowledge of the environment.

De Greene (1973) has pointed out that there is a historical tendency to overempha-

* I would like to thank Mr. A. Tordesillas for supplying this quotation.

size the effects of scientific and technological advances on human progress and to underemphasize social and economic factors. Many theoretical and technical innovations which eventually led to great advances were originally developed purely out of intellectual curiosity or were stumbled upon or came about as a spin-off from some other enterprise. C. S. Smith (1981) has reached similar conclusions and emphasized the importance of religion and aesthetics as prime movers of technological advance. A brief historical review is appropriate to illustrate some relationships between scientific advance and human progress in the light of these notions.

The geometry necessary for understanding ellipses and parabolas was worked out by the Greek mathematician Apollonius in the third century B.C. purely as a theoretical exercise, as was most Greek mathematics and philosophy. One thousand years later, in medieval times, the opinion about most Greek mathematics and philosophy was that it was of little use, and for most of history, the social, economic, and intellectual climate did not stimulate thinkers to find uses for it. In fact, Apollonius's work bore fruit only in the sixteenth and seventeenth centuries when it turned out to be essential to understand the behavior of the planets and of projectiles such as artillery. Boole was a nineteenth-century country clergyman whose invention of Boolean algebra turned out to be important in twentieth-century computer science. Finally, we can consider the advances in microelectronics and miniaturization which were essential for the success of the U.S. space program but also had a profound effect on the design of many consumer products as unexpected spin-offs from the main enterprise.

The tendency to overemphasize the role of science is well illustrated by the controversy over the invention of the miners' safety lamp in the nineteenth century. The problem of explosions in mines caused by the ignition of gases by naked lights was solved independently by the eminent British chemist Sir Humphrey Davy and by George Stephenson, the pioneer of the railways in Britain. Stephenson received little education and achieved only a rudimentary level of literacy at the age of 17. Not surprisingly, Davy was initially credited with the invention of the lamp (which was named after him). Stephenson's sponsors soon laid their claim (the "Geordie" lamp, developed by Stephenson, had been in use for some time) and Stephenson eventually received recognition.

It can be argued that a knowledge of science was not needed to discover the principle behind the safety lamp or to implement it in a prototype. It may well have been the case that Davy, thanks to his knowledge of chemistry, formal education, and scientific vocabulary, would have been better able to explain why the lamp worked. As is often the case, the superiority of science over intuition may be overestimated when explanation is confused with invention.

For much of history, the scientific and technical knowledge which was available to society was not used to improve the design of worksystems. The ancient Greeks who invented mathematics and philosophy also developed pumps, astronomical calculators (preempting Babbage by almost two thousand years), and a primitive steam engine (Heron's "aerolabe"), but these were regarded as curiosities and their industrial potential was never exploited. Plato described the principle of work specialization and its advantages as follows:

. . . more things are made and they are made better and more easily when each man does the work to which he is suited.

But the necessary integration of this idea with the application of science and technology in work design had to wait more than two thousand years. Some historians attribute this neglect to the social and economic effects of the use of slave labor in the ancient world, which meant that free citizens had no interest in physical work and looked upon it with disdain. It is interesting to note that one of the foundations of F. W. Taylor's scientific management was the insistence that increased output, lower product prices, *and* higher wages went together as part of the scientific management of industrial enterprises.

When deforestation caused the ancient Greeks to use stone as a building material, they were slow to exploit the new possibilities for construction it offered. Although they were aware of the advantages of the arch and the vault over the column and the lintel, they continued to build in stone using designs appropriate to construction in wood (the Parthenon is a famous example). On those occasions when scientific and technological skills were exploited, such as in civil engineering, the design of viaducts, town planning, etc., the results were impressive, judging by the many extant constructions.

Necessity is said to be the mother of invention, but it also appears to be the force which drives the large-scale application of different disciplines in an integrated way to achieve a common goal. The assembly line is sometimes credited to Henry Ford, but in the fifteenth century, the Venetian republic, then the main European maritime power, had a line for refitting its ships. Vessels were floated along the line while groups of workers at fixed stations performed specific operations.

According to Flinn (1966), the main factors which gave rise to the industrial revolution in Britain (1780–1860) were social and economic rather than scientific:

1 Steady economic and population growth in the preceding centuries and an energy crisis resulting in the substitution of coal for wood.

2 Increased agricultural activity due to improved land use and crop rotation. The size of the nonagricultural workforce grew. For much of British history prior to this time, the labor of two people was needed to provide enough food for three. This placed severe limitations on the size of the nonagricultural workforce.

3 Better access to education, particularly for the middle and lower classes.

4 Improved economic organization on a national scale.

5 Emergence of a belief in work as an end in itself and the moral duty of thrift, moderation, and prudence.

The Napoleonic wars accelerated the trend toward industrialization even further by causing an increase in the price of fodder for horses. Mine owners looked to the development of railways using steam locomotion as a cheaper method of transporting coal.

This is not to suggest that developments in physics and chemistry due to Galileo, Newton, Boyle, and others were not important. Watt, the developer of the steam engine, is known to have had a keen interest in physics, particularly the gas laws. The Royal Society was established in Britain in 1660 to guide and strengthen scientific inquiry and the Royal Society of Arts, created in 1754, offered prizes for solutions to technical problems. Although these institutions can be seen as symptomatic of an intellectual climate receptive to scientific inquiry and the solution of practical problems, very few of the really important technical developments of the era arose out of these efforts. In fact, much of the technology (as well as the science) needed to start an industrial revolution had been in existence for centuries and just needed adapting to the new applications. For example, the techniques for fabricating cylinders for steam engines were adapted from those developed previously to bore cannon. The casting of cannon, in turn, made use of expertise gained centuries previously in the casting of church bells. What is significant, however, is that the spirit of inquiry which had always existed was directed to the design of worksystems and the solutions of practical problems. Scientists and thinkers were willing to "get their hands dirty" and find ways of applying their scientific knowledge for practical purposes.

As has already been stated, one of the characteristics of the development of industry over the last hundred years or so has been the application of more and more sciences to the design and analysis of worksystems in an attempt to improve them. Even apparently mundane occurrences can be better understood by the systematic application of basic principles and theories, but only if people are willing to do so. There is currently concern about global problems such as population growth, possible ill effects of accumulation in the atmosphere of carbon dioxide from the combustion of fossil fuels, and the future availability and affordability of energy sources. These concerns may give rise to profound changes in the technological base of advanced societies and the way this technology is used.

Much scientific knowledge about humans is available. It is, and will continue to be, the responsibility of the ergonomist to ensure that, where possible, this knowledge is applied to the study, design, and improvement of worksystems.

2

ANATOMY, POSTURE, AND BODY MECHANICS

It is easy to overlook the fact that the human body is a mechanical system which obeys physical laws. Many of the postural and balance control mechanisms, essential for even the most basic activities, operate outside of conscious awareness. Only when these mechanisms break down—as when we slip or lose balance—are we reminded of our physical limitations. An understanding of these limitations is fundamental to practically all applications of ergonomics.

The skeleton plays the major supportive role in the body. It can be likened to the scaffolding to which all other parts are attached, directly or indirectly. The functions of the skeletal and muscular systems are summarized in Table 2-1.

Like any other mechanical system, the body may be stable or unstable and is able to withstand a limited range of physical stresses. Stresses may be imposed both internally or externally and may be acute or chronic. The function of the ergonomist in the study of such stresses is to use principles of anatomy and biomechanics to design the working environment in order to minimize undue stress, preserve the health of the workforce, and improve task performance.

A useful starting point in the discussion of mechanical loading of the body is to distinguish between **postural stress** and **task-induced stress.** According to Grieve and Pheasant (1982), postural stress is the term used to denote the mechanical load on the body by virtue of its posture. Posture may be defined as the average orientation of the body parts, with respect to each other, over time. Task-induced stress is that which results from the performance of the task itself. An astronaut in space under conditions of zero gravity is "weightless" and experiences minimal postural stress. When using a wrench to tighten a nut, the astronaut would be subject to task-related stress caused by gripping the wrench and generating a torque to tighten the nut while stabilizing the body

31

TABLE 2-1   FUNCTIONS OF THE SKELETAL AND MUSCULAR SYSTEMS

Skeletal system	Muscular system
1. Support	1. Produce movement of the body or body parts
2. Protection (the skull protects the brain and the rib cage protects the heart and lungs)	2. Maintain posture
3. Movement (muscles are attached to bone and when they contract, movement is produced by lever action of bones and joints)	3. Produce heat (muscle cells produce heat as a by-product and are an important mechanism for maintaining body temperature)
4. Homopoiesis (certain bones produce red blood cells in their marrow)	

parts (wrist, elbow, shoulder, etc.) to counteract the force reaction at the wrench-nut interface.

Under the influence of gravity, postural stress becomes an important part of the total mechanical stress. In order for the body to be stable, the combined center of gravity (COG) of the various body parts must fall within a base of support (the contact area between the body and the supporting surface). When one is standing, the weight of the body must be transmitted to the floor through the base of support described by the position of the feet (Figure 2-1). The alignment of the body parts must be maintained to ensure continuing stability, and it is in the maintenance of posture that much stress arises.

The posture most free of stress is reclining. Nachemson (1966) found that the pressure inside the intervertebral disks was lowest in this posture compared with others. Most of the muscles of the body which maintain posture in other body positions are relaxed in the reclining position.

SOME BASIC BODY MECHANICS

The basic limiting condition for postural stability is that the combined COG of the various body parts is within the base of support described by the position of the feet (assuming no other external means of support). The main parts of the axial skeleton must be positioned vertically above the base of support. Ideally, the lines of action of the masses of the body parts should pass through or close to the relatively incompressible bones of the skeleton (Figure 2-2). The jointed skeleton thus supports the body parts and is itself stabilized by the action of muscles and ligaments which serve merely to correct momentary displacements of the mass centers from their bony supports. In a rather crude analogy, the skeleton can be likened to an articulated tent pole with guy ropes (postural muscles) on every side. The fabric of the tent corresponds to the soft tissues of the body. Any displacement of the COG of the structure in a given direction leads to tension in the guy ropes on the opposite side. Ligaments can be likened to the springs and rubber fittings which stabilize the articulations of the tent pole, and tendons to the ends of the guy ropes where they insert into the poles.

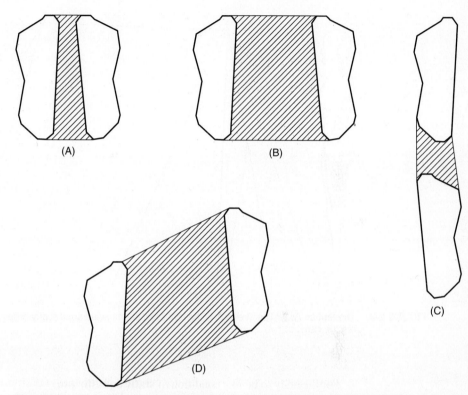

FIGURE 2-1   The stability of the body parts depends on the shape of the base of support described by the position of the feet. (A) Unstable; (B) fairly stable in all directions; (C) stable anteroposteriorly; and (D) laterally stable.

Stability and Support

A stable posture can be maintained only if the various body parts are supported and maintained in an appropriate relation to the base of support, such as the feet or the squab of the seat (Figures 2-1 and 2-3). The size of the base of support determines not only the stability of the body but also the postures which can be adopted.

For example, when from a standing position a person leans forward as if to touch the toes, it can readily be observed that the pelvis moves rearward to compensate for the forward displacement of the COG of the upper body (Figure 2-4). It is impossible to carry out this movement without falling over if there is insufficient space at the rear to allow for compensatory projection of the buttocks unless one foot is placed in front of the other to extend the base of support forward. This can readily be demonstrated by standing with the back and heels against a wall and attempting to pick an object off the floor in front without moving the feet away from the wall. This demonstration shows why it is important to provide **sufficient space** around standing operators and why plenty of **room for the feet** is essential if losses of balance are to be avoided.

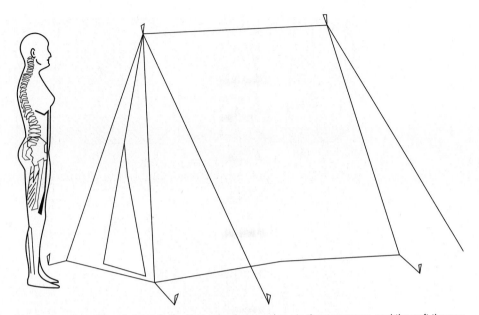

FIGURE 2-2   The tent analogy. The skeleton is the tent pole, the muscles are the guy ropes, and the soft tissues are the canvas.

For the body to be in a condition of **static equilibrium:**

1 Upward forces (from floor) must equal downward forces (body weight plus any objects held).

2 Forward forces (e.g., bending forward) must equal backward forces (extension of back muscles).

3 Clockwise torques (e.g., from asymmetric loads) must equal counterclockwise torques (back and hip muscles).

Ideally, the skeleton should play the major role in supporting the various body parts since this is its function. However, muscles, ligaments, and soft tissues can also play a role but at a cost of increased energy expenditure, discomfort, or risk of soft tissue injury.

When a person is leaning forward in the manner shown in Figure 2-4, a stable posture can be maintained indefinitely, although the posture itself is uncomfortable. The discomfort is a result of the strain being placed on the posterior spinal ligaments and lumbar intervertebral disks—the upper body load is no longer supported by axial compression of spinal structures but by tension in ligaments and asymmetric compression ("wedging") of the intervertebral disks (Figure 2-5).

This demonstrates that a stable posture can be stressful if support of body mass depends on soft tissues rather than bone. Ligaments are able to resist high tensile forces, particularly if these forces are exerted in the direction of their constituent fibers. Ligaments play a major role in protecting joints by limiting the range of joint movement and

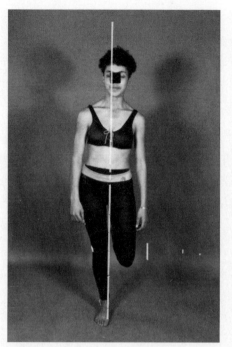

FIGURE 2-3 Standing posture and foot position. The size and shape of the base of support determines the range of postures it is possible to adopt without falling over.

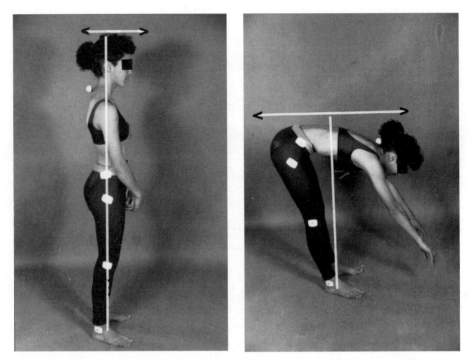

FIGURE 2-4   When the support base is constrained, compensatory movements of which we are usually unaware are needed to maintain stability during movement.

by resisting sudden displacements which might damage the joint. However, injuries can occur if ligaments are exposed to sudden forces when prestressed by extreme joint positions or by complex movements. This is one of the reasons ergonomists stress the importance of the posture of the hands, wrists, elbows, and trunk when tools or controls are operated or when loads are lifted. **Poor equipment design which forces the adoption of extreme joint positions when one is holding an object predisposes the joint to injury.** Good design enables equipment to be used with the joints in the middle of their range of movement.

When a person leans forward in the manner described above but arches the back in an effort to prevent the spine from losing its shape, the posture which is produced is still stable, but a different physiological cost is incurred. In this position, the muscles of the back must carry out static work to maintain the shape of the spine against the pull of gravity which is causing it to flex. **Static muscle contractions lead rapidly to fatigue.**

In some cases, soft tissues can support some of the mass of the body parts. It has been suggested that the lumbar spine is supported by increased intra-abdominal pressure during manual materials handling. The workings of this "abdominal mechanism" and the manner in which it supports the spine are still under debate, however (Gracovetsky and Farfan, 1986; Aspden, 1987). Further discussion of the abdominal mechanism in lifting may be found in Chapter 5.

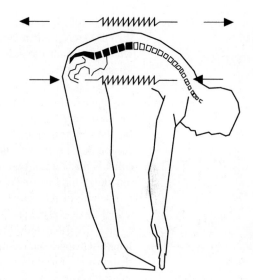

FIGURE 2-5   In this position, postural stress occurs in the form of compression of abdominal contents and intervertebral disks and stretching of the posterior spinal ligaments.

SOME ASPECTS OF MUSCLE FUNCTION

The function of skeletal muscles is to exert tension between the bony points to which they are attached. Tension is exerted when a muscle changes from its resting to its active state in response to impulses from the central nervous system. The maximum tension a muscle can exert depends on its maximum cross-sectional area and also its length (as described below).

The term "muscle contraction" refers to the physiologically active state of the muscle, rather than its physical shortening. Muscles are able to contract eccentrically, isometrically, and concentrically:

1 *Eccentric contractions.* The muscle lengthens while contracting.
2 *Isometric contractions.* The muscle length remains constant during contraction.
3 *Concentric contractions.* The muscle shortens while contracting.

These different contractions can be illustrated by considering the action of the elbow extensors (such as the triceps muscles) of a person doing push-ups. When these muscles contract concentrically, the body is raised off the floor because the shortening of the muscles causes the elbows to extend. They contract eccentrically when lowering the body to the floor. The eccentric contraction acts like a brake. It controls the rate at which the muscles increase their length and at which the elbows flex. This contraction enables the body to be lowered smoothly to the floor in a controlled way. If the person pauses halfway, isometric contraction is required to counteract the downward pull of gravity and maintain the position of the body in space.

A close relationship exists between the length of a muscle and the tension it can exert. If a muscle is removed from the body, it assumes a resting length which depends on its own internal properties. If the muscle is placed in an apparatus suitable for manipulating its length and for measuring the tension required to maintain the muscle at a given length, a length-tension curve may be plotted. As the muscle is artificially lengthened, increased tension is required to overcome the elastic resistance of the connective tissue which surrounds the individual muscle cells and holds the muscle together (Grieve and Pheasant, 1982).

Muscles can exert tension both actively and passively by their resistance to being lengthened. One of the most well-known examples of passive resistance occurs when the hamstring muscles begin to stretch as a person attempts to touch the toes while keeping the knees straight. A muscle's resistance to passive stretching can have a limiting effect on the movement of a joint. In Figure 2-6, it can be seen that forward bending of the trunk is made up of two movements—flexion of the hips and flexion of the spine. The amount of hip flexion depends on the resistance to lengthening of the hamstring muscles (stretching of the hamstrings is experienced as pain behind the knee). The amount of spinal flexion depends on the resistance of the ligaments of the spine.

If an isolated muscle is stimulated electrically to cause it to contract, the active length-tension relationship can be investigated (Gordon et al., 1966). Experiments demonstrate that the tension **actively** produced depends on the length of the muscle when it is stimulated to contract—**muscles have an optimum length at which they are capable of exerting their maximum tension** (Figure 2-7).

In living organisms, muscles form part of a larger system of joints and ligaments, and the tension they exert is transmitted via their bony attachments. This tension causes a turning force to be exerted around the joint which the muscle spans. The magnitude of the force depends not only on the length of the muscle but also on the geometry of the joint and the mechanical advantage of the muscles at a particular joint angle. Angle-torque curves can be plotted for different muscle-joint systems to identify those joint angles at which maximum torque can be exerted. Thus, manual tasks can be simulated and "strong" and "weak" body postures identified. For example, the flexion torque around the elbow joint is greatest in the middle portion of the joint range because the elbow flexors (such as the biceps muscles) have their greatest mechanical advantage in this position.

Control of Muscle Function

Skeletal muscle contains afferent and efferent fibers from the central nervous system. Some of these form complex feedback loops. Receptors in the muscle body provide information about the length of the muscle and its tension (Figure 2-8). This information is conducted to the spinal cord and to higher levels of the central nervous system. The knee-jerk reflex is a well-known example of a feedback loop which is made possible by a neural circuit between the **stretch receptors in the muscle spindles** and **motor neurones in the spinal cord.** When the muscle is lengthened, the stretch reflex comes into operation, causing the muscle to contract—that is, the reflex opposes the lengthening of the muscle. **Muscle tension** is detected by an organ known as the **golgi tendon organ,** which is found at the point where small bundles of muscle fibers fuse with tendon fibers. The golgi tendon organ forms a negative feedback system which prevents damaging levels of tension from building up in the muscle and possibly damaging the muscle fibers or the insertions of the tendons.

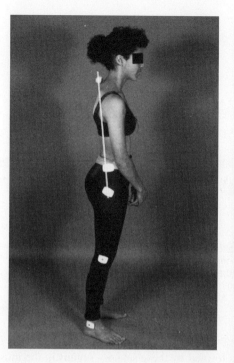

FIGURE 2-6 Analysis of body movement. Many body movements involve coordinated activity of several joints. Movements at the hip are limited by the maximum length of the hip muscles. When one is bending forward, only about one-third of the movement takes place as hip flexion. The rest takes place in the spine. If the knees are bent, the hamstring muscles are shortened and allow more hip flexion. When one is leaning backward in standing, most of the movement takes place as lumbar hyperextension because the hip flexors do not permit much hip extension.

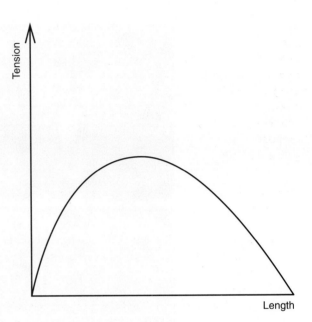

FIGURE 2-7 Active length-tension relationship. A longer muscle generates more tension when it contracts until its physiological length is exceeded.

The neural circuits formed by these sensory organs in the muscles provide the central nervous system with information about the state of the muscle and the posture of joints. This information is known as **proprioceptive feedback** and is essential for all but the most primitive of movements. Skeletal muscle also contains nerve endings which, when stimulated, lead to the sensation of pain. These pain receptors can be stimulated by chemical, thermal, or mechanical stressors. Nerve impulses induced by any of these causes can give rise to reflex muscle contraction via neural circuits in the spinal cord. This contraction can further exacerbate the cause of the pain and lead to a positive feedback situation in which the **muscle goes into spasm**—a topic which is returned to in a later chapter.

For present purposes it can be concluded that muscles are not just simple mechanical actuators like motors or springs—they are reactive tissues. Their tension, length, and thermal, mechanical, and chemical state are integrated and controlled by means of complex neural circuitry at different levels of the central nervous system.

Fatigue and Discomfort

A major goal of research into posture is to develop principles for the design of working environments which impose low postural stress on workers. The use of these design principles, it is hoped, will reduce the incidence of fatigue and discomfort in the workplace. Discomfort is difficult to define because it has both objective and subjective elements. Branton (1969) suggested that discomfort results in an "urge to move" caused by a number of physical and physiological factors. Pressure on soft tissues can cause **ischemia** (depletion of the local blood supply to the tissues), resulting in a shortage of

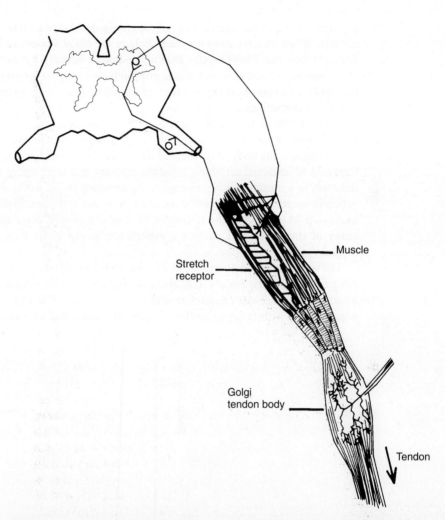

Stretch
receptor

Muscle

Golgi
tendon body

Tendon

FIGURE 2-8 Positive and negative feedback comes from receptors in the muscles and a feedback loop at the spinal cord. A contraction-relaxation reflex exists between antagonistic muscles; when one contracts, the other relaxes to allow movement at the joint. The reflex depends on the feedback from the muscle spindles and golgi tendon bodies, respectively.

oxygen and a buildup of carbon dioxide and waste products such as lactic acid. This condition is known to lead to pain or discomfort.

Active muscles convert glucose and oxygen to carbon dioxide and water, liberating energy in the process. They require a regular blood supply to replenish fuel and remove the waste products. During rhythmic exercises involving eccentric and concentric contractions, blood flow is facilitated by the pumping action of the muscle. Fatigue may ultimately occur as a result of the depletion of metabolites (glucose or glycogen which is stored in the muscle and may be readily converted to glucose). Depletion of muscle glycogen is one of the main causes of fatigue in endurance athletes such as marathon runners. When muscles contract, they occlude the blood vessels within them and thus diminish their own blood supply. **During sustained isometric contractions, the muscle is starved of oxygen and waste products accumulate as oxygen-independent metabolic processes take place.** Discomfort and fatigue occur rapidly during sustained isometric contractions for this reason.

Kroemer (1970) has shown that muscles fatigue rapidly under conditions of static loading even at low workload (Figure 2-9). Skeletal muscle can be regarded as the largest organ in the body. It is essential for all activities involving voluntary movement. Repeated or sustained activities, rapid movements, and large forces can stimulate pain receptors in a muscle. Such movements are common in the workplace, during participation in sports, and in the performance of everyday activities. **Since skeletal muscle makes up 40 percent of the tissues of the body, it should come as no surprise that many of the aches and pains we experience in our daily lives are of muscular origin.**

The tension exerted by muscles can serve to cause movement of a body part in carrying out some work activity. It is also essential for resisting externally imposed forces which might otherwise destabilize and damage a joint. The loss of muscle function which occurs with fatigue therefore implies increased risk of musculoskeletal injury.

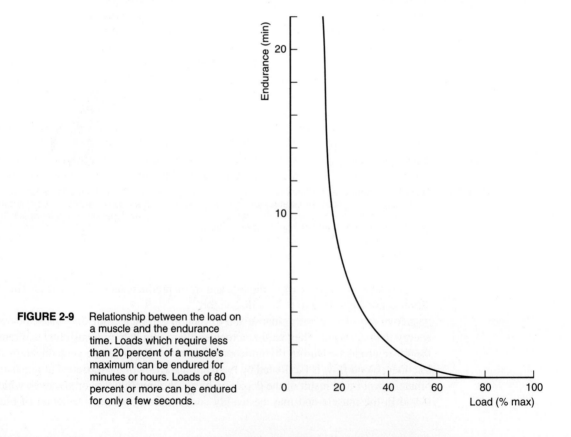

FIGURE 2-9   Relationship between the load on a muscle and the endurance time. Loads which require less than 20 percent of a muscle's maximum can be endured for minutes or hours. Loads of 80 percent or more can be endured for only a few seconds.

This indicates why a consideration of work-rest cycles, work durations, and forces is an essential part of the risk-assessment process in job evaluation and injury prevention.

Electromyography

Discomfort is a subjective experience which can result from a combination of physiological and psychological processes, including muscle fatigue. Muscle fatigue is a physiological phenomenon which can be observed directly using techniques such as electromyography (EMG). Muscle cells are arranged in functional units known as "motor units." Each motor unit is connected to a single nerve fiber and all of the cells in that unit fire in response to an impulse from the fiber on an on-off basis. The number of muscle cells per motor unit varies considerably between muscles—calf (soleus) muscle motor units contain many hundreds of muscle cells, whereas those in the eye muscles contain as few as five. The latter muscles are capable of much finer, coordinated movement because of their superior innervation.

Muscle tension can be varied by increasing the number of motor units firing at a single time or by increasing the firing rate. Electrical activity in muscles can be detected by using either surface electrodes placed on the skin overlying the muscle or needle electrodes inserted into the muscle body. Soderberg and Cook (1984) provide a useful introduction to electromyography. EMG activity from fresh muscle resembles electrical noise—a smooth or "buzzy" trace usually appears on the oscilloscope. It has long been known that the frequency content of the myoelectric signal shifts toward the lower frequencies as a sustained contraction progresses to exhaustion. Stuhlen and De Luca (1982) reviewed theories of muscle fatigue and concluded that compression of the myoelectric spectrum occurs as a result of a decrease in the conduction velocity caused by a decrease in muscle pH due to the accumulation of lactic and pyruvic acids. They used the median frequency of the myoelectric signal as an index of muscle fatigue. The initial median frequency (of the fresh muscle) was taken as a cutoff and the ratio between the root mean square (rms) voltages below and above the median was used to monitor the progress of fatigue. This parameter increases in magnitude as the muscle fatigues. In severely fatigued muscles, the availability of fresh motor units becomes limited and visible tremor sets in (Figure 2-10).

Electromyography is used by ergonomists to detect workspace and task factors which cause unnecessary or rapid muscular fatigue. It complements subjective techniques in which workers are asked to indicate on body diagrams or by questionnaire the location and severity of musculoskeletal pain.

ANATOMY OF THE SPINE AND PELVIS RELATED TO POSTURE

The spine and pelvis support the weight of the body parts above them and transmit the load to the legs via the hip joints. They are also involved in movement. Almost all movements of the torso and head involve the spine and pelvis in varying degrees. The posture of the trunk may be analyzed in terms of the average orientation and alignment of the spinal segments and pelvis. Figure 2-11 depicts the spine and pelvis viewed frontally and sagittally.

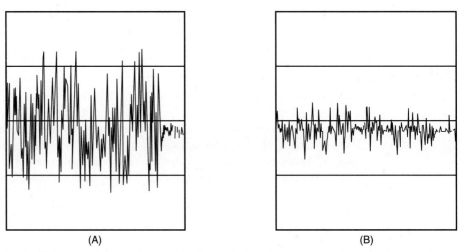

(A)   (B)

FIGURE 2-10   EMG activity from (A) fatigued and (B) fresh muscle. The "buzzy" trace from fatigued muscle is due to desynchronization of motor unit firing.

The Spine

Quadrupedal animals and human babies have a single spinal curve running dorsally from pelvis to head. The thorax and abdomen hang from the spine and exert tension which is resisted by the spinal ligaments, the apophyseal (facet) joints, and the back muscles. In adult humans, the spine is shaped such that it is close to or below the COG of the superincumbent body parts which are supported axially—the effect of weight bearing in the standing posture is to compress the spine (Adams and Hutton, 1980). This compression is resisted by the vertebral bodies and the intervertebral disks. The **cervical** and **lumbar** spines are convex anteriorly—a spinal posture known as **lordosis.** It is the presence of these lordotic curves which positions the spine close to or directly below the superincumbent body parts. The effect is to reduce the energy requirements of the maintenance of the erect posture and place the lumbar motion segments in an advantageous posture for resisting compression (Klausen and Rasmussen, 1968; Adams and Hutton, 1980, 1983; Corlett and Eklund, 1986). The thoracic spine is concave anteriorly and is strengthened and supported by the ribs and associated muscles.

The term "spinal column," although universally accepted, is something of a misnomer; "spinal spring" might be more appropriate—the S shape of the spine of a person standing erect gives the entire structure a springlike quality such that it is better able to absorb sudden impacts, transmitted along its axis, than if it were a straight column. The loss of the S shape in sitting may be one reason drivers of trucks and farm vehicles who are exposed to vibration in the vertical plane are so prone to back trouble.

The cervical and lumbar spines are not fixed in lordosis. Each vertebral body is joined to its superior and inferior counterpart by muscles, ligaments, and joints. The spine takes part in functional movements of the body—part of the postural adaptation required to carry out many activities takes place in the lumbar and cervical spines (Figure 2-6).

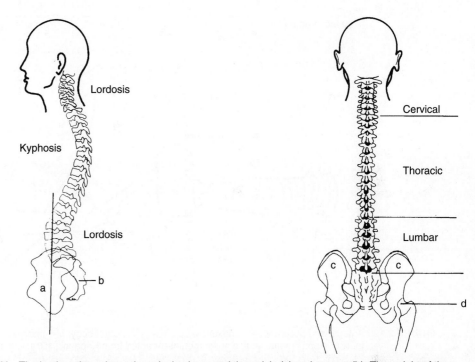

FIGURE 2-11   The lumbar, thoracic, and cervical spines, and the pelvis (a) and sacrum (b). The weight of the upper body is transmitted through the lumbar spine and the iliac bones of the pelvis (c) to the hip joints (d) and legs.

The spine can be considered simplistically to consist of three anatomically distinct but functionally interrelated columns (Figure 2-12). The anterior column, consisting of the vertebral bodies, intervertebral disks, and anterior and posterior longitudinal ligaments, is the main support structure of the axial skeleton. It resists the compressive stress of the superincumbent body parts. The two identical posterior columns are positioned astride the neural arch (which forms a bony cavity through which passes the spinal cord), and consist of the facet joints and the associated bony projections, ligaments, and muscles. The posterior elements of the spine act as jointed columns which control the movement of the complete spine and provide attachment points for the back muscles. The vertebral bodies and their related structures increase in size from the top to the bottom of the spine in accordance with the increased load that they must bear.

The intervertebral disks act as shock absorbers and limit and stabilize the articulation of the vertebral bodies. Each disk consists of concentric layers of cartilage whose fibers are arranged obliquely in a manner similar to a cross-ply tire (Figure 2-13). The layers of cartilage enclose a central cavity which contains a protein-mineral solution. Positive osmotic pressure ensures that water is always tending to enter the disk. Thus, the disks are **prestressed** to withstand loading (in a manner analogous to reinforced concrete beams used in the construction of modern buildings). The fluid-filled nucleus pulposus

Anterior elements   Posterior elements

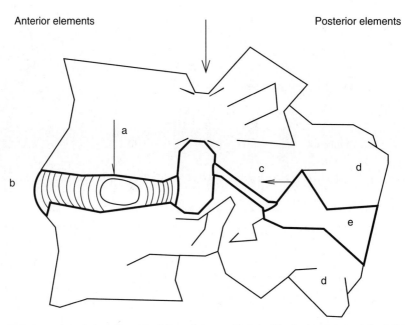

FIGURE 2-12   Anterior and posterior elements of the spine. a = Vertebral body; b = intervertebral disk; c = facet joint; d = bony projections; e = ligaments. (Adapted from Kapandji, 1974, with permission.)

cannot only be likened to a form of reinforcement for the disk, but according to Kapandji (1974), the nucleus pulposus functions as a swivel joint.

Intervertebral disks exhibit viscoelastic behavior. Forces of rapid onset are resisted in an elastic manner—the disk deforms initially, then returns rapidly to its original shape when the force is removed. Under continuous loading, however, the disk exhibits a type of viscous deformation known as "creep." Creep occurs as a result of loading above or below a threshold level. Under compressive loading, the disk narrows as fluid is expelled and the superior and inferior vertebral bodies move closer together (Eklund and Corlett, 1984). Under traction (stretching or pulling forces), fluid moves into the disk and the disk space becomes wider (Bridger et al., 1990).

Klingenstierna and Pope (1987) have shown that vibration increases the rate of creep in the spinal motion segments. Kasra et al. (1992) have shown that loss of fluid from the nucleus reduces its incompressibility, which manifests itself as a decrease in the resonant frequency and an increase in the response amplitude of the motion segments when they are exposed to compressive forces. An already degenerated disk will respond in this way when exposed to vibration and experience greater stresses and strains than a healthy disk—the likelihood of further disk degeneration will be greater and the facet joints will be exposed to greater loading (Shirazi-Adl, 1992). Boshuizen et al. (1992) report a very high prevalence of back pain among forklift truck drivers exposed to vibrations of 0.8 m/sec² (68 percent compared with 25 percent in comparable control subjects). Of particular interest is the fact that the drivers were relatively young (under 35 years of age) and the risk of getting pain was greatest in the first 5 years of driving. Older drivers had a lower risk of pain (presumably, older drivers with pain had found alternative employ-

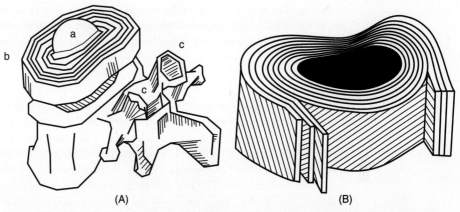

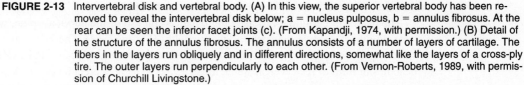

(A)   (B)

FIGURE 2-13   Intervertebral disk and vertebral body. (A) In this view, the superior vertebral body has been removed to reveal the intervertebral disk below; a = nucleus pulposus, b = annulus fibrosus. At the
rear can be seen the inferior facet joints (c). (From Kapandji, 1974, with permission.) (B) Detail of
the structure of the annulus fibrosus. The annulus consists of a number of layers of cartilage. The
fibers in the layers run obliquely and in different directions, somewhat like the layers of a cross-ply
tire. The outer layers run perpendicularly to each other. (From Vernon-Roberts, 1989, with permission of Churchill Livingstone.)

ment). Apart from the vibration, the drivers often sat with the trunk twisted while looking backward, which increases the load on the facet joints. Taken together, all these findings point to a pattern of loading which will cause chronic pain and hasten the onset of
degenerative diseases of the spine.

Clearly, driving tasks which expose the driver to vertical vibrations should not be
combined with lifting tasks. Cabs should be fitted with mirrors to obviate the need to
look behind, and vibration should be reduced. Drivers of goods delivery vehicles should
not have to load and unload the goods or should be provided with lifting and carrying
aids.

The narrowing and expansion of the disk spaces is natural and occurs as a resut of the
forces exerted on the spine during everyday activities. Since there are 24 vertebral bodies, all with disks between them, the shrinkage and expansion of the disk spaces results
in measurable changes in stature—most people are about 1 percent taller when they
wake up in the morning than when they go to bed at night for this reason (de Puky,
1935). Stature change varies exponentially with loading time—almost 50 percent of the
stature gained after a night's sleep is lost in the first half hour after rising. Grieco (1986)
suggests that, since the disks have no direct blood supply, the daily ingress and egress of
fluid due to variations in loading is the mechanism whereby nutritional exchange with
the surrounding tissues takes place. Postures which exert static loads on the body will
interfere with this mechanism and are hypothesized by Grieco to accelerate the degeneration of the disks. **Sitting for 8 hours per day is regarded as a health hazard according to this view.**

Stature change also occurs with age—after about 30 years of age, the intervertebral
disks degenerate, developing microtears and scar tissue; fluid is lost more readily; the
disk space narrows permanently; and the spinal motion segments lose stability. It is not

surprising that most occupationally induced low back pain occurs in middle-aged people. In the elderly, disk degeneration reaches a stage where, together with other degenerative processes, the spine is restabilized but with a corresponding loss of mobility.

The Cervical Spine

The cervical spine has several functions—principally, to support the weight of the head and to provide a conduit for neural structures and attachment points for the muscles which control the position of the head. It consists of seven vertebrae designed to permit complex movements of the head. The lower five vertebrae of the cervical spine have the same general structure as those in the rest of the spine and are surrounded by anterior and posterior ligaments. The cervical spine consists of vertebral bodies and intervertebral disks, facet joints, bony processes for the attachment of ligaments and muscles, and the intervertebral foramen through which passes the spinal cord.

The head can be thought of as being balanced on top of the cervical spine, with the fulcrum directly above the first cervical vertebra (Figure 2-14). The head can be considered to be in balance when a person looks directly forward. Because the COG of the head lies in front of the cervical spine, the head has to be held erect by contraction of the posterior neck muscles. These powerful muscles are true postural muscles—they are essential to the maintenance of the erect posture in Homo sapiens and constantly work to prevent the head from falling forward as a result of gravity. The role of the posterior neck

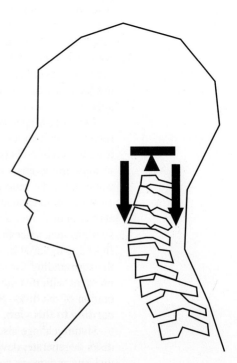

FIGURE 2-14   Balancing of the head on the cervical spine.

muscles in the maintenance of posture becomes clear when it is recalled how a sitting person's head droops forward onto the chest when the person is overcome by sleep. It can be appreciated that in ordinary standing and sitting postures, the structures of the cervical spine are prestressed by the need to maintain the head in an erect position. They are therefore prone to overexertion by any additional stresses imposed by work tasks. As can be seen from Figure 2-14, in the balanced position, a cervical lordosis is present.

Movements of the head are accomplished by the muscles attached to it and to the surrounding parts of the skeleton. The arrangements of the muscles, like the movements of the head, are complex and will be discussed only in greatly simplified form. The deep, short muscles of the neck serve to stabilize the individual vertebrae, whereas the longer, more superficial muscles produce movements of the spine and head as a whole. The posterior muscles, which extend the neck, are stronger than the anterior muscles, which flex the neck, because the latter are assisted by gravity, whereas the former have to work against gravity.

The erector muscles of the neck produce extension of the head and neck if they contract bilaterally (i.e., together). If they contract unilaterally (on one side only), lateral flexion and rotation of the head is produced. The trapezius muscle plays a very important role in many work activities. Because of its oblique orientation, it produces extension, lateral flexion, and rotation of the head toward the side of contraction. It is also involved in elevating the shoulders (Figure 2-15).

Degeneration of the Cervical Spine

According to Kapandji (1974), prolapse of the intervertebral disks of the cervical spine is rare. However, the disks can certainly degenerate, as can the intervertebral joints, and this degeneration can cause irritation of the nerve roots in the cervical spine. Pain in the neck and shoulder may result. Degeneration of the cervical spine, sometimes known as **cervical spondylosis,** can have serious consequences. Compression of the spinal cord at the level of the cervical spine can take place, resulting in weakness and wasting of the upper limbs, which may then spread to the lower limbs.

As is the case with the lumbar spine, some of the degeneration of the cervical spine is part of the natural process of aging. According to Barton et al. (1992), by the age of 65, 90 percent of the population has radiological evidence of cervical spondylosis. Cervical spine degeneration is a potential cause of neck pain because of the mechanical changes which occur as a result of age-related degenerative processes.

Static flexion of the cervical spine increases the moment arm of the head according to the sine of the angle of flexion. This increases the load on the soft tissues in the cervical region, and the posterior neck muscles are placed under increased static load in order to maintain the forward-flexed head in equilibrium with gravity. According to de Wall et al. (1991), the increased static load on these muscles may cause pressure ischemia and starve the muscle tissues of fuel and oxygen. Pain in the neck and shoulders may result, causing **muscle spasm** (reflex contraction of the muscles). This condition, in turn, may exacerbate the pain and lead to a vicious circle. The forward-flexed position may subject the cervical intervertebral disks to increased compression and the posterior ligaments to increased tension. Poor workspace design, if it requires workers to

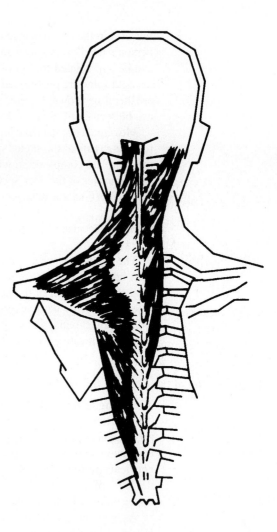

FIGURE 2-15   Rear view of neck and shoulder musculature.

adopt flexed cervical postures, may be a cause of reversible pain or may amplify pain
due to existing degenerative changes.

It is very important to identify occupational factors that increase the mechanical load
on the cervical spine through an increase in flexion and to remove these factors by redesigning the workstation or the task.

The Pelvis

The pelvis is a ring-shaped structure made up of three bones, the sacrum and the two innominate bones. The sacrum extends from the lumbar spine and consists of a number of
fused vertebrae. The three bones are held together in a ring shape by ligaments (Figure
2-16). The innominate bones are themselves made from the fusion of three other bones,
the ilium, the ischium, and the pubis. The pubis lies at the anterior part of the pelvis. It

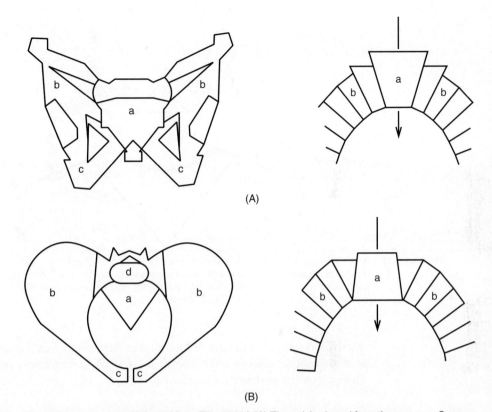

(A)

(B)

FIGURE 2-16   The pelvis as an arch. (Adapted from Tile, 1984.) (A) The pelvis viewed from the rear. a = Sacrum; b = ilium; c = ischium. The sacrum acts like a true keystone in this plane. (B) The pelvis viewed from above. a = Sacrum, b = ilium, c = pubis, d = position of the intervertebral disk between the first sacral and fifth lumbar vertebrae. Under load, the sacrum tends to move forward, like an inverted keystone in an arch. It has to be held in place by strong ligaments.

joins the other bones together, completing the ring shape and acting like a strut to prevent the pelvis from collapsing under weight bearing (Tile, 1984). The posterior structures of the pelvis, the sacrum, and the ilia, carry out the actual weight-bearing function.

The pelvis can be likened to an arch which transfers the load of superincumbent body parts to the femoral heads in standing and to the ischial tuberosities (part of the two ischia) in sitting. When viewed from the rear (Figure 2-16A), the sacrum resembles the keystone of the arch. The load from above is transmitted through the innominates to the femoral heads. However, when viewed from above, the sacrum has the wrong shape for a keystone—it tends to slide forward, out of the arch (Figure 2-16B). Under weight bearing, the tendency for the sacrum to slide forward anteriorly is resisted by the strong ligaments between the sacrum and the ilia. It is these posterior sacroiliac ligaments which stabilize the joint between the sacrum and the ilia. DonTigny (1985) has pointed out that standing postures increase the tendency for the sacrum to be anteriorly displaced,

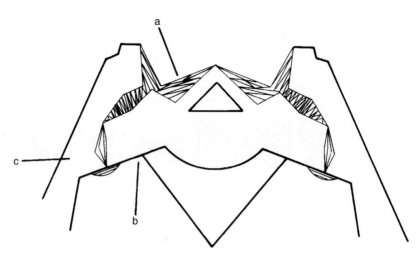

FIGURE 2-17   View of the sacroiliac joint from above. a = Ligaments, b = sacrum, c = pelvis. According to Tile (1984) the ligaments act like the cables of a suspension bridge, preventing the sacrum from slipping forward. If the joint is deformed by loading, the ligaments can be pinched by bone, causing pain in the very low back, usually on one side.

thereby increasing the tension in the sacroiliac ligaments. Small displacements of the sacrum can occur, causing soft tissues to be "pinched," which causes pain (Figure 2-17). This pain can be mistaken for low back pain.

The Lumbopelvic Mechanism

The lumbar spine arises from the sacrum, and the degree of lumbar lordosis depends on the sacral angle which, in turn, depends on the tilt of the pelvis (Figure 2-18). The relationship between the posture of the pelvis and that of the lumbar spine was eloquently illustrated by Forrester-Brown (1930) as follows:

> The simplest countryman understands that one cannot put on the top story of a house until one has built the ground floor and foundations; yet medical men are constantly trying to alter the position of the upper bricks of the spinal column without adjusting the base on which they stand.

Some ergonomists could be accused of making the same mistake when trying to design seats to prevent excessive lumbar flexion. Parents' admonitions to young children to "sit up straight and don't slouch" may be equally mistaken.

When looking at the posture of the spine at work, one must consider the factors which determine the position of the pelvis, such as the slope of the seat or the position of the feet on the floor. The pelvis can be represented as in Figure 2-19, with the hip joint regarded as a fulcrum.

Many muscles attached to the pelvis can be considered as guy ropes or stays which fixate the pelvis onto the heads of the femora. These muscles can exert moments on the pelvis which cause change in pelvic tilt (even though this is not their main function). The

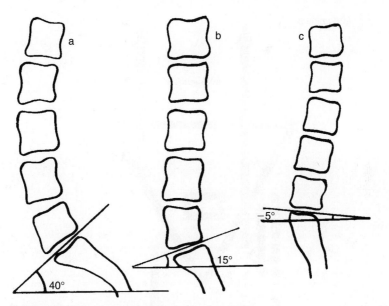

FIGURE 2-18   Relationship between sacral angle and lumbar angle.

hamstring, gluteal, iliopsoas, erectores spinae muscles and other muscles, together with the ligaments of the hip joint, are part of the lumbopelvic system. The tilt of the pelvis in the anterior-posterior plane depends on equilibrium of the moments and countermoments exerted by the antagonistic muscles in the system.

When a person is standing, the line of gravity falls slightly behind the center of the hip joint, which causes the pelvis to automatically tend to tilt backward. This position relieves the abdominal muscles of a postural role and explains why these muscles are relaxed in standing (this applies to normal standing; when one is carrying a load on the back or walking down a steep hill, the abdominals do play a role).

An understanding of the functions of the one- and two-joint hip flexors (iliopsoas and rectus femoris) and the one- and two-joint hip extensors (gluteus maximus and hamstrings) is of relevance to ergonomics.

Posture and the Pelvis

Contraction of the hip flexors on a fixed femur (as in standing) tends to tilt the pelvis forward, increasing the sacral angle. A complex system of reflexes exists to maintain the head in an erect position when this happens. Forward tilting of the pelvis tends to cause the whole of the trunk and head to tilt forward, but the reflexes act to prevent this by hyperextending the lumbar spine to restore the erect position of the head. In a standing position, anterior pelvic tilt is accompanied by excessive lumbar curvature (hyperlordosis). Similarly, contraction of the hip extensors and abdominals causes posterior tilting of the pelvis and compensatory flattening (flexion) of the lumbar curve to keep the head erect. Posterior pelvic tilt is accompanied by diminished lumbar lordosis.

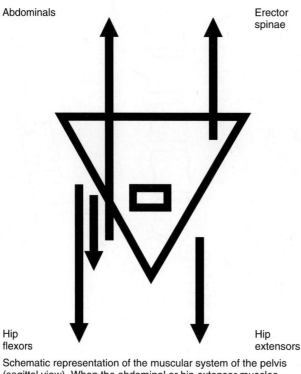

Abdominals

Erector
spinae

Hip
flexors

Hip
extensors

FIGURE 2-19   Schematic representation of the muscular system of the pelvis
(sagittal view). When the abdominal or hip extensor muscles
shorten, the pelvis tilts backward. The result is a flattening of the
lumbar spine to maintain the trunk erect. When the hip flexors or
erector spinae muscles shorten, the pelvis tilts forward. This action is accompanied by a compensatory increase in the lumbar
lordosis.

In extreme cases, individuals can develop postural faults as a result of the development of imbalances between the muscles in this system. When a muscle is weakened and
its antagonist is not, a state of imbalance is said to exist. The stronger muscle then shortens and the weaker lengthens, resulting in postural deformity. Deformities can also
occur if weak muscles are no longer able to adequately oppose gravity.

When a healthy person adopts a comfortable standing position, the pelvis is in an anterior position and there is a distinct lumbar lordosis (Figure 2-20). In seated positions,
the hip joint flexes and the hamstrings and gluteal muscles lengthen and the iliopsoas
and hip flexors shorten. The resulting muscle imbalance causes a new lumbopelvic posture characterized by posterior pelvic tilt and a loss of lumbar curvature. The change is
proportional to the amount of hip flexion required to sit which, in turn, depends on the
design of the seat (Figure 2-20).

In most people, a pronounced lumbar lordosis is really apparent only in standing and
is an unphysiological posture for the spine in most other body positions (Bridger et al.,

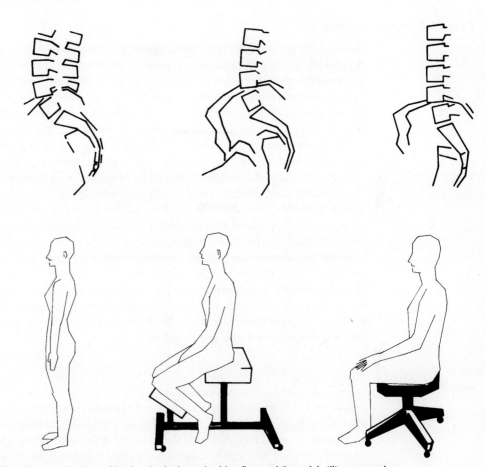

FIGURE 2-20   Progressive loss of lumbar lordosis as the hips flex and the pelvis tilts rearward.

1992). Although the literature on seating sometimes suggests that certain chair designs can "preserve" or "restore" the lumbar lordosis, this claim is incorrect because it ignores the fundamental change in pelvic posture which occurs when a person sits down and which no amount of lumbar support can overcome. Modern office chairs with pronounced lumbar supports may restore a small degree of lordosis, but their main function is to prevent the lumbar spine from flexing.

MUSCULOSKELETAL PROBLEMS IN SITTING AND STANDING

When a person is working in a standing or sitting position, the spinal motion segments, particularly the lumbar motion segments, are susceptible to **extreme postures.** The risk of injury to muscle-joint systems is greatest when they are in extreme postures and least when around the midpoint of the range of motion.

Extreme Postures and Pain

Genaidy and Karwowski (1993) investigated the discomfort associated with joint postures which deviated from the neutral position. Their subjects moved, in turn, each body part from the neutral position to the end of its range of motion, held the posture for 30 seconds, then rated the discomfort associated with the posture on a 10-point scale. This investigation enabled a number of extreme postures to be identified which were high in perceived stress and therefore risk factors for the onset of acute musculoskeletal pain.

For the shoulder joint, high levels of discomfort were perceived when the arms were elevated in any direction away from the body. For the elbow, supination was perceived to be the most stressful posture, followed by pronation, with flexion and extension the least stressful. Lateral bending of the neck was perceived to be more stressful than neck flexion, extension, or rotation. In standing, lumbar extension was perceived to be more uncomfortable than lateral lumbar bending or rotation. In both standing and sitting, flexion of the back was the least uncomfortable of the extreme postures—presumably because the erectores spinae muscles are relaxed when the spine is in full flexion (Floyd and Silver, 1955). All extreme hip movements in standing were associated with high levels of discomfort, abduction being the worst, followed by flexion, extension, adduction, and medial and lateral rotation. Having to hold the legs in extreme positions while standing seems to be particularly problematic. If these extreme postures are designed out, the risk of musculoskeletal pain can be reduced.

Low Back Pain

Pain in the lower regions of the back is one of the most common sources of work-related discomfort. It can also occur as a result of many everyday activities, such as car driving, housework, or gardening. Although the anatomy of the spine is well known, finding the cause of low back pain can be a much more elusive problem for clinicians. Pain is unlikely to arise from the intervertebral disks themselves since they do not contain nerve endings in the adult. Similar reasoning rules out pain from the capsules of the apophyseal joints. Likely sources of pain are the posterior ligaments and other soft tissues. These may be irritated by mechanical trauma due to damage to or degeneration of bony structures. Nerve root compression can also be a source of pain.

Kumar (1990) has shown that mechanical load is a risk factor for low back pain. He used a two-dimensional, static mathematical model of spinal loading to estimate the shear and compression forces at the lumbosacral and thoracolumbar joints. These forces were found to be higher in workers with self-reported pain than in workers without pain.

Back Pain and Muscular Fatigue

It has been shown that the lumbar muscles of chronic low back pain sufferers fatigue more rapidly than those of nonsufferers. Presumably, pain occurs both directly, as a result of stimulation of pain receptors in the muscles due to the biochemical changes which accompany fatigue, and indirectly as a result of the increased load on soft tissues in the lumbar spine itself. Roy et al. (1989) used median frequency analysis of the EMG signal to investigate lumbar muscle fatigue in chronic low back pain sufferers and in normals. They found that the median frequency parameters could be used to reliably classify 91 percent of the low back sufferers and 84 percent of the healthy subjects. In a second study, Roy et al. (1990) showed that the median frequency analysis could distinguish between low back sufferers and pain-free individuals in a group of elite rowers. Finally, Klein et al. (1991) compared the median frequency technique with conventional clinical diagnostic tests for low back pain (range of motion of the lumbar spine and maximum voluntary contraction). The conventional tests correctly identified 57 percent of the pain sufferers and 63 percent of pain-free individuals, whereas median frequency analysis correctly identified 88 percent of pain sufferers and 100 percent of pain-free individuals. Other researchers (e.g., Biedermann et al., 1991) have obtained similar findings using median frequency analysis to investigate lumbar muscle fatigue. It seems that a lack of back muscle **endurance,** rather than a lack of **strength,** is the key characteristic of the type of chronic low back sufferers who suffer no other obvious physical abnormalities or pathological conditions. Van Dieen et al. (1993) have shown that the greatest shift in erectores spinae median frequency is at the L5 level. Although there is great intersubject variability, spectral analysis of back muscle EMG activity can be used to evaluate the risks associated with different lifting tasks.

These findings help explain why chronic low back pain sufferers are at risk in tasks which involve repetitive lifting, carrying of weights in front of the body, leaning forward or with trunk extended—all these activities require sustained activity of the back extensors.

Other Causes of Back Pain

However, pains in the low back can occur for entirely different reasons and from unrelated structures such as the kidneys. Colds and flu may cause complaints of pain in the back. Back pain is a complex problem (Waddell, 1982) and detailed investigation of back problems is best left to expert clinicians. If a worker complains of back pain, the ergonomist's natural inclination is to search for causes in the workplace. In the case of strenuous jobs, this approach may be appropriate, but in less obvious cases, nonoccupational and potentially more serious causes should also be considered.

Low back and shoulder girdle pain are major problems in the industrialized world. Frequently, the pain is of an acute form and is due to muscular fatigue. This type of pain usually subsides within hours or days if the sufferer rests. Many researchers now acknowledge that health problems are often exacerbated, if not caused, by habitual daily activities. It is known that low back problems have a higher incidence among certain groups of individuals, such as professional drivers. Magora (1972) carried out an epidemiological survey to investigate the incidence of low back pain in relation to the occupational requirements for sitting, standing, and lifting. Low back symptoms were higher among those with uniform occupational requirements than those whose daily activities were more varied (who were able to alternate between standing and sitting). Gilad and Kirschenbaum (1988) investigated back pain across a broad spectrum of jobs. More back pain was found in groups who worked in unusual body positions or with the trunk flexed laterally or forward in standing or sitting. Keyserlig et al. (1988) came to similar conclusions after investigating back problems in an automobile assembly plant. Persistent back pain was associated with forward and lateral flexion and twisting (axial rotation) of the spine. Pain prevalence increased substantially if a nonneutral posture was held for more than 10 percent of the work cycle, suggesting that such postures be designed out of the work cycle or minimized.

Although the etiology of musculoskeletal problems involves several factors, it is known that pain can be caused or exacerbated by excessive loading of joints and muscles. This can occur not only as a result of traumatic events but also because of sustained exposure to particular working postures. Nachemson (1966) used a needle transducer to measure the hydrostatic pressure in the third lumbar intervertebral disk. Disk pressure was found to be higher in sitting than in lying down but was reduced when the sitter reclined against a backrest. Other researchers (e.g., Adams and Hutton, 1985; Andersson, 1986; Keegan, 1953; Schierhout et al., 1992) have also presented data which support the notion that it is not *whether* we stand or sit that causes undue postural stress but *how.*

Spinal Problems in Standing

Low back pain is common in standing workers and a number of authors have suggested reasons for this fact. Adams and Hutton (1983) have suggested that the function of the lumbar facet (apophyseal) joints is to resist the intervertebral shear force, whereas the function of the disk is to resist the compressive force (Figure 2-21).

In extended postures (e.g., when one is standing with a pronounced lumbar lordosis), the facet joints may begin to take on some of the compressive load. If the lumbar intervertebral disks are degenerated, the space between adjacent vertebrae decreases and the load on the facet joints increases even more. Adams and Hutton (1985) suggest that excessive facet joint loading may be a causal factor in the incidence of **osteoarthritis.** Bough et al. (1990) have shown that degeneration of the facet joints is a source of low back and sciatic pain.

Both Adams and Hutton (1985) and Yang and King (1984) suggest that **excessive loading of the facet joints stresses the soft tissues around the joint and causes low**

FIGURE 2-21 Function of intervertebral disk (a) and facet joints (b). The disk resists the compressive load, and the facets resist the intervertebral shear force.

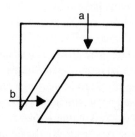
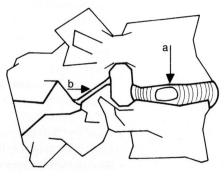

back pain. For these reasons, **excessive lumbar lordosis should be avoided when standing.** Extrapolating from this, one can determine that any workspace or task factors that require workers to arch the back greater than they would normally do should be designed out. Reaching overhead to operate controls or access work objects is an example of a factor requiring modification.

Low back pain can also be caused by **muscular fatigue** if a standing person has to work with the trunk inclined forward (when washing dishes or ironing, for example). This posture puts static load on the low back muscles, which rapidly fatigue.

For work that requires standing, the workspace must be designed to prevent workers from having to stand with excessive lumbar lordosis or having to adopt forward-flexed working positions.

Spinal Problems in Sitting

Sitting has a number of advantages over standing as a working position. The main one is that static, low-level activity of the soleus and tibialis anterior muscles is required in standing and these muscles can, eventually, fatigue. Because drainage of the lower limbs has to be done against gravity, pooling of blood in the lower limbs may occur if a person stands still for long periods. Venous return depends on the pumping action which occurs only during rhythmic contraction of the leg muscles, which occurs when people move or during postural sway. Venous pooling in the lower limbs causes swelling at the ankles. In extreme circumstances, the reduced flow of blood back to the heart may cause a drop in blood pressure and the person may faint. The hydrostatic head which has to be overcome to return blood to the heart from the lower limbs is less in sitting than in standing if the seat is correctly designed.

However, the sitting posture has its own problems. In a study at the Eastman Kodak Company in New York, 35 percent of sedentary workers visited the medical department with complaints of low back pain over a ten-year period. People with existing low back problems often cannot tolerate the sitting position for more than a few hours over the workday.

Keegan (1953) was one of the first authors to discuss the anatomy of the sitting position in relation to the problem of low back pain in sedentary work. In the standing position, the iliopsoas muscle is lengthened. In this position it stabilizes the pelvis in an anteriorly tilted position and the lumbar spine in lordosis. In the sitting position, the increased pull of the hamstrings and gluteals against the weakened hip flexors causes the pelvis to tilt rearward and the lumbar lordosis to be lost. Loss of the lumbar lordosis occurs **reflexively,** as a way of compensating for the rearward tilting of the pelvis which occurs in the sitting position. As the pelvis tilts rearward when a person sits down, the lumbar spine flexes to keep the trunk and head erect.

When the lumbar spine is flexed, the front part of the intervertebral disks is compressed and the rear part stretched. This "wedging" action causes the disks to be extruded rearward, pressurizing the posterior spinal ligaments and possibly the nerve roots. This may result in a sensation of pain in the low back and referred pain in the legs (sciatic pain), respectively (Figure 2-22).

It is generally accepted that flexed sitting postures should be avoided because of the

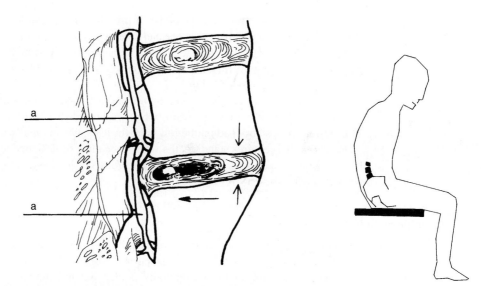

FIGURE 2-22 Anterior wedging of the intervertebral disk occurs in the slumped sitting position (a = posterior ligaments). Soft tissues between the anterior and posterior elements of the spine may be pressurized, resulting in pain. (Adapted from Keegan, 1953, with permission of the publisher.)

posterior protrusion of the disks caused by anterior wedging strain. Mandal (1981, 1991) has criticized the use of the "90-degree" concept in seat design for this reason, stating that the slumped, flexed posture is inevitable if seats are designed in this way (Figure 2-23).

A number of researchers have investigated the effects of hip and knee flexion, pelvic tilt, and lumbar flexion in order to identify hip and knee angles which do not cause extreme lumbar postures (Brunswic, 1984; Eklund and Liew, 1991; Bridger et al., 1992). Brunswic concluded that if the knees are flexed by 110 degrees, seat tilts between 5 degrees rearward to 25 degrees forward were acceptable. If the knees were flexed by 70 degrees or less, forward tilt of at least 10 degrees was required. Bridger et al. found that kneeling, a posture used in developing countries and by children in western society (Hewes, 1957), placed the lumbar spine almost at the midpoint of its range of movement and was optimal from this point of view if not from that of the knees.

These findings provide some support for the use of chairs with forward-sloping seats, or high chairs of the sit-stand variety (Corlett and Eklund, 1986). They also suggest that seated operators should be able to operate foot pedals with their knees in a flexed position (at least 70 degrees). If necessary, the seat should be adjustable to enable those with short legs to move the seat closer to the pedals. Certainly, **seated workers should never be required to fully extend a leg to operate foot pedals.**

Risk Factors for Musculoskeletal Disorders

Any factors which reduce the physical strength of body parts will increase the risk of injury. Much is now known about the strength of spinal motion segments and their com-

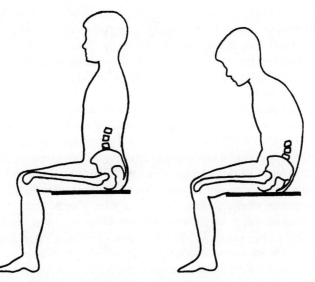

FIGURE 2-23   Most people are unable to sit erect in the 90-degree posture for long periods and soon adopt a slumped posture.

ponent disks and intervertebral bodies, as a result of the in vitro studies of Yamada (1970), Adams and Hutton (1980, 1985, for example), and others. These studies typically involve removing motion segments from cadavers and testing them to failure in a compression testing machine. Variations on the method include first placing the motion segment into a flexed posture before compressing it or combining compression with axial rotation. The load (in newtons) at which failure occurs is recorded and the specimen is removed from the machine and examined to determine where and how failure has occurred (the vertebral body itself can fail or the bony end plates can fail; the disk can also fail).

Detailed discussion of spinal compression tolerance limits is beyond the scope of this volume and the reader is referred to Genaidy et al. (1993) for more discussion. However, the data obtained from mechanical testing of spinal motion segments enable tolerable loads to be identified which will not cause failure of lumbar motion segments. The spinal compression tolerance limit (SCTL) is the maximum compressive load that a specified motion segment can be exposed to without failure. In practical situations, manual handling tasks can be evaluated using biomechanical models (see Chaffin and Anderson, 1984) to estimate the compression load. If the estimated load exceeds the SCTL, then the tasks must be redesigned, either by reducing the load or the load moment.

According to Genaidy et al., several factors reduce the SCTL and therefore increase the risk of injury. SCTL is greatest in 20- to 29-year-olds, declining by 22 percent in the next 10 years, 26 percent in the next 10, and 42 percent in the next 10. At 60 or more years of age the SCTL has declined by 53 percent. Female SCTLs are approximately 67 percent of male values. SCTLs are lower when spinal motion segments are loaded in complex ways, as when compression and bending are combined. A hyperflexed lumbar spine has a lower SCTL than a flexed lumbar spine (Adams and Hutton, 1985). In a flexed position, however, the compressive load on the thicker part of the disk (the anterior annulus) is increased and the load on the facet joints is reduced. Physical activity seems to strengthen both the vertebral bodies and the intervertebral disks. Women, older workers, and those unaccustomed to lifting should not be expected to carry out forceful exertions at work.

In addition to tissue strength and loading, the dynamic aspects of task performance are of interest in relation to the prevalence of musculoskeletal disorders. Marras and Schoenmarklin (1993) investigated the wrist motions of workers engaged in jobs entailing high or low risk of contracting carpal tunnel syndrome. Wrist position was characterized in terms of flexion/extension, pronation/supination, and ulnar/radial deviation. The angular velocity and accelerations of the wrist when moving in the corresponding spatial planes were also recorded. No significant differences were found between the high- and low-risk groups in **wrist posture.** However, **wrist movements** in the high-risk group had greater velocity and acceleration. The findings were interpreted by the authors in terms of Newton's second law of motion (force = mass $\times$ acceleration). In order to produce greater accelerations of the wrist, larger muscle forces are required which are transmitted to the bones via the tendons. The tendons will also be exposed to greater friction by contact with surrounding stuctures.

In a similar vein, Marras et al. (1993) investigated trunk motion in relation to the risk of low back disorder in a variety of jobs. As might be expected, high-risk jobs were associated with high load moments and lifting frequencies and large trunk flexion angles. In addition, lateral trunk velocity and twisting trunk velocity were also associated with high-risk jobs—the faster these movements were, the greater was the risk of injury. This finding can be explained by recalling that a major function of muscles is to stabilize joints. When a body part has to be moved rapidly, in a controlled way, greater coactivation of synergistic and antagonistic muscles is needed to stabilize the joints involved. Since many of the muscles involved will be working against one another, the result is a magnification of the joint loading (Marras and Mirka, 1992).

It seems that many bodily structures will be at risk of injury when exposed to repetitive movements requiring rapid acceleration of limbs or fast responses. A general principle for the prevention of disorders would seem to be to reduce the repetition rate and not just the number of repetitions or work cycles in a job. One way to do this is **by combining the microelements of tasks into larger units.** Further details of the mechanisms of tissue response to mechanical loading resulting in injury can be found in Chapters 5 and 6. A method for characterizing postural stress and repetitive motion can be found in Radwin and Lin (1993). A checklist for evaluating risk factors for upper extremity disorders can be found in Keyserling et al. (1993).

BEHAVIORAL ASPECTS OF POSTURE
Body Links

Anyone who has ever attempted to hold an unconscious person upright will readily appreciate that the body is like a floppy chain of interlinked segments which can move in unpredictable ways if not stabilized. Stabilization can take place internally—via low-level muscular activity—and externally—by the person's interacting with objects in the environment. In the absence of external stabilization, tonic (low-level) muscle activity is required to maintain posture and poise. Although we are not normally aware of this activity and do not perceive it as arduous work, it can lead to discomfort if sustained for long periods. Behaviors such as folding the arms and crossing the legs are postural strategies which turn open chains of body links into approximate closed chains stabilized by friction. Closed chains have fewer degrees of freedom for movement and move in more predictable ways when subjected to destabilizing forces. The net effect of such postural control strategies is to reduce the postural load on the muscles involved. This may be perceived as being "comfortable."

Dempster (1955) viewed the body as an open-chain system of links. Each joint of the body has a freedom for angular motion in one or more directions. A complex linkage such as that between the shoulder, arm, and hand has many degrees of freedom of movement, and power transmission is impossible without accessory stabilization of the joints by muscle action. For example, supination of the wrist may be required to turn a door handle, but this is possible only if the elbow and shoulder joints are stabilized and can counter the reaction at the hand-handle interface.

We are not normally aware of the postural reflexes which exist to control muscle-joint systems even though these reflexes are essential for the performance of nearly all our daily activities. Individual muscles almost never work in isolation to produce a movement. Rather, **synergistic** recruitment of muscles takes place in which contraction of the prime mover is accompanied by contraction of surrounding muscles to position and stabilize the joints. In very few activities do the trunk and limb muscles exert direct forces on the environment—rather, they maintain joint postures such that shifts in body weight can be transmitted via the chain of body links to exert a force. The use of body weight when kneading dough is a simple example of the importance of joint stabilization which enables shifts in body weight to be used to carry out a task.

The same principle also underlies the performance of almost all highly skilled activities in which very large forces are exerted or where body parts or external objects undergo rapid acceleration through pivoting actions. Throwing a discus and swinging a golf club or a tennis racket are excellent examples of this principle (Figure 2-24).

Preparation for Action

In tennis and golf and in the use of hand tools such as hammers, very large forces can be exerted by skilled performers with little apparent effort. One of the key features of this behavior is the efficient stabilization of joint postures and the use of a **backswing** involving eccentric contraction of muscles and stretching of tendons. This backswing can be likened to an archer's pulling back the bowstring before releasing the arrow. In the case of club and racket sports and hand tools, it is the hands, rather than the arrow, which are released when sufficient tension has built up.

In repetitive and particularly in cyclic movements, the function of tendons and ligaments is to store energy. When a person is walking, the stretching and release of the Achilles tendon throughout the gait cycle stores some of the energy when the foot is

FIGURE 2-24   The body link concept in action. Most activities which result in the application of large forces or rapid acceleration of body parts are caused by shifts in body weight and the pivoting action of the links in the open chain of body segments.

planted on the ground, which energy is released during the push-off phase. The deformation of the arch of the foot at these times serves a similar function. The combined effect is to increase the energy efficiency of walking. This dual action illustrates that tendons and ligaments play an **active** role in many activities and that their continued ability to withstand tension and compression is **essential** if these activities are to be carried out efficiently. Excessive tension and compression or **a regime of loading cycles** which exceeds the body's repair capability can occur in industrial jobs as well as in sports if tasks or tools are badly designed. The outcome is pain and diminished capability. Further discussion of musculoskeletal problems of this nature may be found in Chapter 5.

The body-link concept is often of value in posture analysis since the muscular workload of many tasks can be reduced if the operator stabilizes a body part by forming temporary, approximate closed chains with his or her environment. For example, upper body weight can be used to stabilize a forearm and hand resting on a work surface. The hand which is stabilized in this manner can then grasp a component while the other hand uses a tool to work on the component.

CURRENT AND FUTURE RESEARCH DIRECTIONS

Basic modeling of spine biomechanics in health and disease continues to be one of the main areas of fundamental research of relevance to ergonomics. Better models which account for the loading patterns of spinal motion segments in different activities and which enable loads to be predicted will be of obvious value in drafting safety standards for industry. Together with a better understanding of the processes of spinal degeneration, these models will support a more scientific approach to the prevention of spinal degeneration and disease.

It is often stated that degeneration of the lumbar spine is a disease of western industrial society rather than an inevitable consequence of aging. Few cross-cultural studies of spinal degeneration have been carried out. Those that have suggest this area is a promising one for future research.

Those wishing to learn more about spine biomechanics are directed to the book by Chaffin and Anderson in the Bibliography and the research of Nachemson. The paper by Gracovetsky and Farfan (1986) provides an interesting point of departure for further study.

SUMMARY

Evaluation of the physical workplace requires a basic knowledge of human anatomy and body mechanics. If a physical task is to be carried out in a safe and comfortable manner, a number of physical requirements must be met. First, the body must be stable. Stability depends on the relationship between the body parts and the base of support provided by the feet, the seat, and any other surfaces in the workplace which can be used to support the body weight. The design of a workspace can determine the range of stable postures which can be adopted and can be evaluated from this point of view. In vehicles, destabilizing external forces due to motion may also have to be accounted for. If a task requires a posture to be held for any length of time, posture analysis is necessary. A starting point for such an analysis is to determine the mechanisms by which the posture is maintained, whether static muscular effort is required, whether ligaments are being strained, and whether parts of the work surface such as the backrest of a chair or the work surface itself is providing support. Knowledge of the anatomy of the spine and pelvis is particularly valuable here as is an understanding of the mechanisms of physical fatigue.

Finally, it is essential to observe a worker's postural behavior. Skilled workers use postural strategies both to minimize fatigue and to enable them to exert large forces efficiently.

ESSAYS AND EXERCISES

1 One of the best ways to develop an understanding of anatomy is to practice drawing anatomical structures. Copy the anatomical drawings in this book and label the various structures using the appropriate terms.
2 Go to the anatomy section of your library. Obtain books on anatomy and try to improve on your drawings.
3 Observe people carrying out the following or similar activities:
   a Students taking an exam
   b A bus driver behind the wheel

    c A person washing dishes
    d A computer programmer at work
    e An archer aiming at a target
    f A gardener digging a hole
    g A mother holding a young child
    h A carpenter operating a lathe

Draw stick figures to indicate the position of the main body parts and the base of support at the feet. Indicate where you think the main areas of static and dynamic body load are by considering how the person maintains the posture. Try to visualize the load on underlying structures and indicate whether the task load is greater than the postural load.

FOOTNOTE: Origins and Antiquity of the Bipedal Posture in Homo sapiens

The upright posture and bipedal gait of humans has a long evolutionary history. Charteris et al. (1982) attempted to reconstruct the gait of *Australopithecus afarensis* (a small hominid weighing about 30 kilograms (kg) and with a cranial capacity comparable to a chimpanzee) from 3.7-million-year-old fossilized footprints found at Laetoli, East Africa. Their estimates of stride length, cadence, and velocity lead them to conclude that Australopithecus was fully bipedal.

Figure 2-25 depicts the bipedal posture adopted by a human and by a chimpanzee. The COG of the chimp's body lies anterior to its lumbar spine and hip joint, whereas in the human, it lies just above the hip joint. When the position of the COG of the various body parts is compared, it is apparent that in humans the COG of superincumbent body parts lies above and close to the various joints of the body in relaxed standing. This position of the COG reduces the bending moments on the spine and increases the mechanical advantage of the gluteal muscles in keeping the trunk erect—preventing it from jackknifing forward over the legs (Figure 2-25).

Additionally, because the chimpanzee has limited hip extension and cannot position the femur perpendicular to the ground, the knee joint has to flex slightly to position the feet below the COG of the body. Thus, the joints in the standing chimp are stabilized by isometric muscle activity in the legs and flexion strain of the spine. This anatomical analysis leads to the conclusion that, **in chimpanzees, standing is a position of high postural stress.**

Some of the main anatomical adaptations to bipedalism can be summarized as follows (Robinson, 1972; McHenry and Temerin, 1979; Lovejoy, 1988):

1 A lowering of the relative COG of the body by means of a broadening and flattening of the ilium and sacrum, accompanied by shortening of the arms and lengthening of the legs.

2 A change in the function of the gluteal muscles (Figure 2-26). In apes, gluteus medius and gluteus minimus are major hip extensors used for locomotion. Gluteus maximus is of minor importance. These muscles are attached to the iliac blades and the upper femur. In chimpanzees, the iliac blades are flat and lie across the body in a single plane

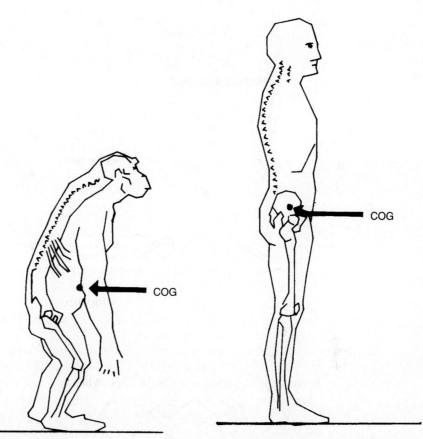

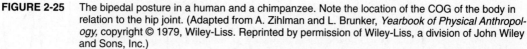

FIGURE 2-25   The bipedal posture in a human and a chimpanzee. Note the location of the COG of the body in relation to the hip joint. (Adapted from A. Zihlman and L. Brunker, *Yearbook of Physical Anthropology,* copyright © 1979, Wiley-Liss. Reprinted by permission of Wiley-Liss, a division of John Wiley and Sons, Inc.)

(Figure 2-26B), whereas in hominids (including Homo sapiens), the ilium is rotated forward around the body (Figure 2-26C).

This rotation displaces the anterior gluteals into a lateral position on the body and converts them into hip abductors. Gluteus maximus remains posteriorly positioned and exhibits hypertrophy. In hominids, its function is no longer locomotion (as it is in apes and most other quadrupedal mammals) but to stabilize the now erect trunk and prevent its jackknifing forward over the legs when a person is walking. The sacrum widens to leave room for the viscera and birth canal.

3 For part of the bipedal gait cycle, the pelvis is supported on only one leg. To stabilize the upper body, the reoriented anterior gluteals act as hip abductors. Viewed frontally, the hip joint acts like a fulcrum with the weight of the upper body and the unsupported leg on one side. This is counteracted by the hip abductors on the other side

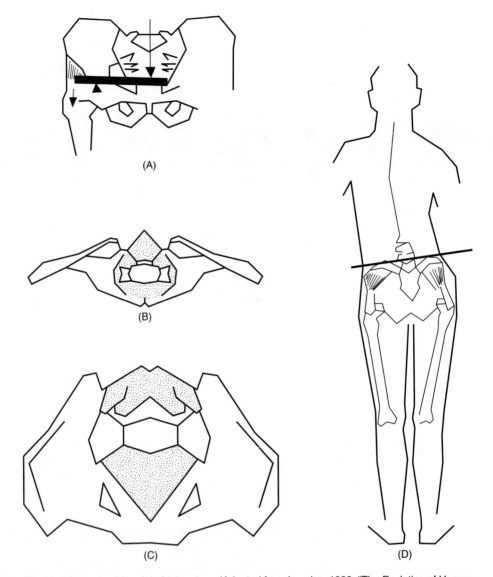

FIGURE 2-26   The hip joint and pelvis related to posture. (Adapted from Lovejoy, 1988, "The Evolution of Human Walking," Copyright © 1988, by Scientific American, Inc. All rights reserved.) (A) Frontal view of a human pelvis and hip joint. The joint acts as a fulcrum during walking. The anterior gluteals act as hip abductors on the supported leg and counteract the adductor moment exerted by the upper torso and swinging leg. This action prevents the trunk from tilting toward the side of the swinging-through leg in walking. (B) Top view of the pelvis of a chimpanzee. The ilia are flat and lie almost in a single plain. The gluteal muscles act as hip extensors in the chimpanzee. (C) Top view of a human pelvis. Note that the ilia are curved, which provides lateral attachment points for the anterior gluteals. This enables these muscles to abduct the hip, as described in (A). (D) "Trendelenburg" posture. Weakness in the anterior right gluteals causes deviation of the pelvis to the right and lateral tilting to the left. The spine exhibits compensatory scoliosis to keep the trunk erect. This type of posture can also be observed when people have to stand on an uneven surface to work.

(Figure 2-26A). In hominids, the femoral neck is lengthened and the ilium is flared outward, away from the body, to increase the lever arm of the hip abductors. People with weak or paralyzed hip abductors exhibit the "Trendelenburg gait," in which the pelvis tilts toward the unsupported leg and the upper body compensates by tilting toward the supported leg.

4 In order for a hominid to stand erect, the trunk must be repositioned from horizontal in the quadrupedal position to vertical in the bipedal position. This position cannot be achieved by simply rotating the pelvis on the vertical femur by 90 degrees such that the trunk is vertical, because extension of the femora required for walking (i.e., extension past the vertical by the trailing leg) is then impeded by the ischium (Figure 2-27). This is one reason the great apes cannot walk on two legs efficiently.

Rather counterintuitively, the evolutionary solution seems to have been to rotate the sacrum forward on the ischium and increase the extension of the lumbar spine. **This is the origin of the lumbar lordosis** which is of so much concern to ergonomists.

The pelvis can be likened to the platform which supports the spine and transmits the weight of the upper body to the ground via the legs. Many muscle groups which in other species provide locomotive power take on new roles in the bipedal posture. The anterior gluteals now stabilize the pelvis; the iliopsoas muscle initiates "swing-through" of the trailing leg in walking and stabilizes the lumbar spine in standing. The hamstrings de-

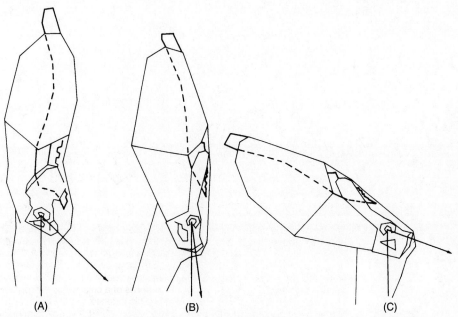

(A)   (B)   (C)

FIGURE 2-27   Orientation of the ischium with respect to the hip joint. (Adapted from Robinson, Copyright © 1972, Chicago University Press, with permission.) (A) Human standing erect. (B) Chimpanzee standing erect. (C) Chimpanzee in its normal quadrupedal position. In (B) the ischium prevents extension of the thigh past the vertical. Comparison of (A) and (C) reveals the fundamental nature of the lumbar lordosis as an adaptation to bipedalism.

celerate the swinging leg prior to heel strike in walking. Only the quadriceps and the plantar flexors (gastrocnemius and soleus) are left to provide a ground reaction force for locomotion.

The erection of the trunk is essential for a bipedal posture and was only partly achieved by backward rotation of the pelvis. The rest was achieved by means of inversion of the normal quadrupedal lumbar curve (concave anteriorly) into a lumbar lordosis (concave posteriorly). The posture of the spine—particularly the lumbar spine—is a topic of great interest in ergonomics. From the above discussion it can be seen that it is only one of a number of other anatomical adaptations which characterize human bipedalism.

3

ANTHROPOMETRIC PRINCIPLES IN WORKSPACE AND EQUIPMENT DESIGN

The word "anthropometry" means measurement of the human body. It is derived from the Greek words *anthropos* ("man") and *metron* ("measure"). Anthropometric data are used in ergonomics to specify the physical dimensions of workspaces, equipment, furniture, and clothing so as to "fit the task to the man" (Grandjean, 1980) and to ensure that physical mismatches between the dimensions of equipment and products and the corresponding user dimensions are avoided.

ANTHROPOMETRY AND ITS USES

As a rule of thumb, if we take the smallest female and the tallest male in a population, the male will be 30 to 40 percent taller, 100 percent heavier, and 500 percent stronger (Grieve and Pheasant, 1982). Clearly, the natural variation of human populations has implications for the way almost all products and devices are designed. Some obvious examples are clothes, furniture, and automobiles. Anthropometric data can be used to optimize the dimensions of a diverse range of items—the length of toothbrush handles, the depth and diameter of screwtops on jars and bottles, the size of tools in tool kits supplied with automobiles, and almost all manual controls, such as those that are found on televisions, videocassette recorders, radios, etc.

Body size and proportion vary greatly between different population and racial groups—a fact which designers must never lose sight of when designing for an international market. A U.S. manufacturer hoping to export to Central and South America or southeast Asia would need to consider in what ways product dimensions optimized for a large U.S. and probably male user group would suit Mexican or Vietnamese users, who belong to one of the smallest population groups in the world. Ashby (1979) illustrated the importance of anthropometric considerations in design as follows:

If a piece of equipment was designed to fit 90% of the male U.S. population, it would fit roughly 90% of Germans, 80% of Frenchmen, 65% of Italians, 45% of Japanese, 25% of Thais and 10% of Vietnamese.

It is usually impracticable and expensive to design products individually to suit the requirements of every user (although this is a recent development in the history of design). Most are mass-produced and designed to fit a wide range of users—the custom tailor, dressmaker, and cobbler are perhaps the only remaining examples of truly user-oriented designers in western industrial societies.

In the design of mass-produced items the task of the ergonomist is first to characterize the way a product is to be used and then to identify the issues which might affect usability—including the constraints which are imposed on the design by the anthropometry of the user population. From this, anthropometric dimensions appropriate to the design of the particular product can be specified. Second, the necessary data from the corresponding consumer/user group are obtained for use in dimensioning either the product itself or its range(s) of adjustability.

Availability of Anthropometric Data

The anthropometry of military populations is usually well documented and is used in the design of everything from cockpits to ranges and sizes of boots and clothing. Anthropometric surveys are expensive to carry out since large numbers of measurements have to be made on sizable samples of people representative of the population under study. Apart from military populations, data are available for U.S., British, and other European groups, as well as Japanese citizens. Pheasant (1986) provides a useful and well-illustrated collection of anthropometric data and a method of estimating unknown anthropometric dimensions from data on stature.

Problems with much of the anthropometric data from the United States and Europe are the age of the data and the lack of standardization across surveys (not all researchers measure the same anthropometric variables). Marras and Kim (1993) note that the first large-scale survey of civilian women in the United States was carried out in 1941 for garment sizing purposes. From 1960 to 1962, the National Center for Health Statistics carried out a survey of 20 anthropometric variables of both men and women. A survey of civilian weights and heights was carried out between 1971 and 1974. In Britain, several civilian surveys have been published since 1950 (see Oborne, 1992), but many of these suffer the same limitations as their U.S. counterparts—the data are often fragmentary, only relevant to the design problems of a particular industry and probably out of date. Marras and Kim (1993) present recent measurements of 12 anthropometric variables from 384 males and 124 females. Abeysekera and Shahnavaz (1989) present data from industrially developing and developed countries and discuss some of the problems of designing for a global marketplace. The International Organization for Standardization has done some work on the standardization of anthropometric measurements (ISO DIS 7250, Technical Committee 159). The ISO list of variables will be of use in the planning of anthropometric surveys.

A certain amount of data on the anthropometry of children exists but is often of limited value for designers, having been obtained by clinicians and physical anthropologists to help define normal growth rates for particular populations (i.e., normal ranges of stature and body mass as a function of age). Even for these purposes, the data can be used only to assess individuals from the same population from which they were sampled.

Types of Anthropometric Data

Structural Anthropometric Data These are measurements of the bodily dimensions of subjects in fixed (static) positions. Measurements are made from one clearly identifiable anatomical landmark to another or to a fixed point in space (e.g., the height of the knuckles above the floor, the height of the popliteal fossa, or back of the knee, above the floor, etc.). Some examples of the use of structural anthropometric data are to specify furniture dimensions and ranges of adjustment and to determine ranges of clothing sizes. Figure 3-1 shows structural variables which are known to be important in the design of vehicles, products, workspaces, and clothing. Figure 3-2 shows examples of vehicle dimensions which would require user anthropometry to be specified. Tables 3-1 to 3-10 present selected anthropometric data from various parts of the world (collated from Ashby, 1979; Pheasant, 1986; Woodson, 1981; Ministry of Science and Technology, 1988).

The reader is advised to study the tables, looking, in particular, for differences in body **proportion** between different groups. For example, there is an approximately 100-millimeter (mm) stature difference between U.S. and Japanese males in the standing position. This decreases to between 5 and 25 mm in the seated position. When considering possible modifications to a vehicle to be exported from one country to the other, the designer might, therefore, first pay attention to the rake of the seat and position of the foot pedals, rather than to the height of the roof.

Functional Anthropometric Data These data are collected to describe the movement of a body part with respect to a fixed reference point. For example, data are available concerning the maximum forward reach of standing subjects. The area swept out by the movement of the hand can be used to describe "workspace envelopes"—zones of easy or maximum reach around an operator. These zones can be used to optimize the layout of controls in panel design. The size and shape of the workspace envelope depends on the degree of bodily constraint imposed on the operator. The size of the workspace envelope increases with the number of unconstrained joints. For example, the area of reach of a seated operator is greater if the spine is unencumbered by a backrest and can flex, extend, and rotate. Standing reach is also greater if the spine is unconstrained and greater still if there is adequate foot space to enable one or both feet to be moved. Somewhat counterintuitively, one way to increase a worker's functional hand reach is to provide more space for the feet.

Generally speaking, fewer functional than structural anthropometric data are available. Although clinicians have long been interested in determining normal ranges of joint movement in healthy individuals to assist in assessment of patients, these data are not always quantified nor even directly applicable to design problems. However, existing functional anthropometric data are useful for designing workspaces and positioning

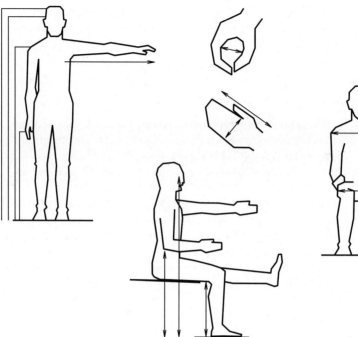
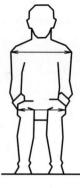
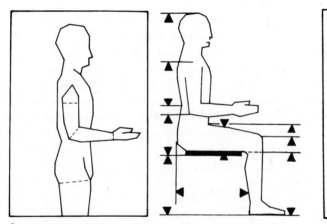
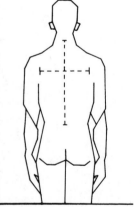

FIGURE 3-1   Some common structural anthropometric variables.

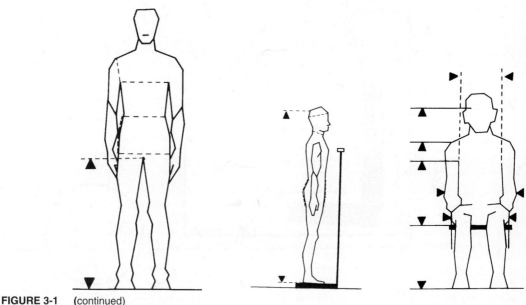

FIGURE 3-1   (continued)

objects within them—particularly in the design of aircraft cockpits, crane cabs, vehicle interiors, and complex control panels in the process industries (Figure 3-3).

Newtonian Anthropometric Data Such data are used in mechanical analysis of the loads on the human body. The body is regarded as an assemblage of linked segments of known length and mass (sometimes expressed as a percentage of stature and body weight). Ranges of the appropriate angles to be subtended by adjacent links are also given to enable suitable ranges of working postures to be defined. This defining enables designers to specify those regions of the workspace in which displays and controls may be most optimally positioned. Newtonian data may be used to compare the loads on the spine from different lifting techniques.

PRINCIPLES OF APPLIED ANTHROPOMETRY IN ERGONOMICS

Anthropometric variables in the healthy population are usually considered to follow a normal distribution, as depicted in Figure 3-4.

The Normal Distribution

For design purposes, two key **parameters** of the normal distribution are the **mean** and the **standard deviation.** The mean is the sum of all the individual measurements divided by the number of measurements. It is a measure of central tendency. The standard deviation is calculated using the difference between each individual measurement and the

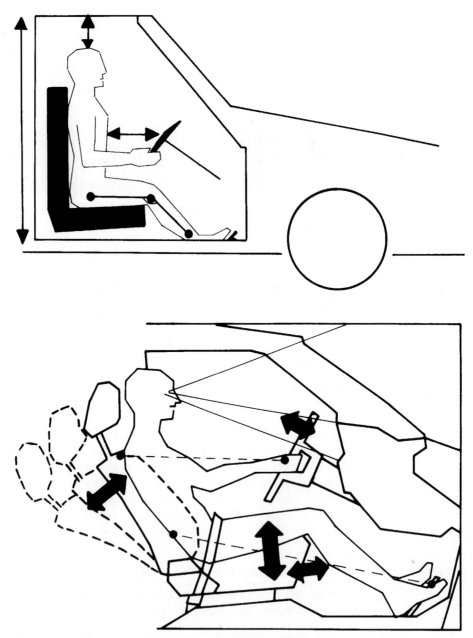

FIGURE 3-2 Some product dimensions which are determined using anthropometric considerations.

TABLE 3-1   STATURE OF SELECTED ADULT POPULATIONS (in millimeters)

Population	Adult males		Adult females	
	5th percentile	95th percentile	5th percentile	95th percentile
Americans (U.S.)	1640	1870	1520	1730
Northern Europeans	1645	1855	1510	1720
Japanese	1560	1750	1450	1610
Brazilians	1595	1810		
Africans	1565	1790		

TABLE 3-2   EYE HEIGHT OF SELECTED ADULT POPULATIONS (in millimeters)

Population	Adult males		Adult females	
	5th percentile	95th percentile	5th percentile	95th percentile
Americans (U.S.)	1595	1825	1420	1630
Northern Europeans	1540	1740	1410	1610
Japanese	1445	1635	1350	1500
Brazilians	1490	1700		
Africans	1445	1670		

TABLE 3-3   ELBOW HEIGHT OF SELECTED ADULT POPULATIONS (in millimeters)

Population	Adult males		Adult females	
	5th percentile	95th percentile	5th percentile	95th percentile
Americans (U.S.)	1020	1190	945	1095
Northern Europeans	1030	1180	910	1050
Japanese	965	1105	895	1015
Brazilians	965	1120		
Africans	975	1145		

TABLE 3-4   FINGERTIP HEIGHT OF SELECTED ADULT POPULATIONS (in millimeters)

Population	Adult males		Adult females	
	5th percentile	95th percentile	5th percentile	95th percentile
Americans (U.S.)	565	695	540	660
Northern Europeans	595	700	510	720
Japanese	565	695	540	660
Brazilians	590	685		
Africans	520	675		

TABLE 3-5   SITTING HEIGHT OF SELECTED ADULT POPULATIONS (in millimeters)

Population	Adult males		Adult females	
	5th percentile	95th percentile	5th percentile	95th percentile
Americans (U.S.)	855	975	800	920
Northern Europeans	865	970	795	895
Japanese	850	950	800	890
Brazilians	825	940		
Africans	780	910		

TABLE 3-6   SITTING EYE HEIGHT OF SELECTED ADULT POPULATIONS (in millimeters)

Population	Adult males		Adult females	
	5th percentile	95th percentile	5th percentile	95th percentile
Americans (U.S.)	740	860	690	810
Northern Europeans	760	845	695	685
Japanese	735	835	690	780
Brazilians	720	830		
Africans	670	790		

TABLE 3-7   ELBOW REST HEIGHT OF SELECTED ADULT POPULATIONS (in millimeters)

Population	Adult males		Adult females	
	5th percentile	95th percentile	5th percentile	95th percentile
Americans (U.S.)	195	295	185	285
Northern Europeans	195	270	165	245
Japanese	220	300	215	285
Brazilians	185	275		
African	175	250		

TABLE 3-8   POPLITEAL HEIGHT OF SELECTED ADULT POPULATIONS (in millimeters)

Population	Adult males		Adult females	
	5th percentile	95th percentile	5th percentile	95th percentile
Americans (U.S.)	395	495	360	450
Northern Europeans	390	460	370	425
Japanese	360	440	325	395
Brazilians	390	465		
Africans	380	460		

TABLE 3-9 BUTTOCK-POPLITEAL LENGTH OF SELECTED ADULT POPULATIONS (in millimeters)

	Adult males		Adult females	
Population	5th percentile	95th percentile	5th percentile	95th percentile
Americans (U.S.)	445	555	440	540
Northern Europeans	455	540	405	470
Japanese	410	510	405	495
Brazilians	435	530		
Africans	425	515		

TABLE 3-10 HIP WIDTH OF SELECTED ADULT POPULATIONS (in millimeters)

	Adult males		Adult females	
Population	5th percentile	95th percentile	5th percentile	95th percentile
Americans (U.S.)	310	410	310	440
Northern Europeans	320	395	320	440
Japanese	280	330	270	340
Brazilians	306	386		
Africans	280	345		

mean. It is a measure of the degree of dispersion in the normal distribution. Thus, the value of the mean determines the position of the normal distribution along the *x*-axis (the horizontal axis). The value of the standard deviation determines the shape of the normal distribution. A small value of the standard deviation indicates that most of the measurements are close to the mean value (the distribution has a high peak which tails off rapidly

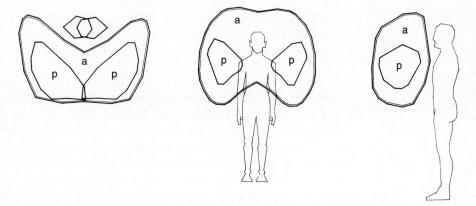

FIGURE 3-3 Functional anthropometric data. The figure shows the shapes of the reach envelopes and the allowable (a) and preferred (p) zones for the placement of controls in a workspace.

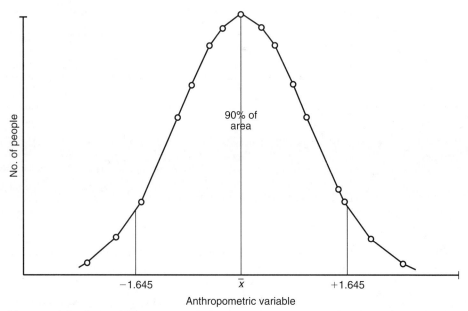

FIGURE 3-4   The normal distribution. Ninety percent of the measurements made on different people will fall in a range whose width is 1.64 standard deviations above and below the mean.

at both sides). A large value of the standard deviation means that the measurements are scattered more distantly from the mean. The distribution has a flatter shape.

In order to estimate the parameters of stature in a population (the mean and standard deviation), one must measure a large sample of people who are representative of that population (e.g., central African females, male Amazonian Indians, U.S. female bus drivers). The formulas given in Figure 3-5 can then be used to calculate the estimates of the mean and standard deviation. Estimates of **population parameters** obtained from calculations on data from samples are known as **sample statistics.**

The distribution of stature in a population may be used to exemplify the statistical approach to design. The assumption that stature is normally distributed in the population, together with the estimates of the mean and standard deviation, enables stature to be characterized in a quantitative manner. An important characteristic of the normal distribution is that it is symmetrical—as many observations lie above the mean as below it (or in the graphic terms of the figure, as many observations lie to the right of the mean as to the left). If a distribution is normally distributed, 50 percent of the scores (and thus the individuals from whom the scores were obtained) lie on either side of the mean.

This simple statistical fact applies to very many variables, yet it is often ignored or misunderstood by both specialists and laypeople alike. It is common to hear statements such as "your child is below average height for his age" or "you are above average weight for your height" both pronounced and perceived in a negative way. The statistical reality is that half of any normally distributed population is either above or below "average." In itself, this fact has no negative or positive connotations. What *is* important is how *far* or how *different* from the mean a particular observation is. Observations which are different from the mean become rarer as the "distance" increases. The distance

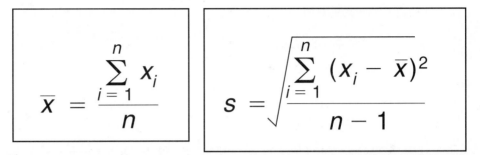

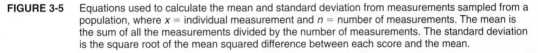

FIGURE 3-5   Equations used to calculate the mean and standard deviation from measurements sampled from a population, where x = individual measurement and n = number of measurements. The mean is the sum of all the measurements divided by the number of measurements. The standard deviation is the square root of the mean squared difference between each score and the mean.

of an observation from the mean is usually measured in **standard deviations**—we can talk about an observation's being a certain number of standard deviations greater or less than the mean.

An important consequence of standard deviations is that it is unwise to design for "Mr. or Ms. Average," a mythical person with the mean stature. As can be seen, very few individuals will be *exactly* of "average" height, and therefore it is necessary to design to accommodate a range of people—preferably those both above and below the mean.

Estimating the Range

The standard deviation contains information about the spread of scores in a sample. It is known, for a normal distribution, that approximately two-thirds of the observations in the population fall within 1 standard deviation above and below the mean. Thus, for a population with a mean stature of 1.75 meters (m) and standard deviation of 0.10 m, approximately two-thirds of the population would be between 1.65 and 1.85 m tall. The remaining third would lie beyond these two extremes.

Using the standard deviation and the mean, one can calculate estimates of stature below which a specified percentage of the population will fall. From the mathematics of the normal distribution, the area under the normal curve at any point along the x-axis can be expressed in terms of the number of standard deviations from the mean. For example, if the standard deviation is multiplied by the constant 1.64 and *subtracted* from the mean, the height below which 5 percent of the population falls is obtained. If 1.64 is *added* to the mean, the height below which 95 percent of the population falls is obtained. These are known as the **5th and 95th percentile heights.** The 1st and 99th percentile heights are obtained when the constant 2.32 is used (Table 3-11).

Using the mean and standard deviation of an anthropometric measurement and a knowledge of the area under the normal curve expressed as standard deviations from the mean, one can estimate ranges of body size which will encompass a greater and greater proportion of individuals in a population. Thus, given the mean and standard deviation of any anthropometric variable, one can compute a range of statures, girths, leg lengths, etc., within which a known percentage of the population will fall.

TABLE 3-11   CONSTANTS USED TO ESTIMATE POPULATION PROPORTIONS

Required percentile	Number of standard deviations to be subtracted from or added to the mean
50th	0
10th or 90th	1.28
5th or 95th	1.64
2.5th or 97.5th	1.96
1st or 99th	2.32

Applying Statistics to Design

Statistical information about body size is not, in itself, directly applicable to a design problem. First, the designer has to analyze in what ways (if any) anthropometric mismatches might occur and then decide which anthropometric data might be appropriate to the problem. In other words, the designer has to develop some clear ideas about what constitutes an appropriate match between user and product dimensions. Next, a suitable percentile has to be chosen. In many design applications, mismatches occur only at one extreme (only very tall or very short people are affected, for example) and the solution is to select either a maximum or a minimum dimension. If the design accommodates people at the appropriate extreme of the anthropometric range, less extreme people will be accommodated.

Minimum Dimensions

A high percentile value of an appropriate anthropometric dimension is chosen. In the design of a doorway, for example, sufficient head room for very tall people has to be provided, and the 95th or 99th percentile (male) stature could be used to specify a minimum height. The doorway should be no lower than this minimum value, and additional allowance would have to be made for the increase in stature caused by items of clothing such as the heels of shoes, protective headgear, etc.

Seat breadth is also determined using a minimum dimension—the width of a seat must be no narrower than the largest hip width in the target population (Figure 3-6). Minimum dimensions are used to specify the placement of controls on machines, door handles, etc. Controls must be sufficiently high off the ground so that tall operators can reach them without stooping, i.e., no lower than the 95th percentile standing knuckle height. In the case of door handles, the maximum vertical reach of a small child might also be considered (to prevent young children from opening doors when unsupervised).

Circulation space must be provided in offices, factories, and storerooms to allow for ingress and egress of personnel and to prevent collisions. In a female or mixed-sex workforce, the body width of a pregnant woman would be used to determine the minimum. About 60 centimeters (cm) of clearance space is needed for passages, the separation of machines, and the distance of furniture from walls or other objects in a room (some anthropometric data of the seated pregnant woman can be found in Culver and Viano, 1990).

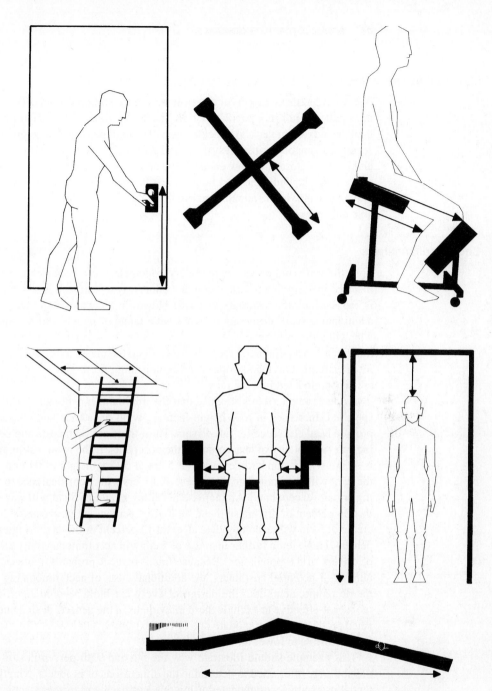

FIGURE 3-6 Some minimum dimensions. The height of a doorway must be no lower than the stature of a tall man (plus an allowance for clothing and shoes). The width of a chair must be no narrower than the hip breadth of a large woman. A toothbrush must be long enough to reach the back molars of someone with a deep mouth. An escape hatch must be wider than the shoulder width and body depth of a large person. A door handle must not be lower than the highest standing knuckle height in a population so that all users can open the door without stooping. The distance from the kneepad to the back of the seat of a "kneeling chair" must exceed the longest buttock-knee lengths in the population of users. The length of a wheel brace must provide sufficient leverage for a weak person to generate sufficient torque to loosen the wheel nuts.

Maximum Dimensions

A low percentile is chosen in determining the maximum height of a door latch so that the smallest adult in a population will be able to reach it. The latch must be no higher than the maximum vertical knuckle reach of a small person. The height of nonadjustable seats used in public transport systems and auditoriums is also determined using this principle—the seat must be low enough so that a short person can rest the feet on the floor when using it. Thus, the seat height must be no higher than the 1st or 5th percentile popliteal height in the population. Figure 3-7 gives examples of some maximum allowable dimensions.

Cost-Benefit Analysis

Sometimes it is not necessary to use anthropometric data because there may be no costs incurred in designing to suit everyone (a doorway or entrance is a hole and the bigger it is, the less building materials are used). However, there are often trade-offs between the additional costs of designing to suit a wide range of people and the number of people who will ultimately benefit. As can be seen from the normal distribution, the majority of individuals in a population are clustered around the mean and attempts to accommodate more extreme individuals soon incur diminishing returns since fewer and fewer people are accommodated into the range.

As an example, let us suppose that the minimum height of a car interior has to be specified. Increases in roof height increase wind resistance and construction cost but provide head clearance for tall drivers. Thus, there are both costs and benefits of building cars with plenty of headroom for the occupants. If the mean sitting height for males is 90 cm with a standard deviation of 5 cm, a ceiling height of 91 cm (measured from the seat) will accommodate 50 percent of drivers with 1 cm clearance to allow for clothing or hair (except hats). An increase in ceiling height of 5 cm will accommodate a further 34 percent of drivers so that a total of 84 percent are now catered for. A further increase of 5 cm will accommodate an extra 14 percent of drivers, bringing the total to 98 percent. However, a further increase of 5 cm will accommodate only an extra 2 percent of drivers in the population. Because this 2 percent probably represents a very small number of potential customers, the additional costs of accommodating their anthropometric requirements into the design of every car built become significant. It may be more cost-effective to exclude these individuals in the generic design but retrofit the finished product to accommodate one or two extremely tall buyers (for example, by making it possible to lower the seat slightly).

This example should illustrate why the 5th and 95th percentiles of anthropometric variables are often used to determine the dimensions of products. Ninety percent of potential users are accommodated using this approach and further sizable alterations will accommodate only a small number of additional users—the point of diminishing returns has been reached.

Anthropometric data must always be used in a cautious manner and with a sound appreciation of the design requirements and the practical considerations. In particular, the designer should try to predict the consequences of a mismatch—how serious they would be and who would be affected. The height of the handle of a door leading to a fire escape in an apartment building dramatizes the seriousness of such mismatches: It is es-

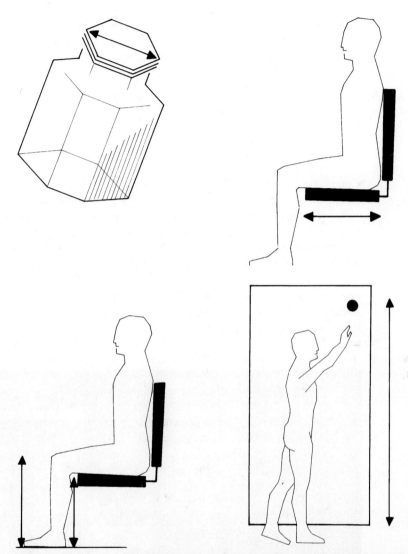

FIGURE 3-7   Some maximum allowable dimensions. A door lock must be no higher than the maximum vertical reach of a small person. Seat heights and depths must not exceed the popliteal height and buttock-knee lengths of small users. Screw-top lids must be wide enough to provide a large contact area with the skin of the hand to provide adequate friction and so that pressure "hot spots" are avoided. However, the lids must not exceed the grip diameter of a small person.

sential that a very wide range of users—including children—be able to reach and operate the handle in an emergency. The design of passenger seats for urban transportation systems is also important—although somewhat more mundane. Because the seats are used regularly by a very wide range of users, even small imperfections will affect the comfort of a very large number of people every day. In the use of anthropometric data, the selection of a suitable cutoff point depends on the consequences of an anthropometric mismatch and the cost of designing for a wide range of people. One of the ergonomist's most important tasks is to predict and evaluate what any mismatches are going to be like. It is not normally sufficient only to specify the required dimensions without considering other aspects such as usability and misuse.

Use of Mannequins

Design aids such as mannequins (jointed two-dimensional representations of the human form used in conjunction with appropriately scaled drawings or models) can help potential problems of anthropometric fit to be visualized.

Mannequins representing 5th, 50th, and 95th percentile users may be used but with caution because of differences in proportion—a person of 5th percentile stature may have 20th percentile reach and 60th percentile girth, which the mannequin will not disclose. It is incumbent on the designer therefore to try to anticipate potential anthropometric mismatches rather than using anthropometric data in a cookbook fashion.

Computerized design aids now exist to facilitate visualization of the physical interaction between users and hardware. SAMMIE (system for aiding man-machine interaction evaluation, SAMMIE CAD Ltd.) is an example of such a system (Figure 3-8). The anthropometry of a three-dimensional model displayed on the computer screen can be manipulated and the consequences evaluated using a computer-generated representation of the product being designed.

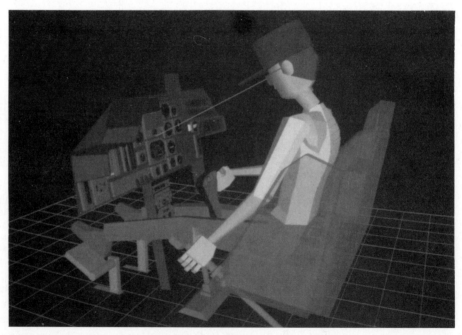

(A)

FIGURE 3-8   Computerized anthropometry. (A) JACK. (JACK is a registered trademark of the University of Pennsylvania. Compliments of the University of Pennsylvania Computer Graphics Research Laboratory.)

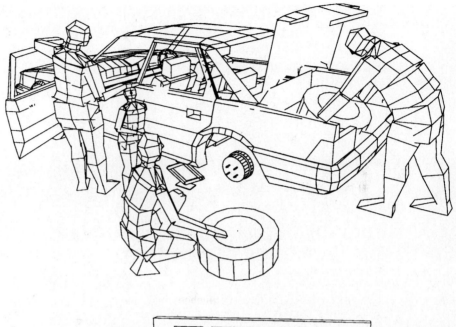

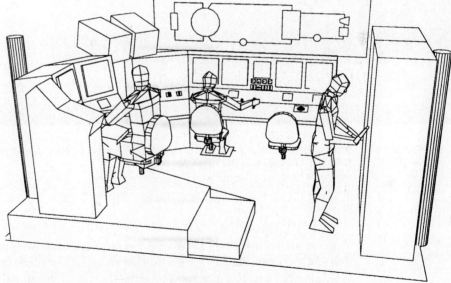

(B)

FIGURE 3-8   (continued) (B) SAMMIE. (SAMMIE CAD Ltd., Loughborough, United Kingdom.)

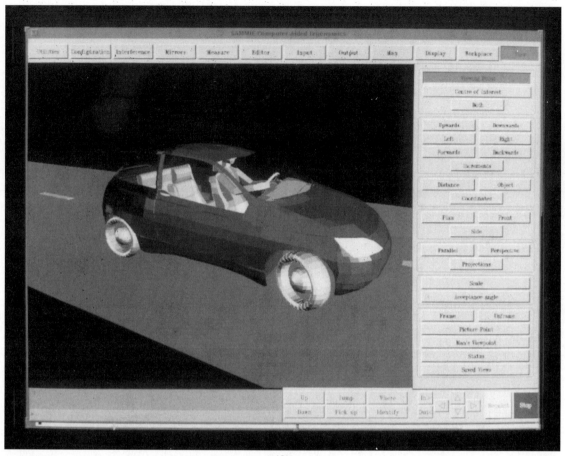

(C)

FIGURE 3-8 (continued) (C) The user-interface of the SAMMIE system for computer-aided design. (SAMMIE CAD Ltd., Loughborough, United Kingdom.)

Several computer-aided design (CAD) systems are available which incorporate model humans whose body dimensions can be manipulated using anthropometric data. The JACK system (Maida, 1993), which has 88 articulated joints of the human body and a 17-segment torso, incorporates data on human body contours and strength limits. Despite the existence of such computer-based anthropometric design aids, the successful application of anthropometric data still requires a sound understanding of the fundamental principles described above.

APPLICATIONS OF ANTHROPOMETRY IN DESIGN

Table 3-12 gives examples of some common anthropometric variables and how they are used in ergonomics. The list is intended to be representative rather than exhaustive, and

TABLE 3-12   DESCRIPTION AND USE OF SOME COMMON ANTHROPOMETRIC VARIABLES

Standing eye height: Height above the ground of the eye of a person standing erect. Can be used as a maximum allowable dimension to locate visual displays for standing operators. The displays should not be higher than the standing eye height of a short operator so that short operators do not need to extend the neck to look at displays.

Standing shoulder height: Height of the acromion above the ground. Used to estimate the height of the center of rotation of the arm above the ground and can help specify the maximum allowable height for controls so that short workers need not elevate the arms above shoulder height to operate a control.

Standing elbow height: Height above the ground of the elbows of a person standing erect. Used to design the maximum allowable bench height for standing workers.

Standing knuckle height: Height of the knuckles above the ground. Used to determine the minimum height of full grip for a standing operator. Operators with high standing knuckle heights should not have to stoop when grasping objects in the workplace.

Standing fingertip height: Height of the tips of the fingers above the ground. Used, as above, to determine the lowest allowable position for controls such as switches.

Sitting height: Distance from the seat to the crown of the head. Can be used to determine ceiling heights in vehicles to provide clearance for users with tall sitting heights.

Sitting elbow height: Height of the elbows of a seated person above the chair. Used to determine armrest heights and work surface heights for seated operators.

Popliteal height: Height of the popliteal fossa (back of the knee) above the ground. The 5th percentile popliteal height may be used to determine the maximum allowable height of nonadjustable seats. The 95th percentile popliteal height may be used to set the highest level of adjustment of height-adjustable seats.

Knee height and thigh depth: Taken together, these variables specify the height above the floor of the upper thigh of a seated person. Can be used to determine the thigh clearance required under a table.

Buttock-popliteal length: Distance from the buttocks to the back of the knee. Used the determine the maximum allowable seat depth so that seat depth does not exceed the buttock-popliteal length of the short users.

Shoulder width: Widest distance across the shoulders. Used to determine the minimum width of narrow doorways, corridors, etc. to provide clearance for those with wide shoulders.

Hip breadth: Widest distance across the hips. Used to determine the space requirements necessary for clearance and, for example, the minimum width of seats to allow clearance for those with wide hips.

Abdominal/chest depth: Widest distance from a wall behind the person to the chest/abdomen in front. Used to determine the minimum clearance required in confined spaces.

Vertical reach (sitting and standing): Highest vertical reach. Used to determine maximum allowable height for overhead controls so that they are reachable by the shortest users.

Grip circumference: Internal circumference of grip from the root of the fingers across the tip and to the palm when the person is grasping an object. Used to specify the maximum circumference of tool handles and other objects to be held in the palm of the hand. Handle circumferences should enable those with small hands to grasp the tool with slight overlap of the thumb and fingers.

Reach: The dimensions of the reach envelope around an operator can be used to locate controls so that seated operators can operate them without having to lean forward away from the backrest or twisting the trunk and standing operators can operate them without forward, backward, or sideways inclination of the trunk. Arm movements should be kept in the normal work area to eliminate reach over 40 cm for repeated actions. These data are applicable to the design of all vehicle cockpits and cabs.

further information may be found in Croney (1980), Clark and Corlett (1984), and Pheasant (1986). Some common approaches to the anthropometric solution of design problems are described below.

DESIGNING FOR EVERYONE

The problem of designing to suit a range of users can be approached in several different ways.

Make Different Sizes

In clothing and school furniture design, a common solution is to design the same product in **several different sizes.** Anthropometric data can be used to determine a minimum number of different sizes (and the dimensions of each size) which will accommodate all users. Mass production or long production runs often bring economies of scale in product design through reduced retooling and stoppages. Mass production usually has economic benefits and demonstrates why it is important to determine the **minimum number** of sizes in a product range which will accommodate most of the users in the population in question.

An unusual and interesting example of this approach comes from some research on the optimum length of chopsticks, which was carried out by Hsu and Wu (1991). Chopsticks can be classified as hand tools, which extend the capabilities of the hand in the following ways:

1 They extend reach.
2 They pick up food.
3 They protect the hand from heat.
4 They protect the food from bacteria or grime on the hands.

Chopsticks operate as a third-class lever system, which means that the longer the chopstick, the greater the excursion (opening and closing) of the tips and the lower the pinch force for the same amount of effort. Longer chopsticks extend the user's reach more than shorter chopsticks and can be used to pick up larger items of food. However, they require greater effort to exert the same pinch force at the tip than shorter chopsticks. Thus, there is a trade-off between two aspects of usability which has to be optimized (the effort required to use the tool versus the reach and pinch diameter it provides) using anthropometric considerations.

The researchers evaluated chopstick lengths varying from 180 to 330 mm. Experimental subjects used the different chopsticks to perform a peanut-picking task, which gave an index of food-pinching efficiency. Objective measures of pinching force and subjective measures of preference were also taken. It was found that the food-pinching efficiency of adults was best with the 240-mm chopstick. For children, 180-mm chopsticks gave the best performance. The authors recommended that, since the operation of chopsticks is more difficult for novice users, families with children should buy chopsticks of both 240 mm and 180 mm. Furthermore, since longer chopsticks cost more than shorter ones, restaurants could consider using 210-mm chopsticks, which were found to be second in terms of usability and preference. The researchers included only young males and primary school pupils in the study. Since females have smaller hands than males and less strength, it would be of interest to determine the optimum length of chopstick for them. A shorter chopstick may have been preferable for females and possibly intermediate between that for men and that for children, which would have provided further support for the author's conclusions. This is a matter for empirical study but serves to illustrate how questions of cost and efficiency interact with the purely ergonomic considerations of designing to suit a wide proportion of the population.

Design Adjustable Products

An alternative approach to product design is to manufacture products whose critical dimensions for use can be **adjusted** by the users themselves. A first step is to determine what the critical dimensions for use are. The next step is to design the mechanism of adjustability with the emphasis on ease of operation. Finally, some instructions or a training program may be necessary to explain to users the need to adjust the product and how to adjust it correctly.

In seated work, for example, the height of the seat and desk are critical dimensions for seated comfort. The seat height should be no higher than the popliteal height of the user so that both feet can be rested firmly on the floor to support the weight of the lower legs (otherwise the soft tissues on the underside of the thigh take the weight and blood circulation is impeded as a result of compression of these tissues). Secondly, the desk height (or middle row of keys on a keyboard) should coincide with the user's sitting elbow height. Since popliteal height and elbow height do not correlate strongly in practice (Verbeek, 1991), adjustable seat and desk heights are needed.

Shute and Starr (1984) investigated the effects of adjustable furniture on visual display unit (VDU) users. On-the-job discomfort was reduced when either adjustable chairs or desks were used in place of nonadjustable furniture. The greatest reductions in discomfort were found when the two adjustable items were used in combination.

One problem with adjustability is that users may not use the adjustment facility if they do not *expect* a product to be adjustable or if they do not *understand* the reason for incorporating adjustability into the product. Verbeek (1991) investigated the effect of an instruction program for office workers on the anthropometric fit between users and their chair-desk workstations. Before the program a survey of chair-desk settings in an office revealed mean deviations from the ideal of 71 mm for seat height and 70 mm for desk height. A model of "correct" sitting was used as a criterion to evaluate the chair-desk setting. After the program, these deviations were reduced by 11 mm and 18 mm, respectively. However, only 7 percent of users adjusted their seat heights as advised and only 13 percent adjusted their desk heights as advised. It was concluded that this meager result was due to practical difficulties, the unaesthetic appearance of adjacent desks' having different heights and the suspect validity of the model of correct sitting which had been used to specify the method of adjustment (further discussion of concepts of correct sitting can be found in the following chapter).

Anthropometry and Personal Space

Personal space can be defined as the area immediately around the body. Argyle (1975) describes personal space in the context of territorial behavior. Many animals display this type of behavior. They regard a certain area of space as their exclusive preserve. The area immediately around an individual's body is usually regarded in this way—two important issues in "psychoanthropometry" are the volume of space regarded as personal territory and the consequences of an invasion of this space by others.

Large individual and cultural differences exist in the volume of space immediately around the body which is regarded as personal. Violent prisoners and schizophrenics have larger personal space than nonviolent prisoners and nonschizophrenics. Argyle reports the results of studies which indicate that Arabs stand closer together than Europeans and North Americans, with Latin Americans and Asians intermediate.

Invasion of personal space appears to be stressful, as indicated by objective measures of psychological stress such as the galvanic skin response (GSR). However, the degree of stress depends on the context; invasion of personal space in a library, for example, is much more stressful than in a crowded train or elevator. Personal space is another important consideration in design in addition to the purely dimensional ones. Design decisions regarding the size and spacing of seats in public areas, the proximity of desks, and so on need to take account of people's personal space requirements and the particular **social context.**

In the workplace, a minimum separation of desks or benches of approximately 1.2 meters is thought to be necessary.

The reader is referred to Altman (1975) and Argyle (1975) for further discussion of these issues.

CURRENT AND FUTURE RESEARCH DIRECTIONS

Many of these directions are discussed in more detail in the Footnote at the end of this chapter. In the light of the increasingly international nature of business and trade, an urgent area of research is to fill in the gaps in the world anthropometric database. Data are lacking on the anthropometry of many populations, particularly in Africa.

Also of interest is the increase in body size of populations which occurs as countries develop industrially. This increase is known to have happened in western industrial countries and Japan and more recently has been happening in southern Europe and parts of South America. In developing countries, there may be differences in the anthropometry of urban versus rural peoples.

Available data are being incorporated into computer-based design aids. Several packages have been developed, all of which assist in integrating data with three-dimensional representations of the human form to improve the visualization and solution of problems of anthropometric fit. Further development of these aids will no doubt take place.

SUMMARY

Anthropometric data provide the designer with quantitative guidelines for dimensioning workspaces. However, a number of precautions are needed if data are to be used correctly:

1 Define the user population and use data obtained from measurements made on that population.

2 Consider factors that might interfere with the assumption of normal distribution of scores. For example, in some countries, stature may be negatively skewed because many individuals do not attain their potential stature because of disease or malnutrition.

3 Remember that many anthropometric variables are measured using seminude subjects. Allowance for clothing is often necessary when designing for real users. Centimeter accuracy is usually appropriate because the effect of clothing on the estimates of user anthropometry can never be accurately predicted. These considerations are particularly important when using data on stature and leg length—allowances for heel heights of 5 cm or more may be needed, depending on the user population and current fashions of dress. The effect of clothing also depends on climate—the colder the climate, the more bulky the clothing and the greater the importance of allowing for this factor in design.

In practical design situations, the data required for a particular body dimension or population may not be available. However, techniques are available for estimating unknown dimensions and the literature on anthropometry can still be of use to the designer in drawing attention to the various body dimensions which need to be considered in the design and the types of human-machine mismatches which could occur. Anthropometry can provide the designer with a very useful perspective on usability issues at the very early stages of the design process. Later, the designer might then take a more empirical approach and test out prototypes using a small sample of users from the extremes of the anthropometric range.

ESSAYS AND EXERCISES

1 Using the data in this chapter, specify seat and work surface dimensions to fit 95 percent of VDT users in two different populations. Comment on any design and equipment procurement implications of this exercise.

2 Carry out an anthropometric evaluation of either a commercial kitchen or of a number of domestic kitchens. Measure and comment on the following:
  a The heights of all major work surfaces, including the sink
  b The heights of all cupboards and shelves
  c The reach requirements of all major work and storage areas
  d The amount of space for movement and foot position
  Interview users to find out how they spend their time in the kitchen. List the most to the least common tasks and note any problems and difficulties experienced. Use all the above information to suggest improvements.

3 Write a report which explains to a design engineer why anthropometric data should be used in the design of human-machine systems. Give examples from everyday life to support your arguments.

FOOTNOTE: Anthropometry and Continuing Human Evolution

In any discussion of human body size, it is worth mentioning the well-documented increases in growth and development which have occurred during the last 150 years. In the United States and Britain, increases in mean stature of 1 cm every 10 years have been observed, although the trend now appears to have leveled off (see Pheasant, 1986, for more details on the secular increase in stature).

According to Konig et al. (1980), this growth "acceleration" is a worldwide phenomenon and is not restricted to adult stature—it is apparent even in the fetal stage of development. Neonates have increased 5 to 6 cm in length and 3 to 5 percent in birth weight over the last 100 years. Additionally, the age of onset of puberty has decreased by about 2 and 3 years for boys and girls, respectively, and the age of menopause has increased by about 3 years. Konig also remarks that acceleration has not proceeded uniformly. From about 1830 to 1930 the average height of juveniles increased by about 0.5 cm per decade, whereas from 1930 onward it increased by up to 5 cm per decade.

A number of explanations for this increase in body size have been put forward. The United States and western Europe have seen dramatic improvements in diet and sanitation over the last 100 to 150 years, particularly in the cities. Proper proportions of protein, fats, and carbohydrates in diet, together with a more generally reliable supply of vitamins and better sanitation have improved the likelihood of people's attaining their genetic potential. The same phenomena have been observed more recently in other parts of the world which have developed industrially.

There is support for the hypothesis that the environment influences the anthropometry of a population—certainly that poor living conditions can retard development. Research carried out on genetically similar Indian populations shows greater stature among groups who subsist mainly on unrefined wheat flour compared with those who subsist mainly on polished rice (many nutrients are removed from rice during the polishing process). Other research indicates that there are differences in stature between socioeconomic groups, with increased stature in the better-off groups. It is interesting to note that the appearance of people over 183 cm (6 ft) tall is not a recent occurrence in history. According to documents written at the time, Henry VIII, King of England in the sixteenth century, and Charlemagne, Holy Roman Emperor in the eighth to ninth centuries, were both well over 6 ft tall.

Konig comments that the secular trend in human development has been most marked in large cities rather than rural areas. It can be argued that the improvement in general living conditions has been more marked in the former than in the latter, but this assumption does not necessarily explain the reductions in age of onset of puberty and delayed menopause. Konig has speculated that acceleration is phenotypic—an overcompensation to exposure to retarding processes found in the industrial milieu.

Anthropologists note that several other recent trends may also have caused an increase in body size and faster development. The last 100 to 150 years have seen the migration of large numbers of people from their respective homelands as a result of colonization or immigration. This trend has been facilitated by advances in transport availability and speed. The resulting intermarriage of previously isolated rural communities has been suggested to result in "hybrid vigor." This genetic hypothesis depends, however, on the validity of its assumptions—that the original communities were in any way isolated and inbred and that on arrival in the new homeland, significant outbreeding took place. It should be noted also that there have been many occasions in history when large numbers of people migrated from one place to another.

Finally, Henneberg (1992) notes that the secular increase in stature in some populations seems to be too big to be accounted for by individual responses to environmental factors alone (15 cm in Holland over one century, for example). He argues that in prehistory, natural selection operated by means of differential mortality—only 33 percent of individuals born into a population could expect to live long enough to reproduce, compared with almost 100 percent in developed countries today. Since natural selection is of a stabilizing type, it tends to favor reproduction of "average" individuals. The continuing relaxation of selection pressure through improved health care and better living conditions increases the range of individuals in a given population. In other words, a given population will exhibit increased variability as selection pressure is relaxed.

This implies that the concept of "Mr. or Ms. Average" or an ideal body type will become increasingly invalid and **designers must expect users to be different from themselves.** This incongruity suggests that it will continue to be of importance to measure increasing human variability in order to provide designers with the information needed to enable them to accommodate an increasingly diverse user population.

4

WORKSPACE DESIGN FOR STANDING AND SEATED WORKERS

Standing on two legs is the exception rather than the rule in the animal kingdom. Most animals are adapted to function on four or more legs arranged more or less symmetrically. This configuration is stable because a line drawn from the COG of the animal's body will invariably fall within the large base of support described by the position of its four or more feet. Animals which do stand on two legs normally exhibit anatomical adaptations to overcome the stability problems caused by bipedalism. For example, the long tail of the kangaroo counterbalances its upper body in the fore-aft plane when it moves and acts as a "third leg" when the animal is stationary. In humans, the lumbar lordosis is a specialized adaptation to the erect position which helps minimize the energy requirements of holding the upper body erect for long periods.

Standing is an energy-efficient posture for humans to adopt. As Hellebrandt (1938) has noted, "Standing is cheap in terms of metabolic cost. . . . Normal standing on both legs is almost effortless." Fatigue is unlikely to be the result of the cardiovascular load of the standing posture (Grandjean, 1980). Static loading of ligaments, compression of soft tissue, and venous pooling are more likely causes of fatigue. Standing is the position of choice for many tasks in industry but can lead to discomfort if insufficient rest is provided or if unnecessary postural load is placed on the body. Some advantages of the standing work position are given in Table 4-1.

In most subsistence societies, **people rarely stand still** for any length of time—if not walking or moving, they adopt a variety of resting positions which depend on the particular postural repertoire of the society in question (Hewes, 1957; Fahrni and Trueman, 1965). In fact, the idea that anyone should stand still is physiologically and mechanically unacceptable. To quote Hellebrandt again: "In biological terms, posture is constant continuous adaptation. . . . Standing is in reality movement upon a stationary base."

Anecdotal evidence across many cultures and over time tells us that people who do

96

TABLE 4-1   SOME ADVANTAGES OF THE STANDING WORK POSITION*

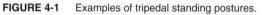

1. Reach is greater in standing than in sitting.
2. Body weight can be used to exert forces.
3. Standing workers require less legroom than seated workers.
4. The legs are very effective at damping vibration.
5. Lumbar disk pressures are lower[†]
6. Standing can be maintained with little muscular activity and requires no attention.[‡]
7. Trunk muscle power is twice as great in standing as in semistanding or sitting.[§]

* From Singleton (1972).
† From Nachemson (1966).
‡ From Hellebrandt (1938).
§ From Cartas et al. (1993).

have to stand for long periods use standing aids, such as the staff of the Nilotic herdsman or the spear of the sentry (Figure 4-1). Reliance on such aids indicates that if standing is to be prolonged, it should be tripedal rather than bipedal—the third leg providing the opportunity to rest each of the others.

In industrial workplaces, it is often not possible to make use of traditional standing

FIGURE 4-1   Examples of tripedal standing postures.

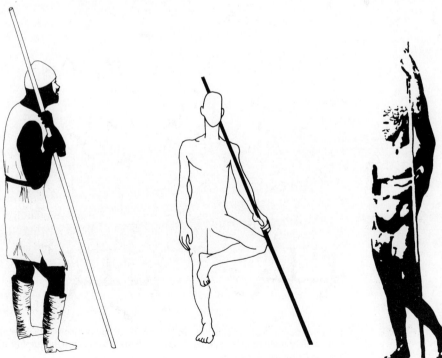

aids or sit-stand seats, and fatigue and discomfort are more likely. Prolonged daily standing is known to be associated with low back pain. Where possible, **jobs which require people to stand still for prolonged periods without some external form of aid or support must be redesigned to allow more movement or to allow the work to be done in a combination of standing and sitting postures.** In the designing of workstations and tasks for standing or seated workers, it is best to avoid positions which place the spine in an extreme posture. Some examples of body positions which require extreme versus midrange lumbar postures can be identified. They are depicted in Figure

(A)

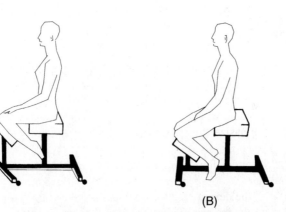

(B)

FIGURE 4-2   Body position and spinal posture are related. (A) Body positions which require extension of the lumbar spine. (B) Body positions which place the lumbar spine toward the midpoint of its range of motion.

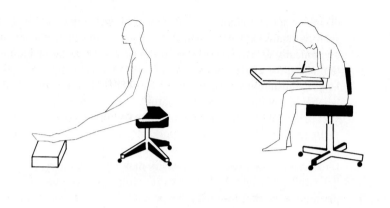

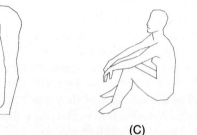

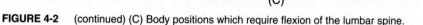

(C)

FIGURE 4-2   (continued) (C) Body positions which require flexion of the lumbar spine.

4-2. The midrange body positions are appropriate models for task and workstation design. The extreme postures should be avoided or adopted only for short periods of time.

CONTRIBUTION OF ERGONOMICS TO WORKSTATION DESIGN

Buyers of office and industrial furniture and equipment undoubtedly make purchasing decisions on the basis of cost followed perhaps by such factors as appearance, status value, faith in a particular company's reputation and products, and popular folklore and personal beliefs about what constitutes good design. Manufacturers, on the other hand, increasingly emphasize the ergonomic features of their products—ergonomically designed backrests, seats, stools, or keyboards. Some key ergonomic considerations in the design of workstations for standing and seated workers are reviewed below.

The application of ergonomics to the design of chairs, desks, benches, etc. influences those aspects of the design with which a user physically interacts. In the case of office chairs, for example, the principal features are the design of the seat pan, backrest, base, and armrests, as well as the height, swivel, and tilt mechanisms. The principal features of desks are the desk height and the desktop itself, including the provision of thigh-clearance space under the front of the desk and provision for sundries and accessories, such as lamps and document holders and cable management in VDT work. An ergonomically designed workstation can be arrived at by including all relevant information about the characteristics of users into the design process. For example, ranges of height adjustability of seats can be specified using data on the distribution of popliteal heights in a population. Seat breadths can be specified as a minimum allowable dimension using data on the distribution of hip breadth in the population.

Bench heights for standing workers can be specified using data on standing elbow height, together with information about the requirements of the task. If great vertical forces have to be exerted, as in pastry making or butchery, the working height should be below elbow height. In lighter work, requiring more precision, a higher work surface permits the worker to stand erect and, at the same time, to stabilize the elbow joints by resting them on the work surface.

Importance of Ergonomics in Workstation Design

The objective of ergonomics in workspace design is to achieve a "transparent" interface between the user and the task such that users are not distracted by the equipment they are using. Distractions may be due to discomfort (e.g., numbness in the buttocks) or to workstation usability problems. For example, the elbow rests of chairs sometimes hit the leading edge of the desk if they are too long, thus restricting access to the desk. In workstations for standing workers, lack of clear space for the feet may impede task-related postural movements. Well-designed workstations should be unobtrusive with respect to task performance. Designers therefore need to consider task requirements as well as the anatomical, physiological, and anthropometric characteristics of users. Usability problems can occur particularly when a change in function or method is not accompanied by workplace redesign.

AN ERGONOMIC APPROACH TO WORKSTATION DESIGN

Good posture is a basic requirement in workspace design. Figure 4-3 presents a framework for characterizing working posture, emphasizing the role of three classes of variables (Table 4-2). This framework stresses that an ergonomically designed piece of furniture cannot be bought off the shelf. Decisions about the appropriateness and relative advantages of different designs can be made only after considering the characteristics of users and the requirements of their jobs (Figure 4-4).

Characteristics of Users and Workspace/Equipment Design

Furniture dimensions can be specified in advance only if the anthropometry of the user population is known (Figure 4-5). Designers typically design to ensure that 90 percent of users will be accommodated. Problems can occur with extremely tall, short, or obese individuals, and special arrangements may need to be made to accommodate them.

Much office furniture is designed around a desk height of approximately 73 cm and assumes provision of a height-adjustable chair. This combination ensures, within limits, that:

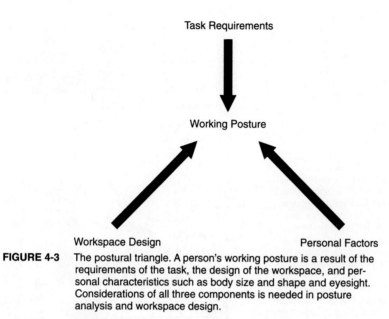

FIGURE 4-3 The postural triangle. A person's working posture is a result of the requirements of the task, the design of the workspace, and personal characteristics such as body size and shape and eyesight. Considerations of all three components is needed in posture analysis and workspace design.

TABLE 4-2   EXAMPLES OF FACTORS WHICH INFLUENCE WORKING POSTURE

Factor	Example
1. User characteristics	Age
	Anthropometry
	Body weight
	Fitness
	Joint mobility (range of movement)
	Existing musculoskeletal problem
	Previous injury or surgery
	Eyesight
	Handedness
	Obesity
2. Task requirements	Visual requirements
	Manual requirements (positional forces)
	Cycle times
	Rest periods
	Paced/unpaced work
3. Design of the workspace	Seat dimensions
	Work surface dimensions
	Seat design
	Workspace dimensions (headroom, legroom, foot room)
	Privacy
	Illumination levels and quality

FIGURE 4-4 Application of the postural triangle to workspace evaluation. The illustration depicts a monk transcribing a text (taken from a stone relief in a central European medieval cathedral). The slumped sitting posture is a result of the low seat, excessive task distance, and the visually demanding task. The right elbow is resting on the thigh to improve the stabilization of the writing hand. The left hand is steadying the book, and the left elbow is resting on the work surface to close the postural chain and reduce load on the left-hand side of the body. The left foot is resting on a footrest. How would you redesign this workspace to improve the posture?

1 Short users can raise their chair heights so that the desk height approximates their sitting elbow height.

2 The desk is not so high that the chair height exceeds the popliteal height of a short person in order to achieve the first criterion.

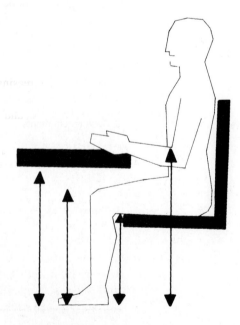

FIGURE 4-5 Critical user and furniture heights.

3 The desk is high enough so that tall users can approximate desk and elbow heights and still have space under the desk for their legs (i.e., they do not have to lower their chairs too much).

Immediately, it can be seen that with fixed desk heights, there are conflicting requirements in trying to accommodate both tall and short users (a very interesting and still valid discussion of these matters can be found in Floyd and Roberts, 1958).

Anthropometric mismatches can have serious consequences for comfort and efficiency because of the way they increase the postural load on the body (Figure 4-6). Short users, for example, may have to raise seat heights beyond popliteal height in order to gain access to the desk. As a consequence, the feet no longer rest firmly on the floor and the floor cannot be used by the legs as a fulcrum for stabilizing and shifting the weight of the upper body. The load of stabilization is not transferred to the muscles of the trunk. The weight of the legs, instead of being borne by the feet, is borne by the underside of the thigh. This position can restrict blood flow and is particularly undesirable for those with varicose veins. Further, the feet can no longer be used to extend the base of support beyond the base of the chair, which makes activities such as reaching and picking up heavy objects more hazardous. Continuous compensatory movements of body parts may be necessary to maintain stability, thus hastening the onset of fatigue.

An alternative to raising the chair is for the short user to work with the elbow below desk height. This position increases the static loading of the upper body, particularly the shoulder girdle, as users are required to maintain the elbows in an elevated position.

Tall users may find desk heights of approximately 70 to 75 cm too low since, even with the chair at its lowest level, the distance between the eyes and the work surface may be too great for comfortable viewing, causing the user to slump over the desk when writing. Figure 4-7 illustrates why it is not sufficient to simply replace an old chair with a newer one in the hope of making a worker "sit up straight." If the anthropometrics are wrong or task requirements dictate a particular posture, no design of seat will correct a stressful posture. The more holistic approach described here is needed.

Lack of foot space may also be a problem for tall workers. Mandal (1981, 1991) has recently called for an increase in desk heights to overcome these problems and to account for the increasing height of the population. Short users can be given **footrests as standard equipment** to ensure anthropometric fit. A footrest should be at least 5 cm high, 40 cm wide, and tilted toward the user by 15 degrees.

More recently, **height-adjustable desks** have been proposed (Ostberg et al., 1984) for use with height-adjustable chairs in order to increase the range of users accommodated by a workstation. The correct way to adjust one of these workstations is as follows (Starr, 1983):

1 Adjust the chair height so that the feet are resting firmly on the floor.
2 Adjust the work surface height for comfortable access to the desktop or keyboard.

Seat depth is another important variable which may affect short users. Chairs with deep seats may look more comfortable, but if the seat depth exceeds the user's buttock-knee length (the distance from the back of the buttocks to the popliteal fossa) the backrest cannot be used properly. Short users are often observed to be perched on the front edge of too-deep seats.

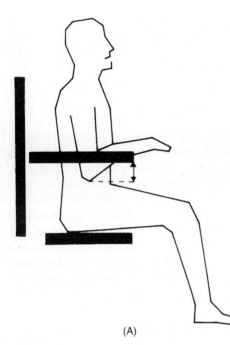

(A)

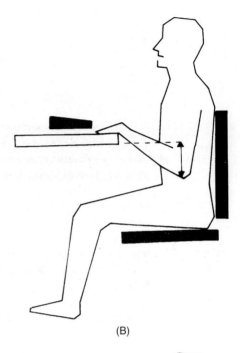

(B)

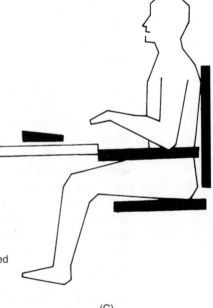

FIGURE 4-6   Common anthropometric mismatches in seated
work. (A) Seat too high, elbow rest too high.
(B) Desk too high (above elbow height). (C)
Task distance too great (elbow rest prevents
access to desk).

(C)

FIGURE 4-7   A "better chair" is not usually the answer to postural problems at the workspace. Consideration of task requirements, user characteristics, and workspace design and layout is essential in the upgrading of existing workstations.

Backrest inclination has been found to influence trunk inclination in sitting, as might be expected. However, the location of the lumbar support with respect to the level of the lumbar spine (L1-L2 or L4-L5) does not seem to be of importance (Andersson, 1986). This finding suggests that a height-adjustable lumbar support is not necessary since the height of a fixed lumbar support can be optimized to contact the lumbar spines of a wide range of users (population differences in stature are due largely to differences in the lengths of the long bones rather than the lumbar vertebrae). The American National Standards Institute (ANSI) recommends the backrest be at least 30.5 cm wide in the lumbar region.

Avoiding Anthropometric Mismatches

In the absence of anthropometric data, engineers, designers, and facilities personnel can avoid anthropometric mismatches by means of a **participative approach. A consumer panel** of workers can work with the design team to evaluate furniture prior to purchase. The panel should consist of managers; dedicated VDT users, and secretaries, clerical workers, and other sedentary workers. Trials to determine "goodness of fit" can be carried out using potentially generic configurations to identify mismatches and to specify the requirements for accessories such as footrests, document holders, etc.

Ergonomically designed workplaces must also be **flexible** if **postural fixity,** with its resultant static loading of the musculoskeletal system, **is to be avoided.** Flexibility implies that the worker can carry out the task, at least some of the time, in more than one working posture with a workspace designed to accommodate both postures. A third, additional, resting posture is also desirable. This is important—a well-designed seat may enable the user to "take the weight off the feet," but it will not take the weight off the spine unless the task permits the user to recline against a tilted backrest at least some of the time.

Task Requirements

All tasks can be considered to have three sets of requirements which can have an influence on workspace design:

1 Visual requirements
2 Postural (effector) requirements
3 Temporal requirements

Visual Requirements The position of the head is a major determinant of the posture of the body and is very strongly influenced by the visual requirements of tasks. This is easy to overlook in workspace design. When the head is erect, the straight-ahead line of sight, parallel to the ground, can be comfortably maintained. However, most people will not maintain the head in this position in order to look at objects on a horizontal work surface if this requires the eyes to be rotated downward by more than 30 degrees. If the main visual area is more than 30 degrees below the straight-ahead line of sight, it is accessed by tilting the head forward. This position places a static load on the neck muscles and displaces the COG of the body anterior to the lumbar spine, causing the characteristic forward slumped posture (Figure 2-23) in which the backrest or lumbar support of the chair is no longer effective.

If objects are placed above the line of sight, the neck is extended to tilt the head backward, which places a static load on the neck muscles. This is an important consideration for determining the location of visual displays for standing workers. **Frequently used displays should not be above the standing or sitting eye height of a short worker.** For VDT users carrying out editing tasks, document holders should be provided and placed orthogonal to the line of sight and adjacent to the VDT screen. Brand and Judd (1993) found that this configuration produced significant reductions in text editing time (of around 15 percent).

Woodson (1981) provides a certain amount of information on the lateral visual field. The eye is sensitive to stimuli up to 95 degrees to the left and right assuming binocular vision. The optimum position for placing objects is, however, 15 degrees either side of the straight-ahead line of sight (Figure 4-8.) These limits describe a visual field in which objects may be placed so that they can be viewed without moving the head from its comfortable erect position. Thus, static loading of neck muscles and other soft tissues in the neck can be avoided if the visual component of the task is kept within a cone from straight-ahead line of sight to 30 degrees below and 15 degrees to the left and right.

It is also important to consider **employee eyesight and the quality of lighting** in the working environment. **If visual requirements exceed worker abilities, poor postures may well result despite careful attention to the furniture design.**

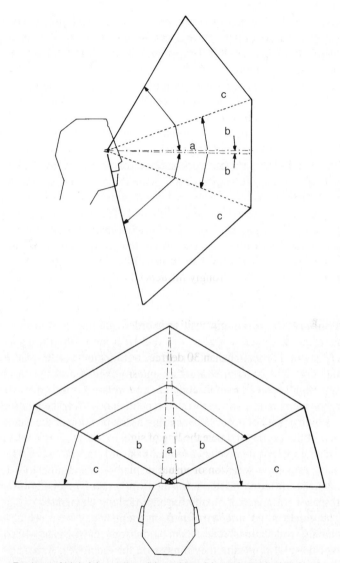

FIGURE 4-8   Regions of high (a), medium (b), and low (c) resolution in the visual field. a and b are preferred regions for placement of objects. Objects should not, however, be placed above the straight-ahead line of sight, particularly at fixed workstations.

Postural Requirements   The position of the hands, arms, and feet is another major determinant of posture and postural load. The working area of a desk or bench should always be close to its front edge and with relatively unobstructed access. In vehicle design the comfort of the driver's seat depends not just on its particular dimensions but its positioning in relation to the foot pedals and manual controls. For both standing and sedentary workers, it has been demonstrated that reduction of the magnitude of required forces and redesign of jobs to avoid excessive reach is of value in the prevention of back problems (Rivas et al., 1984). The preferred zone for work objects lies in the front 40 cm of the bench or desktop.

The slumped, twisted trunk posture of a left-handed person who writes "upside down" illustrates the physical strain which can occur as a result of the postural requirements of a task (Figure 4-9). When working at a conventional desk, these people are usually unable to maintain an erect trunk because of the requirement that the writing hand and forearm be above rather than below the text and thus farther away from the body. Resting the forearm on the work surface requires forward flexion of the trunk. It is thought that many left-handed people write this way to overcome the problem that, for them, if the forearm is kept close to the body, it is impeded by the trunk as it moves from left to right across the page in front of them. Left-handed children may be taught to write by placing the writing paper on the left, rather than in front of the body, thus providing clearance for the writing hand and arm as they move across the page from left to right. This illustrates the importance, more generally, of training people in optimal work techniques.

Temporal Requirements The temporal requirements of tasks are a major consideration in the design of workspaces and exert a moderating influence on the effects of other factors. For example, in a multiuser computer workstation in which individuals spend only 15 to 20 minutes at a time interrogating the system, a sit-stand workstation using a high bench and stool might be appropriate. At the other extreme, certain data-entry jobs require users to carry out a repetitive task over the whole day. These jobs impose a high degree of postural constraint, as the position of the hands and head are fixed by that of the keyboard, documents, and screen. Such highly constrained jobs are associated with increased prevalence of health hazards in VDT work (Eason, 1986). It is particularly important when analyzing physically constraining jobs to characterize task components in terms of frequency and importance of operation and ensure that the workspace is arranged optimally for high-ranking elements.

The **constrained/unconstrained** distinction may be a useful aid in designing or evaluating workstations in an organization. Jobs may be categorized in terms of the degree of postural constraint they impose on the employee. A manager may be relatively unconstrained—free to arrange the layout of items on the desk, to leave the desk or office at will, and to switch from one task to another. Variety may be built into the job (attending meetings elsewhere, for example). A secretary may be relatively constrained, inasmuch as work with typewriters and VDTs will be common, but there will also be other tasks, such as answering the phone or receiving visitors. Workers such as data-entry clerks will often be highly constrained.

Jobs falling into the "highly constrained" category require **maximum flexibility** to be built into the workspace (in terms of adjustable furniture, etc.) to compensate for the lack of flexibility in the design of the job. In addition, current thinking about VDT work now insists that **all VDT jobs must be designed with frequent natural rest breaks (by providing alternative tasks).**

(A)

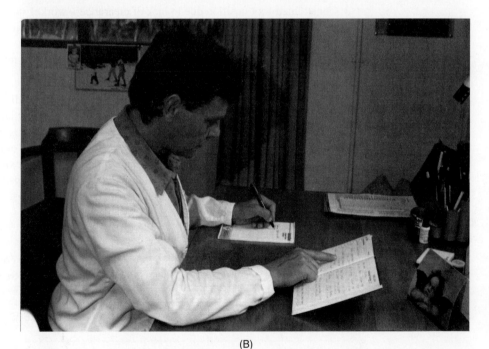

(B)

FIGURE 4-9 "Crab-handed" left-hand writing. Our writing system is designed for the right hand, which drags the pen across the page. For left-handers it is an unergonomic system, since the pen has to be pushed and rightward movement of the left elbow is impeded by the trunk. (A) The left shoulder is abducted to provide clearance for the left elbow as the left hand moves from left to right. The trunk is rotated to the right. (B) If the writing paper is placed to the left of the trunk, the left arm can move to the right as the person writes without being impeded by the trunk. A more symmetrical and erect sitting posture is facilitated by this paper placement. The left wrist is in a more neutral position.

DESIGN FOR STANDING WORKERS

As a rule of thumb, it is often suggested that all objects which are to be used by standing workers should be placed between hip and shoulder height to minimize postural stress caused by stooping or working with the hands and arms elevated. Work surface heights should approximate the standing elbow height of workers, depending on the task. For fine work, a higher work surface is appropriate to reduce the visual distance and allow the worker to stabilize the forearms by resting them on the work surface. For heavy work, a lower work surface is needed to permit the worker to apply great vertical forces by transmitting part of the body weight through the arms. Some recommended work surface heights are given in Table 4-3 (from Ayoub, 1973). These should be taken only as a guide because the actual working height depends on the size of the work objects as well as the height of the surface they are resting on.

Some workspace design faults which increase postural stress in standing workers can be summarized as follows (Figure 4-10):

1 Working with the hands too high and/or too far away—compensatory lumbar lordosis

2 Work surface too low—trunk flexion and back muscle strain

3 Constrained foot position due to lack of clearance—worker standing too far away

4 Working at the corner of the bench—constrained foot position, toes turned out too much

5 Standing with a twisted spine (having to work at the side rather than directly ahead)

Postural constraint in standing workers can be relieved by providing stools to enable workers to rest during quiet periods or to alternate between sitting and standing. Adequate space for the feet should be provided to permit workers to change the position of their feet at will. Clinicians and physiotherapists often suggest that one foot be periodically placed on a footrest or foot rail (White and Panjabi, 1978). Flexing a hip and knee by placing one foot on a raised foot rail is thought to release the iliopsoas muscle on that side, which prevents excessive anterior pelvic tilt and lumbar lordosis.

Evaluation of Standing Aids

Several researchers have investigated the effects of standing aids on comfort and fatigue in standing. Bridger and Orkin (1992) determined the effect of a footrest on pelvic angle

TABLE 4-3   RECOMMENDED WORK SURFACE HEIGHTS FOR STANDING WORKERS (in centimeters)*

Task requirement	Male	Female
Precision work	109–119	103–113
Light assembly work	99–109	87–98
Heavy work	85–101	78–94

*From Ayoub (1973).

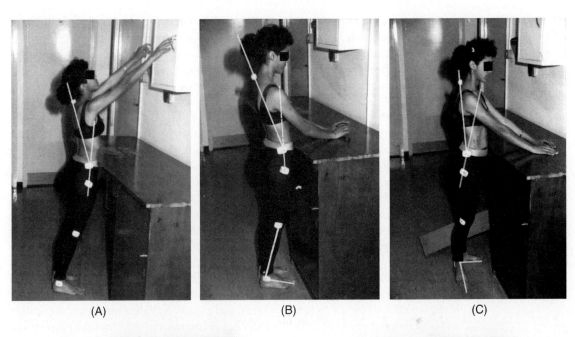

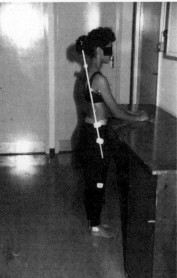

FIGURE 4-10   Illustrations of some causes of excessive lumbar extension in standing work and methods of correction. (A) Working with hands too high or reaching too far. (B) Standing too far away because of a lack of forward foot space. (C) Standing too far away and with feet in an awkward position. (D) Plenty of foot space and correct work height lead to neutral standing posture. (E) A footrest can reduce the standing lordosis and prevent postural fixity of the lumbar spine in standing if used periodically.

in standing. The footrest raised the resting foot 250 mm above the level of the floor and resulted in a net posterior rotation of the pelvis of 4 to 6 degrees (up to 12 degrees in some subjects, which is similar to the pelvic adaptation to a high saddle seat, according to the data of Bendix and Hagberg, 1984). The footrest would appear to be a valid way of reducing lumbopelvic constraint in standing workers and help prevent discomfort in the lumbopelvic region. Rys and Konz (1994) have reviewed recent research on the ergonomics of standing. Use of a 100-mm foot platform by subjects was perceived as more comfortable than normal standing in 9 of 12 body regions. During a 2-hour period of standing, subjects placed one foot on the platform 83 percent of the time, switching their foot from the platform to the floor every 90 seconds on average. The freedom to stand with one foot forward and elevated seems to be an important feature of a well-designed standing workplace. Stewart-Buttle et al. (1993) report that prolonged standing causes significant localized leg muscle fatigue, particularly in the gastrocnemius muscles. Use of a footrest would be expected to relieve some of the load on the resting leg. The provision of a mat to stand on, however, does not reduce lower leg fatigue, although it does reduce discomfort in the lower leg, feet, and back (Rys and Konz, 1994) and muscle fatigue in the erectores spinae muscles. There are many reasons standing workers should not have to stand on hard and sometimes cold concrete floors to work. Mats; wooden, rubber, or plastic platforms; and carpets provide a more yielding surface and better insulation. They may also offer better friction and therefore aid postural stability and help prevent accidents. In a comparison of three different mats, Kontz et al. (1990) found that a mat with 5.8 percent compression was perceived as more comfortable than mats of 7.4 percent and 18.6 percent compression. All were preferred over concrete. Further information can be found in Kim et al. (1994).

DESIGN FOR SEATED WORKERS

In sedentary work in an office, the flexibility requirement implies that the user can adopt both upright, forward-flexed, and reclining postures. Some recommended work surface heights for sedentary work are given in Table 4-4 (from Ayoub, 1973). For work with keyboards, the work surface is often 3 to 6 cm lower than a writing work surface to allow for the thickness of the keyboard. In addition, space must be provided for the sitter's legs. An open area approximately 50 cm wide and 65 cm deep (measured from the sitter's ischial tuberosities) is suggested by Ayoub.

Office Chairs

Some important features of office chairs are summarized in Table 4-5.

A number of researchers have suggested that chairs should be designed with forward-tilted seats (Figure 4-11). These chairs should permit a user to sit with an erect trunk and less posterior pelvic tilting and flattening of the lumbar curve because the tilt of the seat increases the trunk-thigh angle. Comparisons of lumbar angles of people sitting on conventional and forward-sloping chairs indicate that this is the case (Mandal, 1981; Bendix and Beiring-Sorensen, 1983; Brunswic, 1984; Frey and Tecklin, 1986; Bridger, 1988). Drury and Francher (1985) found that many users found a particular implemen-

TABLE 4-4   RECOMMENDED WORK SURFACE HEIGHTS FOR SEDENTARY WORKERS (in centimeters)*

Task requirements	Male	Female
Fine work	99–105	89–95
Precision work	89–94	82–87
Writing	74–78	70–75
Coarse or medium work	69–72	66–70

*From Ayoub (1973).

TABLE 4-5   KEY FEATURES OF CHAIR DESIGN

1. Seats should swivel and have heights adjustable between 38 and 54 cm. Footrests should be provided for short users.

2. Free space for the legs must be provided both underneath the seat to allow the user to flex the knees by 90 degrees or more and underneath the work surface to allow knee extension when the user is reclining.

3. A five-point base is recommended for stability if the chair has castors.

4. The function of the backrest is to stabilize the trunk. A backrest height of approximately 50 cm above the seat is required to provide both lumbar and partial thoracic support.

5. If the backrest reclines, it should do so independently of the seat to provide trunk-thigh angle variation and consequent variation in the distribution of forces acting on the lumbopelvic region.

6. Lumbar support can be achieved either by using extra cushioning to form a lumbar pad or by contouring the backrest. In either case, there must be open space between the lumbar support and the seat pan vertically below it to allow for posterior protrusion of the buttocks.

7. The seat pan must have a slight hollow in the buttock area to prevent the user's pelvis from sliding forward. This keeps the lower back in contact with the backrest when the user is reclining. The leading edge of the seat should curl downward to reduce underthigh pressure.

8. Armrests should be high enough to support the forearms when the user is sitting erect. They should also end well short of the leading edge of the seat so as not to contact the front edge of the desk. If the armrests support the weight of the arms, less load is placed on the lumbar spine.

9. Modern chairs tend to have a thin layer of high-density padding. Layers of thick foam tend to destablize the sitter. The foam can collapse after constant use.

10. Cloth upholstery provides friction to enhance the stability of the sitter.

tation of this seating concept uncomfortable—probably due to the lack of postural variation permitted by the particular design. A similar principle has also been incorporated into a form of high stool for use in industry (Gregg and Corlett, 1988). Few extensive field trials of these "alternative" forms of seating have been reported in the literature, however.

Proper contouring of the backrest of conventional chairs can provide support for the lumbar and thoracic spines in the erect sitting position. A lumbar support acts like a brace which uses the lumbar spine as a lever to tilt the sitter's pelvis forward. Free space

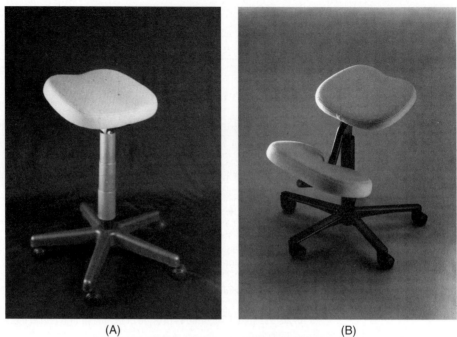

(A)   (B)

(C)

FIGURE 4-11   Alternative seating. (A) Forward-tilted stool. (B) Sloping seat with kneepad. (C) Saddle seat (backrest also functions as a thoracic support). (Courtesy of HAG Inc., Greensboro, North Carolina.)

for the posteriorly protruding sacrum and buttocks must be provided for a lumbar support to be effective (Figure 4-12). Although lumbar supports do reduce posterior pelvic tilting and prevent lumbar flexion, they do not "preserve" the lumbar lordosis as is sometimes stated in the ergonomics literature and in advertisements for office chairs. The amount of lumbar lordosis which is preserved is modest—similar to that of a person sitting in a neutral position such as kneeling (Bridger et al., 1992). The lumbar lordosis is an adaptation to standing and is not an appropriate posture for the lumbar spine of a seated person. Both forward-tilted and conventional chairs should be designed to keep the lumbar spine in a neutral position (i.e., to prevent flexion).

Dynamics of Sitting

Few studies have been carried out on the dynamics of sitting or sitting behavior. Branton (1969) is an exception. He used the body-link concept, described in Chapter 2, to evaluate the comfort of train seats by observing sitting behavior. An open-chain system of body links can behave in unpredictable ways when subject to internal or external forces. The prime function of a seat is to support body mass against the forces of gravity. A second function, which was emphasized by Branton, was to stabilize the open-chain system. In the absence of external stabilization, tonic muscle activity is required, which leads to discomfort if sustained. Behaviors such as folding the arms and crossing the legs can be seen as postural strategies to turn open chains into approximate closed chains stabilized by friction. The comfort of a seat depends, in a dynamic sense, on the extent to which it permits muscular relaxation while stabilizing the open-chain system

FIGURE 4-12   Modern office chair with lumbar and thoracic support suitable for both VDT and non-VDT office work.

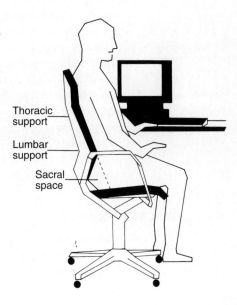

Thoracic support

Lumbar support

Sacral space

of body links. Figure 4-13 shows a chair designed for dynamic sitting behavior by accommodating several different working and resting positions.

WORK SURFACE DESIGN

Designers can improve workstations by considering various aspects of desk and bench design. Some important considerations are the provision of tilt in the work surface and/or the provision of document holders and free space in the working area. Mandal (1981) suggested that desktops should tilt toward the user by about 15 degrees to lessen the visual angle and encourage a more upright posture of the trunk. Zacharkow (1988) provides many interesting illustrations of Victorian school desks having a 15-degree slope for writing and an integral book holder for reading, angled at 45 degrees.

Several studies have indicated that tilted desktops (of 15 or even 10 degrees) do reduce the trunk and neck flexion of seated persons engaged in reading and writing (Bridger, 1988; de Wall et al., 1991) and thus reduce the load on the corresponding parts of the spine (Figure 4-14). Significant effects have been found on subjects seated on both conventional and forward-sloping seats. Porter et al. (1992) have reported similar benefits with a sloping computer desk. Burgess and Neal (1989) found that using a document holder when writing on a flat desk significantly reduced the moment of flexion of the head and neck at the C7-T1 level of the spine and was rated by the subjects as more comfortable than not using one.

For both standing and sedentary workers, work surfaces should be arranged so that the worker does not have to work continually with objects placed at one side or reach excessively to the side. The main working area should be directly in front of the worker's body to minimize any twisting of the trunk when carrying out task-related movements. Twisting strain increases the risk of injury to the intervertebral disks.

Pearcy (1993) has shown that the twisting mobility of the human back is increased in sitting compared with standing either in an upright or a forward-flexed position. The increase occurs because, in the flexed posture, the morphology of the lumbar facet joints permits more axial rotation of the superior vertebral body over the inferior body. The increase is considerable (38 percent more twisting in 90-degree sitting and 44 percent more twisting in long sitting over the whole spine or about 2 degrees more at each intervertebral joint). Because the posterior fibers of the annulus are already stressed when the spine is flexed, the additional stress of twisting may result in very high annular stresses which will predispose the fibers to rupture. Jobs involving the asymmetric handling of loads from a seated position (such as supermarket checkout personnel) would seem to be particularly hazardous. Seated workers who have to resist sudden, external twisting forces (such as catching an object falling from a supermarket conveyor) have a high risk of injury, according to Pearcy.

VISUAL DISPLAY TERMINALS

The design of the workspace for the visual display terminal (VDT) operator has received much attention in recent years (e.g., Grandjean, 1987) because of the importance of attaining a proper relationship between user and workspace to support the requirements

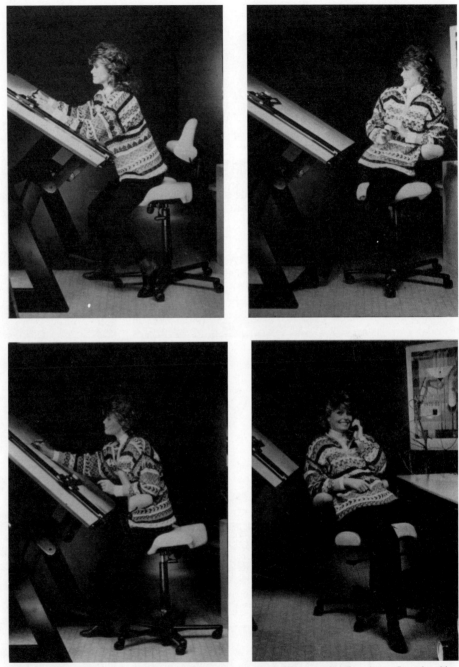

FIGURE 4-13   A flexible chair design (chair can be used like a stool, for reclining, and in "fore-leaner" mode with thoracic support).

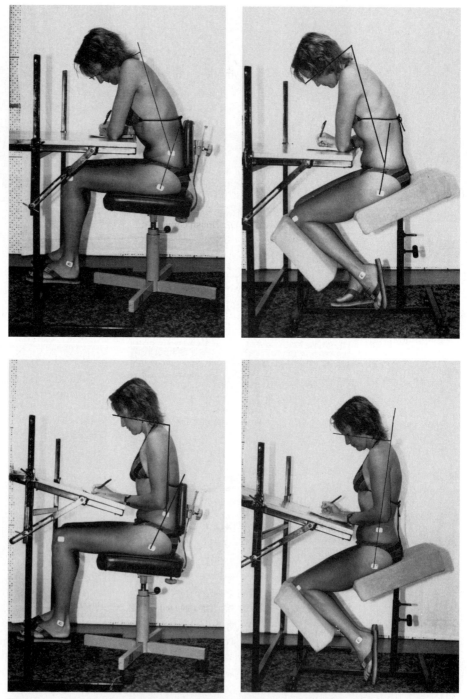

FIGURE 4-14   Tilted work surfaces may be effective when used with both conventional and alternative seats.

for interactive computer work. Figures 4-15 and 4-16 summarize the important considerations in the design of the VDT workspace.

Standards for the design of these workspaces have been drafted or proposed (e.g., ANSI/HFS 100-1988; BS 7179, 1990; BS 7179, 1990; Health and Safety Commission, 1992). The guidelines proposed in the above documents attempt to specify user space requirements and to define an appropriate working environment in functional as well as physical terms. Readers are referred to these documents for detailed information.

Fixed postures and cumulative trauma disorders (CTDs) are the main musculoskeletal problems associated with VDT work. Certain types of VDT work (such as data entry) have requirements somewhere between traditional office work and factory work. Grieco

FIGURE 4-15   Essential furniture considerations in VDT work. (From "Display Screen Equipment—Work Guidance on Regulations," 1992. British Crown Copyright. Reproduced with the permission of the Controller of Her Britannic Majesty's Stationery Office.)

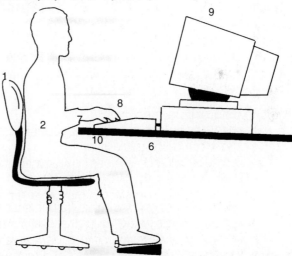

SEATING AND POSTURE FOR TYPICAL OFFICE TASKS

1 SEAT BACK ADJUSTABILITY
2 GOOD LUMBAR SUPPORT
3 SEAT HEIGHT ADJUSTABILITY
4 NO EXCESS PRESSURE ON UNDERSIDE OF THIGHS AND BACKS OF KNEES
5 FOOT SUPPORT IF NEEDED
6 SPACE FOR POSTURAL CHANGE, NO OBSTACLES UNDER DESK
7 FOREARMS APPROXIMATELY HORIZONTAL
8 MINIMAL EXTENSION, FLEXION OR DEVIATION OF WRISTS
9 SCREEN HEIGHT AND ANGLE SHOULD ALLOW COMFORTABLE HEAD POSITION
10 SPACE IN FRONT OF KEYBOARD TO SUPPORT HANDS/WRISTS DURING PASSES IN KEYING

1   ADEQUATE LIGHTING
2   ADEQUATE CONTRAST, NO GLARE OR DISTRACTING REFLECTIONS
3   DISTRACTING NOISE MINIMISED
4   LEG ROOM AND CLEARANCES TO ALLOW POSTURAL CHANGES
5   WINDOW COVERING
6   SOFTWARE: APPROPRIATE TO TASK, ADAPTED TO USER, PROVIDES
    FEEDBACK ON SYSTEM STATUS, NO UNDISCLOSED MONITORING
7   SCREEN: STABLE IMAGE, ADJUSTABLE, READABLE, GLASS/REFLECTION FREE
8   KEYBOARD: USABLE, ADJUSTABLE, DETACHABLE, LEGIBLE
9   WORK SURFACE: ALLOW FLEXIBLE ARRANGEMENTS, SPACIOUS, GLARE FREE
10 WORK CHAIR: ADJUSTABLE
11 FOOTREST

FIGURE 4-16   Other essential design considerations in VDT work. (From "Display Screen Equipment—Work Guidance on Regulations," 1992. British Crown Copyright. Reproduced with the permission of the Controller of Her Britannic Majesty's Stationery Office.) These design considerations are dealt with in other chapters.

(1986) uses the term **"postural fixity"** to describe static postures of the head, neck, and trunk which occur in VDT work. CTDs are usually associated with the highly repetitive elements of VDT tasks such as typing.

In VDT work, the workplace is dominated by the equipment. The position of the hands is determined by the keyboard and that of the head by the screen or documents. In fact, the task requirements of operating a VDT resemble those of car driving rather than writing at a desk (substitute the keyboard for the steering wheel and the screen for the windshield). Grieco has pointed out that static postures are very common in modern office work and has highlighted some potential health hazards of seemingly "doing nothing." According to Grieco, **enforced 8-hour-a-day sitting** should be regarded as an **occupational risk factor** (like manual handling or exposure to vibration). Static postures are often accomplished by means of isometric muscle contractions which deplete local blood supply and hasten fatigue. In data entry, muscular contractions of the trapezius muscles can exceed 20 percent of the muscle's maximum—enough to impede the flow of blood to the muscle. This leads to neck and shoulder pains.

Grieco also notes that since the intervertebral disks have no direct blood supply, the variations in the loading to which they are subjected as a result of activities of daily living provide a useful function in "pumping" intercellular fluid into and out of the disk. This provides nutrients and removes waste products. The daily adoption of fixed postures is thought to be hazardous because it interferes with this nutrient exchange mechanism.

Note that **Grieco's arguments apply even to well-designed workplaces which place low loads on the disks.** Variations in loading above and below a threshold value are thought to be the important factor in ensuring adequate disk nutrition. This hypothesis suggests that **dedicated VDT work is a health hazard and provides a rationale for the inclusion of non-VDT work into an operator's daily schedule of activities.**

VDTs support the activities of many different occupations and professions, and there is no single, correct workspace arrangement to satisfy the requirements of all workers. Therefore detachable keyboards and screens with adjustable tilt and swivel are recommended. Such features permit a copy typist, for example, to work with the screen to one side and the document directly ahead and supported in the same visual plane as the screen by means of a document holder. The screen is the main work area for a programmer and thus should occupy the main visual area of the programmer's desk.

Table 4-6 summarizes some design solutions for increasing the flexibility of VDT workstations. A participatory approach is often the most effective way to decide which product features and accessories are likely to have the greatest cost benefits for different groups of workers in an organization. For example, document holders are often of great benefit to copy typists and data-entry clerks, whereas systems designers may regard increased storage space and shelving as a higher priority.

Table 4-7 presents sample workspace dimensions for U.S. workers (from ANSI/HFS 100-1988).

VDTs and Health

Several researchers have investigated the health of VDT workers in relation to workspace ergonomics. Green and Briggs (1989) found that there was no difference between the workstations of CTD injury sufferers and nonsufferers on the basis of adjustability. Thus, adjustable furniture, in itself, does not guarantee that the musculoskeletal health of workers will be safeguarded. However, sufferers were less satisfied with their equipment and were observed to adopt more inappropriate postures. There was also a perceived need for more training and information on correct workstation adjustment—preferably in verbal rather than literary form. Sauter et al. (1991) collected self-reported data on musculoskeletal discomfort from several hundred VDT users, and aspects of posture and workstation design from 40 users—38 percent of the variance in discomfort

TABLE 4-6   PRODUCT FEATURES AND ACCESSORIES FOR ENHANCING VDT WORKSTATION FLEXIBILITY

Workstation	Feature or accessory
Source documents	Document holder
	Book stand
	Tilted area of work surface
	Shelving
	Bulletin board
VDT screen	Movable screen
	Tilt and swivel adjustment
	Screen holder
Keyboard	Keyboard detached from VDT
	Keyboard drawer
Seating	Footrest
	Armrest
	Lumbar pad
	Narrow backrest (permits trunk lateral flexion and rotation)
	Lumbar and thoracic support
	Recline mechanism
Desk	Height adjustment
	Tilt adjustment
	Extensions for expanding surface area

TABLE 4-7   VDT WORKSPACE DIMENSIONS FOR U.S. USERS (in centimeters)

	95th percentile male	5th percentile female
Seat height*	49	41
Elbow rest height*	29	18
Work surface height*	71	58
Eye height (90-degree sitting)	130	103
Screen height*	Adjust screen so that entire primary viewing area is between 0 and 60 degrees below horizontal plane passing through the eyes.	

*From ANSI/HFS 100-1988.

was explained by the ergonomic factors. In particular, leg discomfort was associated with the use of low, soft seats. Arm discomfort was found to increase with the height of the keyboard above elbow level. Good design of VDT workstations does appear to be of benefit in safeguarding the well-being of office workers.

CURRENT AND FUTURE RESEARCH DIRECTIONS

About 75 percent of workers in developed countries are sedentary, and the basic requirements for the design of seated workspaces are now well understood. However, what is less well understood is the contribution of a sedentary lifestyle to the prevalence of spinal disease and how sedentary jobs should be modified to reduce any risks. Although clear recommendations can be made, more evidence of an epidemiological nature is needed.

A number of alternative seat designs have appeared over the years. Ergonomists usually criticize the design of "kneeling" chairs because they have no backrest and constrain the position of the legs. People who use these chairs by choice counter by saying that the increased base of support enhances lower body stability, which implies that postural adjustments can be made using the large leg muscles rather than the trunk muscles. The forward tilt reduces lumbar flexion strain, so there is no need for a backrest. Both sides lack the support of a body of experimental research to determine the circumstances under which one design will be preferable to another.

Finally, it is remarkable how little research has been done on the standing work position. Unlike that of seated work, the functional anatomy of standing work is not nearly so well understood. Apart from anthropometric guidance and the recommended provision of carpeting or rubber mats to stand on, ergonomics provides the designer with very little advice about the design of the standing worker's workspace. More research in this area is needed.

SUMMARY

The reduction of postural stress is fundamental to workstation design in ergonomics. A multifaceted approach is needed to arrive at appropriate workstation designs for different workers. The requirements of tasks and the characteristics of users need to be considered in relation to the options for workstation design. It is not appropriate to talk about an "ergonomically" designed chair in isolation since the appropriateness of any design depends on its use and the other equipment used to carry out a task. In the design or selection of office and industrial furniture, the workstation-task interaction must be the basic level of analysis.

ESSAYS AND EXERCISES

1 Carry out a survey of the furniture in a modern office. Measure furniture dimensions and relate them to user anthropometry, commenting upon any anthropometric mismatches you observe. Find out if the furniture is adjustable and if users know how to adjust it to optimize their working postures.

   Examine and classify the furniture and accessories used by different categories of staff and different levels of management. Are the different types of furniture used appropriate for the different users and their task requirements? Is new furniture needed? Are adequate accessories provided?

2 Obtain adult anthropometric data from three different parts of the world. Specify work surface and furniture dimensions and heights for displays and controls for each group. Are there geographical differences in workstation dimensions? Can you accommodate both men and women in each group using one set of workspace dimensions?

3 You are the facilities manager of a major international financial institution. You need to spend $50 million to upgrade the company's head and regional offices worldwide. Write

a presentation for top management outlining the company's needs and explaining what constitutes good office furniture and why it is essential in the automated office.

FOOTNOTE: The Unnatural History of Sitting

Homo sapiens is capable of over 1000 comfortable resting positions (Hewes, 1957), but most cultures practice only a subset of these because of differences in custom, climate, clothing, and terrain. The most widely researched posture is undoubtedly standing, because of its theoretical and clinical importance in physical anthropology and comparative anatomy. The 90-degree sitting posture of occidental culture is probably the second most studied posture because of its prevalence in work situations.

The bipedal, erect stance is universal across cultures and can be regarded as fundamental in an anatomical sense. The same cannot be said for sitting in chairs. This posture has its origins in Egypt and ancient Greece as a posture for high-status individuals. The rest of the population appears to have squatted, sometimes on low stools, but usually on the ground. The deep squat is a common working and resting posture worldwide. Millions of people in Africa, many parts of Asia, and Latin America customarily work and rest in this position (the ancient Greeks regarded it as vulgar, depicting its use only by satyrs and other base forms of life, which may explain why squatting is not practiced by countries influenced by Hellenism). The deep squat resembles the habitual resting position of the chimpanzee and is easy for young children of any culture to adopt. Adults not accustomed to it find it uncomfortable—they cannot dorsiflex the ankle joints sufficiently to attain an acute angle between the foot and lower leg, which is necessary to position the COG of the body over the feet. For this reason, they tend to fall over backward when attempting to squat or can squat only when wearing high heels.

Cross-legged sitting is found in north Africa, the Middle East, India, southeast Asia, Indonesia, Korea, Japan, and South America. Convention, clothing, and cold damp floors probably discouraged its use in the United States and northern Europe (although it was the traditional work position of tailors and is still adopted by schoolchildren in these countries). In India, there are many variations. Sen (1984) has summarized some of the benefits of this way of sitting.

A common posture shared by women in Africa, Melanesia, and southeast Asia, and by native American women in northwest America is known as "long sitting" (Figure 4-17). Sitting takes place on the floor with a 90-degree angle between the trunk and the thighs and the knees extended. The trunk is maintained erect. Long sitting is a posture adopted while nursing children or working. It is likely that women who habitually sit in this way have long hamstring muscles; otherwise there would be a tendency for the pelvis to tilt rearward and severely flex the lumbar spine. This doesn't appear to be the case, and the posture merits further investigation since it contradicts most of the recommendations for sitting comfort made by ergonomists in industrialized countries. Those accustomed to sit in this way appear to do so voluntarily and to find it comfortable for long periods.

Sitting on the heels with the knees resting on the floor is a formal posture in Japan and is the position for praying in the Islamic world. In Africa, Mexico, and parts of

FIGURE 4-17   The long sitting position. Used as a working and resting position in many parts of the world, it breaks all the conventional rules of "correct sitting." People who habitually adopt this posture appear to find it comfortable, however.

southeast Asia it is used mainly by women. Kneeling on the floor is generally unacceptable in western society. It is interesting to observe, however, that a piece of furniture (the Balans chair) has recently been designed which allows a kneeling posture to be adopted in a more socially acceptable way. It is often suggested that lumbar spine degeneration is a disease of "Western society" caused by an "unnatural lifestyle." Several cross-cultural investigations of the incidence of spinal pathology have been carried out. Fahrni and Trueman (1965) compared the incidence of disk space narrowing and pathological change in the spines of North Americans and Europeans and a jungle-dwelling tribe in India. In all groups, lumbar degeneration increased with age, as might be expected, but the trends were steeper for the western group. Disk space narrowing was much less marked in the Indians and was hardly age-related at all. The investigators noted that the spines of horses, cows, dogs, camels, and giraffes all degenerate at levels where the mechanical strain is greatest—the cervicothoracic region of camels and giraffes, the lower thoracic level of the basset hound and the dachsund dog, and the apex of the lumbar spine in humans. Studies of the fossilized remains of the giant dinosaur, diplodocus, suggest that degeneration of its spine occurred in the tail (which was used as a prop) as a result of mechanical strain (Blumberg and Sokoloff, 1961).

Fahrni and Trueman, finding that the Indians in their study habitually squatted and avoided extreme lumbar postures, suggested that excessive lumbar lordosis caused by the prevailing postural repertoire might be the cause of accelerated disk degeneration in western society. Although habitual daily activities do influence the incidence of disease, there is no reason to think that the mechanical strain on the spine is greater in people who stand compared with people who sit—what evidence there is suggests the opposite. In addition, not all cross-cultural studies show spinal pathology to be a disease of industrialized countries (e.g., Tower and Pratt, 1990). Anderson (1992) reports studies of central Asian peasants and of subsistence farmers in the remote highlands of Papua New Guinea which show there to be as high a prevalence of mechanical back pain among these communities as there is in the industrially developed world.

Postural behavior and motor habits develop through childhood and are influenced by custom and the particular environment. Milne and Mireau (1979) found that hamstring length decreased throughout childhood and adolescence, but there was a **marked drop** around the age of 6 years, which they attributed to increased chair sitting when the children started school. Preschool children spend long periods sitting on the floor in postures which require a wide range of joint motion. Fisk and Baigent (1981) found that a spinal pathology in young people (Scheuermann's disease) was related to extreme hamstring shortness. They postulated that the hamstring shortness restricted movements of the hip, thereby placing a greater postural adaptation load on the spine in movements requiring flexion of the trunk.

Although a great deal is known about the requirements for workspace design in the western postindustrial milieu, it is clear that much more research is required to elucidate the relationship between the postural behaviors and motor habits which people in a particular culture acquire over the course of their lives and the incidence of discomfort and disease. Whether current design practice is optimal is not known, particularly for implementation in developing countries. Further research in this area will undoubtedly lead to a better understanding of the true limitations and possibilities of human anatomy.

5

THE UPPER BODY
AT WORK

The upper body is taken to include the cervical spine, the head and shoulders, and the elbow and wrist joints. From an anatomical point of view, each of these structures may be dealt with separately, but from an ergonomic point of view, they are best dealt with together. The hands are the main effector organ of the body and are the focus for much of the physical work which is carried out in a job. As was pointed out in an earlier chapter, postural stabilization of the hands and arms is essential if they are to carry out all but the grossest movements in a purposeful way. This stabilization is provided by muscles farther up the kinetic chain, muscles which cross the elbow and shoulder joints and have their origins in the cervical spine and thoracic regions. These muscles may become fatigued if exposed to heavy loads or denied rest.

From an occupational point of view then, the cervical spine, head and shoulders, and elbow and wrist joints can be considered to be interrelated as far as problems of efficiency, design, and discomfort are concerned.

DEVELOPMENT OF WORK-RELATED MUSCULOSKELETAL DISORDERS

Many of the recognized musculoskeletal disorders, such as cervical spondylosis, carpal tunnel syndrome, and tennis elbow, are very common—particularly in older people. The usual pattern is for symptoms to appear before objective signs of degeneration or disease. By 50 years of age, 50 percent of studied populations have had neck, shoulder, or arm pain (Lawrence, 1969). This number increases steadily with advancing age. The age-relatedness of these disorders complicates the task of determining the occupational contribution to the risk of developing a particular disorder because workers presenting with a complaint may have developed it anyway. It is known, though, that certain jobs will cause people with a disorder to experience pain at work—that the symptoms of the disorder will be amplified or exposed by the demands of the job. Thus, a disorder may be more noticeable or troublesome in some occupations than others and workers in these occupations may be more likely to report an injury or take time off from work because of it. This can lead to the mistaken conclusion that the syndrome was caused by the work rather than having been exposed by it. For this reason, modern approaches to the investigation of neck and shoulder problems at work take a multifactorial approach.

Musculoskeletal pain of this type is said to be work-related because it is partially caused by work conditions (Armstrong et al., 1993). Kilbom (1988) proposed a simplified model of the development of work-related occupational musculoskeletal disorders. According to this model, physical task demands combine with the individual's capacity to determine the level of musculoskeletal stress which the individual experiences when carrying out work tasks. Some of the factors which influence an individual's capacity are age (tendons, for example, lose elasticity with age and their fibers are more susceptible to microscopic tears which result in increased cell death), strength, and fitness. Physical stress then combines with psychosocial stress (due to the mental demands of the work, dissatisfaction with the job, and so on) to produce a level of musculoskeletal strain. The perception of pain or fatigue which results from this is modified by other psychosocial factors. For example, a manual worker may perceive a certain degree of discomfort as "part of the job," whereas a computer programmer may not. Since most occupational musculoskeletal pain goes away after a few days' rest (at least in the early stages), rest through absenteeism or job redesign will lead to recovery. Chronic disorders may develop if no action is taken, in which case the only solution may be for the sufferer to seek alternative employment or to continue doing the job but with diminished capabilities.

Following on from the Kilbom model, Armstrong et al. (1993) have developed a model of musculoskeletal disorders which emphasizes exposure, dose, capacity, and response. These elements are summarized in Table 5-1. Exposure refers to work demands such as posture, force, and repetition rate, which have an effect (the dose) on the internal body parts. Metabolic changes in the muscles, stretching of tendons or ligaments, compression of the articular surfaces of joints are examples of what is meant by a dose. The dose may produce a response such as a change in the shape of a tissue, the death of cells, or the accumulation of waste products in the tissues. These primary responses can be accompanied by secondary responses such as pain or a loss of coordination. As can be seen, a response (such as pain) can be a dose which causes another response (e.g., increased muscle contraction).

Capacity refers to the individual worker's ability to cope with the various doses which his or her musculoskeletal system is exposed to. An individual's capacity is not fixed. According to the model, it may change over time as the person ages, or the development of skill may improve the ability to generate great forces with less effort. Training can increase strength or endurance, whereas the development of scar tissue to replace injured muscle tissue may impair strength or endurance. Armstrong et al. point out that muscles can adapt to work demands faster than tendons and that the differential adaptation may lead to reduced (relative) tendon capacity. We might speculate that one of the dangers faced by bodybuilders and others who use illegal anabolic steroids to pro-

TABLE 5-1   KEY ELEMENTS OF ARMSTRONG ET AL.'S MODEL OF THE DEVELOPMENT OF
WORK-RELATED UPPER BODY MUSCULOSKELETAL DISORDERS*

Element	Example	
Exposure	Physical factors:	Workplace layout Tool design Size, shape, weight of work objects
	Work organization:	Cycle times Paced/unpaced work Spacing of rest periods
	Psychosocial factors:	Job dissatisfaction Quality of supervision Future uncertainty
Dose	Mechanical factors:	Tissue forces Tissue deformations
	Physiological factors:	Consumption of substrates Production of metabolites Ion displacements
	Psychological factors:	Anxiety
Primary responses	Physiological factors:	Change in substrate levels Change in metabolite levels Accumulation of waste products Change in pH
	Physical factors:	Change in muscle temperature Tissue deformation Increase in pressure
Secondary responses	Physical factors:	Change in strength Change in mobility
	Psychological factors:	Discomfort
Capacity	Mechanical factors:	Soft tissue strength Bone density/strength
	Physiological factors:	Aerobic capacity Anaerobic capacity Homeostatic control
	Psychological factors:	Self-esteem Tolerance of discomfort Tolerance of stress

*From Armstrong et al. (1993).

duce rapid increases in muscle bulk is injury to the tendons, because tendon strength does not have time to catch up with the increased muscle strength.

It is clear that poor workstation and tool design can increase the discomfort of both healthy and less fit individuals. Ergonomics has an important role to play in identifying stressful jobs and designing out unnecessary musculoskeletal stress. Some common causes of musculoskeletal problems, their relation to work activities, and ergonomic approaches to their prevention or amelioration are described in the following sections.

INJURIES TO THE UPPER BODY AT WORK

The most clear-cut work-related upper body injuries occur as a result of accidents at work, and many of them occur when hand tools are being used. Aghazadeh and Mital (1987) carried out a questionnaire survey to determine the frequency, severity, and cost of hand tool–related injuries in U.S. industry and to identify the main problem areas.

The hand tools most commonly involved in injury were knives, hammers, wrenches, shovels, and ropes and chains. The power tools most commonly involved were saws, drills, grinders, hammers, and welding tools.

Of the main incidents which precipitated an injury, the majority involved the tool's striking the user. This situation was the case with both power and hand tools. However, a significant minority of injuries were caused by overexertion (approximately 25 to 30 percent). The upper extremities were the body area most commonly injured, and the most common injuries were cuts and lacerations, followed by strains and sprains. A **strain** may be defined as overexercise or overexertion of some part of the musculature, whereas a **sprain** is a joint injury in which some of the fibers of a supporting ligament are ruptured although the ligament itself remains intact.

Many power tools can cause strains or sprains because of the reaction forces they exert on the user—particularly if these forces are unexpected or occur suddenly as a result of irregularities at the interface between the tool and the workpiece. Percussive tools, such as paving breakers, exert a reaction force which has to be opposed by the user. In practice, if the tool is well designed and used on a flat surface, the weight of the tool will dampen much of this force. Rotary-powered tools such as drills, sanders, and screwdrivers can exert a reaction torque on the user which may force the wrist into ulnar or radial deviation, causing a strain or sprain. The design of handles for holding power tools in place has received the attention of ergonomists, as is described below. It should not be forgotten, however, that additional handles may need to be fitted to provide the user with sufficient mechanical advantage to overcome the reaction torque of the tool or to carry it. Bone (1993) reports that in the United States, General Motors specifies a maximum allowable torque for freely hand-held power tools and fits torque-arresting arms or slip clutches to more powerful rotary tools. Such modifications reduce the risk of injury and may enhance the usability of the tool.

It appears that there are several different classes of hand tool–related injury and that several different approaches for prevention may be needed. The most common injury would seem to be of a catastrophic nature in which the tool itself suddenly strikes the user, causing a laceration, bruise, or sprain. A second, more pernicious, type of injury involves sprains or strains which appear to result from the handling of the tool itself over longer periods of time. A third type of injury occurs to the skin in the form of blisters due to pressure "hot spots" caused by poor handle design. Attempts to prevent the first type of injury might emphasize training workers to use safe tool-handling techniques and to think ahead—to recognize potentially dangerous situations and to prepare the workplace to minimize the likelihood of unforeseen events. Attempts to prevent the second type of injury might concentrate on the redesign of the tool itself and training workers to recognize the onset of fatigue and to avoid stressful work postures. Handle redesign can prevent the third type of injury, as can increasing the task variety of the job.

Prevention usually requires a multilayered approach involving training, safety propaganda, and workspace design. Of particular relevance to the present discussion is the ergonomics of equipment design in relation to the prevention of strains and sprains. This subject is discussed in a later part of the chapter.

CONTROL OF NECK PROBLEMS AT WORK

Grandjean (1987) concluded that the head and neck should not be flexed forward by more than 15 degrees if undue postural stress is to be avoided. There is considerable evidence that frequent or sustained flexion of the head and neck beyond this point is related to chronic neck and shoulder pain. This pain is exacerbated if the flexion is accompanied by rotation of the head and if the shoulders and arms have to be held in an elevated position at the same time (as is common in certain occupations, such as dentistry and hairdressing).

Tola et al. (1988) compared the prevalence of neck and shoulder symptoms in operators of powered vehicles and compared it with that of carpenters and sedentary workers. The former group of workers was exposed to vibration, static loading due to prolonged sitting, and rotation of the back and neck when reversing. In all groups, symptoms increased with age, but the vehicle operators always had more symptoms than workers of the same age in the other occupational groups. Thus, occupational factors can increase the prevalence of neck symptoms in all age groups, but older workers are more susceptible to occupationally induced symptoms because of the degenerative changes which have already occurred in their necks.

It is known that dentists have a high incidence of cervical spondylosis. Milerad and Ekenvall (1990) compared cervical symptoms in a group of dentists and a group of pharmacists. Although both groups had symptoms of the disorder, the frequency was highest in the dentists and was more often accompanied by shoulder and arm pain. As noted previously, degenerative changes in the cervical spine can interfere with the spinal cord and nerve roots, causing referred pain in the arms.

Vihma et al. (1982) compared sewing machine operators and seamstresses with respect to musculoskeletal complaints. The sewing machine operators spent more than 6 hours per day working in a seated position at their machines, and the work consisted of short, repetitive cycles. The task required neck flexion to look at the horizontal work surface and sustained muscular activity. The machinists reported more recurrent aches and pains in the neck and shoulders than the seamstresses, whose work was less constraining. Similar results have been reported for other occupational groups, such as data-entry clerks (Hunting et al., 1981).

The so-called cervical syndromes or cervicobrachial disorders are influenced by occupational factors ("cervicobrachialgia" is the name given to pain in the neck which radiates to the arm as a result of compression of the nerve roots in the cervical spinal cord). Static neck postures, repeated flexion and/or extension of the neck, forceful movements of the upper body, shoulder elevation, and arm abduction are all possible risk factors. Older workers are considered to be more likely to experience pain of this nature if their jobs expose them to the above mentioned postural stressors.

In practice, the working posture of an individual will depend on the requirements of the task, the design of the workspace, and the individual's personal characteristics—including deeply ingrained motor habits built up over years of work. Each of these factors needs to be taken into account in any attempt to reduce task-induced postural stress on the cervical spine.

The position of the head depends on the visual requirements of the task. These are determined by the work itself and by the worker's eyesight and the level of task illumination. Although many studies suggest that workspace design is an effective method of reducing the load on the cervical spine, these other factors must also be evaluated.

Effects of Ergonomic Intervention

Bendix and Hagberg (1984) compared the trunk posture of subjects sitting at horizontal desks and at sloping desks of 22 and 45 degrees. A more upright trunk posture was adopted when the sloping desk was used. The authors suggested that reading matter should be placed on a sloping surface and writing done on a horizontal surface—i.e., desks should have different sloped surfaces to account for differences in visual and manual requirements of reading and writing. De Wall et al. (1991) investigated the effect of using a 10-degree sloping desk and found average reductions in cervical and thoracic spine load of 15 percent and thoracic spine load of 22 percent.

Schuldt et al. (1986) carried out an electromyographic investigation of neck and shoulder muscle activity. They concluded that the forward-flexed ("slumped") posture imposed the most load. The least load was observed when subjects reclined backward slightly (10 to 15 degrees). Significant increases in the activity of the upper thoracic and cervical spine extensor muscles occurred when subjects abducted their arms (lifted them away from the body at the sides). **Elbow rests would appear to be beneficial for both sitting and standing workers; by stabilizing and supporting the arms, they reduce the load on the shoulder musculature.** (It may be mentioned in passing that by supporting the weight of the arms, elbow rests would also reduce the load on the lumbar spine.)

Jonsson et al. (1988) investigated the development of cervicobrachial disorders in female workers in the electronics industry over a 2-year period. Severe symptoms were present in 11 percent of subjects initially and in 24 percent after 1 year. Symptoms declined in the second year if workers were reallocated to different work and increased if they persisted with the same work. Previous heavy work, high productivity, and previous sick leave were predictors of deterioration. Satisfaction with work tasks and an absence of shoulder elevation were predictors of remaining healthy. Reallocation to different work and spare-time physical activity were predictors of an improvement in

(A)   (B)

(C)

FIGURE 5-1   Postures which place a strain on the neck and shoulders. (A) Continually looking down and to one side.(B) Having to extend the neck to look at the screen (screen too high or user wears bifocal lenses). (C) Slumping over a too-low VDT.

condition. Reduction of the monotony of work and job dissatisfaction seem to be important factors in the control of neck problems at work.

Finally, in an investigation of the design of workstations for computer-aided design and manufacturing workers, de Wall et al. (1992) found that more upright postures were obtained if the monitor was raised so that the middle of the monitor was at eye height rather than at 15 to 25 degrees below eye height as is usually recommended.

Figure 5-1 depicts some postures known to increase the load on the cervical spine.

THE SHOULDER, ELBOW, AND WRIST

Activities which require repetitive movements are also known to be associated with discomfort or disability. Strictly speaking, it is incorrect to make a sharp distinction between repetitive and static work, since most repetitive work includes a static component in the form of postural fixation of the body or of a limb. Pathological changes associated with repetitive jobs may therefore be due to the repetitive component, the static component, or a combination of the two. Any part of the musculoskeletal system may be affected, but most of the recent research has concentrated on problems of the shoulders, elbows, hands, and wrists. A distinction can be made between pain which occurs as a result of excessive task demands and which is reversible if the person is allowed to rest or given alternative work, and pain which occurs as a result of some underlying pathology which may or may not have been caused by the excessive task demands. The latter may require medication, physiotherapy, or surgical repair before the person can resume pain-free working.

Many different factors may cause an individual to experience pain in the upper body at work. According to Kroemer (1989), some of the nonoccupational causes are leisure activities, age, gender, pregnancy, and a cold climate. Force, posture, and repetition are the three main ergonomic variables which are associated with occupational musculoskeletal pain. High-risk jobs are those which require repeated, forceful movements of body parts held at the extremes of their ranges of movement, such as with the wrist flexed, extended, or pronated. The solution is either to reduce the repetition rate or the forces exerted, or to mechanize the job or redesign the tool (as described below) to improve the posture of the body part or give the worker greater mechanical advantage. The speed of movement is particularly important in jobs requiring cyclic or periodic activities. **Repetition frequencies greater than 1 Hz may pose a risk to joints or soft tissues, whereas frequencies less than $\frac{1}{6}$ Hz may cause fatigue through the mechanisms of static loading** (Solomonow and D'Ambrosia, 1987).

Anatomy and Physiology of Pain in the Upper Body

Some pain sites and causes of musculoskeletal pain are reviewed below.

Muscle Pain   Repetitive or prolonged exertion can cause pain through accumulation of waste products in the muscles (cramp). Muscle weakness or spasm may result. Cramp in the hand or forearm is known to be more common in those whose jobs involve prolonged handwriting, typing, or other repetitive movements. Cramp is more likely when extreme postures have to be adopted, since these postures weaken muscle-joint systems. This may increase the number of repetitions and time taken to complete a task. Patkin (1989) reports that cramp can be caused while using badly designed ballpoint pens which require undue pressure to write well. Fountain pen use is compulsory for schoolchildren in some countries because these pens can be used with lower forces.

Muscles can be damaged by sudden high forces, particularly during eccentric contractions (as, for example, when one is trying to hold a falling object or resist a sudden reaction torque of a power tool). This primary response may be accompanied by a feeling of soreness in the muscle which diminishes as the damaged muscle fibers regenerate (Armstrong et al., 1993). In properly designed weight-training programs, the goal is, in fact, to cause such damage because the body responds by increasing the number of contractile elements in the muscle, resulting in improved strength and physique. However, a rest period of at least 48 hours is usually an intrinsic part of strength-training programs—the time needed to allow the exercised muscles to recover. The pattern of activity in many industrial jobs bears little resemblance to that in proper muscle-training programs. One of the main differences is that rest periods are far more frequent in muscle-training regimes than at work, which is why work does not normally have the same beneficial effects as exercise or training. Damage to muscle tissue on a daily basis may exceed the repair capability, leading to a decrease rather than an increase in strength or endurance and to chronic pain in the muscle, or **myalgia.**

In a conscious person, skeletal muscles always have a certain degree of tightness. There is a baseline level of muscle fiber recruitment even during relaxation. This is known as **muscle tone** and it is controlled by the central nervous system and by a feedback system involving the spinal cord and the muscle spindles. Muscle tone is essential for the maintenance of posture. (As an aside, it is noteworthy that one of the disabling factors in cerebral palsy is excess or inadequate muscle tone due to central nervous system damage.)

People under mental stress may, without realizing it, develop increased tension in their muscles which they cannot control. At work, the prestressed body part may be a source of pain even though the task loading is mild. A chronic, stereotyped pattern of recruitment of motor units may be the dose which leads to damage of the muscle tissues.

Tendon Pain   When highly repetitive movements are required, the increase in blood supply to the muscles may be associated with a decrease in blood supply to the adjacent tendons and ligaments of the associated joints. As Hagberg (1987) put it, the muscles "steal" blood from the insertions of the ligaments and tendons. "Policeman's heel" (caused by the repetitive microtrauma of walking long distances every day) is an example. Problems of this kind are sometimes referred to as **insertion syndromes.**

Impaired blood supply to the tendons is thought to be the cause of much occupational shoulder pain, because it increases the rate of cell death within the tendon. This is thought to provide sites in which chalk (calcium carbonate) is deposited. It seems that increased tension in tendons reduces their blood supply, which may explain why static work positions are associated with tendon problems. Armstrong et al. (1993) describe an interesting hypothesis which states that the accumulation of dead cells in tendons can cause an inflammatory response in the tendon by the immune system. Inflammatory responses normally occur when there is an injury such as a cut—the blood supply to the affected region is increased to attack any incoming foreign bodies and the part of the body in question normally feels hot and swollen. According to Armstrong et al., if the person already has an infection such as influenza, the immune system will have been activated and the local inflammatory response described above is more likely to be triggered. This may underlie the popular conception that we are more prone to injury when suffering from colds or flu or other diseases.

Frequent mechanical loading can cause inflammation of tendons, which is known as **tendonitis,** or of the cartilage surrounding a joint. Extreme positions of the wrist can press the flexor tendons of the fingers against the bones of the wrist, increasing the friction in the tendons. Rapid, repetitive movements of hand or fingers can cause the sheaths surrounding tendons to produce excess synovial fluid (Figure 5-2). The resultant swelling causes pain and impedes movement of the tendon in the sheath. This condition is known as **tenosynovitis.** Repeated exposure which leads to trauma can ultimately cause the growth of scar tissue which impedes movement of the tendon in its sheath and thus degrades function. Joint structure may be degraded by the formation of bony spurs

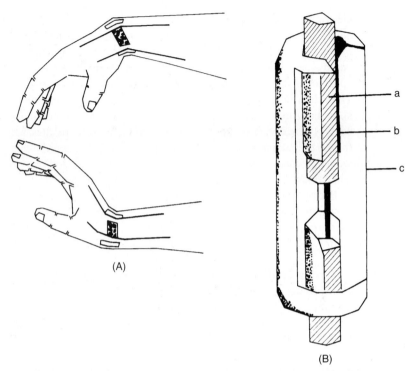

(A)

(B)

FIGURE 5-2   Wrist posture and tendon function. (A) Extreme postures can preload the finger tendons. (B) Simplified view of a flexor tendon of the fingers. The tendon (a) is surrounded by a synovial sleeve consisting of an outer (c) and an inner (b) layer. When the tendon moves, the inner layer glides over the outer layer, lubricated by synovial fluid. If the layers become inflamed or scar tissue builds up, the layers cannot glide smoothly over one another. The tendon then behaves like a rusty brake cable, and smooth, pain-free movement is impossible.

around damaged areas. Reduced mobility, pain, and weakness may result (Chaffin, 1987). Sudden, great forces may cause tendons to separate from bones.

Bursitis   A **bursa** is a sac containing viscous fluid situated at places in tissues where friction would otherwise occur ("bursa" is the Greek word for wine skin and is related to the English word "purse"). There are about 150 bursae in the body and they act like cushions which protect muscles and tendons from rubbing against bones during movements of the body. Overexertion and injury can cause inflammation of the bursae, or **bursitis.** "Housemaid's knee" is a well-known type of occupational bursitis. Bunions are also a form of bursitis which are caused by wearing ill-fitting shoes—friction of the shoe on the bursa on the joint of the big toe causes it to become inflamed. Bursitis can be distinguished from tendonitis anatomically and because of the dull, aching pain that accompanies it—in contrast to the sharper pain of tendonitis.

Neuritis   Repeated or prolonged exertion can cause damage to the nerves supplying a muscle or passing through it. This nerve damage can cause sensations of numbness or tingling (pins and needles) in areas of the body supplied by the nerve. As noted earlier, the model of Armstrong et al. states that the response to a given dose can itself be a dose that leads to a response. In the case of nerves, overexertion can cause increased pressure in a muscle as a result of edema or scar tissue formation. The increased pressure can itself be a dose which results in impaired nerve function. Impaired nerve function, destruction of fibers, or damage resulting in reduced nerve conduction velocity may cause muscle weakness.

All these problems are more likely to occur if the joints are held in an awkward posture (at the extremes of the ranges of movement) since this preloads tendons and ligaments and stretches muscles and nerves.

Bones and Joints   Mechanical trauma seems to be a contributory factor in the development of **osteoarthrosis** in joints. Osteoarthrosis is a noninflammatory disease characterized by degeneration of the articular cartilage, hypertrophy of bone, and changes to the synovial membrane, which causes stiffness and pain in the joints.

When a muscle-joint system is placed in an extreme posture, the muscles on one side of the joint will be lengthened and their antagonists shortened, resulting in a strength imbalance in the antagonistic pair. The ability of the muscles to protect the joint against external forces is degraded, and the joint itself is more easily damaged when the limb is exposed to high forces.

An analysis of joint posture is particularly important in the evaluation of the design of hand tools, particularly in heavy work where the joints may be exposed to high forces. Extreme postures should be avoided to better protect the joints.

Epidemiology of Upper Body Disorders

As will become clear throughout this chapter, there is controversy about the work-relatedness of many musculoskeletal disorders of the upper body. Much of the ergonomics literature seems to be underpinned by the assumption that if a worker experiences pain at work, then it must be the work which caused it. Sceptical clinicians, on the other hand, often attribute this pain to systemic factors or age-related changes.

Epidemiologists have developed a classification scheme for describing the work-relatedness of disorders (e.g., Shilling and Anderson, 1986):

Category 1. The work exposure is a necessary cause of the disorder (as in occupational diseases such as silicosis or lead poisoning).

Category 2. The work exposure is a contributory causal factor but not a necessary one.

Category 3. The work exposure provokes reaction by a latent weakness or aggravates an existing disease.

Category 4. The work exposes the worker to potential dangers which may increase the likelihood of a disease developing (such as alcoholism in liquor industry workers).

Most researchers would place the ailments discussed in this chapter in categories 2 and 3.

There are a number of well-known syndromes which affect particular structures in the upper body and which are caused by the demands of work. They can usually be objectively diagnosed and are accompanied by physical signs of disorder. Stock (1991) carried out a meta-analysis of previous research on upper body disorders at work. A meta-analysis is an analysis in which a researcher, instead of designing a procedure to gather her or his own data to test a hypothesis, uses a number of previous studies and aggregates their findings in the hope of coming to a more general conclusion. Stock found that there was strong evidence of a causal relationship between repetitive, forceful work and the development of disorders of the tendons and tendon sheaths of the hands and wrists and of carpal tunnel syndrome.

Common Pathologies of the Wrist, Elbow, and Shoulder

There are a number of well-known pain syndromes which are sometimes associated with work. **Carpal tunnel syndrome, epicondylitis (tennis elbow), and impingement syndrome are examples.**

Carpal Tunnel Syndrome Carpal tunnel syndrome is a common ailment affecting the wrist and hand. The muscles which flex the fingers lie in the forearm and have long tendons which pass through a narrow opening in the wrist before inserting into the fingers. This opening, known as the carpal tunnel, is also traversed by the nerves and blood vessels of the hand (Figure 5-3).

An increase in the pressure in the carpal tunnel can cause carpal tunnel syndrome if it affects the median nerve. The result is a sensation of tingling and numbness in the palm and fingers. In severe cases, surgery may be required to relieve the pressure. Carpal tunnel syndrome has been reported in jobs requiring rapid finger movements, such as typing, and is found among professional musicians. However, Barton et al. (1992), in their review of the literature, concluded that the majority of cases of carpal tunnel syndrome are not caused by work. Carpal tunnel syndrome can have many nonoccupational causes and is more prevalent in women than in men. It is common during pregnancy be-

FIGURE 5-3   Carpal tunnel. Section through the wrist showing (a) carpal bones, (b) tendons, and (c) the carpal tunnel, containing finger flexor tendons, blood vessels, and nerves. If the wrist is held in extreme postures, the tendons rub against bone when the fingers move and friction is increased. (Thanks to Dr. Douglas A. Bauk of IBM Brazil for supplying the illustration.)

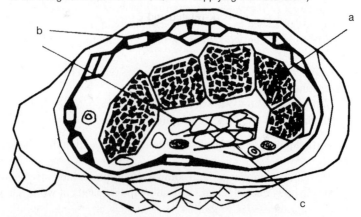

cause the increase in blood volume causes an increase in the volume of extracellular fluid in the body. Since the carpal tunnel is a confined space, this increase may be sufficient to increase carpal tunnel pressure and cause symptoms of carpal tunnel syndrome. The syndrome has also been associated with the taking of oral contraceptives.

Cervical spondylosis and stenosis (narrowing) of cervical structures, as well as nerve entrapment in the arms, may give rise to symptoms which mimic carpal tunnel syndrome. This illustrates that the diagnosis of all arm and hand problems is best left to the experts because pain in these regions may, in fact, originate from other parts of the body.

Loslever and Ranaivosoa (1993) recently investigated the prevalence of carpal tunnel syndrome in light industrial work. The syndrome was found to occur twice as often in both hands as in either the preferred or nonpreferred hand which, they suggest, is evidence that nonoccupational factors are more important than occupational factors. However, the prevalence for both hands was found to correlate positively with measures of wrist flexion and high grip forces. It appears that wrist posture and loading as determined by task and tool design are important cofactors in the etiology of the syndrome. This finding is consistent with the fact that grip strength is considerably reduced when the wrist is flexed rather than extended. It seems reasonable to conclude that redesign of equipment to favor the adoption of a neutral or slightly extended wrist posture may be of value in lowering the incidence of carpal tunnel syndrome at work.

Tennis Elbow The elbow joint may suffer injury following exposure to high forces or repetitive tasks. Somewhat counterintuitively, the elbow can be considered as a weight-bearing joint. The act of grasping and holding objects is possible only if the wrist is stabilized by the muscles of the forearm, many of which originate at the elbow. For example, when the finger flexors contract to enable an object to be grasped, the wrist extensors also have to contract to prevent the wrist itself from flexing. These contractile forces are transmitted to the elbow since the wrist extensors originate from the distal end of the humerus. Clearly, any activity which requires a strong grip to be maintained for long periods will be high in static elbow load.

Overexertion of the extensor muscles of the wrist can lead to a condition known as "tennis elbow" (lateral humeral epicondylitis) (Figure 5-4). In severe cases, the muscle and tendon may separate from the bone. The risk of injury is said to be increased by activities requiring large grasping forces. Dimberg (1987) found that 35 percent of cases of epicondylitis were work-related and 27 percent were due to leisure activities. Tennis accounted for another 8 percent of cases and of the rest, the cause was unknown. Work-related sufferers were found to have jobs higher in elbow stress (e.g., driller, carpenter, polisher) than people who suffer for other reasons or than the workforce as whole.

Dimberg's findings do not support the view that tennis elbow is caused by job content since jobs high in elbow stress may have been more likely only to expose an existing condition or weakness. Many of the cases of tennis elbow in Dimberg's study were of unknown origin. It is clear though that people with symptoms of the disorder will be at increased risk, irrespective of the original cause, if they are required to carry out jobs high in elbow stress.

Impingement Syndrome and Shoulder Pain Most work involving hand tools imposes a combination of repetitive and static loads on the body, loads which usually in-

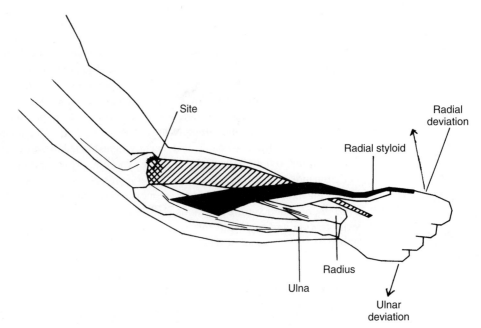

Site

Radial
deviation

Radial styloid

Radius

Ulna

Ulnar
deviation

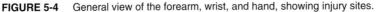

FIGURE 5-4   General view of the forearm, wrist, and hand, showing injury sites.

volve the shoulder, if only indirectly. The shoulder joint is the most mobile in the body and together with its related soft tissues is particularly prone to injury in any activities where the arms are held above the horizontal. Working with the hands above shoulder height is stressful and may increase the risk of developing the so-called impingement syndrome, otherwise known as "swimmer's shoulder," "pitcher's arm" (Wieder, 1992), or "rotator cuff syndrome." The disorder is known to be more common in athletes who use high overhead actions. Sommerich et al. (1993) have recently reviewed the evidence for occupational risk factors in the development of shoulder pain syndromes. The following factors have been identified:

- Awkward or static postures
- Heavy work
- Direct load bearing
- Repetitive arm movements
- Tasks requiring fast arm movements
- Working with the hands above shoulder height
- Lack of rest

Figure 5-5 depicts in broad outline the anatomy of the shoulder joint and related structures. Strictly speaking, movements of the shoulder involve the coordinated movement of several joints. The main one is the scapulohumeral joint, which is usually what is meant when the term "shoulder joint" is used. The shoulder joint is a kind of ball-and-socket joint, but the ball part, the head of the humerus or upper arm bone, represents only a third of the surface of a sphere when it engages the socket. The socket (the glenoid cavity of the scapula, or shoulder blade) is correspondingly shallow.

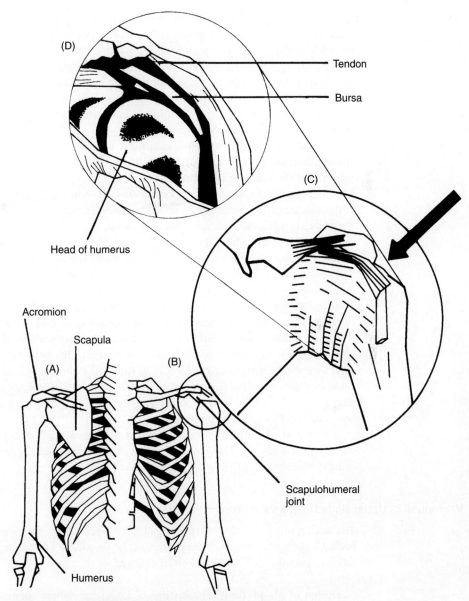

FIGURE 5-5   Shoulder joint. (A) Rear view; (B) front view; (C) expanded view; (D) the scapulohumeral joint itself. Shoulder problems can occur in the narrow space between the head of the humerus, the acromion, and the ligaments, as indicated by the arrow.

In quadrupeds, the fact that the head of the humerus is not enclosed by the bony socket does not matter much because the shoulder is a weight-bearing joint and the head of the humerus is continually pushed against its articular surface. In humans, however, the arms hang down, a position which tends to dislocate the joint, and for this reason, the head of the humerus has to be held in place by ligaments and tonic muscle activity. This explains why the shoulder joint is so easily dislocated. The shoulder joint can be contrasted with the much more stable hip joint where over 50 percent of the femoral head is enclosed by the acetabulum.

That the shoulder joint requires muscle activity to be held in place might alert the ergonomist to its likely susceptibility to rapid fatigue and damage when exposed to static loads or high forces at work. A similar line of argument suggests that one of the simplest ways to reduce occupationally induced shoulder stress in many jobs is through the provision of armrests, slings, or other means of supporting the weight of the arms to enable the shoulder muscles to relax.

Above the humeral head lie the acromion and the coracoacromial ligament, and in between lies a bursa. In this narrow space pass many tendons, nerves, and blood vessels. The space can easily become taken up by the growth of bony spurs, by bleeding, or by soft tissue swelling as a result of overexertion (the subacromial bursa can become inflamed, for example). **If any of these occur the range of motion of the shoulder joint is reduced because the impingement of the subacromial structures causes pain when the joint is moved.** Sufferers cannot raise their arms above shoulder height. It is likely that jobs which require the hands to be chronically elevated above elbow height can cause short-term changes which, over time, may ultimately lead to disorders of the shoulder joint. Shoulder pain can also be caused by localized muscle fatigue—particularly if the arms have to be held above shoulder height for long periods of time (as when one is painting a ceiling or pruning a tree).

Wiker et al. (1989) investigated shoulder posture in relation to the incidence of localized muscle fatigue. They concluded that shoulder muscle fatigue could be prevented if the considerations in Table 5-2 were adhered to.

Prevention of Upper Body Disorders at Work

The design of tools and workspaces can have a profound effect on the posture of the body. Of particular interest is the posture of the shoulders, elbows, and wrists and its relation to pathological musculoskeletal changes.

Design of Hand Tools The control of finger movements depends on many small muscles which can easily become fatigued, particularly during prolonged work with inadequate rest periods and poorly designed tools (for example, when writing an exam). One of the most fundamental problems in hand tool design is to optimize the dimensions of the tool in relation to the hand anthropometry of the population under study.

Ducharme (1977) found that many women working in previously male craft skills in the U.S. Air Force were dissatisfied with the design of the tools they used. Crimping tools, wire strippers, and soldering irons were said to have grips that were too wide or required the use of two hands to operate. Other tools were said to be too heavy and too

TABLE 5-2   METHODS OF REDUCING SHOULDER STRESS

1. If possible, work with the hands near waist level and close to the body.
2. If the hands have to be positioned above shoulder level, their elevation above the shoulders should be no more than 35 degrees. Hand loads should not exceed 0.4 kg, and the posture should be held for no more than 20 seconds for each minute of work.
3. Select taller workers for workplaces which cannot be modified.

awkward to use. Pheasant (1986) suggests, on anthropometric grounds, that the tools were probably inadequate for male use as well but that more males were able to overcome the design deficiencies because of their greater strength. Pheasant and Scriven (1983) report that the car lug wrenches provide inadequate mechanical advantage for both male and female users. Although a greater proportion of females than males are affected, both sexes would benefit from improved designs offering greater mechanical advantage.

Pheasant and O'Neill (1975) investigated handle design in a gripping and turning task (such as using a screwdriver). They found that strength deteriorated when handles greater than 5 cm in diameter were used and that to reduce abrasion of the skin, hand-handle contact should be maximized. Knurled cylinders were found to be superior to smooth cylinders because of the increase in friction at the hand-handle interface. The authors concluded that, for forceful activities, the size of a handle, rather than its shape, was most important. A useful rule of thumb for evaluating handle diameters is that the handle should be of such a size that it permits slight overlap of the thumb and fingers of a worker with small hands. One of the most important criteria for handle design is that sufficient hand-handle contact is provided for. The larger the handle diameter, the bigger the torque that can be applied to it, in principle, but people with small hands must be able to enclose the handle with their fingers. Forty mm seems to be the maximum handle diameter for male users, smaller still if gloves are to be worn. Cylindrical handles are better than handles with finger grooves since these cause pressure hot spots and blistering of the skin of hands they don't fit. Handle lengths should be at least 11.5 cm plus clearance for large (95th percentile) hands. An extra 2.5 cm should be added if gloves need to be allowed for (NIOSH, 1981). Some examples of screwdriver handles designed for one- and two-handed operation are shown in Figure 5-6.

It may also be noted that grip strength depends very largely on the posture of the wrist. When the wrist is extended, the finger flexors are lengthened and can therefore exert more tension, resulting in a stronger grip. When the wrist is flexed, the opposite occurs and grip strength is severely weakened. This underlies the method taught in self-defense classes for disarming an attacker. The hand and wrist holding the weapon are twisted (pronated) and the wrist flexed before any attempt is made to remove the weapon from the attacker's hand. The pronation and flexion of the wrist shortens and therefore weakens the finger flexors. A general requirement of handle design is that the wrist joints should be kept in a neutral position (in the middle of their ranges of movement) when tools are being used. Pheasant (1986) describes how the axis of a handle is at an

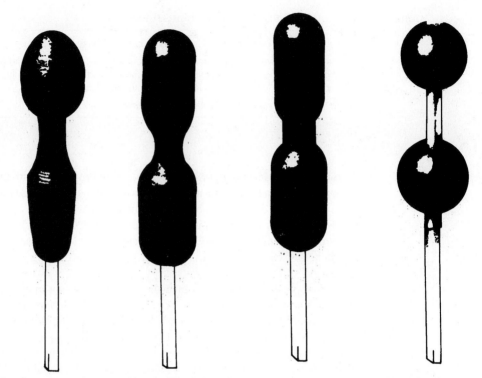

FIGURE 5-6 Some innovative screwdriver handle designs for one- and two-handed operation (designed by the Ergonomic Design Group, Stockholm, Sweden).

angle of 100 to 110 degrees with respect to the forearm when the wrist is in a neutral position (Figure 5-7).

Tools such as saws and pliers can be designed with obliquely set handles to enable the wrist to be maintained in the neutral position (Tichauer, 1978). Tools such as soldering irons can be redesigned using a pistol-grip handle rather than the more traditional

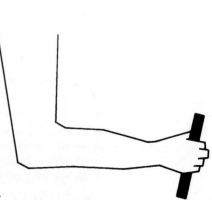

FIGURE 5-7 Angle of grip when the wrist is in the neutral position.

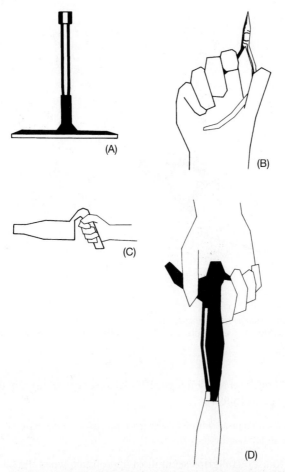

(A)

(B)

(C)

(D)

FIGURE 5-8   Redesigned tools with improved hand-handle interfaces. (A) Socket wrench with T-bar for greater mechanical advantage. (B) Pliers with bent handles to maintain neutral wrist posture. (C) Drill with handle at appropriate wrist angle. (D) Paint scraper with thumb stall to relieve pressure on the palm of the hand and prevent blistering of the skin on the palm.

straight handle for the same reason (Figure 5-8). When one is using a straight-handled tool there is a tendency for the wrist to be bent outward (ulnar deviated). This position stretches the tendons of the forearm muscles on one side, causing them to rub against a bony protrusion on the thumb side of the wrist (known as the "radial styloid"). Repeated exposure can cause the sheath (synovium) within which the tendon runs to become inflamed. This condition is known as **DeQuervain's syndrome.** Inflammation of tendons and tendon sheaths can occur in other parts of the hand and other body structures. In the long term, permanent damage to the tendon and its sheath may result. The buildup of scar tissue in the tendon may ultimately reduce the range of movement of the wrist.

The idea of "bending the handle instead of the wrist" is a valid one and has many potential applications. It has been incorporated into the design of such diverse products as hammers (Knowlton and Gilbert, 1983) and even tennis rackets and table tennis paddles

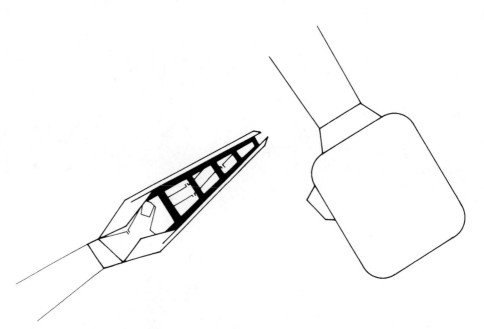

FIGURE 5-9   Table tennis paddle design to minimize wrist strain. The design is based on the notion that a tool is an extension of the hand and therefore encapsulates and becomes part of the hand.

(Figure 5-9) (Chong and McDonough, 1984). Konz (1986) sums up the use of bent handles in tool design as follows:

> . . . When a tool, gripped with a power grip, has its working part extend above the hand, then a curve in the handle may be beneficial. . . . A small bend (5 to 10 deg) seems best.

For power tools, pistol grips seem most appropriate when the task is oriented vertically with respect to the operator (as when one is drilling a hole into a wall). When the task is a horizontal one (as when one is fastening a screw into a horizontal desktop with a power screwdriver), an in-line tool may be better (Figure 5-10).

In the latter case, the forearm is perpendicular to the working part of the tool (the shaft of the screwdriver points vertically downward and the forearm remains horizontal). Power tools tend to be considerably heavier than their nonpowered counterparts, and a potential source of wrist strain comes from the weight of the tool itself—particularly if the handle is placed at one end of the tool rather than in the middle. Having to hold a heavy drill with a pistol grip while positioning the drill bit exemplifies this problem. Wrist loading can be reduced by fitting the handle at the tool's center of mass so that the tool is counterbalanced. Johnson (1988) investigated the design of power screwdrivers in relation to operator effort and concluded that grip diameters should be at least 5 cm. Vinyl sleeves fitted over too-narrow handles were effective in reducing effort. A biomechanical brace was designed to fit over the screwdriver handle and run along the palmer (inside) side of the forearm. The brace transmitted the reaction torque to the user's forearm and enabled the tool to be used with reduced grip force.

Finally, tools can be redesigned and fitted with longer handles, or handle extensions

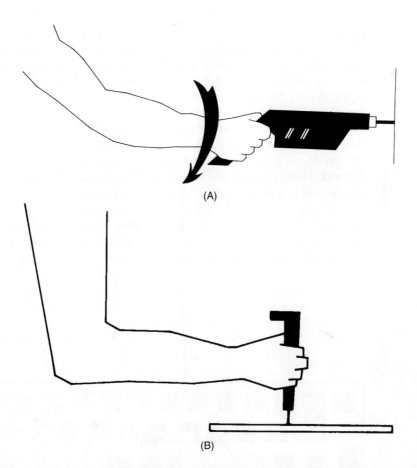

(A)

(B)

FIGURE 5-10   For work at a vertical surface (A), an in-line grip reduces wrist strain. For work at a horizontal surface (B), a perpendicular grip may be better. Note the large moment of ulnar deviation in (A) when
the tool is not engaged with the workpiece. This deviation can be minimized by locating handles
below the mass center of the tool.

can be fitted to increase the worker's vertical reach, obviating the need for the hands to
be raised above shoulder height. Some examples are handle extenders for paint brushes
for painting ceilings and long-handled pruning sheers for pruning the higher branches of
trees.

Keyboard Design   VDT work has received much attention recently as a response to
the prevalence of upper limb CTD in VDT users. It appears that these disorders are more
prevalent in tasks where the keyboard is connected to a VDT rather than a typewriter.
There are several possible explanations for this disparity and it is likely that all of them
can play a part to a greater or lesser degree. With old-fashioned mechanical typewriters,
the typist had to stop to return the carriage manually after each line. Errors had to be corrected manually with erasers or correcting fluid, and the paper had to be changed after
each page had been typed. All these secondary tasks provided changes of posture and

broke up the continuity of the typing task. Brief periods of rest are intrinsic to the operation of mechanical typewriters. With word processors, the secondary tasks are carried out automatically or via special keys on the keyboard, so the work is intrinsically less varied and more likely to cause fatigue. It should also be noted that before the introduction of desktop computers, almost the only people who ever used keyboards were typists who had received special training in keyboard skills (e.g., how to use the fingers most efficiently and how to type without looking at the keyboard). It is likely that a very large proportion of today's desktop computer users have never undergone formal training and do not possess these keyboard skills. Patkin (1989) has suggested that poor motor skill leads to excess cocontraction of muscles and temporary muscle aches which may be mistaken for tenosynovitis (which is rare amongst typists).

Several researchers have focused on the design of keyboards as a means of reducing musculoskeletal problems in keyboard operators. Zipp et al. (1983) investigated the posture of the hands and wrists, noting marked ulnar variation and fatigue (Figure 5-11). They concluded that keyboards should be designed with separate banks of keys (one for each hand), each bank being inclined and contoured to be compatible with the functional anatomy of the hand. Keyboards based on this design are commercially available (Figure 5-12), are perceived as comfortable to use, and give fair performance (Smith et al., 1993). There is a lack of epidemiological evidence concerning their long-term efficacy.

Fernstrom et al. (1994) measured EMG activity from the forearm and shoulder mus-

FIGURE 5-11 Conventional keyboards cause ulnar deviation.

FIGURE 5-12 Separate banks of keys for each hand may relieve wrist strain. (Photograph courtesy of Kinesis Corporation, Bellevue, Washington.)

cles of eight typists using mechanical, electromechanical, and electronic typewriters; a PC keyboard; and a PC keyboard with the keys angled at 20 degrees to the horizontal. The mechanical typewriter required a higher keystroke force, which placed greater strain on the muscles investigated (although the difference was small). The electronic typewriter placed more strain on the right shoulder than did the mechanical typewriter. The unequal strain appeared to be due to the low keystroke force which made it impossible to rest the fingers on the keyboard without typing a character. A keystroke force of about 0.5 newton (N) is recommended. Finally, use of a palmrest while typing was not effective in reducing muscle strain.

Work Design, Education and Training, and Job Design Disorders associated with repetitive work have existed both in and out of industry for many years—biscuit packing, fruit wrapping, and chicken plucking are well-known examples of high-risk tasks. The same disorders are also associated with leisure activities such as playing musical instruments or electronic arcade games, solving Rubic's cube, and running marathons (Brown et al., 1984). Kemp (1985) describes several approaches to the management and prevention of such disorders. Task rotation and scheduling of work-rest cycles to minimize overexposure to the repetitive elements of a task are examples. Careful analysis of individual tasks is essential to identify high-risk areas.

Ergonomists usually advise correct design of tools, equipment, and workspaces for the prevention of musculoskeletal pain in the upper body. Workspaces must be designed to allow flexibility in carrying out a task by means of a wider repertoire of postural behaviours to vary the load on different parts of the body. Training programs to increase a worker's capacity have received little attention in ergonomics, although they are regarded as fundamental in sports science to avoid the very similar types of musculoskeletal problems which can occur when engaging in sport. In addition, sports scientists emphasize the importance of self-preparation immediately before any arduous activity is undertaken. This preparation normally includes proper nutrition and fluid intake and a practiced set of stretching and warm-up exercises.

Table 5-3 summarizes complementary approaches to the reduction of work-related musculoskeletal disorders.

TERMINOLOGICAL PROBLEMS AND THE RSI "EPIDEMIC"

Several terms have been coined to describe work-related musculoskeletal ailments, but none is generally accepted. The term "cumulative trauma disorder" (CTD) is used in the United States. It implies that the disorder is due to the cumulative effects of multiple exposure to stress rather than to disease or degeneration. "Repetition strain injury" (RSI) is used in parts of Europe and Australia. It emphasizes the causative role of repetitive work. "Overuse injury," although it might refer to an injury of any body part, normally refers to an injury of the musculoskeletal system caused by chronic exposure to high loads. None of these terms are accepted diagnostic categories—they refer to syndromes whose underlying pathology is often unknown and are of similar status to terms such as "eyestrain" (which means eye discomfort and viewing difficulties) and "lumbago" (which is derived from the Latin words for low back pain).

TABLE 5-3   SIMPLE MEASURES FOR PREVENTING WORK-RELATED MUSCULOSKELETAL DISORDERS OF THE UPPER LIMBS

Tool design and wrist posture	Workstation design
Bend the handle (5-10 degrees), not the wrist.	Damp jerk and impact forces.
Avoid excessive use of pinch grip (thumb and fingers).	Position work objects so as to eliminate static neck postures.
Maintain neutral wrist posture.	Eliminate static shoulder elevation, or provide elbow rests or slings.
Reduce required grip forces:	Provide vices, clamps, or other aids to minimize the need for sustained holding of work objects.
Use high-friction materials for handles.	
Longer handles increase mechanical advantage.	
Add handles for carrying tools and for resisting reaction torque.	
Damp vibration from power tools.	
Encourage use of large muscle groups.	

Task/job design	Management of working conditions and worker exposure
Limit repetitive movements to 2000 per hour or fewer.	Avoid repetitive work in cold conditions.
Eliminate highly repetitive (under 30-second cycle time) jobs.	Use ergonomic criteria when making equipment procurement decisions.
Design out movements requiring rapid acceleration of body parts.	Redesign tasks before a discomfort becomes a disorder.
Rotate workers between high-repetition and low-repetition tasks.	Eliminate excessive overtime working.
Avoid mental stress by eliminating:	
Unrealistic deadlines or production targets.	
Machine pacing with fast cycles.	
Excessive supervision.	
Piecework systems of remuneration.	
Design in microbreaks (2–10 seconds rest every few minutes).	
Increase task variety (e.g., integrate routine maintenance, inspection, and record keeping with basic production tasks).	

Readers unfamiliar with these terms should note with caution that they are not used in a consistent way in the literature. Sometimes they are used to describe pain which cannot be diagnosed precisely and which does not resemble pain from one of the recognized causes, such as those described above. When used in this way they usually mean "occupational arm pain with no identifiable pathological basis." However, other researchers *do* include recognized syndromes in the category of RSI or CTD.

In the 1980s, a sudden increase in upper limb pain was reported in several countries, particularly Australia. Investigations revealed that some of the pain could be traced to well-known clinical entities such as those described above. Objective signs such as inflammation accompanied the pain. However, in many cases the pain appeared to have no organic origin and the symptoms did not follow well-known patterns.

Mullaly and Grigg (1988) proposed three categories of repetition strain injury, or RSI. Type 1 consists of well-known clinical entities, and type 2 of chronic pain without degeneration, inflammation, or any sign of neurological or systemic disease. In type 3 an initial type 1 lesion acts as focus for pain which is amplified by psychological processes. Types 2 and 3 appear to be a collective name for occupational arm pain, and many clinicians have cast doubts concerning the existence of physical pathology.

Psychological processes seem to be of great importance in the development and course of types 2 and 3. In particular, type 2 does not appear to involve tissue damage and is reversible. The existence of a strong psychological component suggests that ergonomic intervention based solely on the analysis of posture and workstation layout would be insufficient on its own. Psychological and organizational analysis would also be necessary.

The inclusion of well-known clinical entities under the category type 1 repetition strain injury is potentially problematic for the reasons described above. Some of the recognized syndromes are not exclusively caused by the repeated strain or cumulative trauma of work activities, and it may be difficult to demonstrate precisely the role of occupational factors in the development of a disorder. The practice of including recognized syndromes in the RSI category gives a false impression of accuracy and invites the misclassification of disorders and the overestimation of the contribution of occupational factors in the etiology of musculoskeletal disease. It is known, for example, that cigarette smoking and other lifestyle-related activities contribute to the development of these ailments, but the contribution of these other factors is not implicit in terminology such as RSI or CTD.

Mullaly and Grigg's typology does have some value in highlighting that pain in the upper body can have different causes and can be amplified by psychological factors such as job dissatisfaction.

Barton et al. (1992), among many others, have criticized the use of the term "RSI" to describe pain in the arms or hands. First, the term implies the existence of a physical condition with an identifiable physical cause, whereas much of what is reported as RSI does not correspond to well-known syndromes in terms of either the symptoms which are reported or the location of pain. Physical signs such as inflammation are often not identified or reported either. Second, the word "strain" is not used in its true mechanical sense. A strain injury occurs as a result of excess physical stress being imposed on a part of the body and can be objectively confirmed by observing the damage to or change in the affected body part (in this sense, a broken leg or a sprained ankle is a strain injury). This is usually not the case with RSI. Third, an injury is usually thought to occur as a result of a single event in which both cause and effect are clear and are closely related in space and time. However, RSI is used to describe pain which occurs as a result of carrying out repetitive work over long periods (perhaps the word "wear" rather than "injury" is really what is meant). For these reasons Barton et al. suggest that the RSI terminology is obsolete and that it be discarded.

A New Theory of RSI

Several researchers have suggested that much of the diffuse and difficult-to-diagnose arm pain which is currently labeled "RSI" may really be peripheral nerve pain (Quinter and Elvey, 1993). The pain is the result of a disorder of the pain receptors and occurs independently of the person's mental state. Unlike neuritis, there is no inflammation of the nerve itself. Disorders of nerve function are known as **neuropathies.** Chronic abnormal inputs from peripherally damaged nerves can sensitize nerve cells within the spinal cord, resulting in the hypersensitivity to painful stimuli characteristic of RSI. According to the theory, RSI is really a form of peripheral neuropathy in which "damaged nerves can come to contribute actively to chronic pain by injecting abnormal discharge into the nervous system and by amplifying and distorting naturally generated signals" (Quinter and Elvey, 1993).

Butler (1991) points out that nerves are "bloodthirsty" structures—the nervous system constitutes 2 percent of body weight and consumes 20 percent of the oxygen in the blood. The nervous system is also the most extended and connected system in the body. This means that nerves have to accommodate postural movements. When, for example, a person flexes the elbow, the nerves on the flexion side of the joint shorten and may be pinched by other tissues, and nerves on the opposite side are stretched. It is conceivable that both these accommodations may interfere with the nerve's blood supply. According to Butler, movements at joints such as the ankle can increase the tension in the nerves in distant parts of the body. For example, the angle through which the straight leg of a person lying supine can be flexed depends on whether the person's neck is flexed. Cervical flexion pretenses the nervous system and hastens the onset of pain during the straight-leg-raising maneuver. In these situations, the nerve is susceptible to injury, not just at the joint which moves, but at any point along its length where **adverse neural tension** occurs. Vulnerable areas are where nerves pass through tunnels, where the nervous system branches and is less able to glide over surrounding tissues, where the nerve is relatively fixed (at some points in the spine, for example), and where the nerve passes close to unyielding surfaces. It is interesting to speculate whether the flexed cervical postures of many office workers predispose them to pain in the upper limbs through the mechanism of adverse neural tension.

Another puzzling feature of some activity-related upper limb pain is its task specificity. The person may complain of pain only when carrying out small repetitive motions such as typing or playing the piano but be perfectly capable of carrying out similar activities which require a wider range of joint movement. Butler (1991) uses the term **activity specific mechanosensitivity** to describe this phenomenon. It is hypothesized to occur when, during movement, a small region of scarred nerve tissue moves, *in a particular direction,* against a damaged or pathological surface such as bony outgrowth.

CURRENT AND FUTURE RESEARCH DIRECTIONS

There has been a great increase in reported musculoskeletal disease in recent years. This increase has created a demand for better-designed production facilities and equipment (Eklund and Frievalds, 1993). However, before being able to introduce clear guidelines for the control of upper body musculoskeletal injuries in industry, researchers would find it beneficial to understand more about their epidemiology and the contribution of occupational factors to their development.

The fundamental problem facing researchers is to find ways of quantifying worker exposure to occupational stressors, to correctly classify the musculoskeletal outcomes of this exposure, and to demonstrate a relationship between the two. Only from such data can a rational approach to intervention be formulated. Some recent studies with these strengths (e.g., Magnusson et al., 1990) have failed to show strong links between ergonomic exposure and musculoskeletal outcomes, suggesting that ergonomists must pay attention to psychosocial and work organization issues as well as workspace and equipment design.

There is currently a need for better longitudinal and cross-sectional studies of the prevalence of these disorders. Even for recognized disorders such as carpal tunnel syndrome, the diagnostic criteria used by clinicians are not well standardized or are applied incorrectly. Previous studies have used indexes as crude as job title to estimate exposure, but dose cannot be directly inferred from such information. Recent developments in inexpensive and nonintrusive goniometers and other devices for exposure measurement seem promising. Stock (1991) has concluded that the more refined the definition and measurement of ergonomic exposures, the more able researchers are to demonstrate cause and effect where it exists.

It is hoped that these developments will provide data for the design of low-risk jobs and tasks. At present, there is a lack of data for optimizing the design of work-rest cycles and for setting limits on force exertion and repetition rates.

SUMMARY

Although mechanization and automation have reduced the amount of heavy work in society, they have introduced new "light" jobs which are very repetitive and place the load on the smaller muscle groups.

The development of work-related musculoskeletal disorders depends on the interplay of several classes of variables. The characteristics of the person, including age and skill level, can interact with the requirements of tasks and the design of tools, leading to excessive demands being placed on the musculoskeletal system. Neck and shoulder strain can be reduced at work by appropriate design of the visual requirements of tasks. The use of sloping work tops has been shown to be effective. Any beneficial effects are likely to be greater in older workers where the incidence of cervical spondylosis is higher and symptoms are more likely to be amplified by task-induced stress.

Muscle and tendon problems such as cramp, tenosynovitis, and tendonitis have been shown to be associated with highly repetitive activities. Musculoskeletal stress can be reduced and the efficiency of task performance increased by careful task and tool design. Researchers have recently become interested in problems of the wrist, elbow, and shoulder, and a number of well-documented syndromes exist which are often, but not always, associated with work activities. Several different terms have been used to describe these problems and most are unsatisfactory. Armstrong et al. (1993) use the term "work-related neck and upper limb musculoskeletal disorders," which, despite its verbosity, seems to correctly describe the problem in accordance with the current state of knowledge and has the distinct advantage of being underpinned by a conceptual model of the problem and its associated etiology.

Despite the theoretical and methodological problems of carrying out research in this area, clear implications can be identified. Unnecessary load on the upper body can and should be reduced by better design of workstations and tasks. There is some evidence (e.g., Chatterjee, 1992) that ergonomic intervention to reduce stress can produce large reductions in the incidence rates of upper extremity disorders.

Excessive stress may arise from several sources, including:

1 Repeated, rapid movements such as using a screwdriver for long periods or a keyboard for data entry
2 Fixed positioning of unsupported limbs for long periods such as working with the hands above the head
3 Forceful movements such as turning a heavy crank several times a day

Stress from each of these sources can be reduced by changing the design of jobs, workspaces, and tools, changes that will make the occupational manifestation of painful syndromes less likely.

Further information can be found in Putz-Anderson (1988) and Pheasant (1992).

ESSAYS AND EXERCISES

1 Carry out a survey of the handle design of a variety of commercially available hand tools for use in the home. Pay particular attention to:
  a Handle circumference and length
  b Handle shape and contour
  c Mechanical advantage of handle
  d Position of handle with respect to tool body
  Obtain anthropometric data on hand dimensions and decide how satisfactory is the design of these tools.
2 Obtain a number of commercially available screwdrivers and a large piece of hard wood. Using friends or colleagues as subjects, time how long it takes them to screw in a standard-size screw with each screwdriver. Use five screws per screwdriver, per person, and take the average time for each person-screwdriver combination (use a different screwdriver after each screw to balance out fatigue effects). Can any differences be attributed to handle or blade design? Is previous experience in the use of hand tools important?
3 You are consultant to the production manager of a large motor company. Draft guidelines to assist the production department with the procurement of hand tools.

FOOTNOTE: Sleeping Posture—A Contributory Factor in Neck and Arm Pain?

Several attempts have been made to account for occupational arm pain of unknown origin. Smythe (1988) proposed that occupational arm pain of the RSI types 2 and 3 variety is referred from the neck. This means that the site of the pain is, in fact, healthy and the injured or diseased area is asymptomatic. Researchers using body diagrams and hoping to link regions of upper body pain to the ergonomics of workspaces will appreciate that this possibility would invalidate some of their findings in the absence of other information about employee health status. Although pain referred from the neck may be caused by occupationally induced stress, Smythe proposed that it will be exacerbated by activities outside of work, including nonrestorative sleep. Smythe made an interesting comparison between the anatomy of the thorax in humans and in quadrupeds. In humans, the shoulders are set out at the sides of the body and braced in this position by the

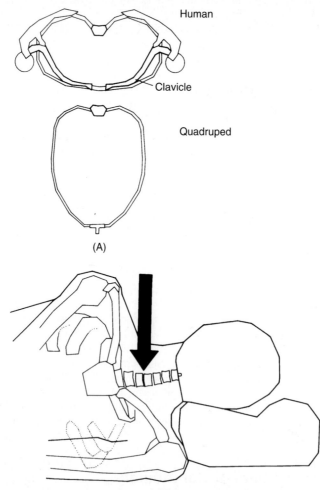

FIGURE 5-13   Sleeping posture and upper limb pain. Thoracic cavity in human and quadruped. In the human, the chest is flat and the clavicles brace the shoulders at the sides. Humans are not well adapted to sleeping on the side because the lower cervical vertebrae lack support. This may prevent recovery from work- or leisure-induced damage. (Reproduced with the permission of Professor H. Symthe and the *Journal of Rheumatology.*)

clavicles (Figure 5-13). Humans are not well adapted to sleeping on the side because the lower cervical vertebrae lack support in this position. Unless the sleeper wears neck ruffs or rolls or uses a special pillow, the neck can be placed under strain. Since most people do not do this, those who are exposed to chronic neck strain at work are also exposed to chronic strain at night when they are asleep. The soft tissues of the neck are never able to recover from the strains of work, and the result is chronic pain which does not respond permanently to simple methods of treatment.

The idea that nonrestorative sleep underlies or exacerbates much of what is reported as occupational arm pain is controversial and open to debate. However, it is of value in taking the emphasis away from the workplace—reminding the ergonomist that the body does not cease to function outside of working hours and that leisure activities can be equally stressful as or more stressful than work activities. Many of the problems discussed in the preceding chapters can also be caused and certainly be exacerbated by activities of daily living outside of work. The ergonomist should keep this fact in mind when evaluating working conditions to determine possible causes of musculoskeletal pain at work.

6

DESIGN OF MANUAL
HANDLING TASKS

A large proportion of the accidents which occur in industry involve the manual handling of goods. In the United States, a report by the National Institute for Occupational Safety and Health (NIOSH, 1981) stated that back pain was attributed to overexertion by 60 percent of back sufferers. About 500,000 workers in the United States suffer some type of overexertion injury per year. Approximately 60 percent of the overexertion injury claims involve lifting and 20 percent, pushing or pulling.

The problem occurs worldwide. According to a recent consultative document by the British Health and Safety Commission, more than 25 percent of accidents involve handling goods in one way or another (Health and Safety Commission, 1991). The seriousness and extent of the problem is well illustrated in the same document where it is estimated that a 10 percent reduction in manual handling injuries would save the British economy some £170 million per annum.

The relation between low back injury and occupation is supported by the findings of epidemiological surveys. Magora (1972) found that low back symptoms were more common in workers who *regularly* lifted weights of 3 kg or more than in those who *sometimes* lifted such weights. It is interesting to note that low back symptoms were even more common in those who *rarely* lifted weights. These individuals may have had sedentary jobs, which is another risk factor for low back problems, or it may be the case that people who rarely lift weights have poor technique. Kelsey (1975) investigated the epidemiology of a common low back injury—the acute herniated lumbar intervertebral disk. A number of demographic factors were associated with increased risk of injury (Table 6-1).

Obesity and occupations involving lifting, pushing, pulling, or carrying were not related. However, lifting was often cited as a **precipitating event**—more so than falling, car accidents, and standing. The evidence seems to suggest that occupational manual

TABLE 6-1   FACTORS ASSOCIATED WITH INCREASED RISK OF BACK INJURY

1 Suburban residence

2 Sedentary occupation

3 Driving motor vehicles (especially professionally)

4 Number of children (in women)

5 Stature (in women)

handling of goods increases the risk of low back symptoms and of back injury in people unaccustomed to lifting.

ANATOMY AND BIOMECHANICS OF MANUAL HANDLING

The spine itself lacks intrinsic stability. Tests carried out on spines (vertebral bodies and associated ligaments) removed from cadavers indicate that the spinal column itself is surprisingly weak—it can support a load of only 2 kg placed on top of the first thoracic vertebra (Nachemson, 1968). How then can the spine perform its function as the main support structure of the axial skeleton?

The spine cannot support the weight of the trunk unless the individual vertebrae are stabilized. In a living person, the whole spine is surrounded by many layers of muscle as well as by the contents of the abdomen and chest cavities. We can consider the spine together with the trunk muscles and other soft tissues as the weight-bearing unit.

Klausen (1965) believed that the short, deep muscles of the back stabilized the individual joints of the spine, holding them together and enabling the spine to function as a rodlike structure. The more superficial muscles stabilized the structure as a whole. The shape of the spine changes when a load is carried because carrying requires an increase in the activity of the back muscles, which results in a greater compressive force being imposed on the intervertebral disks. The lumbar disks become deformed mostly in the anterior part, and the net result of this deformation is a decrease in the degree of curvature of the lumbar lordosis.

The position of the stabilized spine, as a whole, depends on contraction of the more superficial long muscles of the back. Activity in these muscles increases when a person leans forward. When a person leans backward, activity in the anterior abdominals increases and is accompanied by a decrease in back muscle activity (see Figure 2-6). We can consider the spine to be like a tent pole held in place by the antagonistic pull of its surrounding muscle groups.

The abdominal and thoracic muscles play a major role in stabilizing the spine when a weight is lifted, according to Morris et al. (1961). In relaxed standing, these muscles exhibit little activity. When a person leans forward to lift a weight, a moment of flexion is placed on the spine. The heavier the weight, the greater the flexion strain. The back muscles contract to resist the flexion (by exerting a moment of extension about the spine), and this contraction is accompanied by antagonistic contraction of the abdominal and thoracic muscles. These muscles pressurize the contents of the abdomen and thorax, converting them into approximate hydraulic and pneumatic splints which are thought to oppose the flexion moment (Figure 6-1).

The precise role of intra-abdominal pressure is unclear, although it may resist some of the flexion strain and relieve tension in the back muscles. Aspden (1989) has suggested that intra-abdominal pressure compresses the convex surface of the lumbar lordosis. Aspden regards the lumbar lordosis as an arch which, when compressed by intra-abdominal pressure, becomes stiffer (in the same way that placing a load on a masonry arch increases its stability). Whatever the mechanism, the reflex increase in intra-abdominal pressure when a person lifts a load appears to be a normal response to the increased load on the spine.

There is a commonly held view that, when attempting to lift objects, the back must be kept straight. Keeping the back straight can be interpreted as maintaining a lumbar lordosis—or at least, avoiding flexion of the lumbar spine.

Back Injuries and Lifting and Carrying

According to Grieve and Pheasant (1982), the trunk can fail in three ways when a weight is lifted:

1 The muscles and ligaments of the back can fail under excessive tension.

2 The intervertebral disk may herniate as the nucleus is extruded under excessive compression.

3 The abdominal contents may be extruded through the abdominal cavity as a result of excessive intra-abdominal pressure.

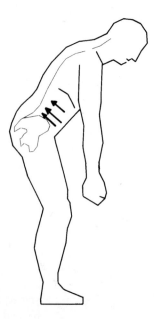

FIGURE 6-1   Abdominal mechanism in lifting.

These injuries are often referred to colloquially as muscle strains or tears, slipped disks, and hernias.

The term "slipped disk" is particularly misleading, so it is appropriate to consider the true nature of this injury in more detail. When the trunk is flexed, as in stooping to lift a weight, the lumbar intervertebral disks undergo anterior wedging by the vertebral bodies above and below them. Material from the nucleus pulposus may be extruded posteriorly, causing the outer layers of the disk to bulge rearward. When one is lifting a load, the spine is subjected to compression which narrows the disk space. When the spine is extended, some nuclear material may become pinched posteriorly, stretching the posterior ligaments and pressurizing nerve roots. This condition is known as a slipped disk (Figure 6-2). It is readily apparent that disks do not slip. Although nuclear material does move posteriorly, the injury is one of prolapse of the annular fibers rather than slippage of the body of the disk itself.

A catastrophic injury such as a disk prolapse is not caused simply by a sudden event such as lifting a heavy weight. It is usually the end product of years of degeneration of the disk and surrounding structures. Over time, the layers of cartilage in the annulus fibrous develop microtears and weaknesses, the nucleus may lose fluid, and the disks may become thinner, causing the vertebral bodies above and below the disk to move closer together. Parts of the vertebral bodies may then be subject to increased load as the degenerated disks take less of the load. Because bone adapts to the mechanical demands placed upon it (being laid down where needed and reabsorbed where not needed), bony spurs (or osteophytes) may grow around the intervertebral foramen. Excessive loading of the facet joints may also cause problems. This process of degeneration may ultimately

FIGURE 6-2   Basic mechanism of a slipped disk. The annular fibers rupture and the nuclear material is extruded posteriorly. (Adapted from Keegan, 1953, with permission.)

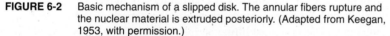

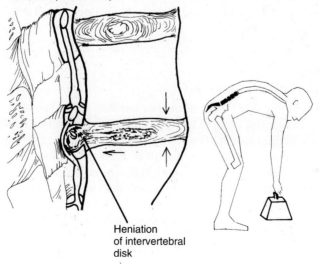

Heniation
of intervertebral
disk

result in herniation of an intervertebral disk—either posteriorly or in other directions (Vernon-Roberts, 1989).

A prolapsed intervertebral disk usually causes severe pain and confines the sufferer to bed (severely limiting mobility) for a considerable time. Because of the viscoelastic nature of the intervertebral disk, the prolapsed material does not return to its original position as soon as the load is released. Sometimes treatments such as lumbar traction can assist in returning the prolapsed material and also relieve pain in the process. Because the intervertebral disks have no direct blood supply in the adult, recovery from injury is slow, and it cannot be said that the damaged tissues ever heal completely, as previously healthy tissue is replaced by scar tissue (Ring, 1981). It is important to note that **an injury such as a slipped disk can permanently reduce a person's ability to carry out strenuous work or leisure activities.** For this reason, the prevention of work-related back injury must be given a high priority since such a prevention policy has great potential to reduce the overall amount of disability in society, with all the corresponding human and economic benefits this reduction entails.

Ring has also pointed out that chronic back sufferers become accustomed to the idiosyncratic features of their own spines and can recognize those tasks and activities likely to cause injury. They are thus able to develop coping skills and avoidance strategies to minimize the risk of injury. In emergency situations or when under stress or distracted as a result of information overload, the person with a weak back may be unable to invoke the appropriate motor strategy or to take the customary precautions when carrying out a potentially hazardous activity. Serious back injury may result, which may explain why epidemiological studies so often find that activities such as lifting are the precipitating cause of back injuries. It is the responsibility of the industrial engineer to ensure that plenty of time is allowed for workers to carry out manual handling tasks and that they are not distracted or pressurized by conflicting task demands.

Back injury prevention programs emphasize the importance of proper work design and of training workers to view lifting and manual handling tasks as problem-solving exercises in which tasks have to be properly analyzed and planned to minimize risk.

Correct and Incorrect Lifting?

Despite any idiosyncratic considerations, it is possible to generalize and to identify methods for the safe design of manual handling tasks. Poor lifting technique is often cited as a major risk factor in lifting-related low back injury. The stoop-lifting technique in which the legs are kept straight and the person flexes at the hips and spine causes a large flexion moment to be exerted about the lumbar spine. This moment is due to the body parts themselves as well as the load. The crouch lift is usually regarded as safer from the point of view of the spine because, with the trunk more erect and the load closer to the body, the flexion moment is reduced (Figure 6-3).

In some situations and with some objects, the crouch lift loses its advantage because the object has to be held farther away from the body. In such cases the stoop lift may be safer. In fact, Bejjani et al. (1984) questioned the principle of bending the knees to save the back, pointing out that more than 2 percent of men and 6.6 percent of women in the

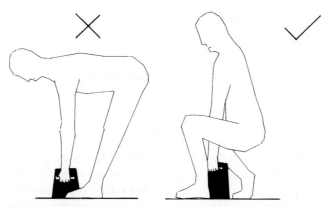

FIGURE 6-3   Correct and incorrect lifting. The load should normally be kept close to the body.

United States suffer some form of knee arthritis. They developed a two-dimensional model of lifting and found an inverse relationship between forces acting on the knee and forces acting on the back in lifting. They argue that successful lifting strategies should be aimed at minimizing the average forces acting on the total musculoskeletal system, which implies that correct lifting strategies need to take worker anthropometry and load characteristics into account.

As in the case of sitting, it is not just a question of *whether* people handle heavy objects but of *how* they do it and *how* the objects are designed. This introduces questions of task design and technique. The main contribution of ergonomics to the reduction of hazards in manual handling is to redesign tasks and to identify techniques and specify workloads which are safe.

DESIGN OF MANUAL HANDLING TASKS

NIOSH has produced a work practices guide for the design of manual handling tasks and an equation for determining safe loads (see below). In Europe, a new directive for the design of these tasks has recently been issued (1990) and the U.K. Health and Safety Commission consultative document (Health and Safety Commission, 1991) provides interesting proposals for the design of manual handling tasks. Three principles of industrial medicine (first, remove the threat; second, remove the operator; third, protect the operator) are usefully paraphrased in the context of manual handling (Table 6-2).

The human-machine model described in the introduction and used in the previous chapter on seating can be adapted to the discussion of manual handling.

LIFTING

A safe posture for lifting loads can be arrived at from a consideration of the basic principles of body mechanics which were described in previous chapters. The COG of the body parts and load must be kept within the base of support described by the position of

TABLE 6-2 GENERAL PRINCIPLES FOR THE CONTROL OF MANUAL HANDLING HAZARDS IN THE WORKPLACE

1 Avoid hazardous manual handling as much as possible.

2 Assess any hazardous operations; then redesign the task to obviate the need to move the load, or automate or mechanize the process.

3 Reduce the risk by providing mechanical assistance, redesigning the load itself, or redesigning the workspace.

the feet. The body must be erect so that the weight of the body parts above the lumbar spine is transmitted through the spine and bending moments are minimized. Space for the feet is essential, as is adequate space for the postural correction movements which take place when a person changes position (e.g., from standing to squatting and vice versa).

Task Requirements

Some factors which increase postural stress in lifting and carrying are given in Table 6-3. Lifting from a seated position deserves special consideration. The flexion moment about the lumbar spine is greater in sitting than in standing (Nachemson, 1966; Brunswic, 1984) and thus the spine can be considered to be prestressed. In sitting, the leg muscles cannot be used to generate a ground reaction force and more is required of the muscles of the trunk. Stooping from a seated position is particularly hazardous. It is curious that, in the light of these facts, furniture designers still design desks with the largest drawer (to contain the heaviest objects) at the bottom, close to the floor.

The characteristics of the load itself should also be considered. Weight is not the only consideration; the size of the object is also important. In a sense, 20 kg of lead is lighter than 20 kg of feathers since, being smaller, it can be held closer to the body. Containers for one- or two-handed handling should be designed as small as possible so that the load is kept close to the body. The provision of handles is important—well-designed handles can facilitate secure grasping of the load and prevent sudden movements of the combined person-load COG. Conversely, slippery materials or unstable loads (lacking rigidity) can increase the risk of injury.

The dynamics of the lift also deserve some consideration. Spinal loading appears to be greater when the lift involves high trunk velocities or acceleration. Slow, smooth

TABLE 6-3 SOME TASK FACTORS WHICH EXACERBATE POSTURAL STRESS IN MANUAL HANDLING

1 Having to grasp or hold the load at a distance from the trunk

2 Having to twist the trunk while supporting or lifting a load

3 Having to lift or lower objects placed below knee or above shoulder height

4 Having to lift or move the load through great vertical or horizontal distances

5 Having to hold or carry the load for long periods

6 Having to lift or carry frequently

7 Having to lift while seated

movements seem best if excessive back loading is to be avoided. Having to lift under time pressure, during emergencies, or to meet unrealistic production targets must be avoided.

Personal Characteristics of Workers

In the design of lifting tasks it is essential to consider the characteristics of the worker population. Because of the requirements for task design described above, straightforward anthropometric considerations are not as useful in the design of lifting tasks as they are in seating, although a tall workforce will be better able to handle objects stacked higher, but not those stacked lower, and the opposite will be true for short workers. The following personal considerations are important.

People whose occupations typically do not require them to lift objects seem to be particularly prone to injury (Magora, 1972). It is undesirable to expect office workers, drivers, or professionals to lift heavy objects either occasionally or in emergencies. There is still some controversy over whether the level of fitness of workers is related to the probability of their suffering a back injury. It has been suggested that hyper- or hypomobility of the lumbar spine may be a risk factor for back problems, but empirical studies have failed to show a clear relationship between spinal flexibility and future risk of back trouble. Flexibility would not appear to be useful for screening out workers at high risk of injury (Battie et al., 1990; Burton, 1991).

Although women are generally less capable of handling heavy loads than men, sex is not always a useful consideration since there is considerable overlap in the abilities of males and females in the workforce. Age should be considered since it is known that muscular strength declines from middle age onward. Middle-aged people are particularly prone to low back problems because of the instability of their lumbar motion segments.

If the workforce contains a large proportion of women, some special considerations are relevant and generally speaking, manual lifting of heavy loads should be avoided because it increases the stress on the trunk. According to Hayne (1981), menstrual pain may incapacitate some women and the increased intra-abdominal pressure when lifting may increase menstrual flow. It may also increase the risk of miscarriage in the first few months of pregnancy. Menopausal women often suffer from **osteoporosis** (demineralization of bone leading to a loss of bone strength) particularly in the spine and pelvis. This condition increases the risk of trunk failure when the sufferer is lifting a load. A detailed discussion of the many other factors affecting women at work can be found in Chavalitsakulchai and Shahnavaz (1990).

Given that lifting is known to be a hazardous activity and that there are many good reasons why women, in particular, should not have to lift heavy weights, the special case of back pain in the nursing profession, consisting primarily of women, deserves some consideration. Nursing is one of the few jobs where women have to lift heavy (and awkward) weights (i.e., patients) on a regular basis. Stubbs et al. (1983a) reported the findings of an epidemiological study of back pain in 3912 nurses. One in six nurses attributed their back pain to a patient-handling incident, 750,000 working days were lost annually because of back pain (16 percent of all sick leave), and 78 percent of the pain was in the lower back.

Another personal characteristic which is relevant to the design of manual handling tasks is employees' level of knowledge of safe handling procedures. It is usually suggested that workers should be trained to use correct lifting techniques, to identify hazardous situations, and to prepare themselves for executing a lift (e.g., preparing the area of the lift and the area where the object is to be placed, removing obstructions if carrying is involved, deciding whether to lift the object in stages). However, there is still some controversy as to the effectiveness of training programs. Van Akerveeken (1985) reports that there is little evidence that training in correct lifting techniques is effective in reducing the incidence of occupational back injuries. Stubbs et al. (1983b) found little relationship between the point prevalence of back pain among nurses and the time spent training in safe lifting techniques. They concluded that **if the lifting task is intrinsically unsafe** (which is the case in the manual handling of patients), **no amount of training will solve the problem. A better approach therefore would be to do away with the lifting task or redesign it to make it safe.**

Most workers do not deliberately set out to injure themselves. Injuries occur when they are least expected—when conscious awareness of a hazard is minimal and knowledge of correct techniques is least likely to be used. For these and other reasons, several researchers (including Snook, 1978) have stressed the importance of engineering and design solutions aimed at reducing the incidence of back injury.

Workspace Design

Many aspects of workspace design can increase the risk of injury from lifting. Some of the more important considerations are given in Table 6-4.

Design of Lifting Tasks

Ayoub (1982) has summarized many of the guidelines for the design of lifting tasks. These are presented, in modified form, in Tables 6-5, 6-6, and 6-7 with permission of the *Journal of Occupational Medicine.*

Attempts have been made, over many years, to determine the maximum load that workers will find acceptable under a particular set of lifting conditions, such as the frequency of the lift, the distance of the load from the body, and whether the load has handles. Much notable work in this area is from Snook and his colleagues (e.g., Snook, 1978; Ciriello et al., 1993). Snook developed a psychophysical methodology in which

TABLE 6-4   SOME WORKSPACE FACTORS WHICH EXACERBATE POSTURAL STRESS IN MANUAL HANDLING

1 *Confined spaces.* The ability to exert forces decreases when space is restricted. Restricted headroom is a good example (Grieve and Pheasant, 1982). Less use of the legs is possible, which increases the load on the trunk muscles.

2 *Height of object.* Only items placed between the knee and elbow height should be lifted. Conveyors, shelves, and palletizing systems should be designed to allow for this factor.

3 *Flooring.* Space for the feet should be provided both underneath the load and around the worker. Slippery floors should be avoided.

TABLE 6-5   HOW TO MINIMIZE THE WEIGHT TO BE HANDLED

1 Assign the job to more than one person.

2 Use smaller containers.

3 If possible, mechanize the process.

4 Use machines, rather than employees, to transfer loads between surfaces.

5 Change the job from lifting to lowering, from lowering to carrying, from carrying to pulling, and from pulling to pushing.

6 Use handles, hooks, or similar features to enable workers to get a firm grip on objects to be lifted.*

7 Reduce the weight of containers used to transfer objects.

8 Balance and stabilize the contents of containers to avoid sudden shifts in load during a lift.

9 Design containers so that they can be held close to the body.

10 Treat work surfaces to allow for ease of movement of containers.

*Maximum acceptable weights are about 16 percent lower if no handles are used (Ciriello et al., 1993).

TABLE 6-6   HOW TO MINIMIZE REACH AND LIFT DISTANCES

1 Increase height at which lift is initiated; decrease height at which it terminates.

2 Stack objects no higher than shoulder height.

3 Store heavy components on shelves between shoulder and knuckle height.

4 Avoid deep shelves.

5 Avoid side-to-side lifting from seated position.

6 Provide access space around components to cut down on the need for manual repositioning.

7 Fit storage bins or containers with spring-loaded bottoms.

8 Use sloped surfaces to gravity-feed items to the point of lifting.

9 Provide free space around and under the work surface to increase functional reach.

TABLE 6-7   HOW TO INCREASE THE TIME AVAILABLE FOR LIFTING

1 Increase the time by relaxing the standard time for the job.

2 Reduce the frequency of lifts.

3 Introduce job rotation to parcel out lifting between workers.

4 Introduce appropriate work-rest cycles.

workers are given control over one of a set of lifting task variables—usually the weight of the object. The other variables are controlled by the experimenters, and for each combination of these variables, the worker adjusts the weight to what is perceived to be a maximally acceptable level for the conditions. Snook's approach seems to have yielded valid data. Ciriello et al. (1993) report that if a manual handling task is acceptable to less than 75 percent of the population, the probability of compensatable back pain is 3 times greater than if the job is acceptable to more than 75 percent of the population. Tables 6-8 and 6-9 depict the weights which are below the maximum acceptable weights for males and females under a variety of lifting conditions.

TABLE 6-8 WEIGHTS ACCEPTABLE BY 75 PERCENT OF MALE INDUSTRIAL POPULATION (in kilograms)*

Width†	Distance‡	Floor-to-knuckle height One lift every						Knuckle-to-shoulder height One lift every						Shoulder-to-arm height One lift every					
		5 s	9 s	14 s	1 min	5 min	8 hr	5 s	9 s	14 s	1 min	5 min	8 hr	5 s	9 s	14 s	1 min	5 min	8 hr
75	76	10	14	15	18	25	29	12	16	18	17	21	24	9	12	14	16	20	23
	51	11	14	16	19	26	31	13	17	19	20	24	27	10	14	15	18	22	25
	25	13	17	19	21	29	34	15	20	22	23	28	32	11	16	18	21	26	30
49	76	12	15	17	21	28	34	12	16	18	17	21	28	9	12	14	16	20	23
	51	12	16	18	22	30	35	13	17	19	20	24	27	10	14	15	18	22	25
	25	14	18	21	24	33	39	15	20	22	23	28	32	11	16	18	21	26	30
36	76	13	17	20	23	31	37	13	17	19	18	23	26	9	13	15	17	21	24
	51	14	18	20	24	32	38	16	18	20	21	26	29	10	15	16	19	24	27
	25	16	21	24	27	37	43	16	21	24	24	30	35	12	17	19	23	28	32

*From Snook (1978).
†Width of object in centimeters. Note: Horizontal hand location is at least 15 + width/2.
‡Vertical distance of lift in centimeters.

TABLE 6-9 WEIGHTS ACCEPTABLE BY 75 PERCENT OF FEMALE INDUSTRIAL POPULATION (in kilograms)*

Width†	Distance‡	Floor-to-knuckle height One lift every						Knuckle-to-shoulder height One lift every						Shoulder-to-arm height One lift every					
		5 s	9 s	14 s	1 min	5 min	8 hr	5 s	9 s	14 s	1 min	5 min	8 hr	5 s	9 s	14 s	1 min	5 min	8 hr
	76	8	10	11	13	17	20	8	11	11	11	14	15	5	9	9	10	12	14
75	51	8	10	12	13	18	21	9	12	12	12	15	17	6	10	11	11	14	15
	25	9	12	14	15	20	24	11	14	14	15	18	20	7	11	13	13	16	18
	76	9	11	13	15	20	24	8	11	11	11	14	15	5	9	9	10	12	14
49	51	9	12	13	15	21	25	9	12	12	12	15	17	6	10	11	11	14	15
	25	10	14	15	17	23	28	11	14	14	15	18	20	7	11	13	13	16	18
	76	10	13	14	16	22	26	9	12	12	12	14	17	6	9	10	11	13	15
36	51	10	13	15	17	23	27	9	12	12	13	17	19	6	10	11	12	15	17
	25	12	16	18	19	26	31	11	15	15	16	19	22	8	12	14	14	17	20

*From Snook (1978).
† Width of object in centimeters. Note: Horizontal hand location is at least 15 + width/2.
‡ Vertical distance of lift in centimeters.

In many situations, workers have to carry out a combination of lifting, carrying, and lowering. Snook and Ciriello (1991) report that the maximum acceptable weights for these combination tasks are lower than for carrying alone, but not significantly different from lifting or lowering performed separately. Combination tasks are more strenuous, as is reflected by higher heart rates for people carrying them out. Detailed information may be found in Snook and Ciriello (1991).

The NIOSH Approach to the Design and Evaluation of Lifting Tasks

Probably the most comprehensive approach to the design of lifting tasks is that of NIOSH in the United States. NIOSH has developed an equation for calculating the recommended weight limit (RWL) for a specific lifting task which a worker could perform for a specified period without an increased risk of low back pain. The equation has been determined empirically and specifies a weight limit as a function of the values of specified task variables. The original equation was developed in 1981. It has been recently updated to cover a wider range of tasks, including asymmetrical lifting tasks.

Three criteria have been used to develop the equation (Table 6-10). The use of three criteria is essential if a wide range of tasks are to be evaluated because different tasks impose different loads on workers. For example, infrequent handling of heavy, awkard objects may be limited by biomechanical rather than physiological factors. Repetitive lifting of light objects may be limited by metabolic stress and local muscle fatigue. Workers' perceptions of their capability are also a limiting factor. Sometimes, the different criteria will produce conflicting RWLs, in which case the lowest is used.

The approach taken by NIOSH has been to recommend a **maximum load under ideal lifting conditions.** The RWL$_{max}$ is taken to be a load of 23 kg, lifted in the sagittal plane from a height of 75 cm above the floor, and held 25 cm in front of the body. The load is to be lifted no more than 25 cm vertically, there is to be a good coupling between the load and the lifter (which is achievable using handles or occurs as a result of the shape of the load itself), and the load is to be lifted only occasionally. Thus, the conditions for lifting a maximum load are specified.

Six coeffcients have been developed which reduce the RWL to account for task factors that cause departures from the ideal situation. The RWL$_{max}$ is multiplied by the coefficients (which are less than 1) to arrive at a new RWL for the specified conditions. The values of the coefficients have been determined using biomechanical models of spinal loading and the findings of epidemiological and psychophysical studies. The 1991 NIOSH equation is presented in Table 6-11 together with definitions of the coefficients and the terminology used in the equation (Figure 6-4).

TABLE 6-10   CRITERIA USED IN THE NIOSH LIFTING EQUATION

Consideration	Criterion	Cutoff value
Biomechanical	Maximum disk compression	3.4 kN
Physiological	Maximum energy expenditure	2.2–4.7 kcal/min
Psychophysical	Maximum acceptable weight for 75% of females and 99% of males	

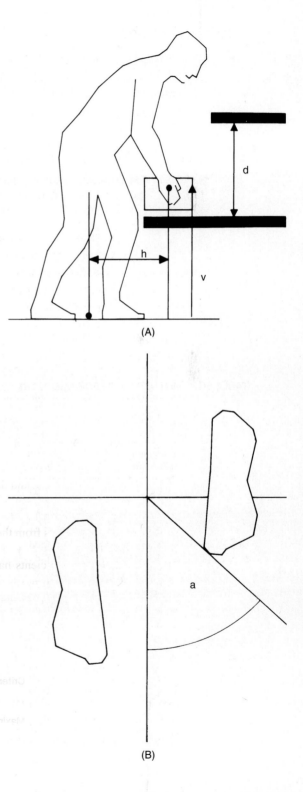

FIGURE 6-4 NIOSH approach to lifting task evaluation. (A) h = distance of load from midpoint, v = height of hands above floor, d = distance through which load is lifted. (B) a = angle of asymmetry of lift.

TABLE 6-11   1991 NIOSH LIFTING EQUATION

RWL = LC × HM × VM × DM × AM × FM × CM

where LC = load constant of 23 kg
      HM = horizontal multiplier = 25/H
      VM = vertical multiplier = 1 − (0.003IV − 75I)
      DM = distance multiplier = 0.82 + (4.5/D)
      AM = asymmetric multiplier = 1 − 0.0032A
      FM = frequency multiplier (from Table 6-12)
      CM = coupling multiplier (from Table 6-13)

and

      H = horizontal distance of the hands from midpoint (0) of the ankles
      V = vertical distance of the hands from the floor
      D = distance through which the load is lifted
      A = angle of asymmetry (see Figure 6-4)
      F = frequency of lifting (lifts/min every 1, 2, or 8 hours)

TABLE 6-12   VALUES OF FM FOR USE IN THE 1991 NIOSH EQUATION FOR DETERMINING RWL

| | Work duration | | | | | |
| | ≦1 h | | ≦2 h | | ≦ 8 h | |
Frequency (lifts/min)	V < 75	V ≥ 75	V < 75	V ≥ 75	V < 75	V ≥ 75
0.2	1.00	1.00	0.95	0.95	0.85	0.85
0.5	0.97	0.97	0.92	0.92	0.81	0.81
1	0.94	0.94	0.88	0.88	0.81	0.81
2	0.91	0.91	0.84	0.84	0.65	0.65
3	0.88	0.88	0.79	0.79	0.55	0.55
4	0.84	0.84	0.72	0.72	0.45	0.45
5	0.80	0.80	0.60	0.60	0.35	0.35
6	0.75	0.75	0.50	0.50	0.27	0.27
7	0.70	0.70	0.42	0.42	0.22	0.22
8	0.60	0.60	0.35	0.35	0.18	0.18
9	0.52	0.52	0.30	0.30	0.00	0.15
10	0.45	0.45	0.26	0.26	0.00	0.13
11	0.41	0.41	0.00	0.23	0.00	0.00
12	0.37	0.37	0.00	0.21	0.00	0.00
13	0.00	0.34	0.00	0.00	0.00	0.00
14	0.00	0.31	0.00	0.00	0.00	0.00
15	0.00	0.28	0.00	0.00	0.00	0.00
>15	0.00	0.00	0.00	0.00	0.00	0.00

*V = vertical distance of the hands from the floor (in centimeters).

TABLE 6-13   VALUES OF CM FOR USE IN THE 1991 NIOSH EQUATION FOR DETERMINING RWL

Coupling	V < 75*	V ≧ 75
Good	1.00	1.00
Fair	0.95	0.95
Poor	0.90	0.90

* V = vertical distance of the hands from the floor (in centimeters).

The use of the NIOSH lifting equation can be illustrated as follows: If the load had to be held at a distance H of 50 cm in front of the body, the RWL_{max} would be halved. If the load had to be lifted from the floor rather than the ideal of 75 cm above the floor, it would have to be reduced to 77.5 percent of RWL_{max}. Increasing the distance of the lift from 25 cm to 50 cm would call for a reduction to 91 percent of RWL_{max}. Lifting asymmetrically at 90 degrees to the sagittal plane would call for a reduction of RWL_{max} by approximately 30 percent.

The ratio between the load actually lifted and the RWL is known as the **lifting index.** An index of less than 1 is believed not to increase the risk of injury. An index of three or more indicates that many workers will be at increased risk and the task should be redesigned.

NIOSH's goal in developing the equation is to specify controls on industrial lifting which will protect healthy workers. It is not valid to apply the equation to set limits for those with already damaged backs.

Further details of the NIOSH method can be found in the publications of the institute (e.g., PB91-226274) as well as in Waters et al. (1993). These sources should serve to illustrate NIOSH's admirably rigorous approach to the pervasive problem of occupational lifting. The approach has the distinct advantage of lending itself to implementation as an interactive decision aid for the industrial engineer or designer.

Lifting Aids and Injury Prevention Programs

There has been a recent upsurge in interest in the possible industrial applications of the use of weight-lifting belts in the performance of tasks which require workers to lift objects. Competitive weight lifters have used these belts, which fit tightly over the low back and abdomen, for many years. Theoretically, they are supposed to reduce the load on the lumbar spine by augmenting intra-abdominal pressure. Redell et al. (1992) evaluated a weight-lifting belt and back injury prevention training in a group of airline baggage handlers. There was no evidence that use of the belt when handling baggage reduced the incidence of lumbar injuries or lost workdays. In fact, the injury rate was higher after discontinuing use of the belt. This finding suggests that the belt may actually decondition the trunk musculature and predispose workers to back injury in their leisure time. More work needs to be done on this area, however. Shah (1993) reports that in Nepal, most people who lift and carry heavy weights wrap a 5-m length of cloth (called a patuka) around their waist before work, which may act like a spinal orthosis.

Back pain is lower in patuka wearers than nonwearers. Wearing the patuka does not seem to decondition the trunk musculature even over a period of many years.

Garg and Owen (1992) evaluated the effects of an ergonomic intervention program on the incidence of back injuries among nursing personnel in a nursing home. The program consisted of identifying patient handling tasks perceived to be most stressful and carrying out an ergonomic evaluation of the stressful tasks, conducting a laboratory study to select patient handling devices which would reduce the physical stress on the handler, and training nursing personnel in the use of these devices. Modifications were also made to toilets and showers to improve ease of handling. Two patient handling devices were introduced to facilitate patient transfer—a mechanical hoist for lifting patients out of bed and a "walking belt" (a method of putting "handles" onto the patient, which consisted of a belt worn by the patient with loops for the patient handler to grasp). The two devices were found to be acceptable by a high percentage of potential users. The back injury incident rate dropped from 87 per 200,000 work hours before the intervention to 47 per 200,000 after the intervention.

It would appear then that back injury prevention programs can be successful if they are based on the redesign of tasks to reduce the risk of injury and on the introduction of appropriate mechanical aids. However, a systematic approach to the analysis of problem areas together with the evaluation of the usability of potential lifting aids and the ease of implementation of solutions is needed.

CARRYING

The discussion of carrying tasks follows naturally from that of lifting. However, no discussion of carrying can be complete without a discussion of postural stability (balance) and of walking itself.

Postural Stability and Postural Control

The erect human body is a tall structure with a narrow base of support—its COG lies at more than half its total height. In terms of physics, the body is at the mercy of environmental perturbations. In terms of physiology, however, an adult with fully developed postural reflexes is very stable, given that movement is not restricted. Posture is an active process which depends on the functioning of a number of neurologically separate systems of reflexes (Martin, 1967).

The antigravity reflexes are centered in the hindbrain (pons and medulla) and cause automatic bracing of a limb when it is loaded by body weight. Although they are essential for the maintenance of an upright posture, they do not assist in the maintenance of equilibrium. A second set of postural reflexes (located in the basal ganglia of the midbrain) controls the posture of the various body parts in relation to each other (as in the postural fixation of limbs, for example) and of the whole body itself. The cerebellum is also involved in posture in the coordination of body movements.

The reflex control of posture can be distinguished from low-level reflex arcs, such as the knee-jerk reflex, and from voluntary movement. Mechanisms exist to protect the body against mechanical instability using feedback from the body itself to maintain balance. Feedback is obtained from both the semicircular canals in the inner ear and from sense organs in the muscles themselves. People with impaired somatic or labrynthine feedback appear to exhibit little instability when standing on a stable base with visual feedback. However, when their vision is impaired, they exhibit particular disabilities (such as the inability to maintain the head erect when blindfolded, Martin, 1967). This finding demonstrates that although vision is of secondary importance in the maintenance of body equilibrium, it can play an important role under certain conditions.

In healthy subjects, under less drastic circumstances, the erect posture is character-ized by postural sway. Small perturbations of the COG of the body are countered by an ankle strategy (Duncan et al., 1990). The ankle strategy underlies normal sway when someone is standing still on an unmoving surface. A hip strategy is used to correct large perturbations or when the position of the feet is constrained, for example, by lack of space. The hip strategy can be seen when someone is balancing on a narrow surface (such as a gymnast on the beam who momentarily loses balance). For posterior dis-placements of the COG of the body (where the line of gravity moves toward the heels), tibialis anterior and then quadriceps femoris contract in the ankle strategy. In the hip strategy, the paraspinal and hamstring muscles contract to thrust the pelvis forward, thereby compensating for the rearward displacement of the upper body and maintaining equilibrium.

Slips, Trips, and Falls: Catastrophic Failure of the Erect Position

Losses of balance resulting in slips, trips, and falls are a major cause of work-related in-jury. According to the United Kingdom Medical Commission on Accident Prevention, over 40 percent of industrial lost-time accidents are due to this cause alone (Porritt, 1985). Apart from trauma such as broken bones and concussion and bruising caused by the impact of body parts against unyielding surfaces, some of the worst back injuries are precipitated by sudden impacts transmitted through the musculoskeletal system when normal postural control mechanisms break down. Because we are not normally aware of the workings of our own postural control mechanisms, we often fail to appreciate **the enormous forces which are unleashed when we are slipping, tripping, or falling,** ex-cept in exceptional circumstances—the shock of discovering that there was one more stair to descend when one is walking down the stairs while reading a newspaper is a dra-matic example. The impact is sufficient to cause fracture of the neck of the femur in el-derly subjects, according to Citron (1985).

Walking itself is a form of controlled falling (Figure 6-5). The body teeters on the brink of catastrophe as each leg swings through to save the ever-falling mass. **Trips** occur when the swinging leg is prevented from reaching its destination, and **slips,** when there is insufficient friction between the weight-bearing leg and the floor. Both problems are a result of a violation of a person's expectations or assumptions by some aspect of the workplace—steps, ramps, or ridges may be in unexpected places; there may be dif-ferences in the depth or height of the steps in a flight of stairs; or there may be loose car-peting or unexpectedly wet or highly polished floors. All these factors can increase the risk of slipping, tripping, and falling.

Several researchers have attempted to establish optimum dimensions for the design

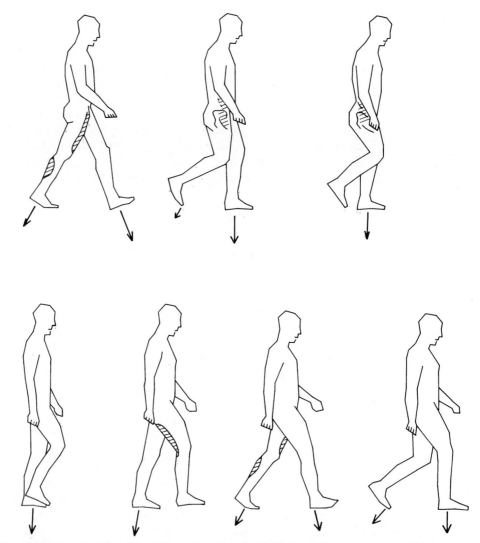

FIGURE 6-5   Mechanism of walking. The calf muscles plantarflex the foot and the quadriceps straighten the knee at toe-off to propel the body forward. The weight shifts to the left foot. The iliopsoas muscle pulls the trailing right leg forward and its knee bends passively. The hamstring muscles decelerate the swinging right leg so that the heel can be planted on the ground. The body weight passes along the length of the foot from the heel at heel strike to the big toe at toe-off—two critical stages of force transmission at the foot-floor interface. It is critical for stability that the swinging leg overtake the body and be planted ahead of it and that the heel not move as it strikes the ground. Throughout the gait cycle, the reaction force must be approximately in line with the supporting leg(s) or the walker will lose balance. (Adapted from the photographs of Eadweard Muybridge, 1884.)

of stairs. Irvine et al. (1990) concluded that riser heights (the height of the step) below 152 mm and above 203 mm should be avoided, and run depths (the distance from the front to the back of a step) less than 254 mm and more than 330 mm should also be avoided.

The Foot-Floor Interface

One of the most important factors influencing the incidence of slipping and tripping is the design of the foot-floor interface. Floor materials such as concrete, steel, earth, tile, and rubber matting differ in their frictional properties, as do shoe materials such as synthetic rubber, natural leather, etc. The static friction between the shoe and the floor depends on the frictional properties of the respective materials and on the area of contact between them. High heels are more hazardous than flat soles for this reason. Rubber and synthetic soles have higher friction on dry floors than leather shoes but not on wet floors. Higher friction is required on slopes, which suggests that, in addition to warning signs, slopes be designed with high-friction materials or lateral ribbing. Concrete floors can be laid with ribbing perpendicular to the intended direction of travel. NIOSH recommends shoes and work surfaces be matched to result in a coefficient of static friction of 0.5 and that there be smooth transitions between areas with different surface frictional properties. Swenson et al. (1992) estimate that the threshold coefficient of friction for walking is between 0.2 and 0.4. Good housekeeping is needed to clear up spills, replace worn carpeting or missing tiles, and repair loose floorboards or cracked or pockmarked concrete surfaces.

Changes in floor materials or slope can be signaled with warning signs. Yellow-on-black stripes are often used to indicate hazards and can be painted on the floor at or close to the transition zone or along the length of ramps. English (1994) makes the very important point that possibly the major determinant of slips and falls is the person's knowledge of the surface. People are perfectly capable of walking on ice *if they know about it* and of carrying out complex movements on the deliberately slippery surfaces found in bowling centers and on dance floors.

Agnew and Suruda (1993) report that from 1980 to 1986 there were 43,505 fatal work injuries to U.S. males, 4179 of which were from falls. Falls from ladders accounted for 20 percent of fatal falls in workers over 55 years compared with 9 percent of all fatal falls. Older workers seem less able to survive falls, as indicated by a lower impact energy associated with the fall. English notes that the slowing of reflexes and increased skeletal fragility which occurs with age makes corrective postural strategies less effective during a slip and intensifies the injuries associated with nonfatal falls in older people. Postmenopausal, osteoporotic women are particularly at risk.

Other Factors Influencing Postural Stability

Many factors influence postural stability and sway in healthy subjects (Ekdahl et al., 1989). Age has been related to sway in some studies but not in others. Pyykko et al. (1990) investigated the postural stability of 23 subjects over 85 years of age. When compared with a control group, the elderly subjects had significantly higher sway velocities even during nonperturbed conditions. Visual deprivation was found to have a large effect on the elderly subjects, contributing to about 50 percent of postural stability. The authors concluded that postural control in elderly subjects was reduced as a result of deterioration of somatic feedback from stretch reflexes in the muscle spindles. It appears then that as we get older, we rely more on visual control of posture. Visual feedback is slower than intrinsic feedback from the muscles, a factor that may explain the susceptibility to falling of older people.

Approximately 6000 people die annually in the United Kingdom as a result of injuries caused by slips, trips, and falls. Two-thirds of these are elderly. The importance of visual feedback for postural control in the elderly has clear implications for the design of the visual environment in retirement villages, hospitals, and the workplace—particularly in those countries undergoing demographic aging. **Good lighting in stairwells, hallways, and so on is clearly a high priority, as is a consistent approach to the placement of light switches inside and outside buildings.**

Ekdahl et al. also observed that females have better balance than males—possibly because of the fact that their body COG is lower. Of the various balance tests investigated by these authors, standing on one leg while blindfolded was the most difficult (impossible for subjects over 55 years of age) and standing on one leg with a blindfold was more difficult than standing with the feet together with a blindfold. These results are not surprising in view of the reduction in the size of the base of support and increased postural load on the supporting leg. Standing on one leg appears to increase the relative importance of visual feedback for postural control—even in young subjects—and has interesting practical implications. For example, **operating foot pedals from a standing position would appear to be contraindicated, particularly when lighting is poor.**

The Design of Carrying Tasks

It is generally believed that loads should be positioned as close to the body as possible in order to minimize energy expenditure during carrying. As soon as the load is placed away from the body (in the horizontal plane), destabilizing forces will be exerted which have to be counteracted by static contraction of appropriate muscle groups on the contralateral side of the body. During sustained carrying, this muscle contraction will be a likely source of local discomfort and fatigue.

Carrying increases the load on the body in two ways. First, the increase in weight increases the physiological cost of walking and the load on the muscles of the legs which propel the body forward (the plantarflexors of the ankle joint and the quadriceps). Second, the method by which the load is held or attached to the body can be an additional source of postural stress. Fatigue, discomfort, or injury may arise from either or both sources of loading. The second source is of particular interest in ergonomics since, through the design of more efficient carrying methods, the total postural load can, in principle, be reduced. From an energy-expenditure viewpoint, **an efficient method of load carriage is one that imposes little additional postural load over and above the extra energy cost per kilogram of load** (Soule and Goldman, 1969).

Klausen (1965) investigated several methods of holding a 40-kg load and measured electromyographic (EMG) activity in various muscle groups (Figure 6-6). He also com-

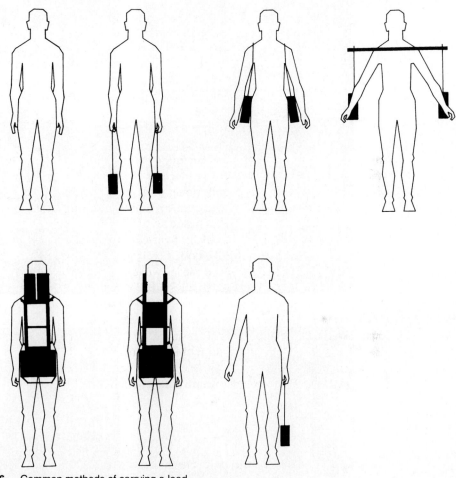

FIGURE 6-6   Common methods of carrying a load.

pared spinal and pelvic posture in free standing and in holding a load and noted any load-induced changes. Holding 20 kg in each hand was found to increase the electrical activity in the thoracic portion of the back muscles. It also placed a static load on the muscles controlling flexion of the fingers used to grip the weights. Holding the weights by means of a yoke over the shoulders caused a slight forward inclination of the spine and stooping, and was accompanied by an increase in back muscle EMG to counter the forward shift in the line of gravity of the superincumbent body parts.

The effects of carrying the load high and low on the back were also investigated using a carrying frame. When the load was carried high on the back, similar results to that using the yoke were observed. The body was inclined forward, and back muscle activity increased to exert an extensor moment on the trunk to counteract the flexor moment introduced by the load. When the load was carried low on the back, the subjects still adopted a forward-inclined trunk posture—presumably to maintain the combined COG

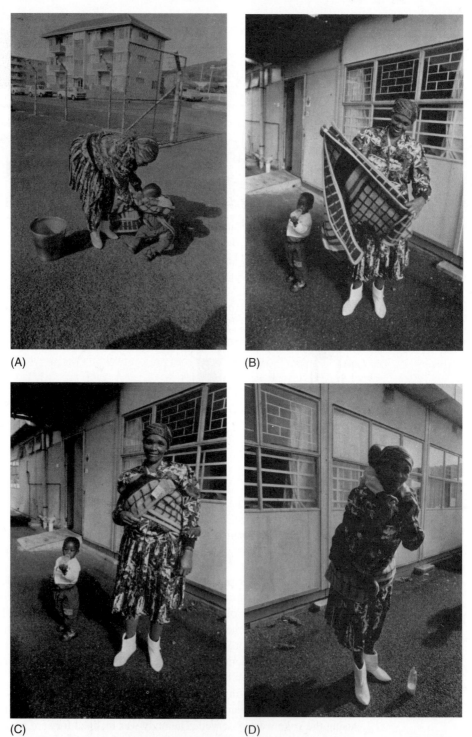

(A)

(B)

(C)

(D)

FIGURE 6-7 Traditional method of carrying in Africa. (A) The problem is to carry both the baby and the bucket simultaneously. (B) A blanket is folded and (C) tied around the waist. (D) The baby is hoisted over the shoulders

(E)

(F)

(G)

FIGURE 6-7 *continued* (E) and secured using the blanket. (F) Ready for carrying. (G) The bucket can now be lifted onto the head, leaving both hands free.

of the load plus body weight over the feet. However, decreased EMG activity was observed in the back muscles, but increased activity was observed in the iliopsoas muscles. The interpretation of these findings was that carrying the load low on the back introduced an extensor moment on the trunk (i.e., it tended to pull the trunk backward). Although this would reduce activity of the long back muscles, the forward-inclined position of the trunk would be maintained only if increased activity in the iliopsoas muscles occurred to counteract the backward pull of the load. Thus, carrying a load low on the back would seem to be a relatively efficient method which decreases activity in the back muscles and, in turn, would be expected to reduced the stress on the lumbar spine.

Cook and Neumann (1987) investigated the effects of load placement on low back muscle EMG during carrying and compared it with EMG activity during walking without an external load. They concluded that carrying with a backpack required the least additional muscular effort of the methods investigated and that carrying a weight in front of the body required the most effort. Carrying a weight in one hand induced asymmetrical loading on the spine, decreasing back muscle activity on the side of the load and increasing it on the unloaded side.

It seems that backpacks provide a low-stress method of carrying loads if the load is carried low in the pack. Soule and Goldman (1969) concluded that the energy cost per kilogram of load carried efficiently on the back is identical to the energy cost per kilogram of body weight; i.e., no additional postural load is imposed. Alternatively, carrying loads in one hand or in front of the body should be avoided because it increases the load on the back muscles. (Neumann and Cook, 1985, cite a special case in which asymmetrical load carriage may be advantageous. Workers with osteoarthritis of a hip joint may benefit by carrying the load on the same side of the body as the bad hip because carrying a load on one side of the body reduces the forces acting on the hip joint of that side, particularly when it is the weight-bearing hip, during the swing-through phase of gait. This is because the load assists in abducting the hip joint on the weight-bearing side.)

Similar reasoning suggests that when one is using carrying aids such as trolleys, **pushing rather than pulling** would also minimize any additional back muscle load and therefore be preferable. When one is pushing a trolley, the reaction tends to extend the trunk and is resisted by the abdominal muscles and the hip flexors. When one is pulling a trolley, the reaction tends to flex the trunk and is resisted by the back muscles. Laboratory investigations confirm the superiority of pushing compared with pulling (Lee et al., 1991). The compressive force on the L5-S1 disk was 2 or 3 times greater in pulling than in pushing. Body weight itself contributes more to the compressive load on the disk when one is pulling a load than when one is pushing it.

Carrying a load in front of the body may also obscure the carrier's view of the floor in front and of potential hazards such as steps or obstructions, thereby increasing the risk of slips, trips, and falls. Thus, for many applications, loads should be carried low on the back and placed symmetrically. It is noteworthy that in Africa, women traditionally carry young children low on their backs in the manner shown in Figure 6-7.

Finally, it is worthwhile to consider the practice of carrying loads on the head, which is the way most people in the world carry heavy objects. In many industrial societies head load carrying is no longer practiced, but it can be readily observed in rural areas of southern Europe and in many developing countries, where it is used mainly by women to carry heavy objects such as containers full of water, firewood, and cumbersome objects such as large baskets. Datta and Ramanathan (1971), in an investigation of different methods of load carriage, found that carrying 14 kg on the head was approximately half as demanding as carrying 7 kg in each hand in terms of oxygen consumption. Soule and Goldman (1969) reported that head load carrying was only slightly more demanding than carrying a load on the torso at the same speed. The extra cost was due to the increased shoulder and neck muscle activity to stabilize the head. The limiting factor in head load carrying appears to be the mechanical loading tolerated by the neck musculature rather than the total energy cost. Head load carrying appears to be one of the more efficient of all methods of load carriage.

CURRENT AND FUTURE RESEARCH DIRECTIONS

Much progress has been made in the drafting of guidelines to reduce the risk of workers' developing back pain or injury. It should be noted that the 1993 NIOSH lifting equation (the state-of-the-art equation) is likely to be further improved in the future as research continues (e.g., Ciriello et al., 1993). Further research is also required to demonstrate the effects of industry's adopting these controls and to express them in cost-benefit terms.

It is known that work in confined spaces and asymmetric loading increase the risks associated with heavy work. Several groups are investigating spinal loading under these conditions, including Gallagher and his colleagues at the U.S. Bureau of Mines.

The development of new devices, such as the lumbar motion monitor, enables researchers and designers to quantify lumbar movements of workers and relate them to task requirements and workspace design. These devices will enable more accurate estimates to be made of worker exposure to spinal stress and can be used to identify and redesign high-risk jobs.

Little research has been carried out on the prevalence of work-related back injury in developing countries. This area would appear to be an important one for future research since the economies of these countries rely on manual work to a far greater degree than industrially developed countries. Of particular interest are the following research topics:

1 The appropriateness for these countries of adopting controls used in developed countries
2 Biomechanical and postural evaluation of traditional work practices
3 Development of low-cost solutions and manual handling aids

SUMMARY

Modern thinking about the design of manual handling tasks emphasizes that a detailed analysis of all aspects of the task is necessary. It is no longer considered adequate to rely on simple specifications for maximum acceptable loads. For example, the design of the load, its size, and its density are as important, within limits, as its weight. The design of the workspace and the environment in which the work is to be carried out are also important considerations, as are data on the characteristics of the workers themselves. A more detailed treatment of many of these issues may be found in Chaffin and Andersson (1984).

ESSAYS AND EXERCISES

1 A worker has to lift a load symmetrically at a distance of 40 cm from the body from a height of 30 cm to a height of 90 cm. The lift is carried out once per minute over an 8-hour day. Use the NIOSH equation to determine the RWL. What is the effect on the RWL of the following?
 a Lifting asymmetrically at an angle of 50 degrees to stance
 b Putting handles on the load
2 Discuss the factors that cause the trunk to fail during heavy work.
3 Using a portable weighing device, survey the loads lifted and carried by people in non-industrial settings. Some examples are:
 a Shopping bags
 b Suitcases
 c Babies and young children
 d Sports equipment
 Would people be allowed to handle such loads in industry?

FOOTNOTE: Weight Lifters

In a discussion of spinal loading and injury, it is easy to convey the impression that the spine is a weak structure. It is often believed that the weakness arises because in the erect human, the spine acts like a weight-bearing column, whereas in quadrupeds, it acts more like a suspension bridge. Some fundamental design fault in the spine is said to put workers at the mercy of heavy manual work. In fact, in healthy individuals, the spine is an extremely strong structure easily capable of withstanding loads above body weight for short periods.

According to Farfan (1978), the lumbar lordosis and other adaptations provide humans with an overabundance of power for attaining an upright, bipedal posture. Most primates, on the other hand, lack the power to attain the erect posture for any length of time and cannot withstand the forces placed on their spines in activities such as lifting and throwing.

Competitive weight lifters are able to lift weights of more than twice their body weight over their heads. This ability is achieved through strength training but depends on excellent technique and timing. In lifts such as the clean and jerk the weight lifter exerts explosive power to accelerate the weight off the floor and squats rapidly to catch it at chest height. The muscles of the lower limbs are used primarily to power the lift. Power lifters attempt to lift maximum weights when carrying out the squat lift (squatting with a weight over the shoulders) and the dead lift (lifting a weight off the floor to thigh height). Weights of over 300 kg are regularly achieved, although in these exercises the movements are slow and controlled.

The key to injury-free lifting in both exercises is that the muscles of the legs and shoulders are used to lift the weight—*not* the muscles controlling flexion or extension of the spine. Anatomical analysis of lifting indicates that most of the movement is **hip extension, not lumbar extension.** It is the hamstrings and gluteal muscles that bring the trunk of the bent-over lifter to an erect position. As Gallagher and Hamrick (1991) explain, the gluteal muscles can generate an extensor moment about 5 to 7 times greater than the lumbar erectores spinae, which is why trying to lift with the back is both inefficient and hazardous.

Noe et al. (1992) compared the lifting technique of weight lifters and control subjects executing a dead lift (lifting a barbell from floor to knuckle height). The weight lifters were observed to exert maximum force at 50 percent of maximum lift height compared with 67 percent for the controls. Both groups initiated the lift using the quadriceps muscles to extend the knee. However, the weight lifters also used the gluteal muscles early in the lift, thereby extending both hip and knee simultaneously. This technique resulted in better force production and better stabilization of the pelvis and trunk. The weight lifters also made better use of the quadriceps throughout the lifting cycle, whereas, in the control subjects, quadriceps muscle activity declined in the latter half of the lift to be taken over by erector spinae activity. Noe at al.'s findings highlight the importance of good technique and of using several different muscle groups in lifting, and the need to strengthen all of them in training and work-hardening programs.

Traffimow et al. (1993) report that when subjects' quadriceps muscles are deliberately prefatigued in an experimental situation, their lifting technique changes from the squat lift to the stoop lift, trunk angular velocity increases, and there is less movement at the hip and knee. Rehabilitation of back-injured workers should include the quadriceps muscles and emphasize endurance as well as strength training of all muscles involved in lifting. Once again, a general principle of physical ergonomics is emphasized by these findings:

To reduce the risk of injury, design work-rest cycles to reduce the incidence of localized muscle fatigue.

7

PHYSIOLOGY, WORKLOAD, AND WORK CAPACITY

Muscles, through the tension they exert when contracting, make physical work possible. There are three types of muscle in the body:

1 Smooth muscle
2 Cardiac muscle
3 Skeletal muscle

Smooth muscle is found in the intestines and makes possible the movements essential for the digestion of food (peristalsis). It is also found in the walls of blood vessels where it is involved in the regulation of blood pressure and blood flow. It is not normally considered to be under conscious control. Cardiac muscle has a special structure and constitutes the bulk of the heart. The present discussion is limited to work involving skeletal muscle—muscle which is connected to the bones of the skeleton and passes over joints, enabling the bones to act like levers when the muscle contracts.

The energy required for muscle contraction is obtained from phosphate compounds in the muscle tissue. These compounds are formed from the breakdown of food. A brief review of muscle structure and function is appropriate (more detailed accounts can be found in Astrand and Rodahl, 1977; Withey, 1982; and Noakes, 1992).

MUSCLES: STRUCTURE AND FUNCTION

Some basic aspects of muscle function were discussed in Chapter 2. This topic will now be returned to, with the focus on muscle structure and function and its relation to energy metabolism. The contraction of most skeletal muscles is under conscious control, and the rate and strength of contraction can be varied at will. A complete muscle is made up of many bundles of fibers (cells) arranged side by side and covered by connective tissue

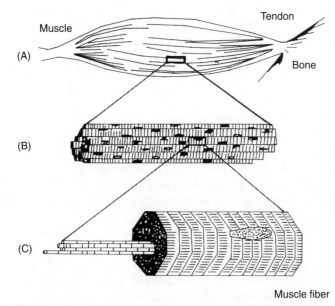

FIGURE 7-1   Basic structure of skeletal muscle. (Adapted from Noakes, 1992.)

sheaths. Nerves and blood vessels are located in the connective tissue. Each muscle fiber consists of many smaller myofibrils (Figure 7-1).

A myofibril is split up into a number of sarcomeres arranged in series (Figure 7-2). A sarcomere consists of many filaments layered over each other in alternating bands. There are two types of filaments. Thick filaments consist of about 300 myosin molecules. Thin filaments consist of a globular protein called actin. The filaments are the true contractile elements of a muscle. Muscles can be likened to bundles of string all joined together. Each individual string (muscle cell) is made up of fibers (myofibrils), each of which is constructed from many alternating bands of actin and myosin. The whole structure is bathed in intra- and extracellular fluid and is permeated by blood vessels and nerves.

The mechanism of muscle contraction consists of the actin filaments' sliding over the myosin filaments. Since the actin and myosin filaments are arranged in overlapping, alternating bands like a multilayered sandwich, sliding of the former over the latter causes the sarcomeres to shorten. Note that the filaments themselves do not shorten (Figure 7-3). The primary stimulus for muscle contraction is the release of calcium ions stored in

FIGURE 7-2   A myofibril.

FIGURE 7-3   Sliding filament theory of muscle contraction.

the sarcoplasm. The calcium ions bind to the actin, which increases the affinity of myosin for actin.

The fundamental physiology of muscle contraction is still not completely understood and much of it is beyond the scope of the present discussion. However, the main steps in the process are summarized in Table 7-1.

According to Guyton (1991), the duration of the calcium pulse (from release to reabsorption of calcium ions) is about one-twentieth of a second. Thus, in order for continuous muscle contraction to occur, a new impulse must reach the muscle after the calcium pump has removed the calcium ions. Smooth movements of the body are, in reality, the result of many individual contractions and relaxations of contractile elements throughout the muscle.

The amount of muscle contraction which can take place depends on the total change in length of all the sarcomeres in the muscle. As was discussed in an earlier chapter, muscle contraction is a biochemical process and need not result in shortening of the muscle itself. It is more accurate to say that muscle contraction consists of the actin filaments' *tending* to slide over the myosin filaments which will result in shortening of the muscle under appropriate circumstances. Further details of this sliding filament theory of muscle contraction and how it underpins the length-tension relationship observed in muscle contraction can be found in the classic paper by Gordon et al. (1966).

ENERGY FOR ACTION

Energy for muscle contraction (and for many other bodily processes) comes from the breakdown of a substance known as ATP (adenosine triphosphate). By breaking one of

TABLE 7-1   STEPS LEADING TO THE CONTRACTION OF SKELETAL MUSCLE

1 A nerve impulse travels along a motor nerve to the motor nerve end plates in the muscle fibers.

2 A small amount of neurotransmitter is secreted and travels to the membrane of the muscle fiber, initiating an action potential in the muscle fiber.

3 The action potential travels along and permeates the muscle fiber until it reaches a structure known as the **sarcoplasmic reticulum.**

4 The sarcoplasmic reticulum, which acts like a store of calcium ions, releases calcium ions into the myofibrils. This increases the attraction between the actin and myosin filaments, causing the sliding action of one filament over another.

5 The calcium ions are then actively pumped back into the sarcoplasmic reticulum. ATP is needed for this. Muscle contraction ceases.

6 Further nerve impulses cause the process to repeat.

the phosphate bonds, ATP is converted to ADP (adenosine diphosphate) and energy is made available inside the cell. Astrand and Rodahl (1977) liken ATP to a rechargeable battery pack—a short-term store of directly available energy. In order for the cell to continue functioning, the ADP must be reconverted to ATP so that energy can continue to be made available when required. A second phosphate compound known as creatine phosphate acts like a backup energy store to recharge the ADP to ATP.

In order for muscle filaments to slide backward and forward over one another, energy is required. In fact, the actin and myosin filaments are always mutually attracted and energy is required to break this attraction and allow the muscle to lengthen again. During the relaxation phase of muscle contraction, ATP is needed to weaken the attraction between the actin and myosin filaments. Endurance athletes sometimes experience muscle cramps when a muscle becomes fatigued during an event (Noakes, 1992) because the ATP needed to break the actin-myosin attraction is depleted.

The creatine phosphate and ATP stores are of very limited capacity and are reduced within seconds or minutes. The creatine phosphate system can provide energy for maximal muscle activity for about 8 to 10 seconds and therefore has to be continuously replenished. The replenishment takes place in structures inside the cells, known as mitochondria. If the ATP and creatine phosphate stores are depleted, the muscle will go into rigor (shortly after death, a state of rigor mortis sets in through the loss of all ATP).

In muscles, the mitochondria are placed close to the filaments. Carbohydrates and fatty acids (derived from fat) are broken down—ultimately into carbon dioxide and water—and energy is released to form ATP. The basic reaction for the liberation of energy involves the oxidation of glucose, as shown below:

$$C_6H_{12}O_6 + 6O_2 \rightarrow 6CO_2 + 6H_2O + energy$$

Carbon dioxide and water (so-called metabolic water) are the products of this chemical reaction.

The carbohydrates and fatty acids used by the mitochondria are derived from the food we eat. They are obtained either from the bloodstream or from stores in the muscle. Glucose is stored as glycogen in the muscles and liver. It is an important store of energy in endurance sports. Oxygen is obtained from the air ventilating the lungs and is transported to all parts of the body by the blood. The body has a very limited capacity to store oxygen, although muscles do contain a substance known as myoglobin, which is a short-term oxygen store.

In an electrical analogy, if the bloodstream is likened to an electrical reticulation network, the mitochondria can be considered transformers, plugged in to the mains, which keep the creatine phosphate and ATP battery packs fully charged.

Oxygen-Dependent and Oxygen-Independent Systems

If insufficient oxygen is available, the mitochondria are no longer able to convert the carbohydrate and fatty acids to produce energy and a second, oxygen-independent, system becomes the main source of ATP in which enzymes in the intracellular fluid surrounding the myofilaments and mitochondria use glycogen and blood sugar (but not fat) to produce energy in the absence of oxygen. This backup system operates at all times, even at rest, but it is inefficient and produces much less energy per glucose molecule than the oxygen-dependent system. Furthermore, it produces waste products which cause the acidity of the muscle cells to increase. This, in turn, reduces the affinity between actin and myosin filaments, which, in turn, weakens the muscle. Oxygen is required to remove these waste products. If it is not available when the person is working, the waste products will accumulate. Under these circumstances it is said that the person has built up an **oxygen debt**, which must be paid back when work ceases.

As was described in an earlier chapter, most tasks involve a mixture of static and dynamic work and the latter can usually be sustained longer than the former. Table 7-2 summarizes some of the physiological differences between static and dynamic work.

Implications

The oxygen-independent system can form ATP molecules about 2.5 times as fast as the oxygen-dependent system but can provide energy for only about 1.3 minutes of maximum muscle activity. It is a valuable system which enables work to be carried out at a high rate for short periods of time interspersed with rest. Clearly, it is unrealistic to expect workers to carry out sudden, highly demanding tasks for more than a minute or so without rest. Therefore in plans for high-workload emergency situations, backup must be provided to enable rapid alternation of work teams. The oxygen-dependent system can function for as long as nutrients are available.

Efficiency of Muscle Contraction

The muscular system uses oxygen to convert chemical energy from foodstuffs (stored in the tissues or delivered by the bloodstream) into mechanical energy via the sliding filament mechanism. This process has a certain efficiency. According to Edholm (1967), muscles are, at best, only 20 percent efficient in converting chemical reactants to mechanical output, and heat is the main by-product of the process. This heat is produced largely as a result of the viscosity of body parts and of overcoming friction of blood flowing through capillaries and tendons sliding over joints. Metabolic heat production has important implications for the design of work in extreme climates, as will be seen in a later chapter. The body's system for converting chemical into mechanical energy can malfunction for a number of reasons (Table 7-3).

TABLE 7-2   DIFFERENCES BETWEEN STATIC AND DYNAMIC WORK*

Static work	Dynamic work
Sustained muscular contraction	Repetitive muscle contraction-relaxation cycle
Reduced muscle blood flow	Increased muscle blood flow
No increase in muscle oxygen consumption	Increased muscle oxygen consumption.
Oxygen-independent energy production	Oxygen-dependent energy production
Muscle glycogen $\rightarrow$ lactate	Muscle glycogen $\rightarrow$ CO_2 + H_2O; muscle uptakes glucose + fatty acids from blood

*From Noakes (1992).

TABLE 7-3   LIMITING FACTORS ON MUSCLE CONTRACTION

1 Demand for energy exceeds supply. This situation can occur when chemical fuel stored in the muscle is exhausted and the rate of replenishment of oxygen or glucose is inadequate. The supply of fuel to working muscle depends on the capabilities of the circulatory system.

2 Mechanical capabilities are exceeded. The force output of a muscle is finite and depends on the number of contractile elements (the maximum strength of a muscle is proportional to its cross-sectional area). A task may exceed the muscle's mechanical capacity. Strength training increases the cross-sectional area of a muscle and the ability to recruit muscle fibers to carry out muscular work.

3 Accumulation of waste products impairs muscle function. This situation can occur during static muscle activity, as described in previous chapters.

4 The rate of heat production exceeds the body's thermoregulatory capacity. Excess heat can no longer be dissipated, and body temperature increases. Cardiovascular capacity is impaired.

THE CARDIOVASCULAR SYSTEM

The cardiovascular system performs a number of functions. It collects oxygen from the lungs and food from the gut and delivers them to the cells of the body. It also collects waste products and delivers them to the excretory organs. Hormones secreted by the endocrine glands in various parts of the body are transported to target cells in other organs by the blood. Finally, the cardiovascular system is involved in thermoregulation—the control of body temperature.

An average adult has about 5 liters (L) of blood. Approximately 2.75 L consist of plasma and 2.25 L of blood cells. There are three main types of blood cells. Red cells (erythrocytes) transport oxygen, white cells (leukocytes) fight infection, and platelets assist in blood clotting. The red cells are formed in bone marrow and contain a red pigment (heme) which is bound to a protein molecule (globin). Hemoglobin combines with oxygen to form oxyhemoglobin—the form in which oxygen is carried in the blood. Plasma consists of 90 percent water and 10 percent solutes—food, proteins, and salts.

The heart is a four-chambered pump about the size of a person's fist. The right atrium receives blood returning from the body and passes it to the right ventricle, where it is pumped to the lungs, where gaseous exchange takes place. Hemoglobin absorbs oxygen (a process known as oxygenation) and the oxygenated blood returns to the left atrium of the heart via the pulmonary vein. It passes to the left ventricle, where it is pumped via the aorta to the tissues of the body through a series of ever-branching blood vessels. Most tissues of the body are permeated by capillaries—tiny blood vessels with internal diameters not much larger than a red blood cell. A schematic representation of the circulatory system is given in Figure 7-4.

The system pumps oxygenated blood under pressure so that it permeates the tissues of the body. Oxygen and food are delivered to the tissue, and carbon dioxide and waste products are removed. After passing through the bed of capillaries which permeates the tissues, the deoxygenated blood drains into the veins (in a manner similar to the flow of small streams or tributaries into a river). The veins contain one-way valves which enable the blood flow in only one direction—back to the heart. Muscle contractions which occur as a result of everyday activities are usually sufficient to act as a "muscle pump"

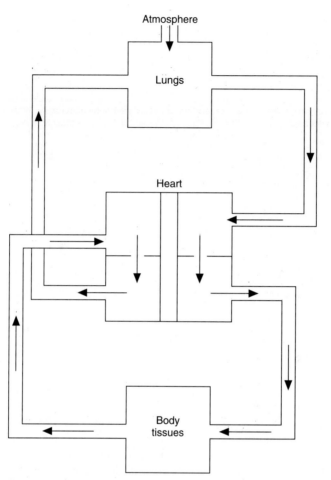

FIGURE 7-4   A model of the circulatory system.

which, by compressing the veins, forces the blood back along the veins until it reaches the heart.

The output of the heart increases from around 5 L of blood per minute when at rest to 25 L per minute during heavy work. As workload increases, the volume of blood pumped per stroke (heartbeat) can be increased as well as the heart rate itself. Furthermore, the flow of blood to the body can be increased on a selective basis by constricting and dilating different blood vessels. During physical work, more blood is directed to the muscles and to the skin (to dissipate the excess heat which is a by-product of muscular contraction). After a meal, extra blood is directed to the intestines. The selective control of blood flow to the muscles is controlled by the nervous system. Changes in acidity and the local concentration of metabolites provide feedback to the neural systems which control blood flow.

THE RESPIRATORY SYSTEM

The respiratory system functions as a gas exchanger. It supplies the body with oxygen and removes carbon dioxide. It consists of the nose, pharynx, larynx, trachea, bronchi, and lungs. The nose filters incoming air, warms and humidifies it, and tests for substances which may be harmful to the respiratory tract. The nasal epithelium is lined with mucus. Small particles in the turbulent incoming air become trapped in the mucus which flows under the action of cilia to the pharynx, where it is swallowed. Before reaching the trachea, incoming air is warmed almost to body temperature and humidified to almost 100 percent.

The pharynx (throat) is a muscular tube which serves as the entrance, or "hallway," for the respiratory and digestive tracts. It plays an important role in the forming of vowel sounds in speech (phonation). The larynx lies at the upper end of the trachea and just below the pharynx. It is part of the open passage through which air travels on its way to the lungs and protects the lungs against the ingress of solids. It is also the organ of voice production.

The trachea (or windpipe) is a tube of cartilage and muscle about 2.5 cm in diameter extending from the larynx (in the neck) to the thoracic cavity (or chest), where it divides into two smaller tubes—the primary bronchi. The bronchi divide many times into increasingly smaller tubes which finally terminate in microscopic sacs, or alveoli. The alveoli are in contact with capillaries containing blood pumped from the heart via the pulmonary artery, and it is through this contact that gaseous exchange takes place.

The respiratory system can be likened to an interface which brings the cardiovascular system into contact with the atmosphere.

PHYSICAL WORK CAPACITY

It was stated earlier that the worker can be thought of as both a user and a source of energy. Food and oxygen are required for the basic metabolic processes necessary for life—the maintenance and repair of cells, the supply of blood to the tissues, etc. The **basal metabolic rate (BMR)** is the rate of energy consumption necessary to maintain life. Individuals differ in their BMR. It is about twice as high in a child as in an adult. The decrease in BMR with age is one of the reasons for adult "middle-age spread." The BMR can drop by about 20 percent in chronically malnourished individuals.

Physical work capacity refers to a worker's capacity for energy output. This capacity will depend primarily on the energy available to the worker in the form of food and oxygen and the sum of the energy provided by oxygen-dependent and oxygen-independent processes. The rate of energy consumption during physical work is the sum of the basal energy consumption and the metabolic cost of the work in terms of energy consumption.

For continuous work at moderate intensities, oxygen-dependent processes usually make the major contribution to energy output. For each liter of oxygen consumed about 4.8 kilocalories (kcal) of energy is released. Work capacity depends on the ability to take up oxygen and deliver it to the cells for use in the oxidation of foodstuffs. The ability to work at a high rate is associated with a high oxygen uptake.

Maximum Oxygen Uptake

Exercise physiologists and sports scientists have used the term "VO$_2$ max" to describe an individual's capacity to utilize oxygen (aerobic capacity). VO$_2$ max has traditionally been estimated by having subjects run on a treadmill or pedal a bicycle ergometer while their oxygen uptake is measured. The running or cycling speed is increased in an incremental manner and oxygen uptake is measured approximately every 3 to 5 minutes after the subject has adapted to each new work rate. As might be expected, it is observed that oxygen uptake increases as the work rate is increased. The relationship is approximately linear, as is illustrated in Figure 7-5.

Clearly, oxygen consumption and heart rate cannot continue increasing indefinitely. In any work situation, a point is reached where the person cannot increase the work rate any more. There are limits to performance of all human activities, be they running, cycling, or loading boxes onto a conveyor belt in a factory. For many years it has been believed that the factor which limits a person's work rate is the inability of the heart and lungs to supply oxygen to the muscles at a sufficiently fast rate to meet the energy requirements of the work. At this point, the person will reach his or her maximum work rate. Individual differences in maximum work rate can be traced back, according to this view, to differences in the ability to supply oxygen to the muscles.

Strictly speaking, the view that oxygen uptake limits work capacity can be criticized on the grounds that it is equally valid to say that a high oxygen uptake is merely a con-

FIGURE 7-5   The relationship between workload and oxygen uptake.

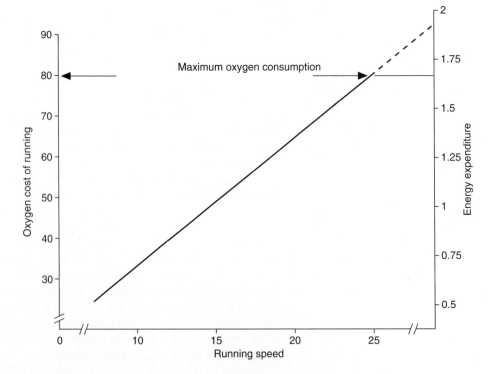

sequence of the ability to work at a high rate. Cause and effect are more difficult to distinguish than they seem to be initially. However, the idea that limitations on oxygen supply place a limit on maximum work rate underlies the classic concept of VO_2 max which has been used in much research. It states that a point is reached where increases in work rate are no longer accompanied by increases in oxygen uptake—the individual is assumed to have reached her or his maximum level of oxygen uptake and cannot sustain a harder pace of work. Further increases in work rate can be achieved only if energy is provided by oxygen-independent processes. These processes can be maintained only for a short period because they cause waste products to accumulate in the blood and tissues. Shortly after a person reaches a work rate which exceeds his or her VO_2 max, performance will decline dramatically.

There is great interindividual variability in VO_2 max, which is thought to be due largely to genetic factors. Training can improve an individual's VO_2 max, although only about 10 to 15 percent.

Noakes (1988) has reviewed the evidence for the idea that the maximum work rate depends on the capacity to deliver oxygen to the muscles. He notes that in many of the individuals tested, the heart rate–VO_2 relationship does not reach a clear plateau—even though the person exercising reaches a point of maximum performance. Noakes has suggested that the limitation on maximum performance may not be the ability to supply oxygen to the muscle tissues but a failure of muscle contractility (or muscle power). This view implies that in groups of athletes, VO_2 max may not be the best method of predicting maximum performance.

Removal of Waste Products

In addition to the capacity to utilize oxygen, sustained performance at a high level depends on the ability to remove waste products of metabolism from the tissues. Endurance athletes do not perform at VO_2 max intensities but at a lower level (about 80 percent in marathon runners, according to Reilly, 1991). Successful performance depends on the ability to remove lactate from the body tissues as well as on aerobic capacity. The accumulation of lactate in the blood of an athlete signals the upper limit to which endurance exercise can be maintained (Reilly, 1991). Unlike VO_2 max, the blood lactate response can be improved considerably by training.

Successful performance of endurance tasks also depends on the ability to dissipate heat and the ability to utilize fat as a source of fuel.

VO_2 max and Industrial Work

Despite these cautionary notes about the meaning of VO_2 max, an understanding of the classical concept is essential in order to understand much of the existing literature on industrial physiology. It remains the case that the consumption of large quantities of oxygen is associated with high levels of performance in all endurance sports. Successful cross-country skiers, distance runners, and cyclists always have high VO_2 max levels of 70 to 80 milliliters of oxygen per kilogram of body weight. This can be compared to an average value of around 40 mL per kg per minute. If maximum aerobic capacity is thought of as a type of anthropometric variable, it can be said that these individuals lie at the extreme of the anthropometric range.

It is generally believed that individuals can work continuously over an 8-hour shift at a rate of 30 to 50 percent of their maximum capacity. When it is remembered that there is a distribution of VO_2 max in the population, it is clear that tasks must be designed using VO_2 max in a way similar to that of other anthropometric variables. That is to say, jobs must be designed so that they can be carried out by the less fit members of the workforce. For example, suppose two otherwise identical workers have whole body VO_2 max values of 6 L and 2 L of oxygen per minute, respectively, and the task requires the expenditure of 2 L of oxygen per minute. The first worker will have a greater work capacity because he or she will be able to work comfortably all day at only 33 percent of VO_2 max. Energy will be met by oxygen-dependent processes, which means that this worker will be able to work in a physiologically steady state and obtain energy from the oxidation of fat as well as from carbohydrates (recalling that oxygen is required to enable fatty acids to be metabolized in the mitochondria). The second worker will be required to work at 100 percent of VO_2 max and will tend to utilize oxygen-independent processes. He or she will be able to work only for short periods and will require frequent rest breaks to metabolize the accumulated metabolic waste products. The second worker will also be dependent on carbohydrates as a source of fuel.

It will be readily appreciated that in heavy manual work which is renumerated on a piecework basis (i.e., in which the amount each worker earns depends on how much she or he produces), individuals with a high VO_2 max will be at an advantage since they will require fewer rest breaks. Astrand and Rodahl (1977) report that this appears to be the case among lumberjacks—high earners tend to have a higher VO_2 max.

Various attempts have been made to classify the severity of work in terms of oxygen uptake, heart rate, and energy expenditure. Table 7-4 (adapted from Astrand and Rodahl, 1977) is a representative example. Some examples of activities with different oxygen requirements are given in Table 7-5.

NIOSH (1981) has published data concerning the maximum aerobic capacity of U.S. workers. The 50th percentile male and female capacities, in terms of energy output, are approximately 15 to 10.5 kcal per minute, respectively. The lower 5th percentile ca-

TABLE 7-4   SEVERITY OF WORK IN TERMS OF VO_2, HEART RATE, AND ENERGY EXPENDITURE*

Work Severity	VO_2 (L/min)	Heart rate (beats/min)	Energy expenditure (kcal/min)
Light work	< 0.5	< 90	< 2.5
Moderate work	0.5–1.0	90–110	2.5–5.0
Heavy work	1.0–1.5	110–130	5.0–7.5
Very heavy work	1.5–2.0	130–150	7.5–10.0
Extremely heavy work	> 2.0	150–170	> 10.0

*Adapted from Astrand and Rodahl (1977).

TABLE 7-5 OXYGEN UPTAKE AND PHYSICAL ACTIVITY

---

1 *Rest.* Basal metabolism requires approximately 0.25 L of oxygen per minute.

2 *Sedentary work.* Office work, for example, requires an oxygen uptake of only a little over resting levels (0.3–0.4 L per minute).

3 *Housework.* Housework includes several moderate to heavy tasks (requiring about 1 L per minute of oxygen). It is unusual for high work rates to be maintained for any length of time though.

4 *Light industry.* Oxygen uptakes from 0.4–1 L per minute are required.

5 *Manual labor.* Oxygen uptake may vary from 1–4 L per minute. The workload can depend greatly on the tools and methods (for example, the design of a shovel, the method of load carriage).

6 *Sports.* Endurance sports can have a very high oxygen requirements. Oxygen uptakes of over 5 L per minute have been recorded in cross-country skiers, for example.

---

pacities are 12.5 and 8 kcal per minute. For continuous work, NIOSH states that energy expenditure should not exceed a value of 33 percent of an individual's maximum capacity (or 5 kcal per minute for men and 3.5 kcal per minute for women) over an 8-hour shift. Referring to Table 7-5, one can see that a 50th percentile male worker would be able to carry out heavy work for 8 hours, but not very heavy work. A 5th percentile worker would be able to carry out only moderate work for 8 hours. Determining the energy requirements of different jobs makes it possible to identify jobs which can be performed only by the fitter members of the workforce and, where possible, redesign these jobs to lower the required energy expenditure, thus making them available to a wider range of workers.

For occasional lifting tasks (1 hour or less per day) NIOSH recommends that energy expenditure rates not exceed 9 kcal per minute for physically fit men and 6.5 kcal per minute for physically fit women. The upper limit of energy expenditure is approximately 16 kcal per minute for men and 11 kcal per minute for women for 4 minutes.

FACTORS AFFECTING WORK CAPACITY

Many factors can influence a person's capacity to carry out physical work. Fatigue, due to lack of carbohydrates or fluids or the accumulation of waste products, is an obvious example. Personal characteristics and environmental factors are also fundamental. Table 7-6 summarizes some of the more common personal and environmental factors.

A wide variety of constitutional, lifestyle-related, and psychological factors can influence work capacity. Some of the more common personal factors are summarized below.

Personal Factors

Body weight Body weight (particularly the percentage of body tissue which is composed of fat) influences all activities in which the worker has to move his or her own body (e.g., walking, cycling, climbing ladders or stairs). In exercise physiology and sports science it is usually more meaningful to express VO_2 max in relative terms by ex-

TABLE 7-6   FACTORS AFFECTING PHYSICAL WORK CAPACITY

Personal	Environmental
Age	Atmospheric pollution
Body weight	Indoor air quality
Gender	Ventilation
Alcohol consumption	Altitude
Tobacco smoking	Noise
Active/nonactive lifestyle	Extreme heat or cold
Training/sport	
Nutritional status	
Motivation	

pressing a person's oxygen consumption in terms of method body weight (liters of oxygen per minute per kilogram of body weight). This method takes into account the fact that the leaner runner, for example, will be at an advantage over the plumper rival—all other things being equal. In this sense, it is possible to increase one's relative VO_2 max by shedding excess kilograms of fat.

Age Age has a significant effect on work capacity. VO_2 max declines gradually after 20 years of age. A 60-year-old has an aerobic capacity of about 70 percent of that of a 25-year-old. This decline is due to a reduction in cardiac output. Current thinking stresses that the fundamental aging phenomenon is due to a loss of muscle function. Since the heart is essentially a muscle, this explains the loss of aerobic capacity with age.

Sex Women have a lower VO_2 max than men and usually have a higher percentage of body fat. They also have less hemoglobin than men. The cardiac output per liter of oxygen uptake is higher in women than in men. For a woman, the heart must therefore pump more oxygenated blood than for a man in order to deliver 1 L of oxygen to the tissues.

Alcohol Consumption Alcohol may increase cardiac output in submaximal work, thereby reducing cardiac efficiency. It also affects liver function and can cause a predisposition to hypoglycemia (low blood sugar).

Tobacco Smoking Tobacco smoke contains about 4 percent by volume carbon monoxide (CO). CO has an affinity for hemoglobin (combining to form carboxyhemoglobin) 200 times as powerful as oxygen. Smoking therefore reduces work capacity by reducing the oxygen-carrying capacity of the blood. It also causes chronic damage to the respiratory system, which impairs the ventilation of the lungs and the transfer of oxygen from the air to the blood. Tobacco smoke also contains a very large number of toxic and carcinogenic chemicals which are likely to have a generally depressing effect on the physical capacity of smokers.

Recent evidence suggests that nonsmokers who work in the same room as smokers may suffer some of the same effects as smokers themselves by breathing in the smoke-laden air. Of particular importance is the sidestream smoke which is emitted from the burning tip of the cigarette. Sidestream smoke contains a higher proportion of toxic substances and gases then exhaled smoke because it has not been prefiltered by the smoker's lung tissues.

Training Work capacity can be enhanced by physical training (to increase a worker's VO_2 max) and job training in more efficient work methods (to obtain more output per liter of oxygen consumed by the worker or to enable the worker to safely exert greater forces by using better techniques). Specific training regimes can be developed to strengthen particular parts of the musculoskeletal system, with the goal of improving performance or preventing injury. Strength training requires that the body part in question be exercised at near maximal levels. Over a period of several months, the muscle fibers increase in size as a result of an increase in the number of myofibrils, and an increase in strength is observed.

Nutritional Status and General Health A balanced diet is important to ensure adequate amounts of necessary foodstuffs and to minimize the accumulation of excess body fat. Excess body fat lowers a person's relative VO_2 max, as was described above.

In developed countries, many people eat a diet high in saturated fats. Such a diet causes raised plasma concentrations of cholesterol. Tiny crystals of cholesterol are deposited on the inside walls of the arteries, which eventually leads to a disease of the arteries known as **atherosclerosis.** The continued accumulation of cholesterol forms deposits called plaques which eventually reduce the cross-sectional area of the artery and thus impede blood flow. Additionally, the arteries lose flexibility (atherosclerosis is sometimes called hardening of the arteries for this reason). These changes in the structure of arteries can impede the flow of blood to the muscles and to the heart itself, resulting in decreased performance and increased risk of heart attacks.

In developing countries a lack of carbohydrate in the diet can reduce work capacity or cause other complications which affect work capacity. This is discussed in more detail in the next chapter.

Motivation Motivation is an extremely important determinant of work capacity. For present purposes, it may be noted that a worker's level of motivation may be affected by intrinsic factors such as personality, personal and career goals, need for achievement at work, and so on, as well as extrinsic factors such as work organization, method of remuneration, and the availability of alternative forms of employment. Piecework systems (where the worker is paid according to how much is produced) may motivate workers to work at an increased rate but have been associated with increased risk of accidents and of developing musculoskeletal ailments. Motivation is discussed in more detail in other chapters of this book.

Many environmental factors can degrade work capacity. Some of them are under management control and can be manipulated to reduce any adverse effects, as is described next.

Environmental Factors

Air Pollution Air pollution may increase the resistance to air flow of the respiratory airways and, in the long term, cause damage to the lungs, permanently reducing the worker's capabilities. If the source of the pollution involves the combustion of organic compounds, carbon monoxide may be one of the by-products and decrements in work capacity may result. When the concentration of carbon monoxide exceeds 6.5 parts per million, it begins to accumulate in the blood during submaximal exercise. Because carbon monoxide concentrations from 37 to 54 parts per million occur in urban traffic, the work capacity of people doing heavy manual work in urban areas may well be degraded.

Climate Extreme environments can have significant effects on work capacity, as is described in a following chapter.

Noise Noise is a stressor which can elevate the heart rate and reduce cardiac efficiency.

Altitude The capacity for sustained work is reduced at high altitude because the partial pressure of oxygen is lower (the air becomes "thinner" with increasing altitude). Less oxygen is available per unit volume of air and thus the functional VO_2 max is reduced. Some adaptation to work at altitude can take place after continual exposure for several weeks. Short-term, high-intensity activities (such as lifting a heavy weight) are not influenced by altitude because their performance does not involve oxygen-dependent processes (Kroemer, 1991). In practice, the capacity for maximum muscular work remains unchanged up to an altitude of 1500 m above sea level. Above 1500 m, maximal work capacity decreases by about 10 percent per 1000 m. Submaximal work capacity is not affected below 3000 m above sea level but does require a larger percentage of the total work capacity compared with the percentage at sea level.

Protective Clothing and Equipment Protective clothing and equipment can affect work capacity in several ways, depending on the application. Bulky clothing may limit the range of motion of certain joints and hamper work movements. It may decrease stride length and therefore reduce walking and running speed and efficiency. Heavy clothing assemblies or devices which have to be carried will reduce work output for the same level of effort in activities such as climbing or carrying. Protective clothing assemblies which trap heat may increase heart rate for the same level of effort, thus reducing work capacity or hastening fatigue.

Breathing equipment almost always imposes an additional physiological cost on the user because of the resistance of air flow caused by pipes, valves, etc. The particular cost depends on the application. In the nuclear power industry, the main consideration is to protect workers from dust by providing a separate breathing system for the provision of oxygen and the removal of carbon dioxide. Firefighters need protection from noxious gases, such as carbon monoxide, and the provision of oxygen which the fire tends to consume. Pilots of high-altitude aircraft need a breathing mixture enriched with oxygen because the ambient partial pressure of oxygen at high altitude is not insufficient to "push" oxygen through the alveoli. In closed breathing systems, only oxygen is provided and the wearer rebreaths his or her own air after it has passed through a carbon dioxide absorber such as soda lime or barium hydroxide. A disadvantage of these systems is that water vapor can accumulate—in the cold it can freeze and block the system. Open systems supply the wearer with fresh air, and expired air is released to atmosphere via a one-way valve. In this respect, open systems waste oxygen and are inefficient.

Breathing equipment can affect performance in several ways. First, it must provide the user with an adequate "lungful" of air at each breath, which varies depending on the conditions. At rest, 10 to 15 L of air are needed per minute. During heavy exercise, 40 to 50 L per minute may be needed. The dynamic aspects of air flow are also important. During speech, high acceleration and peak flows are observed which the system must cater for (up to 200 L per minute). Resistance to inspiration can be exacerbated by any factors which increase turbulence in the flow of incoming air—rough inner surfaces of pipes, sudden bends, joints, or changes in bore. Resistance to expiration depends on the valve characteristics. Under heavy resistance, the breathing patterns may be characterized by panting—shallow breaths repeated frequently. Panting brings with it the threat of **hypercapnia** (accumulation of carbon dioxide in the blood). The acidity of the blood increases and the heart rate increases.

Certain designs of breathing equipment impose a physiological cost to the user because of the "dead space" through which air has to be breathed. An infinitely long pipe would eventually cause suffocation as the user continually rebreathed his or her own air. Shorter lengths (200 mm to 2 m) can have significant effects on physiological variables both at rest and at work, as part of the expired air is rebreathed. Dead space increases heart rate, tidal volume, partial pressure of carbon dioxide, and lowers work capacity.

Sulotto et al. (1994) investigated the energy expenditure of railway workers wearing an air-purifying respirator in which inspired air first passed through filters designed to remove organic vapors and dust. Exercise using the respirator was characterized by reduced frequency of breathing, ventilation rate, oxygen uptake and maximal oxygen uptake, and carbon dioxide production. The mean drop in maximum oxygen uptake was about 9 percent. The recommended work load for an 8-hour day of 40 percent of maximum aerobic capacity needed to be reduced to compensate for the lowered maximum value when the respirator was used. Similarly, for unfit people, a level of 30 percent of the maximum uptake when wearing the respirator was recommended for an 8-hour shift. A maximum inspiratory mouth pressure of 20 mmH$_2$O was recommended for respirators designed for work of this intensity and duration.

CURRENT AND FUTURE RESEARCH DIRECTIONS

A great deal of contemporary research on physical work capacity is taking place in the fields of exercise physiology and sports science. Aerobic capacity is not the only variable which underlies success in endurance events. Biomechanical factors such as running efficiency also play a part. Identification of these factors can enable an athlete's potential to be predicted. The factors which affect cardiovascular capacity in the general population are now quite well understood. Catastrophic failure of the cardiovascular system can be avoided if the individual is not exposed to risk factors such as tobacco smoking, unbalanced diet, lack of exercise, and stress. General recommendations for exercise, diet, etc. can be given, but further research will lead to more specific recommendations for maintaining physical condition.

Although there is a certain amount of data on the aerobic capacities of U.S. workers, the population is undergoing demographic aging and these data may well need revision. Generally, more data concerning the work capacities of U.S. workers and workers in other countries—particularly in industrially developing countries—are needed. Such data will enable ergonomists and industrial engineers to make better decisions about what constitutes an acceptable work rate for a given task when designing production facilities worldwide.

SUMMARY

Physiological mechanisms set limits to the worker's capacity for physical work. Sustained work at a low level requires the operation of oxygen-dependent processes and the removal of waste products and heat. Vigorous work can be carried out for short periods using oxygen-independent processes, but accumulation of waste products and increase in the acidity of muscle tissue cause weakness and make frequent rest periods essential.

ESSAYS AND EXERCISES

1 Discuss what is meant by "fatigue." How would an ergonomist go about investigating complaints of fatigue in manual work?
2 People's occupational and leisure preferences are a reflection of their physiology. Discuss.
3 What practical control measures can an organization take to ensure that its employees' work capacity is not degraded?

FOOTNOTE: Limits of Physical Work Capacity

The ability to exert extremely great forces depends on muscle mass and the utilization of energy stores in the muscle itself. At high exercise intensities (above 75 percent of VO_2 max) more use is made of muscle glycogen stores. This cannot be sustained for long periods for the reasons described above.

According to Edholm (1967), 1 kg of muscle can produce about 0.3 horsepower (hp) (0.746 kilowatts). Thus, a 70-kg man with approximately 31.5 kg of muscle could, in theory, produce 10 hp in a single burst. In practice, champion cyclists can produce a maximum of about 2 hp and sustain this for 10 seconds. One L of oxygen yields approximately 0.1 hp and, as described above, work requiring more than 2 L of oxygen per minute is classified as extremely heavy—as, for example, running a marathon.

Successful marathon runners always have a high VO_2 max compared with the rest of the population. They usually have a low percentage of body fat. It has been shown that as the duration of exercise increases, the ability to obtain energy from the oxidation of fat becomes more important (Noakes, 1992). Endurance training appears to enable a prospective athlete to obtain more of her or his energy requirements from fat rather than from carbohydrates because of the massive increase in mitochondria which occurs in muscle after 8 to 12 weeks of endurance training. This means that the trained athlete can exercise for longer before becoming exhausted as a result of the depletion of carbohydrate stores in the muscles and liver.

Diet is another important factor in determining work capacity. Endurance sports provide another example of the dramatic effects of carbohydrate loading. A high-carbohydrate (and low-fat) diet eaten in the 3 days before the athletic event increases the amount of glycogen stored in the muscles and liver and therefore enhances performance. Carbohydrate ingested during the activity may also delay the onset of fatigue.

8

INDUSTRIAL APPLICATIONS OF PHYSIOLOGY

Industrial systems require large investments of capital in order to be established and maintained. The improvement of productivity to hasten a return on investment has always been a priority. A main goal has been to determine an acceptable **work rate** for a given job. Industrial engineers have developed methods for designing manual jobs in a systematic way (Barnes, 1963, is a classic text). These techniques enable them to specify **time standards,** or standard times for the completion of tasks, and to describe the physical load of tasks by means of **performance rating.** This has provided management with a way of organizing work in which standard levels of production can be defined and actual output can be monitored to ensure that the standards are being maintained. These methods rely on observation of worker behavior by the industrial engineer rather than objective measurement of physiological variables.

In this chapter, the application of physiological methods in industry will be described. In addition to the measurement of workload itself, physiological methods also offer the possibility of investigating mental stress and dealing with wider issues such as nutrition and employees' level of fitness for a given job.

Physiological methods can be applied in industry to evaluate the physical demands of jobs in terms of **energy expenditure.** In principle, any increase in oxygen uptake over and above that required for basal metabolism can be used as an index of the physiological cost to an individual of performing the work. When an individual begins a work task from rest, heart rate and oxygen consumption increase to meet the new demands. Because this response is not instantaneous, the immediate requirements for energy are met by local (i.e., muscular) energy stores. Similarly, when an individual ceases working, heart rate and oxygen consumption slowly return to their initial levels (Figure 8-1). Even though the individual is now resting, extra oxygen is still required to replenish the mus-

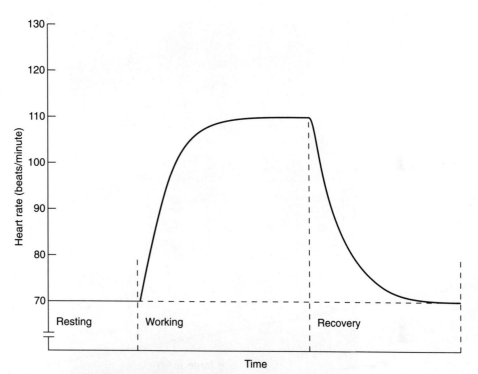

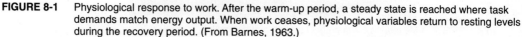

FIGURE 8-1   Physiological response to work. After the warm-up period, a steady state is reached where task demands match energy output. When work ceases, physiological variables return to resting levels during the recovery period. (From Barnes, 1963.)

cle stores and to oxidize the waste products of oxygen-independent processes—to pay back the oxygen debt.

In many industrial tasks there is a warm-up period in which physiological processes adjust to meet the new demands. The warm-up is followed by a period of steady-state work and a recovery period (Manenica and Corlett, 1977). The **total physiological cost** of work includes the energy expenditure during both the working and the recovery periods (until physiological processes return to their resting levels). Thus, in order to assess the physical demands of work using physiological methods, one must begin measurement with the subject completely at rest. Measurement continues throughout the work cycle and into the recovery period until the physiological variables return to the resting levels measured previously.

MEASUREMENT OF THE PHYSIOLOGICAL COST OF WORK

The classic method of determining energy expenditure at work involves the measurement of oxygen uptake using the Douglas bag (Figure 8-2). The subject inhales air from the atmosphere and exhales it through a mask connected by tubing to a large bag (known as a Douglas bag). As the subject carries out a specified task, the initially empty bag is

FIGURE 8-2   Measuring oxygen consumption of a cart puller ("thelawalla") using a Douglas bag to collect the expired air. (Photograph courtesy of Professor S.R. Datta.)

filled with expired air. After about 50 L of air has been collected, the task is terminated or the expired air is diverted to a second empty bag. The volume of air in the filled bag is calculated and its gaseous composition analyzed (electronic gas analyzers are often used nowadays).

The oxygen content of the air in the Douglas bag can be compared with that of the atmosphere to determine the amount of oxygen metabolized by the subject. If the time taken for the subject to fill the Douglas bag is known, the subject's rate of oxygen uptake can be calculated. From the rate of oxygen uptake, the rate of energy expenditure can be calculated. Further details can be found in standard work physiology texts such as Astrand and Rodahl (1977). The energy costs of some common daily activities are given in Table 8-1.

The Douglas bag method is well established but can be inconvenient and interfere with task performance because of the bulky nature of the equipment required. Recently, more compact instruments have been developed to facilitate the measurement of oxygen consumption of moving subjects. The oxylog, for example, measures the volume of inspired air as it passes through a turbine flow meter in a mask which is placed over the subject's nose and face. Some of the expired air passes through the oxylog itself, where

TABLE 8-1   ENERGY COSTS OF SOME COMMON DAILY ACTIVITIES

Activity	Energy cost (kcal/min)
Sitting	0.3
Kneeling	0.5
Squatting	0.5
Standing	0.6
Stooping	0.8
Walking unladen	2.1
Walking with a 10-kg load	3.6
Cycling at 16 km per hour	5.2
Climbing stairs (30 degrees of incline)	13.7

its oxygen content is measured. Harrison et al. (1982) report that the oxylog is sufficiently accurate to yield reliable estimates of oxygen uptake in the working environment and is a viable alternative to the Douglas bag.

Indirect Measures of Energy Expenditure

The rate at which the heart beats is known to increase as a function of workload and oxygen uptake. Because heart rate is easier to measure than oxygen uptake, it is often used as an indirect measurement of energy expenditure.

Heart rate can be likened to a signal which integrates the total stress on the body and can be used as an index of the physiological cost of work. However, as was described in the previous chapter, maximal oxygen uptake varies between individuals. Individuals can have the same rate but completely different levels of oxygen consumption. Thus, on its own, heart rate cannot be used to estimate the energy requirements of a job.

To evaluate physiological workload using heart rate, one must determine, for each worker, the relationship between heart rate and oxygen uptake. Both variables have to be measured simultaneously in the laboratory at a number of different submaximal workloads. This process is known as **calibrating** the heart rate–VO_2 relationship for a worker. Since the relationship between the two variables is linear, a worker's heart rate, when it is subsequently measured in the field, can be converted into an estimate of oxygen uptake by reference to the laboratory data. Estimates of energy expenditure during work can then be calculated from the oxygen-consumption data.

Several researchers have validated this method of using heart rate data to estimate the energy expenditure of previously calibrated subjects (Spurr et al., 1988; Ceesay et al., 1989) by comparing the heart rate estimates with estimates obtained using whole-body calorimetry (which is the "gold standard" for estimating energy expenditure). **Close correlations between these two methods of estimation suggest that heart rate measurement of previously calibrated subjects can give valid estimates of energy expenditure.**

Subjective Measures of Physical Effort

Physiological measures of workload all require instrumentation of one form or another. The idea of developing subjective measures of physical effort is worthwhile from a cost-benefit and practical viewpoint. The Borg RPE Scale (Borg, 1985) is a well-known rating scale for such measures (Table 8-2). Workers are asked to rate the level of exertion they perceive when carrying out a task on a scale from 6 to 20 (corresponding to minimum and maximum heart rates of 60 and 200). The ratings can be used in conjunction with objective measures. High positive correlations between heart rate and RPE are usually found.

It is more debatable whether such scales can be used entirely on their own. All subjective measurements are prone to distortion by what psychologists call experimenter effects—bias caused by the interviewee's or investigator's perception of the situation or of the other person. Recent research (Levine, 1991) in pain reporting has shown that not only are there gender differences in pain reporting but that male subjects report less pain to female than to male interviewers. Mital et al. (1993) investigated workload and fatigue in highly trained subjects. They found a discrepancy between ratings of perceived exertion and objective measures with the subjects underestimating the actual workload.

Subjective measures offer the investigator a convenient and simple pencil-and-paper instrument for measuring workload but should be used with caution and with objective measures of careful analysis of work demands.

APPLIED PHYSIOLOGY IN THE WORKPLACE

Physiological methods have been used to evaluate physical workload in a variety of jobs both in industrialized and developing countries.

Investigations of Forestry Work

Tomlinson and Mannenica (1977) compared physiological methods of evaluating the workload of forestry workers with the observational methods traditionally used in production engineering. They hypothesized that the physiological methods would be superior because:

1 They take into account the physiological capacities of individual workers and are therefore more valid.

2 They are sensitive to extra stress over and above that due to the work itself (e.g., stress due to heat or humidity).

3 They are more objective and more reliable since, if correctly used, they do not depend on the judgment of the person making the measurements.

The workload of a given task cannot always be inferred directly from measures of output. In unstructured manual work (such as forestry work) it is not always valid to assume that the cost of work will bear a one-to-one relationship with the observed rate of working or the amount produced because the **effort** involved in carrying out a particular task may vary depending on the particular circumstances. **Physiological methods** reflect the effort that the worker puts into the worksystem rather than the output of the sys-

TABLE 8-2 THE BORG RPE SCALE

Rating	Interpretation of rating
6	No exertion at all
7	
8	Extremely light
9	Very light
10	
11	Light
12	
13	Somewhat hard
14	
15	Hard
16	
17	Very hard
18	
19	Extremely hard
20	Maximal exertion

tem itself—they **are indexes of the effect of work on the worker rather than the effect of the worker on the output of the worksystem.**

In the field, direct measurement of oxygen uptake is not always possible because the apparatus (ace masks, air hoses, etc.) may encumber the worker. The researchers therefore calibrated the experimental subjects (a group of forestry workers) in the laboratory before taking measurements in the field. The subjects carried out a standard task in the laboratory at different submaximal levels while oxygen consumption and heart rate were measured simultaneously. For each subject, the oxygen consumption was measured over a range of heart rates. In the field, when only heart rate was measured, oxygen consumption could be estimated by extrapolation from the laboratory data.

Maximum oxygen consumption (VO_2 max) was estimated using the data on the heart rate–oxygen-consumption relationship. Each subject's maximum heart rate was estimated using this formula:

$$\text{Max heart rate (beats per minute)} = 200 - 0.65 \text{ age (years)}$$

Thus, a subject of 45 years of age would have a maximum heart rate of 170 beats per minute. From this number, a value of VO_2 max could be obtained by extrapolation from the laboratory heart rate–VO_2 data.

The researchers then measured the heart rates of the forestry workers while they were working—felling trees with a chain saw and trimming branches off the felled trees. The heart rate data were used to estimate oxygen consumption when workers were carrying out the work.

It is generally held that individuals can work at around 40 percent of their VO$_2$ max for 8 hours without suffering undue fatigue. Each subject's oxygen consumption when carrying out forestry work was compared with a value of 40 percent of his VO$_2$ max. It was concluded that the physiological cost to the workers was greater than that which is appropriate for an 8-hour day.

These findings indicated that:

1 The tasks should not be performed continuously without rest.

2 An appropriate work-rest cycle should be established so that the average workload (both work and rest) does not exceed 40 percent of VO$_2$ max.

The researchers then compared the physiological estimates of workload with work study officers' ratings of workload. These ratings were obtained partly on the basis of production requirements and partly on the basis of the work study officers' perceptions of worker effort. The findings indicated, first, a lack of consistency among the ratings of the work study officers themselves and, second, a lack of consistency between the physiological measures and the ratings—with the raters tending to underestimate the workload.

Subsequent research has confirmed that forestry work imposes a high workload on operators. Kukkonen-Harjula and Rauramaa (1984) measured the oxygen consumption and heart rates of lumberjacks in a variety of tasks. Oxygen consumptions of between 1.9 and 2.2 L per minute were observed. Even though the workers were found to have high values of VO$_2$ max, frequent rest pauses would be required.

Calculation of Rest Periods in Manual Work

Murrell (1965) proposed that rest periods can be calculated according to the empirical formula given below:

$$\text{Rest allowance} = w(b - s) / b - 0.3$$

where w = length of the working period
  b = oxygen uptake
  s = standard uptake for continuous work

We can use the forestry workers as an example: If a worker spends 0.5 hour felling a tree at an oxygen uptake of 2.64 L per minute and the standard is taken to be 1 L per minute, the rest allowance is given by:

$$\text{Rest allowance} = 0.5(2.64 - 1) / 2.64 - 0.3$$
$$= 0.35 \text{ hours}$$

Workload in Developing Countries

Developing countries rely very heavily on manual labor. Workload measurement is therefore important to identify unduly heavy tasks, to evaluate traditional work methods, and to arrive at more efficient methods of work. Numerous studies have been carried out in India. Sen et al. (1983) investigated the workload of tea leaf pluckers. They concluded that the leaf-plucking task itself was a light job in terms of energy expenditure and moderately light in terms of heart rate, although severe local fatigue of the finger and forearm muscles was common. A secondary carrying task was found to significantly add to the daily energy expenditure, and it was recommended that the amount of carrying of full baskets be reduced by increasing the number of tea leaf weighing stations.

The authors also evaluated the effectiveness of a hat to shade the worker from the sun. It was found to significantly reduce heart rate.

In another investigation, Datta et al. (1983) used the Douglas bag method to investigate the energy cost of pulling indigenous handcarts (Figure 8-2). Even pulling an empty handcart was found to constitute moderately heavy to heavy work. Loaded handcarts required very heavy and unduly heavy work.

Oberoi et al. (1983) compared the energy expenditure needed when washing clothes manually and when using a machine. Very few Indian households used a machine, according to the authors, and it was therefore appropriate to identify those situations in which high energy expenditure was required by the manual methods. Washing clothes in a squatting position was found to impose the highest energy expenditure; the lowest energy expenditures were recorded when subjects sat on a low stool or when they stood. The latter two work positions imposed no greater workload than that required to wash clothes by machine. Certain supposedly labor-saving domestic devices may, in fact, impose a higher workload than is readily appreciated because of the tasks required to set up and operate the device.

Some interesting studies of domestic work may be found in Grandjean (1973). An example of one of the many practical conclusions is that vacuum cleaners are a major labor-saving device (they save time and effort), whereas dishwashers save time but not effort (because of the added physiological cost of stooping to load and unload the dishwasher).

These studies illustrate some of the ways in which physiological methods can be used to evaluate and redesign manual tasks of all kinds. The methods can also be used to optimize the design of tools used to carry out such tasks. An interesting evaluation of the effects of blade size and weight in the design of shovels can be found in Frievalds and Kim (1990).

Evaluation of Nonphysical Stress

The above research involves the *direct* application of physiology to the analysis of work in situations where performing the work has *direct* physiological effects on the worker. These effects can be measured and used as indexes of physical workload. In developed countries, much industrial work is light—it has no direct consequences for the worker, at least none greater than many other activities of daily living carried out outside of work. However, physiological methods can also be applied to the investigation of light work, particularly to detect the presence of mental stress. It is known that the levels of several physiological variables, including heart rate, increase when a person is under mental stress.

Khaleque (1981) investigated heart rate, perceived effort, and job satisfaction in a group of female subjects in a cigar factory. The subjects were operators of cigar wrapping machines and the work involved simple, repetitive movements according to a 3-second work cycle paced by the machine. Machine-paced work such as this is generally held to be stressful, particularly in older subjects.

Job satisfaction was measured using a questionnaire (the Brayfield-Rothe scale) and perceived effort was measured using the Borg scale. Heart rate was measured using an optical pulse wave detector and appropriate circuitry. In this method, a clip is attached to the subject's ear lobe so that light from a small lamp on one side of the clip is detected by a photoelectric cell on the other side, the ear lobe being interposed between the lamp and the detector. With each pulse wave, the translucency of the ear lobe changes (as the flow of blood through the ear changes). This change is detected as a change in the current produced by the photoelectric cell.

The subjects were split into two groups according to their level of job satisfaction. The satisfied group were found to have an average heart rate of 81 beats per minute and the dissatisfied group, an average of 91 beats per minute. Perceived effort did not differ between the two groups, nor was there a relationship between perceived effort and heart rate. The author concluded that emotional factors probably played a role in elevating the heart rate of the dissatisfied workers. Job dissatisfaction is assumed to lead to negative emotional responses which increase heart rate. Clearly, ergonomists must not ignore the emotional needs of workers when they design jobs.

Physiological methods have long been used in the aviation industry and in military applications as indexes of mental stress. When a person is subjected to a high mental workload or to an emotionally intense experience, the level of cortical arousal (arousal of the cortex of the brain) is hypothesized to increase. There are many physiological correlates of the increased arousal, including increased heart rate, changes in the electrical resistance of the skin (the galvanic skin response), and changes in the concentration of certain hormones in the blood (particularly the stress hormones such as adrenalin). These physiological measures can be used to detect the presence of mental stress (some of these methods are reviewed in more detail in a later chapter).

Although many jobs require great concentration, it is not necessarily the case that they impose negative emotional stress on the worker. Becker et al. (1983) measured heart rate and oxygen consumption of surgeons while operating. The heart rate–oxygen-consumption relationship was measured using a cycle ergometer, and each surgeon's maximum oxygen uptake was predicted. Heart rate was also measured with the subjects at rest, when sleeping, and when doing ward and operating theater work. Oxygen consumption during operations was found to be low (0.26 to 0.41 L per minute). Heart rate during surgery was commensurate with the oxygen consumption and with the isometric arm and other work required during the performance of surgical tasks. The authors were therefore able to conclude that there was little evidence of increased heart rate due to mental stress.

Heart rate and heart rate variability have long been used to assess the workload of pilots. Jorna (1993) presented data on heart rate and heart rate variability of pilots undergoing conversion training to a new aircraft. When coming into land, novice pilots' heart rates increased steadily, but heart rate variability (instantaneous interbeat intervals) decreased. At the completion of training, the same pilots experienced lower heart rates when landing but greater heart rate variability. Pilots undergoing examination had very high heart rates and almost no heart rate variability. Roscoe (1993) concluded that, of the available physiological variables for assessing pilot workload, heart rate seems the most useful. Heart rate profiles of pilots approaching and landing at two airports differing in difficulty showed similar patterns of a steady increase during the approach followed by a decrease after landing. The overall heart rate was higher when landing at the "difficult" airport. A reduction in heart rate was found when pilots landed under autopilot rather than manually. Thus, heart rate and heart rate variability can be used to distinguish between different mental states bought about by task demands in tasks with low physical workloads.

Under stressful conditions, breathing exceeds metabolic requirements. According to Schleifer and Ley (1994), this causes a drop in the carbon dioxide concentration in the end-tidal expired air (the last air to be expired when one is breathing out). Stress-induced hyperventilation decreases the concentration of carbon dioxide in the blood, producing a number of physiological effects. Vasoconstriction of blood vessels reduces both the flow of blood to the heart and brain and the amount of oxygen that hemoglobin in the red blood cells can release to the tissues. This condition is accompanied by feelings of dizziness and heart palpitations. Theoretically, end-tidal carbon dioxide measurement should be a powerful way of measuring nonphysical stress. Unlike heart and ventilation rate measurements, which increase both under stress and during exercise, end-tidal pCO_2 measurements decrease when the person is under stress but not when physical work is carried out (breathing does not exceed physical work requirements under these conditions). Schleifer and Ley compared these stress indexes during self-relaxation, relaxation using a stress-management technique known as progressive relaxation, and during data-entry work at a VDT. As predicted, end-tidal pCO_2 more closely tracked self-reported mood states than the other variables. It seems to be a useful means of investigating the stress of VDT work and might be used as biofeedback in stress-management programs.

FITNESS FOR WORK

The motivation for the research cited above has been to fit the job to the worker—to ensure compatibility between the physical demands of the work and the physiological limitations of the worker. A related question is the assessment of a worker's fitness for doing the work at all.

Fitness is a subject about which even the experts disagree. Performance of a job depends on the level of skill and the motivation of the worker as well as on physical fitness. Definitions of fitness can be problematic. In sports and exercise physiology, fitness is clearly event-specific. The cardiovascular robustness of the marathon runner can be contrasted with the explosive power of the weight lifter and the poise and flexibility of the gymnast. Even concepts such as VO_2 max have an event-specific quality to them. Reilly (1991) points out that VO_2 measurements on athletes are best tailored to the particular endurance event. Long-distance runners are best tested on the treadmill; cyclists, on the bicycle ergometer; and athletes whose sport involves mainly upper body activity, on arm ergometers.

Fitness and Health

Fitness, according to Reilly, is a multivariate concept—although a distinction can be drawn between fitness and health. Medical doctors tend to pronounce a patient healthy if they can find no evidence of disease. In fact, many athletes, while having a high level of fitness for their chosen sport, may actually be very unhealthy and suffer from frequent bouts of colds and flu as well as musculoskeletal and other ailments (often a sign of an overenthusiastic training schedule). Conversely, few healthy people are fit to run a marathon or to lift heavy weights without risk of injury unless they have undergone a prior program of training.

In jobs which require prolonged manual labor, assessment of cardiovascular capacity can be used to determine a worker's level of fitness for work. VO_2 max can be measured directly or indirectly (from heart rate and oxygen consumption data at three or more submaximal workloads). Alternatively, one of a number of simple step tests can be used. These tests have been developed to enable fitness to be assessed objectively and without the need for sophisticated equipment. It appears to be the case that physically fitter individuals have a lower heart rate when resting and when working at a predetermined level than the less fit. Furthermore, their heart rate returns to its resting level more quickly after a period of exercise.

In the gold mining industry a step test has been developed in which subjects step on and off a step 24 times per minute for 9 minutes. The heart rate is then measured and the worker is placed into one of three categories. Category A men (heart rate less than 120 beats per minute) can be allocated to strenuous work, especially in hot conditions, whereas category B men (121 to 140 beats per minute) are given less arduous tasks and are not expected to work in hot environments. In several industries and occupations (e.g., fire fighting, mining) it is not possible to redesign the working environment or equipment so as to "fit the task to the men"; thus physical selection tests have a valuable role to play in ensuring compatibility between worker characteristics and task demands.

Physical Fitness and Everyday Life

Physical fitness is also influenced by the physical demands of everyday activities. Bandyopadhyay and Chattopadhyay (1981) used a step test in a study of the physical fitness of Indian college students. They found that the physical fitness score obtained using the test was higher in physically active than physically inactive students and that resting heart rate was negatively correlated with the fitness index (fitter people had lower heart rates). They also compared the fitness levels obtained in their study with those of studies carried out in North America, Africa, Japan, Australia, and Britain. The Indian subjects were found to be less fit, and the researchers suggested that nutritional and genetic factors might be the reason.

The relationship between physical fitness and customary daily activity appears to be well established. Ballal et al. (1982) investigated the physical condition of young adults in the Sudan. Soldiers, physical education students, and urban workers were found to be in better physical condition than rural villagers and medical students. The differences could not be attributed to body size, body composition, or sex. The authors suggested that the results were due to differences in the level of customary activity. The two most unfit groups were in other ways different: The villagers were smaller and less muscular than the other groups as a result of disadvantageous features of their environments. The medical students had a greater proportion of body fat and low levels of physical activity, possibly because of their more affluent lifestyle.

Workload, Physical Fitness, and Health

The goal of ergonomics in the design of the physical component of jobs is to minimize unnecessary and possibly harmful stress. However, this goal is not to be interpreted to mean that all stress is bad and should be minimized. There is a growing body of evidence suggesting that people who do physically demanding jobs are fitter and healthier than their less active colleagues. Conversely, the risk of fatal heart attack is greater in those with less demanding jobs. According to Paffenbarger (quoted from Noakes, 1992):

> Recognition of the role of exercise in health is changing as robots and automation now perform most of the laborious tasks that used to be done by muscle power. A cartoon in an American newspaper recently quipped, "Time was when most men who finished a day's work needed rest. Now they need exercise!" It is encouraging to see this new wisdom grow in popular acceptance.

In an early study of the relationship between occupation and fitness, Morris et al. (1953) compared the incidence of heart disease in bus conductors and bus drivers and in mailmen and sedentary postal clerks. In both groups, the incidence of heart disease was significantly lower in the active rather than the sedentary group. The bus drivers were more overweight, had higher blood pressure and cholesterol, and smoked more than the conductors. To complicate interpretation though, it appeared that these differences had existed before the individuals entered employment, suggesting that there was an element of **self-selection.** That is, people who are more active anyway choose active occupations and vice versa.

In a second study, Morris et al. (1973) investigated 16,882 British government employees. All of them had sedentary occupations and were comparable in terms of coronary risk factors. The group was then subdivided into those who did or did not participate in vigorous leisure-time activities. **The heart attack rate of the active group was one-third less than that of the inactive group.** Even smokers who exercised were less at risk than smokers who did not exercise.

In the United States, Paffenbarger et al. (1984) studied the risk of fatal heart attack in San Francisco longshoremen. Longshoremen who performed heavy manual labor were less at risk than less active colleagues. The risk of fatal heart attack was up to 50 percent lower in the most active compared with the inactive groups. Figure 8-3 summarizes these findings.

It seems fairly clear then that daily energy expenditure is an important factor in safeguarding the health of the population. Those in strenuous manual jobs will get some de-

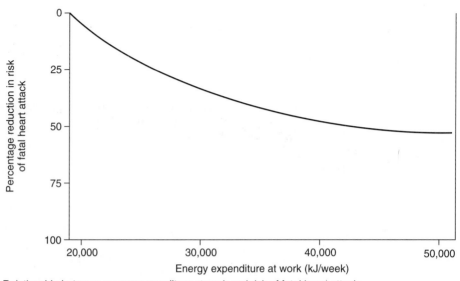

FIGURE 8-3   Relationship between energy expenditure at work and risk of fatal heart attack.

gree of protection by virtue of their work. However, the trend is for physical workload to be reduced, which suggests that Paffenbarger's view is valid. Table 8-3 gives the daily energy expenditure of a variety of occupations.

Exercise and Physical Fitness

Interest in the general fitness of the working population has increased over the last few years, and it has been suggested that a fitter workforce will be more productive and have lower absenteeism. This idea is based on the notion that increased fitness will "inoculate" workers against physical and nonphysical job demands by improving the ability to cope physiologically.

Hardman et al. (1989) investigated the effects of a program of brisk walking on a group of previously sedentary middle-aged women. The women followed a progressive program of brisk walking, building up to a pace of 1.87 m per second after 12 months. They averaged 16 to 17 kilometers (km) per week, which took about 155 minutes and was approximately 60 percent of their predicted VO_2 max. The program was found to enhance their tolerance for exercise and their metabolic and cardiovascular responses to it. Favorable changes in the ratio of total cholesterol to high-density lipoprotein cholesterol concentration were also observed when the walkers were compared with a control group. It would appear that for otherwise sedentary and inactive individuals even **regular brisk walking** may be beneficial. **There may be some justification for designing buildings and work environments to encourage this type of activity and even to attempt to incorporate it into otherwise low-activity jobs.**

Cox et al. (1981) evaluated the effects of an employee fitness program in two Canadian insurance companies (an experimental and a control company). Program partici-

TABLE 8-3   ENERGY COST OF SOME COMMON OCCUPATIONS*

Occupation	Kilocalories per day	
	Men	Women
Bookkeeper	2400	2000
Secretary	2700	2250
Bus driver	3000	2500
Letter carrier	3000	2500
Machine operator	3300	2750
Stonemason	3600	3000
Laborer	3900	3250
Carpenter	3900	3250
Ballet dancer	3900	3250
Coal miner	4200	
Lumberjack	4500	
Athlete	4800	4250

*From Woodson (1981).

pants in the experimental company exhibited increases in VO_2 max, a loss of body fat, and more positive attitudes toward their employment. Labor turnover and absenteeism were lower in the experimental than in the control company. The authors concluded that a potential 1 percent reduction in company payroll costs was possible through the implementation of the program, although the findings of the study could not be attributed directly to increased fitness.

Work-Hardening Programs

Work-hardening programs are a type of physical training program aimed at improving employees' physical capacity for a particular task or set of tasks, with the aim of reducing the risk of physical injury. According to Guo et al. (1992), to be effective, work-hardening programs must:

• Be of greater intensity than that experienced on the job.
• Be based on the techniques of exercise physiology described in the literature (e.g., Noakes, 1992).
• Use exercises which closely resemble the movements made on the job.
• Be evaluated using tests which simulate the activities carried out during training.

In many industries, work hardening takes place informally. Management frequently expect a break-in period of a few weeks before a new employee can work at the required pace. This is particularly true of jobs involving heavy or repetitive work.

Evidence from a variety of sources suggests that low back sufferers have less trunk flexor and extensor strength, diminished spinal extension, overpowered flexors, and tight hamstring muscles (e.g., Beiring-Sorensen, 1984; Pope et al., 1984). Animal and human studies have shown that endurance training causes an increase in muscle stiffness (the ratio of tension developed in a muscle when it is stretched to the amount it lengthens) as a result of increased cross-linking of collagen fibers in the muscle (Herbert, 1988). Animal studies have also shown an increase in the amount of stiffness and collagen in the muscles of elderly and sedentary rats. A sedentary lifestyle appears to reduce hamstring distensibility in the general population (Milne and Mireau, 1979). A lack of exercise combined with age-related changes in muscle structure may reduce flexibility, predisposing workers to injury. Badly designed exercise programs which increase endurance or strength but not flexibility may not have the expected beneficial effects. At the other extreme, it is known that joint hypermobility (which seems to be the result of hereditary factors rather than training in successful ballet dancers, according to Grahame and Jenkins, 1972), is a risk factor for joint dislocation and premature osteoarthrosis.

The three main components of a work-hardening program are **strength, flexibility, and endurance.** Depending on the requirements of the task in question, the program must aim to achieve an **appropriate balance** between these components. This is particularly important when the training is part of a rehabilitation program for previously injured workers.

Guo et al. (1992) compared the effects of flexibility and strength-flexibility training programs on middle-aged maintenance employees. Subjects trained 5 days per week for 30 minutes per day over 4 weeks. Improved physical capacity was obtained using flexibility exercises only by progressively increasing either the time spent in a particular posture or the number of repetitions of the exercise. Increases of 28 percent were found for truncal rotation and low back flexibility and 39 percent for dynamic strength. The strength-flexibility program resulted in increases of 52 percent for dynamic strength, 150 percent for endurance time, and 27 percent for low back flexibility. However, during a 4-week follow-up period in which the exercises were performed twice per week, most of these gains were lost. It would seem that once a degree of work hardening has been achieved, more regular training than twice-a-week sessions is needed to maintain it. Finally, it is thought that abdominal muscle exercises are of value in improving the stabilization of the lumbar spine and thus protecting workers from injury. Richardson et al. (1990) suggest that those exercises involving rotation seem best since they induce a pattern of muscle stimulation involving cocontraction of the lumbar muscles.

Maintaining Fitness at Work: VDT Users

Exercise programs for VDT users have been developed for use at work. Their aim is to be complementary to the design of the job so as to increase physical variety and thus counteract the stresses of static VDT work. A review of physical exercise for VDT users can be found in Lee et al. (1992). Office exercise programs must be usable. Since many VDT workers occupy open-plan offices, they may be reluctant to perform exercises which are conspicuous or embarrassing. Lee et al. found that many of the recommended exercises were inappropriate either because they actually reproduced the demands of VDT work, were contraindicated for physical conditions such as osteoporosis, or were dangerous. Any exercise which requires use of office furniture as a prop is potentially dangerous since the person or the equipment may not be as stable as expected.

The general principles of VDT exercise programs, according to Lee et al., are the following:

- Chronically shortened and tense muscles must be stretched and relaxed.
- The spine must be mobilized to reduce and redistribute compressive forces.
- Chronically stretched or relaxed muscles must be contracted or strengthened to maintain posture.
- Exercises must involve the lower limbs to promote venous return.

VDT users often sit in the upright, slumped posture described in a previous chapter. The neck is extended and there is static tension in the extensors. These muscles require stretching and relaxation in flexion. The scapular elevators are also chronically tensed and need stretching and relaxation. The lumbar and thoracic extensors are passively stretched and need activation—by actively extending the lumbar spine. The anterior thoracic muscles are relaxed and shortened and require stretching (possibly by grasping both hands behind the back). The forearm flexors are chronically tensed and shortened and require lengthening and relaxation.

CURRENT AND FUTURE RESEARCH DIRECTIONS

Escalating health care costs in developed countries have provided a stimulus for the investigation of factors influencing the health of the working population. Since work constitutes the major portion of an individual's waking life, it seems logical to evaluate the activities carried out at work and to suggest ways of incorporating health-promoting features into working life. Fitness programs and in-house gymnasiums are obvious examples.

Before the advent of the automobile and public transport systems it was common for people to walk 10 to 20 miles per day. Nowadays, many people probably walk less than this in a week. Further studies of the influence of activities of daily living on health and fitness would be of interest.

As is discussed in the Footnote to this chapter, the research issues which center on work and health in developing countries are quite different and much work remains to be done.

SUMMARY

Physical workload can be evaluated in terms of the physiological cost to the worker of carrying out the work. Oxygen consumption and heart rate are objective measures of workload. Since people differ in their capacity for aerobic work, it is appropriate to consider the relative workload of a task—an individual's oxygen consumption while working can be expressed as a percentage of VO_2 max. Workloads of 30 to 40 percent of VO_2 max can be sustained for an 8-hour shift. The task of the ergonomist is to design the work so that it can be performed safely and without undue fatigue by the largest possible number of people in the workforce. This task implies designing for individuals with lower aerobic capacity (such as older workers). If it is not possible to accommodate such individuals, a physiological approach is needed to specify a minimum standard of fitness for the job.

Heart rate is influenced by factors other than workload. This situation is both an advantage and a disadvantage for the experimenter. If heart rate is used to estimate oxygen consumption by the extrapolation method, erroneous estimates will be obtained if extraneous variables increase the heart rate over and above the level needed to supply oxygen to the tissues. Some factors which can cause such an elevation include mental stress, drinking tea or coffee, working directly after a heavy meal, and working in the heat. By the same token, heart rate may be of more use as an index of the total workload because it *is* affected by variables such as mental stress and heat, which do increase the load on the worker. It is in this sense that heart rate integrates the total stress of a job.

Despite the global trend to replace manual work with mechanized and automated systems, physiological aspects of work design are of great importance in view of the modern interest in eliminating avoidable ill health. In low-technology jobs the problems of avoiding excessive physical stress remain. In high-technology jobs, the problem is to avoid mental overstress and find ways of increasing the physical activity of the predominantly sedentary workers.

ESSAYS AND EXERCISES

1 Try out the step test described in the chapter on yourself. Calculate your heart rate, measured immediately after the test, in beats per minute. The height of the step is determined as follows:

$$\text{Step height (cm)} = \frac{542 \times 2.54}{\text{Your body mass (kg)}}$$

Step for 9 minutes at a rate of 24 steps per minute. Which fitness category do you fall into?

2 Choose one of the activities below. Lie down for 5 minutes prior to starting the activity; then take your resting heart rate. Carry out the activity for 10 minutes. Measure your pulse for 30 seconds immediately afterward to estimate your heart rate. Use the Borg scale to rate the level of exertion you perceive.

a Jogging
b Walking up a flight of stairs briskly
c Doing squats
d Raising and lowering a 20-kg load from knee height to chest height at a rate of 1 lifting cycle every 5 seconds
e Standing still
f Sitting
g Walking
h Walking the same distance with a 10-kg weight in each hand
i Walking the same distance with a 20-kg backpack
j Walking the same distance with a 20-kg load on your head

Plot the heart rate data (difference between resting and final heart rate) against the ratings of perceived exertion. What conclusions can you draw? Try the experiment again with one or more of the other activities.

3 Distinguish between health and fitness. What measures can organizations take to improve the health and fitness of their employees?

FOOTNOTE: Nutrition, Work Capacity, and Health

At the most extreme levels, the relationship between nutrition, work capacity, and health is a simple one—without adequate food, work capacity will be diminished and the individual will weaken as body mass is lost. Premature death will eventually result. At the other extreme, endurance athletes can improve their performance by nutritional techniques such as carbohydrate loading. Between these two extremes lies a buffer zone where questions about diet and health and the minimum nutritional requirements for a healthy life are still being debated.

The term "malnutrition" refers to problems of overeating and incorrect eating habits as well as to undernutrition. Malnutrition is found in both developed and less developed countries. In developed countries, apart from specific entities such as anorexia nervosa, most of the attention has been paid to problems of excessive food intake or of excessive intake of certain types of food. Malnutrition in these societies is a cofactor (together with other lifestyle-related behaviors such as tobacco smoking) in the causation of heart disease. It is in this way, through its effects on general health, that nutrition can influence the work capacity of people in developed countries.

The nutritional status of most workers in developed countries is usually appropriate to meet the demands of their jobs. Nutritionists in developed countries have developed the concepts of a balanced diet and of recommended daily (or dietary) allowances (RDAs) of protein, fats, and carbohydrates, as well as vitamins and trace elements. However, some controversy remains as to whether these RDAs are essential to maintain health. Bull (1988) states that there are many myths surrounding the specification of diets for sportsmen and sportswomen. The Nutrition Committee of the Canadian Paediatric Society suggests that athletes of all ages require only a normal, balanced diet. Vitamin and mineral supplements are unnecessary since any additional nutrients will be obtained from the extra food eaten to meet the energy requirements of hard training. Among other things, this viewpoint suggests that the enormous quantities of protein eaten by bodybuilders are unnecessary. It can be concluded then that if people are able to eat enough food to meet their calorie requirements, they will almost certainly meet their requirements for other nutrients as well (some well-understood exceptions to this generalization are documented below).

A major problem with the diet eaten by people in industrialized societies, according to Shorland (1988), is nutritional distortion caused by the removal of certain components of plant and animal tissue and the addition of other components (e.g., the removal of bran and other residues to make white flour and the addition of salt to food—often up to 10 times bodily needs).

Banerji (1988), in a review of nutritional problems in third world countries, concluded that the most serious problem was a **deficiency of calories.** Many people in third world countries simply do not get sufficient food to meet their calorie requirements. Under these circumstances, the body uses protein as a source of energy rather than for its main function, which is to maintain body tissue; thus it may be mistakenly concluded that the person is suffering from a dietary deficiency of protein rather than carbohydrate. It is for this reason that a diet sufficient in carbohydrates and/or fats is said to be **protein sparing.**

Inadequate energy intake also leads to fatigue and apathy. The lassitude which results from a calorie-deficient diet will be likely to deprive children of those learning experiences which result from active interaction with the environment. On the other hand, Shorland (1988) suggests that adults are able to satisfy their daily protein requirements as a consequence of satisfying their calorie requirements, i.e., mainly from bread (an average person requires only 30 to 55 grams of protein per day).

NUTRITION, ACCIDENTS, AND FATIGUE

Acute or chronic undernutrition is often a cause of low blood sugar and is accompanied by feelings of tiredness and irritability. Pheasant (1991) has reviewed the evidence for a connection between low blood sugar, low performance, and increased accidents. Studies have shown poorer performance in workers who skip breakfast than those who don't. Midmorning and afternoon snacks seem to help workers sustain performance. Blood sugar is often found to be low in victims of freeway automobile accidents. Other studies have shown that foundry workers have fewer accidents if given a high-energy glucose drink before work. Pheasant concludes that for maximum work efficiency, calorie intake should be spread over the day to include a light breakfast, midmorning and afternoon snacks, lunch, and dinner.

Designers of production facilities in developing countries, where the nutritional status of workers may be poor, might make allowance for this in the design of on-site facilities, particularly if shift work is involved. High-calorie meals and snacks acceptable to local tastes should be provided at no or minimum cost during plant operation.

DIETS AND DIETARY DEFICIENCY

Banerji (1988) states that diseases due to a deficiency of one or other nutrient are rare and occur only in extreme circumstances (e.g., sailors on sailing ships used to get scurvy during long voyages as a result of vitamin C deficiency). If people are able to eat enough food to satisfy their hunger and to meet their daily calorie requirements, they will almost certainly get sufficient protein, fats, vitamins, and minerals. Cassava, which is grown in certain parts of Africa, is one of the few protein-deficient subsistence crops. Corn meal is deficient in one of the essential amino acids (those amino acids which cannot be synthesized in sufficient quantity in the body). In parts of the world where people subsist primarily on corn meal, the protein-deficiency disease **kwashiorkor** is common. Children are particularly prone to this disease, which is characterized by failure to grow, lethargy, and impaired mental abilities.

FOOD SUPPLEMENTS AND WORK CAPACITY

The problem facing people in many third world countries is that they are not able to satisfy their calorie requirements and are forced to subsist at extremely low levels. A number of researchers have investigated the effects of food supplementation on the work output of malnourished workers. Diaz et al. (1989) investigated the work performance of a group of Gambian laborers over a 12-week period during a time of natural food shortage (the wet season). The laborers were split into two groups, one receiving food supplements for the first 6 weeks and the other receiving supplements during the last 6 weeks. Both groups gained weight during the supplementation period and lost weight when there was no supplementation. However, food supplementation had no effect on worker productivity despite the negative energy balance of unsupplemented workers. The fact that the workers were paid on a piecework basis may explain the constant productivity, which was maintained even at the expense of body weight loss. In such harsh situations as these, workers may maintain their level of output at work but reduce the energy devoted to leisure activities to compensate. When this happens, one of the costs of work is reduced leisure activity.

WHAT IS GOOD NUTRITION?

Much remains to be learned about human nutrition and what constitutes a "good diet." Shorland (1988) described cross-cultural studies of diet, health, and longevity in a review of the present state of knowledge about nutrition. Human populations appear to be able to grow on a bewildering variety of diets—some of which differ markedly from others. For example, evidence from the bone remains in caves occupied by Homo erectus indicate that meat made up some 70 percent of the diet (mainly venison, but with the addition of tiger, buffalo, and rhinoceros). Domesticated animals contain a much higher level of fat in the carcass than do these wild animals. Although modern Eskimos have subsisted almost entirely on marine mammals and fish, the fat content of these animals is also markedly different from that of domesticated animals. In ancient Greece and Rome, people subsisted mainly on barley bread and gruel made from barley or millet, with small amounts of olives, figs, cheese, and beans. Bread made from barley, rye, or wheat seems to have been a staple food throughout most of European history. The western "meat-milk" diet appears to have evolved only during the last 100 years as the amount of bread consumed daily has been reduced and replaced with meat and milk, which contains levels of certain types of saturated fat not found in wild animals or in plants. Although some Pacific atoll populations subsist mainly on coconuts, which are rich in saturated fatty acids, these are of a different type from the fatty acids of the western meat-milk diet (lauric as opposed to palmitic acid).

Shorland's (1988) view is that the current state of knowledge of human nutrition is not soundly based. This view is echoed by Labadarios (1992), who suggests that nutrition science has "seriously neglected the lessons to be learnt from studying people who continuously remain healthy, despite consuming apparently outwardly inadequate food intake." The "supernutrition" of western society is said to have hastened maturity and body size such that the young are now ready for adult study and work at an earlier age. This view contrasts with other research suggesting that dietary restriction (in all species tested) increases longevity. The cost of accelerated growth may therefore be premature aging.

9

HEAT, COLD, AND THE DESIGN OF THE PHYSICAL ENVIRONMENT

Chemical transformations used in basal metabolism and muscular work always produce heat. One of the most important developments in animal evolution has been the emergence of warm-blooded animals—animals which use this heat to maintain body temperature within a narrow range. Because the rates of most biochemical reactions are temperature-dependent, it has proved advantageous from an evolutionary point of view to control them by means of a thermoregulatory system.

FUNDAMENTALS OF HUMAN THERMOREGULATION

Humans have a remarkably well-adapted ability to tolerate heat compared with other primates. This statement applies equally to Eskimos as to tropical rain forest dwellers despite any small differences. This is because humans are hairless and have a large proportion of high-capacity sweat glands—known as eccrine glands—in their skin.

Thermal Balance

Thermoregulation is achieved by balancing the two main factors which determine body temperature—the metabolic heat produced and the rate of heat loss. In humans, the thermoregulatory goal is to maintain the temperature of the deep body tissues (the core temperature) within a narrow range, around 36 to 37°C. Core temperatures over 39.5°C are disabling and over 42°, they are usually fatal. The lower acceptable limit is 35.5° and 33° marks the onset of cardiac disturbances. Further drops in core temperature are extremely dangerous, and temperatures as low as 25° are fatal. The temperature of the peripheral body tissues, particularly the skin, can safely vary over a much wider range. From a thermal point of view, the body can be considered to have a warm core where much of its heat is produced. This core is surrounded by a shell of cooler, insulating tissues, particularly subcutaneous fat. It is believed that the hypothalamus is involved in the central control of core temperature.

The principal sources of heat production in the body are the liver and intestines, the brain, the heart, and the working muscles. Muscular work is a major source of heat because the mechanical efficiency of muscles is only about 20 percent. Little heat transfer with the environment takes place via conduction. The body tissues are poor heat conductors. Heat is transferred to the skin from the deep tissues by convection. Blood is an ideal medium of heat convection since its specific heat capacity and thermal conductivity are high. Core temperature can be maintained within its narrow range of values only if there is a state of heat balance with the environment. The fundamental thermodynamic processes involved in heat exchange are described by the equation given in Table 9-1.

Metabolic heat production can be measured directly using a method known as **calorimetry.** A resting person is placed in a sealed chamber, the temperature of which is maintained at a constant level by passing the air in the chamber through a cooling system (pipes in a water bath). The increase in temperature of the water in the bath is directly related to the metabolic heat production of the person in the sealed chamber. Direct calorimetry is suitable only for research purposes; another method, known as indirect calorimetry, is often used in its place. Because metabolic processes for energy production and basal metabolism require oxygen, the oxygen uptake of a person at rest can be measured and used to estimate metabolic energy production (M). The energy equivalent of metabolizing 1 L of oxygen is approximately 4.8 Calories (although it varies depending on whether glucose, fat, or protein is being used as fuel). However, if the person is working or exercising, not all the oxygen consumed will give rise to heat—some will be used to carry out the work. If the amount of physical work (W) can also be measured, then the heat gained as a result of metabolic processes can be calculated (from the total metabolic energy production less the energy used to carry out the work).

Heat may also be gained from the environment or lost to the environment by convection (C) or radiation (R). If the skin of a resting person is warmer than the surrounding air, the air at the skin surface is warmed by the conduction of heat from the skin.

TABLE 9-1   BASIC EQUATION OF HUMAN THERMAL BALANCE

$$S = M - E \pm R \pm C \pm K - W$$

where S = heat gained or lost by the body
            0 when the body is in thermal balance with the environment
       M = metabolic energy production
       E = heat dissipated through evaporation (sweating)
       R = radiant heat to or from the environment
       C = convection to or from the environment
       K = conduction to or from the environment
       W = work accomplished by the worker

Since warm air rises, a flow of air around the person is established and heat is convected away from the person's body. Cool air moves closer to the skin to replace the rising warm air. The air flow is known as a convection current. In practice, a resting person has a layer of slowly moving warm air at the surface of the skin. Movement of one's body or of wind can disrupt this warm layer and increase the rate of heat loss. Depending on the temperature of the air, this heat loss will be perceived as a chilly draft or a cooling breeze.

All objects at temperatures greater than absolute zero emit infrared radiation, which is sometimes referred to as radiant heat. The human body is no exception. If its temperature is greater than that of its surroundings, a radiant heat loss takes place. If the surroundings are hotter than the body, it experiences a net heat gain. Inside buildings, the walls and ceiling are common sources of radiant heat, whereas the sun is the major source out-of-doors. In steel and glass manufacture, furnaces are a major source of radiant heat.

Sweat production and evaporation (E) are a mechanism by which heat is lost to the environment. Since the body tissues are composed largely of water, it is unsurprising that water is lost to the environment by diffusion from the skin, the lungs, and the sweat glands. If the temperature of the surroundings is greater than that of the body, no heat can be lost by convection or radiation, so evaporative heat loss by sweating is essential to maintain thermal balance.

When the body is in a state of thermal balance, the terms on the right-hand side of the Table 9-1 equation cancel each other out. This situation can occur in several ways. For example, in a cold environment, a state of thermal balance can be achieved by either increasing the rate of metabolic heat production (by shivering or by carrying out some fairly strenuous activity) or reducing the heat loss by convection and radiation (by wearing more clothes). Clothing reduces heat loss to the environment by trapping air in the clothing fibers between the garment and the skin. This trapping disrupts the flow of convection currents and maintains a layer of warm air around the skin surface. Clothing also absorbs some of the radiant heat from the body. Both of these effects reduce the temperature gradient between the skin and its immediate surroundings and, therefore, the rate of heat loss. Water practically eliminates the insulating property of clothing because of its high specific heat capacity. In hot conditions (working in a blast furnace, for example) there is a high rate of metabolic heat production as well as heat gain from the environment (by radiation). A state of thermal balance can be obtained by either reducing the rate of metabolic heat production (by reducing the work rate or introducing rest pauses) or protecting the worker from the radiant heat (by means of a protective suit, for example).

The design of clothing assemblies for work in extreme environments is more problematic than it at first might seem. A clothing assembly for cold work, if it traps air too well, may trap sweat as well. If the sweat accumulates, the insulation provided by the clothing will be degraded. Suits to protect workers from radiant heat may also trap air around the worker's body. Since the rate of evaporation of sweat depends on the humidity of the air, evaporative heat loss may steadily decline as the trapped air becomes more saturated with water vapor and less sweat evaporates.

MEASURING THE THERMAL ENVIRONMENT

Dry-Bulb Temperature

In many homes, it is common to see a thermometer placed on the wall of the entrance hall or living room. Most people are familiar with these mercury-in-glass thermometers and with their use in measuring the temperature of the air. These thermometers measure what is known as the dry-bulb (DB) temperature—the temperature of the constituent gases which make up the air. Although dry-bulb temperature indicates the thermal state of the air, other factors have an equally important effect on the heat gained or lost by a worker.

Relative Humidity and Wet-Bulb Temperature

One of the most important factors is the humidity, or water vapor content of the air. At all temperatures above freezing, water tends to evaporate into the air, and the rate of evaporation increases with temperature. At any given temperature, however, a point is reached where no more water vapor can evaporate into the air. At this point, the air is said to be saturated (with water vapor). When the water vapor pressure of the air reaches its maximum for a given temperature, evaporation from a wet object will cease. However, if the air is heated, further evaporation from the wet object will resume—the higher the temperature of the air, the move water vapor it can contain. Thus, saturated water vapor pressure increases with temperature. Conversely, if warm air is cooled, a point will be reached where the water vapor pressure at the original temperature is equal to the saturated water vapor pressure at the new (lower) temperature—this point is known as the dew point, and it is at the dew point that water condenses out of the air.

"Relative humidity" is a term used to describe the water vapor pressure of the air at a given temperature. It is the water vapor pressure at a given temperature expressed as a percentage of the saturated water vapor pressure at that temperature. Figure 9-1 illustrates these terms diagrammatically.

It is common in ergonomics to take three separate measurements of the air temperature. Apart from the dry-bulb temperature, measurements of wet-bulb (WB) temperature and globe temperature (GT) are also taken. Wet-bulb temperature is traditionally measured with a mercury-in-glass thermometer. A wet cloth "sock" is placed over the bulb of the thermometer and the measurement is made after allowing the thermometer and sock to stabilize at the ambient temperature. The wet-bulb temperature depends not only on the dry-bulb temperature but also on the relative humidity of the air. Evaporation of water from the sock cools the bulb of the thermometer. The rate of evaporative cooling depends on the humidity of the surrounding air. The wet-bulb and dry-bulb temperatures can be used to calculate the relative humidity. This calculation is normally done using psychometric charts (e.g., Kerslake, 1982).

When the air is completely saturated (i.e., the relative humidity is 100 percent), no water will evaporate from the bulb of the wet-bulb thermometer, no evaporative heat loss will take place at the bulb, and the wet- and dry-bulb temperatures will be the same. Otherwise, wet-bulb temperature is lower than dry-bulb temperature.

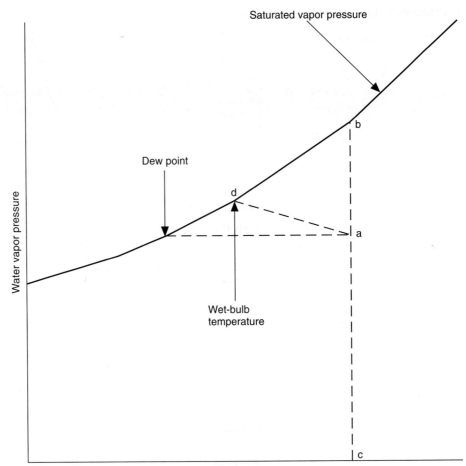

FIGURE 9-1   Relationship between temperature, vapor pressure, dew point, and wet-bulb temperature. The dew point is the temperature at which the vapor pressure at (a) would equal the saturated vapor pressure. Relative humidity at (a) is given by ac/bc. The wet-bulb temperature is the temperature which would be reached if water is evaporated into the air. (From Kerslake, 1982. © Cambridge University Press, 1982. Reprinted with the permission of Cambridge University Press.)

Globe Temperature

Another important measurement of temperature accounts for the effects of radiant heat. It is known as globe temperature. Traditionally, the bulb of a mercury-in-glass thermometer is placed in a metal sphere which is painted matte black. Any radiant heat (from the sun or from hot objects) is absorbed by the sphere and heats up the thermometer. Radiant heat can prove to be a significant source of heat load on the worker (consider the beneficial effects of shade on a hot, sunny day). In many situations, measurements of globe temperature are essential if the true nature of the thermal environment is to be evaluated.

Air Movement and Wind Chill

In an evaluation of the effects of temperature on the worker, it is important to take air movement into account. Air movement moderates the effects of high temperatures and exacerbates the problems of low temperatures (where it causes wind chill). Air movement can be measured using mechanical anemometers (in which the rate of rotation of a vane is proportional to the velocity of the air flowing through it) or by electrical means such as the hot-wire anemometer, where air movement cools a heated wire. The amount of cooling is proportional to the velocity of the air flow.

Heat Stress

Heat stress may be defined as the combination of all those factors both climatic and non-climatic which lead to convective or radiant heat gains by the body or prevent heat dissipation from the body (Leithead and Lind, 1964).

Modern heat stress monitors make use of thermistors (electrical transducers) instead of mercury-in-glass thermometers. Typically, dry-bulb, wet-bulb, and globe temperatures can be measured by the same instrument, and air movement can be taken into account. These devices often provide combined measurements of temperature known as heat stress indexes. The wet-bulb globe temperature (WBGT) is a commonly used heat stress index. Commercial heat stress monitors calculate the WBGT index as follows (for measurements made outdoors and indoors, respectively):

$$\text{WBGT (outdoors)} = 0.7\,\text{WB} + 0.2\,\text{GT} + 0.1\,\text{DB}$$

$$\text{WBGT (indoors)} = 0.7\,\text{WB} + 0.3\,\text{GT}$$

It should be noted that the calculation of the index of heat stress depends very heavily on the wet-bulb temperature. This fact also demonstrates why simple measurements of dry-bulb temperature alone cannot provide an accurate measure of the stress involved in working in a hot environment. Figure 9-2 shows modern instrumentation for measuring heat stress. The reader is referred to Kerslake (1982) for further discussion of the uses and limitations of various heat stress indexes.

THERMOREGULATORY MECHANISMS

A number of physiological mechanisms exist for maintaining heat balance in the human.

Peripheral Vasomotor Tone

Heat production and heat loss can be balanced within a fairly narrow range of skin temperatures by adjustment of peripheral vasomotor tone in the skin circulation. In hot environments, peripheral vasodilation occurs in the skin circulation. The arterioles dilate and capillaries at the skin surface open, blood flow increases, and heat is conducted to the skin surface over a small distance, from where it is dissipated to the environment. In the cold, vasoconstriction of skin arterioles occurs, reducing blood flow in the cutaneous

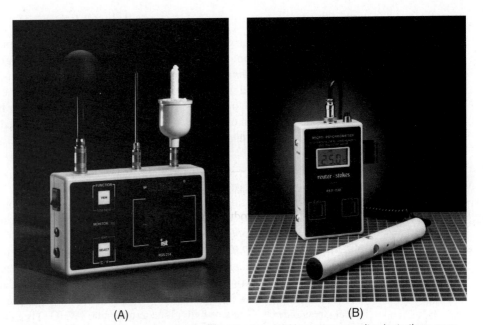

(A)   (B)

FIGURE 9-2   Modern instruments for thermal environment evaluation. (A) Heat stress monitor (note three separate sensors for measuring dry-bulb, wet-bulb, and globe temperatures). (B) Micropsychrometer, which gives direct readings of wet-bulb and dry-bulb temperatures, relative humidity, and dew point. (Photographs supplied by Imaging and Sensing Technology, Ontarior, Canada.)

circulation. This increases the conductive distance, thereby increasing the insulation of deeper body tissues. Less heat is lost at the skin surface.

Hence, the skin can function both as an insulator and as a radiator of heat, depending on the peripheral vasomotor tone. An indication of the effectiveness of this system can be obtained by comparing its insulating capacity with that of clothes. The CLO is a common unit of clothing insulation. A nude person may be said to have a thermal insulation of zero CLO, whereas a person wearing a business suit (plus shirt, underwear, etc.) has 1 CLO (Parsons, 1993). Using this terminology, one can describe the insulation provided by the human vasomotor system as being effective over the range of approximately 0.2 to 0.8 CLO.

Countercurrent Heat Exchange

Countercurrent heat exchange is essentially a method of conserving heat. It involves the exchange of heat between arteries and veins supplying the deep tissues. Arterial blood is precooled before it reaches the extremities, and venous blood is warmed before it returns to the vital organs. The efficiency of this system depends on the anatomical distribution of veins and arteries in the various bodily structures. Some animals have very impressive heat-exchange systems. An efficient heat-exchange system enables penguins, for example, to spend long hours standing on ice at many degrees below freezing and to maintain a very high temperature gradient between their deep body tissues and the soles of their feet.

Sweating

Humans have about 1.6 million high-capacity (eccrine) sweat glands on their skin. Eccrine glands are found only on the hairless regions of other mammals, who sweat principally through low-capacity (apocrine) glands which are associated with hair follicles. This is why humans can lose about 500 grams (g) of sweat per square meter of skin, whereas horses and camels can lose only 100 and 240 g, respectively (Hanna and Brown, 1979). Heat tolerance is well developed in all human populations because of their high-capacity sweat glands.

Sweat, which is produced by the eccrine glands in the skin, consists mainly of water. It is a dilute solution of various electrolytes, principally sodium, potassium, and chloride. Passive diffusion of water through the skin occurs most of the time, as does nonthermal sweating, or loss of water through the sweat glands. This type of sweating is referred to as nonsensible sweating because it is not accompanied by any sensation of sweating (people are not aware that they are sweating). Sweating becomes sensible when the rate of sweat production exceeds approximately 100 g per hour. Sweat rates of 1 to 2 kg per hour are possible. Sweat cools the body when it evaporates (because of the latent heat of vaporization of water). However, in humid environments, evaporation of sweat diminishes and cooling efficiency is lost even though sweat continues to be produced. Another problem in humid environments is a kind of **reverse sweating** in which atmospheric water vapor condenses onto the skin, releasing its latent heat of condensation and warming the skin even further.

Profuse sweating has two important disadvantages:

- Dehydration may occur if more water is lost than is replaced.
- Salt may be lost.

The adrenocortical hormone, aldosterone, is involved in the conservation of sodium, causing it to be actively reabsorbed in the sweat glands and in the kidneys. As the rate of sweat production increases, less sodium ions can be reabsorbed.

People do not voluntarily replace all the lost water when working in the heat, even if an ample supply of fluid is available. In endurance events such as marathon running, successful athletes learn to replace lost water in an appropriate way. **In hot industrial work, plenty of fluid must be made readily available at all times and workers must be encouraged to regularly replace lost fluid to avoid both dehydration and overhydration.**

Shivering

Involuntary shivering is a thermoregulatory mechanism involving active heat production. Groups of motor units act out of phase with one another, and muscles act against their antagonists. Almost no movement occurs, and the result is a high level of heat production.

Voluntary movement also increases heat production but breaks up the insulating layers of air around the body, which increases the rate of heat loss. Both shivering and voluntary movement increase oxygen uptake and cardiac output and can lower a person's capacity to carry out physical work.

WORK IN HOT CLIMATES

Heavy physical work in the heat imposes **conflicting demands on the cardiovascular system.** Peripheral vasodilation requires an increase in blood flow to the skin (up to 10 L per minute). However, working muscle also demands increased blood supply compared with relaxed muscle (up to 25 L per minute). Up to a point, blood flow to the skin and muscles can be increased by diverting blood from the viscera, but this reserve capacity is limited. Since cardiac output cannot exceed venous return and the average maximum cardiac output is about 25 L per minute, cardiac capacity is a limiting factor for muscular work in the heat (Kroemer, 1991). The cardiovascular system may be placed under considerable strain when a person is working in the heat as output rises to meet the demands of both physical work and bodily cooling. A dangerous condition can arise if the heart is no longer able to meet both demands.

Heat Stroke

If a worker becomes dehydrated, sweat production diminishes and the deep body (core) temperature may increase. Rapid elevations in body temperature increase the metabolic rate. Since heat is a by-product of metabolism, the heat load is exacerbated and a **positive feedback situation** may develop. In extreme situations, the thermoregulatory system may be unable to cope. If core temperature rises above 42°C, blood pressure may drop and insufficient blood is pumped to the vital organs, including the heart, kidney, and brain. Under these conditions a worker will collapse with **heat stroke.** Heat stroke (thermoplegia) is a dangerous condition caused by exposure to excessive heat. Sweating stops and the skin becomes dry. The victim experiences dizziness, headache, nausea, muscle cramps, and a dangerously high core temperature. Since the core temperature control system is located in the brain, the elevated temperature may disrupt the thermoregulatory system itself, and even if the worker stops working, core temperature may not return to normal. Rapid cooling of the body using cold sprays or wet sponges can be used to lower the temperature. Many industries take steps to prevent such dangerous situations from occurring, as is described below.

Relative Humidity

Whether workers can tolerate a hot environment depends on several classes of variables (Table 9-2). For example, dry-bulb temperatures of more than 38°C can be tolerated if the relative humidity is less than 20 percent because at such low humidities, the cooling efficiency of sweating is high. However, a relative humidity of 90 percent at 32°C is tolerable only with air movement and if a low level of work activity is required. Work in hot environments can be made more tolerable by introducing job aids or rest pauses to

TABLE 9-2   SOME FACTORS INFLUENCING THE ABILITY TO WORK IN THE HEAT

1 The characteristics of the worker
   Physiological heat tolerance
   Age
   Aerobic capacity
   Degree of acclimatization
2 The thermal environment
   Relative humidity
   Globe temperature (accounts for the effects of radiant heat and shade)
   Wind speed
3 The requirements of the task
   Work rate
   Provision of rest pauses
   Provision of protective clothing

reduce metabolic heat production or by periodically removing the worker from the hot environment. Finally, workers differ in their ability to tolerate heat stress. At one extreme, hyper-heat-tolerant individuals can work in hot environment without the need for prior acclimatization. Heat-intolerant individuals are never able to work safely in the heat. Identifying such individuals is an important part of worker selection in industries such as deep underground mining.

Heat Acclimatization

Heat acclimatization is a physiological process of adaptation rather than a psychological adjustment to life in a hot environment. It involves an increase in the capacity to produce sweat and a decrease in the core temperature threshold value for the initiation of sweating. The maximum rate of sweat production can double from 1 L per hour in an unacclimatized person to 2 L per hour in an acclimatized person. A state of acclimatization is best achieved by exercising in the heat and drinking plenty of fluid.

Increased sweat production enhances evaporative cooling of the skin and thus improves heat transfer from the deep body tissues to the periphery. The risk of dehydration and salt depletion is reduced in an acclimatized person as a result of expanded blood volume and a reduction in the salt concentration of the sweat. **Furthermore, acclimatization reduces the skin's blood-flow requirements, which reduces the cardiovascular load during work in the heat.**

Heat acclimatization occurs naturally, but it may also be induced artificially. In deep mines, manual work has to be carried out in wet-bulb temperatures up to 33°C. In some mines, workers are artificially acclimatized to heat as a requirement of their employment. Surface acclimatization chambers are used and workers exercise (by performing step-ups) in temperatures of 31.5° wet-bulb and 33.5° dry-bulb. Rectal temperatures are monitored and rest periods given in response to temperature increases over 38.3°. Other methods of acclimatization include exercise while wearing a vapor barrier suit (which prevents sweat evaporation) and resting in baths at 41°C.

An individual's ability to produce sweat depends on the climate experienced in the early years of life (Diamond, 1991). It appears that the sweat reflex develops in the first 2.5 years of life (Hanna and Brown, 1979). Although we are all born with a similar number of sweat glands in the skin, the number which develop, or are "switched on," depends on the conditions experienced early in life. People who grew up in a hot environment develop a larger number of functional sweat glands than those who grew up in a cold environment and can therefore tolerate hot conditions more easily.

Although the body can acclimatize to heat, it cannot acclimatize to dehydration. Fluid must be made available at all times even for acclimatized workers.

Personal Factors Affecting Heat Tolerance

Apart from acclimatization, several personal factors can influence a worker's ability to tolerate hot conditions (see Table 9-2).

Age The very young and very old are less tolerant than other age groups. Young children have less sweating capacity than adults. Older men are less able to tolerate high heat stress partly because of their higher skin temperature threshold for the onset of sweating. The increased mortality of the elderly during heat waves is thought to be due to increased cardiovascular load rather than heat stress.

Sex There are no qualitative differences between men and women in their response or acclimatization to heat. There is some evidence that women begin sweating at a higher skin temperature and sweat less than men. Women have a higher proportion of body fat than men, which may also play a role (see below).

Physical Fitness Physical fitness improves heat tolerance because both rely on cardiovascular function and sweat production. Physically fit workers are less stressed by hot conditions even if they are accustomed to a temperate climate.

Body Fat Excess body fat degrades heat tolerance by increasing the mass-to-surface area ratio of the body and reducing cardiovascular fitness. Also, adipose (fat) tissue contains less water than other tissue and has a lower specific heat capacity. Thus, the same heat load will cause a greater temperature increase in obese compared with lean individuals.

Heat Stress Management

Some approaches to the elimination of heat stress are summarized in Table 9-3. Table 9-4 gives threshold WBGT values at which these precautions should be implemented.

TABLE 9-3   SOME BASIC STEPS IN HEAT STRESS MANAGEMENT

1   Reduce high relative humidity by using dehumidifiers.

2   Increase air movement by using fans or air conditioners

3   Suggest that workers remove heavy clothing; issue loose-fitting overalls.

4   Reduce the work rate.

5   Include frequent rest pauses.

6   Introduce job rotation.

7   Carry out outdoor work at cooler times of the day (e.g., early morning).

8   Allow 2 weeks for acclimatization.

9   Enforce rest breaks and provide drinking water or other fluids.

10   Provide shade to reduce radiant heat load (plant trees, build awnings, issue wide-brimmed hats).

11   In factories, build cool spots and refuges to lower worker exposure.

TABLE 9-4   RECOMMENDED THRESHOLD WBGT TEMPERATURES FOR HEAT STRESS MANAGEMENT*

	Air velocity	
Work rate	< 1.5 m/sec	$\geq$ 1.5 m/sec
Light	30.0	32.0
Moderate	27.8	30.5
Heavy	26.0	28.9

*From OSHA.

WORK IN COLD CLIMATES

Core temperature can be maintained in the cold if the person is working and suitable protective clothing is provided. When manual work is not being performed, less metabolic heat is generated and adequate insulation becomes of increasing importance. Metabolic heat production may increase by a factor of 3 when a person is working compared with resting. When the person is resting, therefore, 3 times the amount of clothing insulation is required. Adequate hydration is essential when one is working in a cold environment. Sweating may still take place and moisture is lost as the lungs are ventilated by the cold, dry air. In extreme circumstances, a form of dehydration known as cold dehydration may occur. Cooling of the nasal epithelium, particularly during heavy breathing, may cause nosebleeds.

Core Temperature

If the core temperature of a person in a cold environment does drop below normal, control of core temperature by the central nervous system becomes disrupted. This situation occurs at around 33°C. At 29°C, the hypothalamic control of core temperature breaks down completely. Cooling of the core tissues lowers the metabolic rate and therefore the amount of metabolic heat production. If metabolic heat production cannot be elevated while heat is being lost, a vicious circle of further heat loss and still lower metabolic heat production may develop.

Acclimatization to Cold?

Some animals may be able to acclimatize to cold in a physiological sense, but it is less clear that humans are able to do so. Brown adipose (fat) tissue (known as BAT) has been suggested as the site of heat production in these animals. In addition to shivering, the animal may increase its metabolic heat production by means of chemical transformations taking place in the BAT. In humans, local acclimatization to cold may occur in the extremities. Increased blood flow through the hands can occur after repeated exposure to cold conditions. This preserves manual dexterity but at the cost of increased heat loss from the body as a whole. It has been suggested that people accustomed to cold conditions have an increased basal metabolic rate and the ability to sleep while shivering. It may also be the case that humans exposed to a cold climate increase their daily food intake. This may increase their metabolic rate and the thickness of the subcutaneous fat layer, resulting in a higher rate of heat production and better insulation of the core tissues. The chemical processes involved in the digestion of food are also a source of heat.

In humans, clothing provides a subtropical microclimate of warm, moist air between the skin and clothes. At rest, a skin temperature of about 33°C is perceived as comfortable. Most people put on more clothes in cold conditions in order to maintain this microclimate and, for this reason, are rarely exposed to any stimulus for true acclimatization to the cold. **Behavioral adaptation** to the cold, through experience, is of great importance in humans—wearing correct clothing and "keeping on the move" are examples. It may be noted in passing that for much of history, the wearing of hats by people living in cold climates was the norm, although today, hat wearing seems to have gone out of fashion. Peripheral vasoconstriction takes place in most areas of the body except the head, where up to 25 percent of all heat loss can take place when it is cold (Kroemer, 1991). Because so much body heat can be lost from the head, it should be considered an extremity to be protected as are the hands and feet.

Perception of Cold

The perception of cold also seems to depend on experience. For example, people accustomed to a cold climate may feel comfortable when their deep body tissues are suitably insulated by clothing, despite local cooling of the skin of the cheeks, ears, nose, fingers, and feet. Those unaccustomed to wide disparities in skin temperature between exposed and unexposed parts of the body may confuse "being cold" (i.e., having a low core temperature) with "feeling cold" (i.e., experiencing low skin temperatures on the extremities). In fact, their core temperature remains perfectly normal and in a physiological sense they are not cold.

Cold perception also depends on the **set point of the hypothalamic temperature-**

regulating center. Bacterial infections (which cause chills, for example) can elevate the set point for core temperature thermoregulation. The person therefore feels cold even though the core temperature is normal. Peripheral vasoconstriction takes place and the person shivers. After several hours, the core temperature reaches the new set point and the person again feels neither hot nor cold. When the factor which initially caused the increase in set point is removed, the person feels hot; peripheral vasodilation and profuse sweating occur as the core temperature declines and reaches the original set point.

SKIN TEMPERATURE

The temperature of the skin can vary over a much wider range than that of the deep tissues of the body. Burning takes place at skin temperatures over 45°C. At lower temperatures, sweating may impair functional hand strength by reducing friction at the hand-object interface. Parsons (1993) provides an interesting discussion of the surface temperature of objects in relation to the risk of sustaining burns as a result of accidentally touching the hot object. It is important to distinguish between the surface temperature of an object and its contact temperature—the temperature when it is being touched (Siekmann, 1990). It is the contact temperature which determines whether a burn will take place and which varies depending on the material. Conduction of heat to and from the skin by a solid object depends on the thermal properties of both skin and object—a metal surface at 100°C would feel much hotter than a cork surface. The threshold at which burning would take place can be estimated for different materials assuming a momentary contact time of a quarter of a second (the reaction time to remove the part of the body touching the object). The threshold temperature for wood, for example, is estimated as 187°C compared with 136° for brick and 90° for metal. Similar thinking has lead to the drafting of specifications for the maximum surface temperatures of domestic appliances to minimize the risk of burns (Parsons, 1993).

Siekmann (1990) cites maximum recommended temperatures for touchable surfaces for different contact durations. For example, the maximum surface temperature of steel is 69° and 51°C for contact times of 1 second and 1 minute, respectively. If the steel is coated with lacquer, the recommended maximums are 89° and 51°. The risk of burning is lowered for a given surface temperature if a material with lower thermal conductivity is used. Ceramics, plastics, and, particularly, wood produce lower contact temperatures for a given surface temperature. The maximum surface temperature for wood is 115° and 59°C for contact times of 1 second and 1 minute, respectively. Less dense, dry wood with a roughened surface is an excellent material for making handles for hot objects such as kitchenware.

Wind chill is the decline in effective temperature as a result of air movement and can have consequences for the extremities (hands, nose, cheeks, and ears), particularly if they are directly exposed to the wind. Skin at 20°C is one-sixth as sensitive as skin at 35°. At 5°, pressure and touch receptors do not respond to stimulation. This increases the risk of frostbite as all sensation is lost at lower temperatures. Intracellular fluid freezes at a little below 0°C. Extensive tissue damage is caused by the formation of ice crystals in the cells—damage which may not become apparent until the skin is re-warmed. In extremely cold environments, skin contact with bare metal is particularly hazardous—the skin may freeze immediately and stick to the metal, causing extensive tissue damage.

The reduction of skin sensitivity and loss of manual dexterity which take place at low skin temperatures are important from an industrial viewpoint. Allowance must be made for the fact that tasks may take longer to perform than would otherwise be the case. Controls and equipment should be designed so as to be successfully operated using gross movements of the body.

PROTECTION AGAINST CLIMATIC EXTREMES

In some applications, the working environment cannot be improved nor can the task be redesigned to eliminate worker exposure to extreme conditions. In these circumstances, methods of protecting the worker are required.

Specify Safe Work-Rest Cycles

A number of attempts have been made to identify physiological measures which indicate the presence of heat stress. Welch et al. (1971) measured pulse rate, skin temperature, sweat loss, and rectal temperature of subjects working in hot, humid conditions. Rectal temperature, which is used as an index of core temperature, proved to be the only reliable indicator of the onset of heat exhaustion. Bell et al. (1971) investigated the time taken for subjects to reach a state of imminent heat collapse and specified safe exposure times for men at work in hot environments. Figure 9-3 shows the limits of permissible exposure to hot conditions of workers working at different energy expenditures for a range of work-rest cycles (Dukes-Dobos/NIOSH, 1981).

Design "Cool Spots"

In some situations, thermal comfort can be improved by designing a thermal refuge for operators. Sims et al. (1977) designed a "cool spot" for inspectors in a glass-making factory. Discomfort caused by radiant heat and high air temperatures was reduced using glass and aluminum radiant heat shields and low-velocity cooling air. A similar approach was taken by Ohnaka et al. (1993), who investigated heat stress among workers in the asbestos-removal industry. These workers have to wear protective clothing to prevent the inhalation of asbestos and the contamination of their personal clothing. The protective ensemble, consisting of impermeable clothing and air masks, prevents heat loss and exacerbates the problems of working in hot weather. Several solutions to the problem are possible. First, the working area could be cooled by air-conditioning the work environment. Second, workers could be issued air- or liquid-cooled suits or ice-cooled vests. Third, an air-conditioned cool room could be constructed adjacent to the work area in which rest breaks are taken. The first possibility was not practical because of the temporary nature of the workplace—workers move from one part of a building to another and work on different buildings. The second option reduced mobility and was rejected. The third option was evaluated by measuring heart rate, sweat rate, and rectal temperature of subjects working in hot conditions, in cool conditions, and in hot conditions with rest in a cool room (the "hot-cool" condition). The hot-cool condition was found to re-

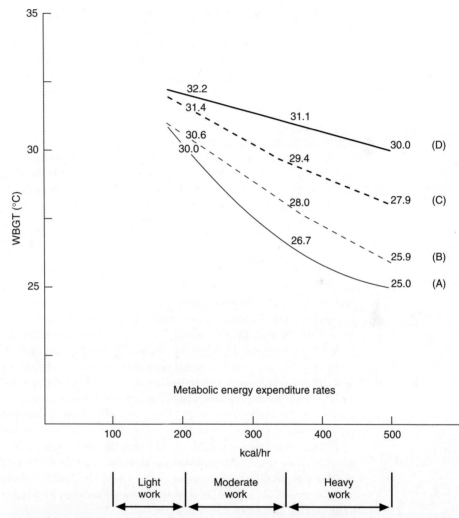

FIGURE 9-3   Lines of permissible exposure to work in hot conditions. (A) 8 hours continuous work. (B) 75 percent work and 25 percent rest each hour. (C) 50 percent work and 50 percent rest each hour. (D) 25 percent work and 75 percent rest each hour. (From Dukes-Dobos/NIOSH 1981 Work Practices Guide for Manual Lifting Tasks.)

duce physiological stress significantly and to reduce discomfort even when subjects were working in the heat. The design of cool spots can be a physiologically valid and practical method of ameliorating the problems of those working in hot conditions.

Issue Protective Clothing

A third method of reducing climatic stress is by means of protective clothing designed to provide an acceptable microclimate between the skin and the clothing assembly. Liquid- and air-cooled suits have been designed for pilots. Air or a special cooling liquid is circulated around the suit to maintain thermal balance. In gold mining, special cooling jackets have been designed to enable selected workers to acclimatize in the work environment itself (Schutte et al., 1982). The workers are selected on the basis of a heat tolerance test carried out on the surface. Suitable workers are given a jacket containing dry ice to wear underground for the first few days. This practice reduces the risk of increased core temperature and enables acclimatization to take place naturally in response to the demands of the task and environment.

In some situations, it is not possible to provide workers with liquid- or air-cooling garments because they interfere with task performance. Constable et al. (1994) investigated the effectiveness of using a cooling garment only during rest periods on the work capacity of subjects walking on a treadmill at 40 percent of their maximum aerobic capacity. Work-rest periods of 30 minutes were used. Subjects walked wearing a chemical protective clothing assembly and the cooling garment underneath. During rest, the cooling was switched on. Compared with walking and resting without the cooling garment, heat storage was significantly attenuated and work capacity doubled. Compared with a control condition in which subjects wore light clothing only, physiological stress was greater. The cooling garment condition resulted in rectal temperature oscillating between 37.5 and 38°C. Without the cooling garment, rectal temperature reached 39°C after 2.5 hours and the experiment was terminated. In the control condition, rectal temperature oscillated between approximately 37.25 and 37.5°. It was concluded that cooling during rest was effective, mostly so in the first few minutes of resting, and would be even more effective if work-rest cycles were reduced to less than 30 minutes.

Engineering solutions, personal protective clothing, and changes in work organization all have a role to play in heat stress management. In subtropical climates, spring and early summer are the seasons when care is needed to prevent heat stress in some industries. Over the winter, workers can lose the acclimatization to hot work acquired in the previous summer and be particularly vulnerable to heat stress on hot spring days.

If the external temperature is greater than the core temperature (about 37°C), additional clothing is required to protect the skin from a net heat gain from the environment. In some hot countries, people traditionally wear long, flowing robes covering the whole body, presumably to maintain a layer of cool air between the skin and the clothing and to protect the skin from solar radiation.

In more severe cases, protection from radiant heat can be achieved using clothing assemblies designed to insulate the worker from the heat. Helmets with a large air space between the shell and the operator can be used in conjunction with thick woolen suits and wooden clogs. Fireproof, reflective materials have also been used in firefighting.

A major source of radiant heat in many parts of the world is the sun. Wide-brimmed hats have been used by rural people for millennia and constitute portable shade—an effective first line of defense against the potentially debilitating effects of intense solar radiation. People who live in hot deserts cover their bodies with long, black, flowing robes. Black materials absorb more radiant heat than white materials but give better protection against sunburn. In hot, humid environments, minimal clothing is worn to maximize evaporative heat loss. If there is also intense sunlight, a conflict arises with the need for protection from solar radiation and the only solution may be to restrict outdoor work to the early morning and late afternoon.

EFFECTS OF CLIMATE ON PERFORMANCE

Climate has a profound effect on the performance of physical tasks. It can also affect mental task performance, but this effect is thought to be indirect and due to the effects of climate on physiological variables, which themselves affect performance. If workers are protected from harsh climates, performance decrements need not occur.

There are two main theories concerning the effects of climate on performance. Arousal theory states that for all tasks there is an optimum level of arousal (readiness to act) at which maximum performance occurs. Climatic extremes can increase the arousal level, whereas overly comfortable conditions can lower it (Meese et al., 1989). If the arousal level deviates significantly from the optimum, performance deteriorates. A competing theory suggests that climatic extremes, particularly cold, have a distracting effect—performance declines because of momentary shifts of attention away from the task toward the environment.

Effects of Heat on Performance

High internal temperatures appear to increase the speed of performance because they accelerate the body's "internal clock." Fox et al. (1967) showed that increases in body temperature accelerated people's perception of the passage of time. Alnutt and Allen (1973) found that when their subjects' body temperatures were raised to 38.5°C, the speed of performance on a reasoning test increased even when skin temperatures were kept at levels regarded as comfortable. Colquhoun and Goldman (1972) investigated the ability of subjects to detect target stimuli against a noisy background at a temperature of 38°C dry-bulb, 33° wet-bulb. It was found that only when work in the heat was accompanied by an actual increase in body temperature was an effect on performance observed. The effect was to increase subjects' confidence in saying they had detected a target rather than to increase the detection rate. In fact, subjects made more false positive errors (saying they had detected a target when there was none). Elevated body temperature may lead to more risky behavior. This research has relevance for quality control and inspector performance in all "hot" industries, such as iron, steel, and glassmaking.

Azer et al. (1972) provide more evidence that hot conditions have significant effects on performance only when they cause a rise in body temperature. In their experiment, performance decrements occurred when subjects worked at 35°C and 75 percent relative humidity, but not at 35° or 37.5° and 50 percent relative humidity. Relative humidity seems to be a key determinant of performance in a variety of tasks. Allen and Fischer (1978) found that if relative humidity was held constant at 40 percent, performance on a simple learning task did not vary over the range of 11 to 28°C. If relative humidity was not controlled, optimum performance was found to occur at 18°C, 35 percent relative humidity. Increases and decreases in temperature lead to performance decrements.

Wyon (1974) investigated the effects of moderate heat stress on typewriting performance by reanalyzing the data from a 1923 report by the New York State Commission on Ventilation and was able to show that people consistently produced more at 20° than 24°C. It seems that mild heat stress can cause drops in performance by lowering arousal and decreasing motivation. More severe heat stress may incapacitate workers as well as

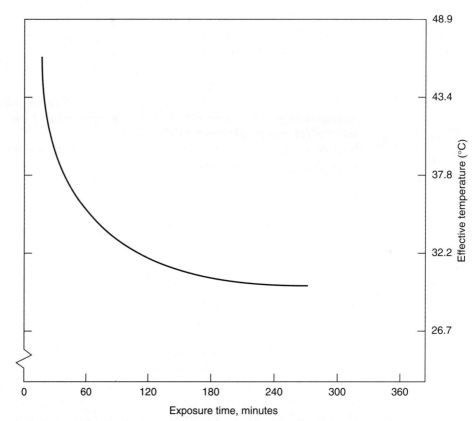

FIGURE 9-4   Upper limit for unimpaired mental performance as a function of exposure time to different effective temperatures. (Adapted from McCormick and Sanders, 1982.)

overarouse them. Figure 9-4 gives data on the upper exposure limits for unimpaired work in hot conditions.

Effects of Cold on Performance

It is often difficult to disentangle the effects of cold on the psychological and physical aspects of task performance. Cold causes peripheral vasoconstriction, shivering, impaired neural conduction, and poorer muscular control—all of which may have effects, such as longer reaction time, which may appear to be of psychological nature.

Meese et al. (1989) investigated the choice reaction time and vigilance of subjects working at temperatures of 6, 12, 18, and 24°C. The subjects, male factory workers, were divided into two groups. Trained workers had received 300 minutes of practice on the task over the previous few weeks in a warm room. Untrained subjects had a short 10- to 15-minute training session immediately prior to their participating in the research. For the trained subjects there was no difference in performance at the different temperatures. For the untrained subjects, improvements in performance (using multiple performance indexes) of between 25 and 100 percent were observed at 24°C compared with 6°C. The effect of cold was to slow reaction time, increase hesitation, and reduce certain errors. Essentially, work in the cold was slower but more cautious. In addition, the performance of the untrained subjects correlated more closely with skin temperature than with heart rate, oral temperature, or air temperature. As with heat, it seems that it is the physiological response to the climatic conditions, rather than the conditions themselves, that determines whether performance decrements will occur. If workers can be adequately protected, they can still function in extreme climates.

Although cold can have a significant effect on performance, previously acquired skill can protect performance against this form of stress. Meese et al. note that their trained subjects acquired skill in a warm environment. Thus, highly skilled workers are better able to maintain their performance in cold conditions even with no previous exposure to the stressor.

COMFORT AND THE INDOOR CLIMATE

One of the most striking effects of technological advance is the relocation of the population from rural areas to cities. Instead of doing manual work in the open air, large numbers of people now work indoors. The workload itself can usually be classified as light or moderate, and the metabolic heat production is therefore low. An immediate benefit has been that indoor workers are protected from many extreme environmental conditions, such as intense sunlight, snow, or rain.

The size of early buildings was limited by the then-available building technology. Several hundred years ago factories were long, single-story buildings in which all workers were close to a window. Office buildings resembled houses and the rooms inside were small. As building techniques improved, larger, taller buildings were possible. Electric lighting enabled deeper buildings to be built with workers distanced from the natural illumination provided by windows. However, lack of ventilation degraded the indoor air quality.

Refrigerated air-conditioning was introduced in the 1930s and enabled radical new designs of buildings, particularly offices, to be constructed. The possibility of controlling the indoor climate bought about by technological developments has lead to an upsurge in interest in environmental design. It is of considerable practical importance to develop specifications for a comfortable indoor climate.

The climatic conditions inside a building depend primarily on the conditions prevailing outside and the people and processes housed within. However, many additional factors can mediate the effects of the exterior conditions on those indoors. Of particular importance is the design of the building itself.

Building Design and the Indoor Climate

Building design determines the amount of solar penetration into the building and the radiant heat gain. The amount of insulation, particularly of the roof, has a large influence on the heat exchange between the building and the environment. The construction materials influence a building's thermal performance via a mechanism known as the **flywheel effect,** which is discussed in more detail below.

Solar heat gain through uninsulated factory roofs and walls is an important cause of overheating. Heat gain is particularly important in developing countries where cost-effective but thermally efficient building concepts are needed. In hot countries, roof insulation is an important first step in combating overheating. Parts of the building which are exposed to direct sunlight can be painted bright colors to reflect solar energy. It has been estimated that reductions in external roof and wall temperatures of almost 20°C are obtained by whitewashing building exteriors. Awnings, arbors, or covered walkways can be built around buildings to provide shade both for people and for the walls. If plentiful, water can be sprayed onto the roof and walls and adjoining courtyards and sidewalks. In developing countries, people sometimes pour water onto a nearby floor before sleeping to lower the air temperature in a room (Hanna and Brown, 1979). Screens may be placed over doorways or windows and may be wet down to cool the incoming air.

Steel roofs in factories can be shaded with tiles, preferably with an air space between the roof and the tiles. Rectangular-shaped buildings can be constructed with their long walls facing north and south. In the summer, the sun shines on the shorter walls in the morning and evening and on the roof during the day. This minimizes the degree of heating of the building by the sun. Conversely, in the winter, when the sun is low in the sky, the long walls and roof are exposed to the sun during the day, which helps warm the building.

Sunlight entering through windows is another major source of heat. In existing buildings, physically constrained workers should not have to work in direct sunlight for any length of time (this restriction also applies in cold countries since sunlight is a source of glare which degrades vision). The orientation of windows with respect to the sun is an important consideration in the design stage of a building. Special glazing, louvers, or shades can be retrofitted to reduce solar penetration. There is a modern trend to build buildings with large windows to save on construction materials. This practice, however, can lead to faster heat loss and gain.

The heat-transfer properties of building materials vary considerably depending on their mass. Lightweight materials store very little heat themselves. Heat transfer through them depends on their thermal conductivities and the temperature differential across them. Heavyweight materials have greater ability to store heat; heat gained during the day warms the material itself before being transferred to the air inside the building. Similarly, at night, heat is lost from the building materials themselves and the interior of the building remains warmer for a longer period of time. The building material itself can be thought of as a buffer or heat store which lies between the internal and external thermal environments. Thus, heavy materials act like a flywheel to smooth out the effects on the indoor climate of daily oscillations in external temperature. To use an extreme example, one could compare the temperature inside a cave, which remains more or less constant throughout the year, with the temperature inside a tent, which changes rapidly throughout the day in response to changes in the external temperature. The magnitude of the flywheel effect does not depend so much on the heat-conducting properties of the materials used in construction as on the mass of material enclosing the interior space. Thus, there is often a trade-off between the cost of constructing a new building and the later cost of controlling its internal temperature within acceptable limits. Money saved on construction materials and construction time in building a more lightweight structure may have to be spent later on additional air-conditioning (except in permanently hot, humid climates where all buildings require artificial cooling to maintain a comfortable indoor temperature). Figure 9-5 shows a renovated building which has a large flywheel effect.

Thermal Comfort in Buildings

The thermal comfort of a factory or office worker depends on there being an average skin temperature (resulting from the combination of climate, clothing, and metabolic heat production) of approximately 33°C (Astrand and Rodahl, 1977), although colder temperatures at the extremities may be tolerated. Large disparities in skin temperature may lead to complaints of discomfort even if the average skin temperature is close to 33°C. Drafts, sunlight falling on an arm or the face, and sitting next to a cold wall are all causes of thermal discomfort due to uneven skin temperature distribution.

Modern approaches to the indoor climate have attempted to specify an acceptable range of conditions for the worker. **Temperatures of 19 to 23°C at relative humidities between 40 and 70 percent** (preferably 50 to 60 percent) **are recommended for sedentary work.** In industry, a slightly lower range (from 18 to 21°) is preferable. In winter, when workers typically wear heavier clothing, the lower part of the temperature range may be more acceptable. In summer, the higher part of the range can be used. At temperatures of 24°C and above, workers may begin to feel lethargic. At temperatures of 18°

FIGURE 9-5   Renovated building. This old prison is now an office building. It has stone walls 1 meter thick at the base and small windows. The large flywheel effect means that it remains cool indoors in hot weather and is warm at night.

or lower, shivering may commence in sedentary or inactive workers unless extra clothing is worn.

The occupants of a building and the machines they use are themselves a source of heat. The body produces approximately 100 watts (about the same as a visual display terminal). When new equipment or processes are introduced into existing facilities (when an office is automated, for example), the additional heat load should be evaluated and increased ventilation or air cooling provided if required, particularly in hot weather.

Perception of Air and Thermal Quality

In the previous sections, individual differences in heat tolerance were noted. Individual differences in temperature preference also exist. Investigations have shown that some people find temperatures as low as 18°C comfortable, whereas others prefer temperatures higher than 23°C. Sundstrom (1986) suggests on the basis of findings such as these that individuals be given a certain amount of control of the temperature of their workplaces (being able to control local ventilation or to use a small fan or heater).

Relative humidity, temperature, and ventilation are the key determinants of how air quality is perceived. **The threshold at which air is perceived as stuffy begins at a relative humidity of 60 percent at 24°C and 80 percent at 18°C.** Dehumidifiers can be used to lower the relative humidity in a building to acceptable levels.

Low relative humidity (less than 30 percent at office temperatures) causes bodily secretions to dry up. Under these conditions, the occupants may complain of dry, blocked noses and eye irritation. Contact lens wearers may experience eye discomfort since proper adhesion of the lens to the eye depends on a continuous supply of lachrymal fluid to maintain a thin, moist film over the cornea. Static electrical charges are also more likely to build up in dry buildings, causing irritating shocks when a grounded conductor is touched. Somewhat counterintuitively, low relative humidity inside houses and offices can be a problem in cold, wet countries if artificially heated buildings are not ventilated adequately (which the occupants may be reluctant to do) and if humidifiers are not provided.

Ventilation

Ventilation is also a determinant of thermal comfort and, more generally, of satisfaction with the indoor environment. The main purpose of ventilation is to provide fresh air and to remove accumulated noxious gases and contaminants. Ventilation helps remove heat generated in a working area by convection and cools the body. It is not always possible to lay down acceptable limits from the point of view of thermal comfort. However, **air speeds** less than approximately **0.1 m per second** will usually cause a sensation of **staleness and stuffiness**—even at relatively low temperatures. Air speeds greater than **0.2 m per second** may be perceived as drafty. In hotter conditions (with a corrected effective temperature of more than 24°C) air speeds of **0.2 to 0.5 m per second** will **aid body cooling,** particularly when the relative humidity is high. It is apparent that whether or not air movement is perceived as an irritating draft or a cool breeze depends on the ambient temperature and the relative humidity.

In hot countries, an important use of ventilation is to cool working areas and their contents by ventilating buildings at night, when the air outside has cooled.

CURRENT AND FUTURE RESEARCH DIRECTIONS

Many aspects of indoor thermal comfort remain relatively unexplored. For most of history, humans have worked out-of-doors during the day or in unheated buildings. Under these circumstances, temperature varies throughout the day, increasing toward the afternoon and declining thereafter. Superimposed on this cycle are minor fluctuations in temperature caused by wind or by the sun's being momentarily obscured by clouds, etc. In modern buildings, however, a fairly constant temperature is maintained throughout the day. Whether people can truly be comfortable in such an environment or whether "thermal boredom" is inevitable is still being debated (e.g., Sundstrom, 1986).

Parsons (1993) has carried out some interesting work to specify safe surface temperatures of appliances (safe in the sense that the risk of burns is reduced if the user accidentally touches the appliance). This work is interesting and useful because it leads to the drafting of very clear specifications and guidelines. However, it could be extended to look at novel ways of providing users with feedback about the state of an appliance. For example, should a domestic kettle change color as it heats up to signal that it is full of boiling water? A more general question is, Can the feedback provided by appliances about their state be augmented in useful ways to enhance safety?

A final, and very important, issue is the question of total indoor environmental quality. This topic is discussed in the accompanying footnote.

SUMMARY

Measurement of dry-bulb, wet-bulb, and globe temperatures and air speed is necessary in order to evaluate the thermal nature of a working environment. The heat or cold stress on the worker depends on these factors and on the workload and insulation value of clothing. Each of these factors can be manipulated to ensure a state of thermal balance between the worker and the surroundings.

Humans are accustomed to being surrounded by a subtropical microclimate of air. An average skin temperature of 33°C is perceived as comfortable by most people. The ability to carry out manual work in extreme environments is limited. Most people are able to acclimatize to hot working conditions within 2 weeks, but the ability to adapt to the cold is much more limited. For these reasons, if work has to be carried out in extreme environments, the following steps should be considered:

1 Remove the operator by mechanizing the task.
2 Change the task or the environment.
3 Protect the operator.

The perception of comfort depends on several factors, and there are individual preferences about what constitutes a comfortable indoor temperature. A temperature range between 19 and 23°C at 40 to 70 percent relative humidity will accommodate most workers. Low relative humidity, still air, and drafts are some of the main causes of thermal discomfort in offices and factories.

A lack of thermal balance between a worker and the surroundings will lead to decrements in the ability to perform many tasks. However, a level of discomfort sufficient to cause complaints will occur long before any serious decrements in task performance or threats to health.

It is impossible to design a perfect thermal environment acceptable to everyone. People differ in basal metabolic rate (which declines with age), in the thickness of their subcutaneous fat layer, and in habits of dress. The existence of this variation among people means that no office or factory will be entirely free of complaints about the temperature. Ranges of temperature to which most people can adapt, however, can be specified.

ESSAYS AND EXERCISES

1 Take dry-bulb, wet-bulb, and globe temperature measurements in the following places (measure the external ambient temperature at the same time):

  a A modern, air-conditioned office
  b A workshop or factory floor
  c A busy commercial kitchen
  d A sauna room
  e A cold room or cold store
  f A park or beach at 8:00 A.M., noon, and 4:00 P.M. on a sunny summer day
  g A park or beach at midday in the shade

Calculate the WBGT temperature and comment on your findings.

2 Measure the WBGT in a modern, air-conditioned office containing 20 employees or more. Take measurements at various parts of the office and at different times of the day. Develop a small questionnaire to assess the occupants' satisfaction with the indoor climate. Suggest possible improvements, and comment on any differences in the level of satisfaction of different people.

3 Write an essay on thermal comfort in the indoor environment.

FOOTNOTE: Air Quality and Sick Buildings

Indoor climate may be defined as the collective whole of all the physical properties of a room which influence a person's heat balance and perception of thermal comfort. However, comfort and well-being do not depend only on the temperature in a room. They are also influenced by the total indoor environment, which includes many additional factors over and above the climatic ones discussed above.

A great deal of attention has been paid recently to the air quality inside buildings. The term "building sickness syndrome" has been coined to describe unduly high absenteeism among the occupants of buildings with ostensibly poor air quality. Particular attention has been paid to sealed, mechanically ventilated buildings which may be either permanently or only temporarily "sick" (Sterling et al., 1983; Scansetti, 1984; Sykes, 1988).

Modern office buildings use sophisticated air-conditioning systems to provide workers with a comfortable indoor climate to work in. In order to conserve energy on heating and cooling outside air, some of the air already in the building is recirculated. Under certain circumstances, this recirculation can cause the concentrations of everyday substances to reach unacceptable levels, which may, in turn, cause health complaints.

All buildings enclose a finite volume of air. If they are not properly ventilated, the air quality is likely to deteriorate. Air quality can be degraded by several classes of contaminants. Carbon dioxide, carbon monoxide, and ozone are examples of inorganic contaminants. Organic contaminants include formaldehyde and other hydrocarbons. Living organisms such as bacteria, fungal spores, and mites can also contaminate the air.

There are several sources of indoor air pollution. The most obvious is the air entering the building from outside. In cities, pollution from factories, car exhaust emissions (carbon dioxide, carbon monoxide, and lead from leaded gasoline) may enter the building. The building itself may be a source of pollution emitted by the construction materials, furniture, and fittings. Business machines and cleaning chemicals are another source. A final source is the occupants themselves. People give off carbon dioxide, water vapor, microorganisms, dead skin cells, unpleasant odors and, sometimes, tobacco smoke. Most people are exposed to some or all of these substances from time to time and experience no problems. Building sickness occurs when unusually high levels occur for some reason. Acute or chronic accumulation of pollutants can occur in any type of building, including residential buildings. Some of the main suggested indoor pollutants are reviewed below.

It has been suggested that photocopiers may degrade air quality because they emit ozone—an unstable molecule made up of three oxygen atoms. Ozone has a half-life of a matter of minutes and is destroyed by contact with most surfaces, including the tissues inside the nose, and by cigarette smoke. An ozone hazard would be unlikely to occur except locally, in a badly ventilated, heavily used photocopy room. Photocopiers might be placed by extractor fans or open windows to prevent the accumulation of ozone, which can cause eye irritation, in the air.

At one time, there was concern that the internal components of visual display terminals might emit carcinogenic gases, such as polychlorinated biphenyls (PCBs). That possibility now seems unlikely.

Humidifier fever is an acute influenza-like disorder which occurs mainly on Mondays at the resumption of work, decreasing in severity across the working week. It is apparently caused by the inhalation of spores and cysts in the air. Some humidifier systems can provide a breeding ground for bacteria and other microorganisms, as can an accumulation of water in cooling systems and air ducts, particularly if the water contains organic substances from other sources. Legionnaire's disease is a dramatic example of this type of problem. Regular system inspection, disinfection, and maintenance can reduce problems of this nature.

Low relative humidity (less than 40 percent) may also increase the risk of respiratory infection (Sykes, 1988). For example, it has been suggested that dry air causes microfissures in the respiratory tract which act as landing sites for infection by airborne bacteria. Several mechanisms have been suggested.

Formaldehyde is a common contaminant of indoor air. It is found in many manufactured wood products, including particle boards, and in ceiling tiles, carpets, and urea-formaldehyde foam. It irritates the skin and the mucous membranes. High levels of formaldehyde "off-gassing" can occur in new buildings, particularly in hot weather, causing eye, nose, and throat irritation.

Much of the dust in occupied buildings consists of dead skin cells. Clothing acts like a microscopic "cheese grater" which removes dead cells from the surface of the skin. Other sources of dust include dry-cleaning and insulation materials. Microscopic particles of glass fiber have reportedly caused conjunctivitis in office workers. Reports of facial rashes among visual display terminal users have sometimes been attributed to dust collecting on the surface of the skin. The cathode ray tube is thought to be able to induce a static electrical charge on the face of the operator, which then attracts oppositely charged dust particles.

Finally, it has been suggested that the air in modern buildings may be deficient in negatively charged small air ions. Air ions are charged forms of the molecules of the various gases found in air. In offices, air ions are destroyed by contact with metal air ducting, dust, smoke, and static electrical charges on visual display unit screens. Although there are marked differences in the concentrations of air ions found in city and country air and inside and outside buildings, it is not known whether artificially increasing the concentration of ions using ion generators does improve air quality. Hawkins (1984) found some evidence that the introduction of negative ion generators into an office reduced the incidence of dizziness, nausea, and headaches. However, in a subsequent study he could not replicate these findings.

Many factors can degrade the air quality inside buildings. Despite sometimes sensationalistic reporting, the sick building syndrome would appear to have some physical basis. Sykes (1988) has summarized some of the main findings (Table 9-5). Increased ventilation can reduce the incidence of symptoms, whereas high, uniform temperatures result in more symptoms. Wanner (1984) cites a minimum ventilation rate of 12 to 15 m^3 of air per person per hour. If manual work is carried out or if cigarette smoking is allowed, 30 to 40 m^3 per person per hour is needed. Dull, uniform lighting combined with a lack of daylight has also been cited as a contributory factor. Although the syndrome is not psychogenic, low morale and job dissatisfaction may amplify employees' negative responses to poor indoor air quality.

When confronted with complaints about poor air quality, the ergonomist might begin with an analysis of temperature, humidity, and air flow, as described earlier in the chapter. If these prove satisfactory, it is possible to test for more exotic forms of contamination. Occupational hygienists have standard equipment to test for the presence of con-

TABLE 9-5   SOME CONCLUSIONS ABOUT SICK BUILDING SYNDROME

1   Symptoms are more common in air-conditioned than naturally ventilated buildings.
2   Clerical staff are more likely than managers to suffer.
3   People with the most symptoms have the least perceived control over their environment.
4   Symptoms are more common in the afternoon than the morning.

taminants such as formaldehyde and other airborne hydrocarbons. There are standards to assist with interpretation of findings. Samples of the air and of dust in the ventilation ducts can be taken and cultured to determine what types of bacteria are present.

One of the most useful tests of indoor air quality, though, is of the concentration of carbon dioxide in a room. Atmospheric air is approximately 78 percent nitrogen, 21 percent oxygen, 0.9 percent argon, and 0.033 percent carbon dioxide. Carbon dioxide buildup occurs whenever ventilation is poor, and even small increases are known to make the air seem stuffy. Occupational hygienists can measure carbon dioxide concentration using commercially available meters. Elevated CO_2 is nearly always indicative of poor ventilation.

10

VISION, LIGHT, AND LIGHTING

Light may be defined as that portion of the electromagnetic spectrum which can be detected by the human visual system. Electromagnetic radiation is characterized in terms of wavelength and frequency. As can be seen from Figure 10-1, the electromagnetic spectrum is extremely wide.

The electromagnetic spectrum is represented using a nonlinear scale—each step is an increase by a factor of 10. It can be seen that the entire frequency range is over 10^{18} Hz long. If the spectrum were presented as a linear scale with a scale division of 1 mm corresponding to 1 Hz, the entire scale would be more than 1 billion km long. The section of the scale corresponding to the visible spectrum would be about 40 cm long. It is apparent that the bandwidth of the electromagnetic radiation to which the human visual system is sensitive is extremely narrow.

VISION AND THE EYE

The eye is a fluid-filled membranous sphere which converts electromagnetic radiation from the environment into nerve impulses. These impulses are transmitted to the brain along the optic nerve. A simplified diagram of the eye is presented in Figure 10-2.

Light enters the eye through a transparent outer covering called the cornea. The cornea plays a major role in refracting the incident light. Further refraction occurs as the light passes through the lens. The pupil can be likened to the aperture of a camera. Its function is to vary the amount of light entering the eye. In bright light, the iris contracts and the diameter of the pupil decreases. When this situation occurs, only the central part of the lens can form an image on the retina. In poor light, the iris expands and a larger area of the lens is used. Because the peripheral regions of the lens focus the light slightly in front of the image formed by the central part (a characteristic of all simple lenses, termed "spherical aberration"), slight blurring occurs when objects are viewed in poor light. Therefore, the ability to discern detail, or visual acuity, is reduced in poor light.

252

Frequency, Hz

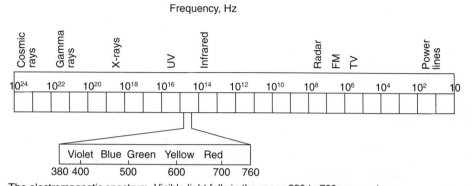

Violet Blue Green Yellow Red

380 400   500   600   700 760

FIGURE 10-1   The electromagnetic spectrum. Visible light falls in the range 380 to 760 nanometers.

In photographic terminology, it can be said that **good lighting increases the depth of field of the eye.** In bright light, the pupil is very small and the eye acts like a pinhole camera. A pinhole camera has infinite depth of field; all objects are focused irrespective of their distance. Increased depth of field reduces the need for optical adjustment of the refractive system when one is looking at objects at different distances. It is in this sense that it can be said that good lighting reduces the load on the visual system. Figure 10-3 illustrates the optics of the depth-of-field phenomenon.

FIGURE 10-2   Basic structure of the eye. (a) = cornea, (b) = pupil, (c) = lens, (d) = ciliary muscle, (e) = suspensory ligaments, (f) = retina, (g) = fovea, (h) = blind spot, (i) = optic nerve.

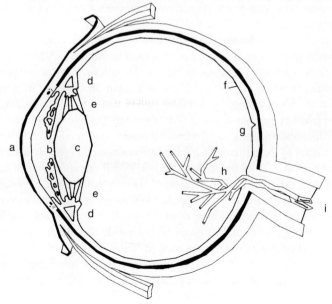

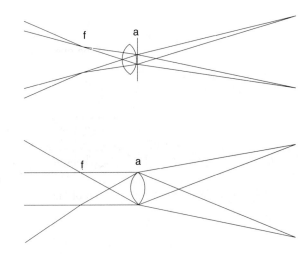

FIGURE 10-3   The depth-of-field phenomenon is one of the reasons we can see better in good rather than poor light. With a large aperture (a), only the image *exactly* at (f) is in sharp focus.

The Refractive Apparatus of the Eye

The lens is held in place by a nonrigid, membranous sling. It divides the eye into two compartments. The smaller anterior compartment contains a watery fluid called the aqueous humor, which is secreted by the ciliary body. The posterior compartment contains the jellylike vitreous humor. The humors help maintain the structural integrity of the eyeball and lens, and they supply the lens with nutrients. The surface of the eye is covered by a transparent membrane known as the **conjunctiva,** which supplies the cornea with nutrients. Dilation of the blood vessels in the conjunctiva as a result of injury or infection causes the characteristic bloodshot appearance known as **conjunctivitis** or pinkeye.

The **cornea,** the **humors,** and the **lens** constitute the refractive apparatus of the eye. The refractive power of the eye is measured in **diopters.** A lens that can focus parallel light rays to a point 1 m from its axis is said to have a refractive power of 1 diopter. A lens that can focus parallel light rays to a point 50 cm from its axis would have a refractive power of 2 diopters. Similarly, a lens that can focus parallel light rays to a point 10 cm from its axis has a power of 10 diopters. The eye is regarded in a simplified way for making calculations of its refractive power. It is considered to have a single lens 17 mm in front of the retina. When focusing on a distant object, it has a refractive power of 59 diopters. About 48 diopters of the eye's total refractive power is due to the cornea rather than the lens. This is because of the large difference between the refractive indexes of the cornea and air as opposed to the smaller difference between the refractive indexes of the lens and the humors of the eye. If the lens is removed from the eye, its refractive power is about 150 diopters, but inside the eye it contributes only about 15 diopters when distant objects are viewed. This explains why removal of the lens of the eye (because of the presence of opacities, or cataracts, in the lens) does not lead to blindness; corrective lenses can be supplied to replace the lost refractive power and compensate for any optical abnormalities.

Blinking

Blinking is a reflex action which occurs every 2 to 10 seconds. It is also a voluntary forced closure of the eye. The function of blinking is to stimulate tear production and flush out any foreign objects (such as dust particles) from the surface of the eye. Tears, a dilute saline solution, lubricate eye movements and are mildly bactericidal. The eyelids can be likened to windshield wipers. At the inner corner of the lower eyelid is the nasolachrymal duct, through which the tears are drained (animals such as seals do not have nasolachrymal ducts, which is why they appear to be weeping continuously). Tasks requiring concentration may reduce the blink rate. This reduction in blinking can cause particles of dust to accumulate on the surface of the eye and can lead to drying and irritation of the surface (particularly if the relative humidity of the air is low) and can cause people to complain of "hot" or "rough" eyes. It has been suggested that this abnormal drop in blink rate is common in work involving VDTs and is a contributory factor in the incidence of VDT-related visual problems. Measurement of the relative humidity of the air in a room would therefore be a useful exercise when attempting to establish the cause of eye complaints by VDT users.

Accommodation

Although the refractive power of the cornea can be regarded as fixed, that of the lens is variable. Variation in lens refractive power enables light from both distant and near objects to be focused sharply onto the retina—a process known as **accommodation.**

Cameras usually have fixed-focal-length lenses. Accommodation, or focusing of a camera, depends on adjusting the distance of the lens from the focal plane (the photographic film). In fish, focusing is done in a way similar to that of a camera—by moving the lens toward or away from the retina. The mammalian eye differs: When near and far objects are viewed, the refractive power of the eye is altered by changing the shape of the lens. When **distant objects** are viewed, the lens assumes a **flatter, disk-shaped appearance** and a longer focal length. When **near objects** are viewed, the lens has a **fat, rounded appearance** and greater refractive power (Figure 10-4).

The optics of accommodation are relatively straightforward, and most visual problems which are due to optical causes can be successfully corrected using appropriate lenses. When one is viewing a distant object, the incident light rays can be considered to be parallel (which they would be if the viewed object were at infinity). The light is then refracted by the cornea and lens to produce an image on the retina. In practice, any object more than about 6 m away can be considered to be at infinity. When one is viewing close objects, the incident light rays are divergent and greater refractive power is required to produce a sharp image on the retina. **In young people,** the refractive power of the lens can increase from 15 to about 29 diopters to bring close objects into focus—the lens has about **14 diopters of accommodation** in these individuals (Guyton, 1991). Although this is quite a small proportion of the eye's total refractive power, it is necessary for seeing clearly both near and middle-distance objects (such as a VDT on a desk and a bulletin board on a wall across a room).

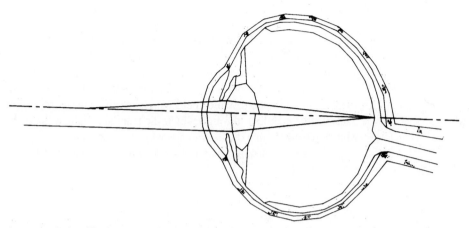

FIGURE 10-4   Accommodation. The lens assumes a spherical shape to focus divergent rays from near objects onto the retina. Parallel rays from distant objects can be brought into focus with a flatter lens.

The mechanism of accommodation is as follows: In a human eye, the lens has a naturally convex shape. It is held in place by a capsule and a muscle known as the **ciliary muscle.** The ciliary muscle is situated around the equator of the lens, to which it is attached by ligaments. When a near object is fixated, the ciliary muscle contracts and moves closer to the lens in a sphincterlike action. This reduces the tension in the ligaments and permits the lens to adopt its natural convex shape (Guyton, 1991), increasing its refractive power. When a distant object is fixated, the ciliary muscle relaxes and moves farther away from the lens. Tension in the capsule is increased by the pull of the ligaments and the lens is pulled into a flatter shape which has less refractive power.

Thus, the ciliary muscle has to contract to accommodate near objects, and it is in this sense that the visual workload can be considered to be greater when one is viewing near rather than distant objects. Visual workload in close tasks can be reduced by permitting microbreaks every few minutes, in which the eyes are refocused on a distant object for a few seconds.

The **near point** of vision is the closest distance at which an object can be brought into sharp focus. A 16-year-old can focus on an object less than 10 cm in front of the eye. However, the lens loses elasticity with age, and in practice, this loss results in a reduction in refractive power. By the age of 60 years, the near point may have receded to 100 cm, which is why older people often need reading glasses or, when reading a newspaper, for example, have to hold the paper at arm's length. By the age of about 50 years, the lens has only about 2 diopters of accommodation left. After this point it can be regarded as completely nonaccommodating—a condition known as **"presbyopia."** It should be noted that presbyopia does not lead to blindness because the eye still has much accommodative power which is supplied by its remaining refractive structures. The result of presbyopia is that the eye becomes focused at a fixed distance—a distance which varies among different people depending on the characteristics and condition of their eyes. Frequently, the fixed viewing distance in the presbyopic eye is intermediate between the previous near and far points, and the person has to wear bifocal lenses, the upper part of which is set for distant vision and the lower part for near vision (required principally for reading). It can be seen that bifocal or trifocal lenses can restore a kind of "stepwise" accommodation to the presbyopic eye. In practice, if workplaces are adequately lit, the depth of field of the eye is increased and the net effect is to lower the requirements for accommodation. **This explains why good lighting is important in all facilities which are used by older people.**

A more detailed discussion of accommodation may be found in Miller (1990).

Visual Defects

In a normal, or **emmetropic,** eye, there is a correct relationship between the axial (antero-posterior) dimensions of the eye and the power of its refractive system. Parallel light rays are focused sharply on the retina (Figure 10-5).

In **myopia,** light rays entering parallel to the optic axis are bought into focus at a point some distance in front of the retina. This can be caused by the eye's being too long antero-posteriorly or can be due to excessive power of the refractive system. Myopia is sometimes referred to as nearsightedness because the near point is closer to the eye in myopic people (for an equal amount of accommodation) than it is to a healthy eye. Myopic individuals cannot bring distant objects into focus. **Temporary myopia** often occurs after a near object has been viewed for a period of time. Accommodation is not instantaneous because the lens requires time to change to a flatter shape when the ciliary muscle relaxes. Myopic individuals can often carry out close tasks such as VDT work

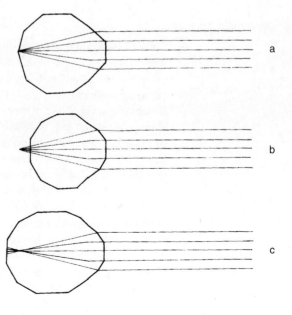

FIGURE 10-5 In the emmetropic eye (a), light rays are focused onto the retina. In hypermetropia (b), the eye lacks refractive power. In myopia (c), there is excessive refractive power.

or sewing with ease but experience difficulties with tasks such as driving, where target objects are more than 5 to 10 m away.

In **hypermetropia,** light rays entering parallel to the optic axis are brought into focus behind the retina. This can be caused by the eye's being too short antero-posteriorly or by insufficient curvature of the refractive surfaces of the eye. Hypermetropia is sometimes referred to as farsightedness because the near point is farther away from the eye (for an equal amount of accommodation) than it is in a healthy eye. Hypermetropic individuals can be said to lack refractive power and may tire quickly when carrying out work in which the viewing distance is short (such as using a VDT).

In **astigmatism,** there is an unequal curvature of the refractive surfaces of the eye; the refractive power is not the same in one plane as in another. When an object of complex shape is viewed, the retinal image may be out of focus in one plane but not in others. Astigmatic individuals often perform quite well when given simple eye tests because the defect is corrected by the depth of focus of the eye. However, they may experience severe difficulties at night or when there is excessive glare.

Each of the above defects is of an optical nature and can be corrected using appropriate lenses. Myopic individuals can be given diverging lenses to reduce the total refractive power of their optical systems. Hypermetropic individuals can be given converging lenses to increase their refractive power. Astigmatic individuals can be given lenses which correct for the unequal curvature of the refractive surfaces of their eyes once the plane(s) where distortion takes place have been identified by optometric examination. Several problems may occur if defective vision is not corrected. First, suboptimal vision may degrade performance. Second, excessive load on the muscles of the refractive system may cause visual fatigue. Third, the worker may adopt stressful body postures to orientate the head in an attempt to see better, and neck and shoulder problems may result.

Chromatic Aberration

"White" light is a mixture of different wavelengths. A defect of all simple lenses is that light of a given wavelength is focused slightly nearer or farther away than light of other wavelengths, a condition known as **chromatic aberration.** Chromatic aberration can cause the outlines of objects to have purple or red fringes (the shortest and longest wavelengths). The refractive power of the eye is about 3 diopters greater for saturated blue than for saturated red, so blue objects are focused in front of red ones and both cannot be in focus at the same time. Purple letters or characters on VDT screens may appear to have fuzzy edges for this reason. In a VDT using less saturated colors, the difference may be only 1 diopter. However, this difference may still be enough to cause startling visual effects, such as chromostereopsis (the illusion of depth).

Since blur is a stimulus for accommodation, **unresolvable blur may destabilize the accommodation mechanism,** making it "hunt" in vain to resolve the blur caused by the different focal points of the different colors. The onset of visual fatigue may be hastened. Reds, oranges, and greens can be viewed without refocusing but cyan or blue cannot be viewed with red (because they are at opposite ends of the visible spectrum). Chromatic aberration is exacerbated when the pupillary diameter is large (as in poor lighting).

For clarity and the avoidance of strange visual effects, the conjoint use of saturated colors or colors from opposite ends of the spectrum should be avoided.

Convergence

The eyes are a small distance apart. When one is looking at a distant object such as a mountain several kilometers away, each eye receives a similar image because the lines of sight of the two eyes are parallel and the distance apart of the two eyes is negligible compared with the distance of the viewed object.

When closer objects are viewed, the eyes **converge** on the object; that is, the lines of sight of each eye meet at the object. Convergence decreases with distance. When viewing close objects, the two eyes view the object from slightly different angles, so the images cast onto the two retinas differ slightly. The position of the eyes in their sockets is controlled by the **vergence system.** Lightly pressing one eye with the finger causes the perception of double images, which is known as **diplopia.** Long hours of close visual work may cause imbalances in the muscles controlling eye movement, a condition known as **phoria,** and increase the perceived effort required to carry out the task.

The Retina

The **retina** is the most complex part of the eye. It consists of a layer of light-sensitive cells connected to nerve fibers and is sometimes considered an extension of the brain (or "a little piece of brain lying within the eyeball"). Unlike photographic film, the retina *actively* processes incoming information before passing it on to the brain via the optic nerve.

Photons of incident light cause chemical changes in the light-sensitive cells, which give rise to nerve impulses. The nerve fibers pass over the cells and converge to make up the optic nerve. The point at which the optic nerve leaves the retina is known as the blind spot. The retina can be likened to an array of electronic light detectors linked in complex ways which act in an on-off fashion when activated by the incident photons.

The retina contains two types of light-sensitive cells, known as rods and cones. There are over 100 million rods and about 6 million cones. Rods are more sensitive to light than cones and are essential for scotopic (night) vision. Bright light bleaches the rods, which renders them ineffective, and the cone, or photopic, system then comes into operation.

Photopic and scotopic vision differ in several other ways. The retina has an uneven distribution of light-sensitive cells. Cones predominate at the central part of the retina (the fovea). Toward the periphery, the concentration of rods increases. Furthermore, each foveal cone cell has its own nerve fiber, whereas several rods may share one fiber. The structure of the retina is exceedingly complex, but it can be concluded that photopic vision is less sensitive to light but more acute than scotopic vision.

Subjectively, we are aware of an area of focal attention in bright light which corresponds to the fovea, or central part of the retina, where there is a high concentration of cones. In dark conditions, the ability to resolve detail is lost because the cones no longer function. The peripheral areas of the retina contain more rods and are sensitive to changes of light and to movement. **This has several implications for display design; for example, warning signals or alarms can be placed in peripheral areas of a panel as long as flashing light or movement is used.**

Retinal Adaptation

Retinal adaptation is the ability of the retina to change its sensitivity in accordance with the ambient lighting conditions. When one walks from a darkened cinema into a sunny street, a temporary feeling of being dazzled is experienced. The diameter of the pupil decreases to reduce the amount of light entering the eye, the rods are bleached, and the photopic system quickly comes into operation. When one goes from a bright area to a dark one, the pupil increases in size and chemical changes take place in the retina as the rods slowly come into operation. **Full adaptation to dark conditions can take up to 20 minutes.** For this reason, it is important not to expose the dark-adapted eye to sudden bright lights, since even brief exposures will degrade scotopic vision for some time afterward. This condition can happen to motorists driving along unlit roads at night when oncoming cars have badly adjusted or undimmed headlights. Extra light is usually provided at the entrances and exits to tunnels for similar reasons—to smooth the transition between light and dark and provide more time for retinal and behavioral adaptation.

Figure 10-6 depicts the sensitivity of the photopic and scotopic systems to light of different wavelengths (the spectral sensitivity of the retina). As can be seen, the scotopic and photopic systems are differentially sensitive to light of different wavelengths. Both are maximally sensitive to light in the middle of the spectrum (which is perceived as blue-green, green, and yellow). Violet and red are less readily sensed, which implies that in order for a red object to appear as subjectively bright as a green one, more illumination is required. It is unfortunate that red is normally used to signal danger since the retina is less sensitive to red than to other colors under the same intensity of illumination.

The scotopic system can be seen to be several orders of magnitude more sensitive than the photopic system to the shorter wavelengths. However, the two systems have a similar sensitivity to the longer wavelengths. This implies that when the eyes are dark-adapted, long-wavelength (red) artificial sources should be used to illuminate objects temporarily. The cones will be sensitive to these wavelengths at an intensity sufficiently low so as not to bleach the rods and degrade their state of dark adaptation. Night guards, sentries, etc. can be issued flashlights containing long wavelengths for use, for example, when inspecting documents at night.

In practice, a stage intermediate to purely photopic or scotopic vision can occur. This stage is known as mesopic vision and has characteristic features of both of the others.

Color Vision

The photopic system can discriminate between light of different wavelengths. Subjectively, this discrimination is experienced as color vision. It may be noted, in passing, that nocturnal animals have many more rods than cones and are almost color blind.

Three types of cone cells with different spectral sensitivities have been identified. Be-

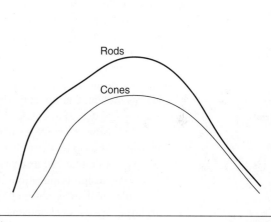

FIGURE 10-6   Sensitivity of rods and cones.

cause the absorption spectrums of the three light-sensitive pigments overlap, a given in-cident wavelength of visible light will cause the three types of cones to respond in vary-ing degree. The result is a **pattern of cone outputs** which is a signature of the particu-lar wavelength. Different wavelengths give rise to different cone output patterns which are interpreted as different colors by the brain. The notion that the cone output patterns are transmitted to the brain where they give rise to the perception of all spectral colors appears straightforward and compatible with our everyday experience. What is experi-enced as white light, when diffracted using a prism, can be seen to consist of light of many different "colors" (i.e., wavelengths). The term **"component process"** has been used to describe this explanation of color vision. For example, red and green light sources can be combined to give a yellow color which matches monochromatic yellow even though the wavelength which gives rise to the perception of yellow is not present. This combining is known as a **metameric match** and is evidence for a component process view of color vision. The cone output pattern caused by the mixed-wavelength stimulus is identical to that caused by the monochromatic source, so the same color is perceived. According to this view, color vision is purely the result of the cone output pat-tern caused by the incident light. Different wavelengths have characteristic cone output patterns, and if a combination of different wavelengths impinges on the retina, a cone output pattern intermediate between the two wavelengths and in proportion to their re-spective intensities will result. This type of reasoning is used to explain how people can distinguish between colors such as bluish green, greeny blue, etc.

Certain visual phenomena are more difficult to explain using simple color-mixing theories. If a bright light is fixated by the eye for several seconds, a retinal afterimage will remain when the viewer looks away. Blue lights produce yellow afterimages and vice versa, and red lights produce green afterimages. It has been suggested that at the retinal level, a component process exists with functionally different cones responding maximally to different wavelengths. The outputs of this system are thought to fire into a so-called **opponent process system.** There are said to be three opponent process systems—a blue-yellow system, a red-green system, and a light-dark system. When long wavelengths are incident, the red-green system increases its firing rate above its spontaneous level, and red is perceived. As the incident wavelength is shortened, the system is inhibited and its firing rate decreases until it drops below the spontaneous level, and green is perceived. Thus, after viewing a bright red light for several seconds, a green afterimage of the object is seen as a result of a sudden drop in the output of the opponent process system. Certain cells in the brain appear to act like opponent process cells, increasing and decreasing their firing rate in this way.

Color vision is a complex phenomenon which can be dealt with here only at a superficial level; however, these theories of color vision can have implications for ergonomics. The notion that **red and green** and **blue and yellow** are related neurologically has interesting design implications. Generally, these **opponent color combinations** should be avoided in display design because of the afterimage problem—particularly in active displays, such as VDT screens, and if the colors are highly saturated. A saturated color is "pure" in that it contains only those wavelengths which give rise to the perception of the particular color—fire engine red is an example. Unsaturated colors are those which have been "diluted" with white. Figure 10-7 illustrates the afterimage problem.

Bright blue letters on a yellow background may make an attention-getting sign but

FIGURE 10-7   Refer to the back cover. Stare at the black dot in the colored area for 30 seconds, then fixate the dot on the white paper of this page. (I would like to thank Dr. Douglas A. Bauk of IBM Brazil for supplying the illustration on the back cover.)

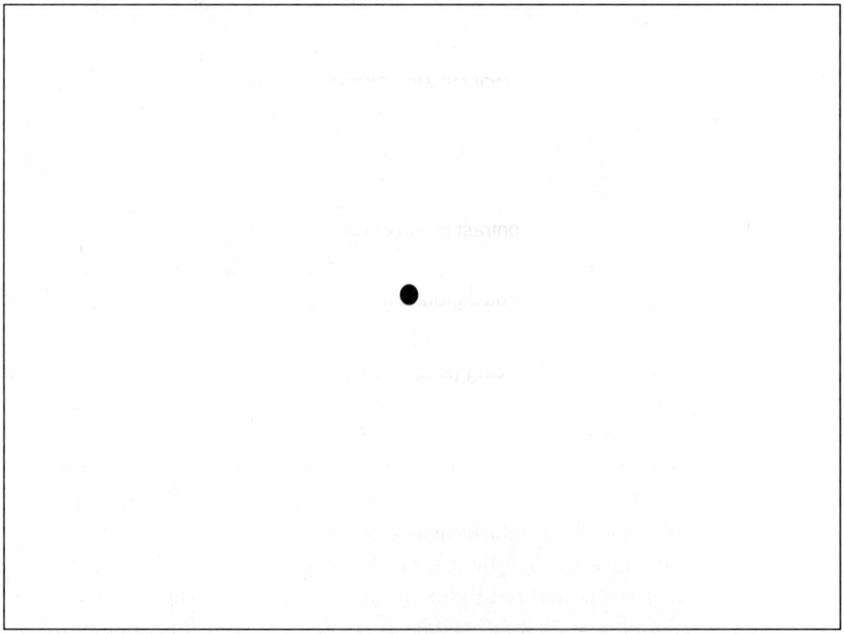

would be unsuitable for use in a visual display screen. Sharp discontinuities between bright colors may produce startling visual effects—red letters on a blue background may "vibrate," for example. Color should not be used for borders or to display messages on the periphery of a display since cone density is very low at the periphery of the retina. The use of flashing lights or a moving cursor would be more effective. Some general guidelines for the use of color in visual displays are summarized in Table 10.1. Further details can be found in Christ (1975).

The dark-adapted eye is more sensitive to shorter-wavelength (blue) than longer-wavelength (red) light and is most sensitive to green. Shorter-wavelength colors should be used to design displays or notices which have to be read in dark conditions. The light-adapted eye is most sensitive to middle-wavelength (yellow-green) light. In the design of passive (reflective) displays for light conditions, maximum color contrast can be obtained by using yellow on black.

Color perception is a complex phenomenon. Not all colors can be duplicated by mixing three primary hues in the laboratory (brown, silver, and gold are examples). In color television sets, three-color mixing is used, yet viewers do perceive these otherwise unattainable colors. In laboratory conditions, it is possible to illuminate a dark object such as a piece of coal so that its luminance is greater than a piece of white paper. However, subjects will say that the paper looks brighter—as long as they can see enough to recognize the objects as paper and coal. This finding demonstrates that higher learning processes are involved in making judgments about color. Some colors are associated with familiar objects (as is indicated by sayings such as "as black as coal"), and judgments are influenced by other factors, such as color intensity and the reflectance of the object.

TABLE 10-1   GUIDELINES FOR COLOR SELECTION*

In general

1 Choose compatible color combinations. Avoid red-green, blue-yellow, green-blue, red-blue pairs.

2 Use high color contrast for character-background pairs.

3 Limit number of colors to four for novice users and seven for experts.

4 Use light blue for background areas only.

5 Use white to code peripheral information.

6 Use redundant coding (shape or typeface as well as color); 6 to 10 percent of males have defective color vision.

For visual display units

1 Luminosity diminishes in the order white, yellow, cyan, green, magenta, red, blue.

2 Use white, yellow, cyan, or green against a dark background.

3 For reverse video use nothing (i.e., black), red, blue, or magenta.

4 Avoid highly saturated colors.

   * Adapted from Durrett and Trezona (1982), van Nes (1986).

Color and Visual Acuity

When stationary objects are being viewed, acuity is least if the object is blue. Long and Garvey (1988) investigated color and dynamic visual acuity. When moving targets are being viewed under photopic conditions, their color does not seem to influence visual acuity. However, dynamic visual acuity under scotopic or mesopic conditions is affected by target color, and blue targets are much more easily resolved than other colors, probably because of the differential sensitivity of the rod system. Long and Garvey suggest that we should be careful not to overgeneralize the usual ergonomic recommendations about color. Under some lighting conditions where there is a moving display or a moving observer, colors such as blue (which are commonly avoided for stationary targets) might be advantageous. There are possible implications here for the design of roadway warning signs which will be used primarily at night.

MEASUREMENT OF LIGHT

The objective measurement of light is essential in the design and evaluation of workplaces. Because the eye adapts to light levels, automatically compensating for any changes in illumination, subjective estimates of the amount of light in a work area are likely to be misleading.

Data concerning the visual response of the eye have been used to define lighting measures. The radiant flux arriving at a surface is weighted according to the eye's sensitivity to each of a number of wavelength intervals. The total incident radiant flux after weighting is known as the **luminous flux.**

The measurement of light is known as photometry. The main photometric units are luminous intensity, luminous flux, luminance, and illuminance. Their definitions are given in Table 10-2.

The SI unit of luminous intensity is the candela (cd). An imaginary point source of luminous intensity 1 cd will emit light in all directions. A source of greater intensity will emit more light. In both cases, we can imagine a sphere of light spreading out from the source. Clearly, the intensity of the source itself does not depend on the distance from which it is viewed. However, the "strength" of the light at the edges of the imaginary sphere will depend on viewing distance.

The surface area of a sphere is 4π steradians of solid angle. A steradian (sr) is defined as the solid angle which encloses a surface on a sphere equivalent to the square of the

TABLE 10-2 MAIN PHOTOMETRIC TERMS

Luminous intensity:	The power of a source or illuminated surface to emit light
Luminous flux:	The "rate of flow" of luminous energy
Luminance:	The light emitted by a surface
Illuminance:	The amount of light falling on a surface
Reflectance:	The ratio of the luminance and illuminance at a surface

radius. By definition, a point source of luminous intensity 1 cd will emit a total luminous flux of 4π lumens (lm), or 1 lm per steradian.

Illuminance refers to the light falling on a surface. If we imagine a point source of 1 cd inside a sphere of 1-m radius, the illumination on the inside of the sphere is defined as 1 lux (lx), or 1 lm per square meter. If we increase the radius of the sphere to 2 m but keep the intensity of the source the same, the illuminance on 1 m^2 of the inner surface will now be 1/4 lx. When the radius of the sphere increases linearly, the surface area increases according to the square of the radius. Illumination therefore follows an inverse square law, with intensity decreasing with the square of the distance from the source.

The luminance of an object depends on the light it emits or reflects toward the eye. It corresponds roughly to brightness, although brightness perception depends on other factors, such as contrast. In the example above, if the inside of the sphere is a perfect reflector of light, then the luminance will be the same as the illuminance. The percentage of the incident light which is reflected by a surface depends on the reflectance of the material. Reflectance is defined as the ratio of luminance to illuminance. White paper has a reflectance of about 95 percent; white cloth, about 65 percent; newspaper, about 55 percent; plain wood, about 45 percent. Matte black paper has reflectance of about 5 percent. More formally, reflectance is given by

$$\text{Reflectance} = \frac{\text{luminance} \times \pi}{\text{illuminance}}$$

Thus, if we know the illuminance on the surfaces in a room (for example, by measuring it with a light meter), we can select materials of appropriate reflectances for each of the surfaces in order to achieve a balance of surface luminances in the room and ensure that the ratio of luminances of adjoining surfaces is not excessive (as described below).

More detailed discussion of photometric terms, the relationships between them, and recommendations for the design of lighting systems can be found in publications of the Illumination Engineering Society (IES) and in Boyce (1982).

Lighting Standards

Much effort has been applied over many years to the drafting of standards for the illumination of workplaces. Standards differ from country to country. Table 10-3 presents recommended illuminances for various work situations. Some other illuminance values which are found in more extreme situations have been included in the table to provide context. U.S. readers should refer to the *IES Lighting Handbook* for up-to-date and detailed recommendations, whereas readers outside the United States may find the *CIBS Code for Interior Lighting* useful. It can readily be seen that the eye can operate under an extremely wide range of illuminance levels—due largely to the differential sensitivity of the photopic and scotopic systems.

Recommended illuminance levels vary both between countries and over time. Before the invention of electric lights, indoor workers depended on daylight, and building dimensions were limited because daylight does not penetrate into a room beyond about 10

TABLE 10-3 EXAMPLES OF RECOMMENDED AND NATURALLY OCCURRING ILLUMINANCES

Area or activity	Illuminance on a horizontal surface (lux)
Clear sky in summer	150,000
Overcast sky in summer	16,000
Performance of extremely low-contrast tasks (certain surgical operations)	10,000
Textile inspection	1,500
Office work (pencil handwriting, poorly reproduced documents)	1,000
Precise assembly work	1,000
Office work (without VDUs)	500
Office work (data entry)	500
Office work (VDU, conversational tasks)	300
Heavy engineering	300
Rough assembly work	200
Minimum illuminance for manual handling tasks (NIOSH)	150
Rarely visited places where little perception of detail is required (e.g., railway platforms)	50
Good street lighting	10
Emergency lighting*	2
Moonlight	0.5

*Data are a representative compilation from many sources. For emergency lighting recommendation, see Jaschinski (1982).

m. Sundstrom (1986) has noted that lighting standards have risen over the years and have been higher in the United States than elsewhere. In the early part of this century, investigations by various companies in the United States and by the Industrial Fatigue Research Board in Britain demonstrated that the performance of visually demanding tasks could be improved by increasing the level of illumination. Artificial lighting became accepted in factories and standards rose under the assumption that "more is better." With the arrival of fluorescent lights, recommended levels could increase further because of the better **luminous efficacy** (lumens per watt of electricity), longer life, and reduced heat production of fluorescents.

Recommended illuminance levels differ between the United States and other countries for several reasons, including differences in design philosophy. U.S. standards have been heavily influenced by the laboratory research work of Blackwell. In other countries a more pragmatic approach has been taken. In Britain, the view has been that daylight be the dominant source of light, with artificial light taking on a supplementary role.

A more recent trend has been to reduce the levels of illumination in workplaces, particularly offices. This trend has occurred partly because of the desire to conserve energy and also as a response to the introduction of VDTs into the workplace. Ergonomists have recommended that illumination levels be lower in VDT offices to avoid glare and reflectance problems (see below). In practice, the choice of an appropriate level of illumination depends not only on the task but also on the distribution of objects in the visual field and their luminances.

Contrast and Glare

The function of the eye is not so much to detect light but to detect **luminance discontinuities** between objects in the visual field. It is the difference in luminosity between an object and its surroundings which makes it visible, rather than the light the object reflects. A person wearing opaque spectacles can detect the presence of light but not the presence of objects.

The retina functions more like an edge detector than a light meter. Special cells (called horizontal cells) are part of a network of cells interposed between the rods and cones and the fibers of the optic nerve. If one part of the retina is stimulated by a bright light, the horizontal cells inhibit the adjacent photoreceptors. This has the effect of increasing the difference in the firing rates of the stimulated and unstimulated parts of the retina and serves to enhance the perceived contrast between the stimulus and its surroundings, which has the effect of sharpening the contours of objects, facilitating their detection (the importance of contrast and contour detection in design is discussed below and in Chapter 13).

The direction of gaze is involuntarily drawn to bright objects in the visual field. This is known as **phototropism.** Jewelry shops usually display their wares on black velvet cloth under bright lights to obtain maximum contrast—the intention being to stimulate a phototropic response in passersby.

One of the most important considerations in the design and evaluation of lighting systems is to ensure appropriate contrast between objects in the visual field. The luminance contrast between two surfaces is given by the difference between the luminances of the brighter and dimmer surfaces expressed as a percentage of the brighter (note that contrast does not depend on the absolute brightness of the surfaces):

$$\text{Contrast} = 100 \times (\text{Lbright} - \text{Ldark})/\text{Lbright}$$

Contrast percentages can also be calculated using data on the reflectance of adjacent surfaces assuming equal illuminance of the surfaces in question or if the actual illuminances are known.

The luminance ratio is the ratio of the luminance of a work area to that of its surroundings. Recommended maximum luminance ratios have been proposed—3:1 between a task and its immediate surroundings to 10:1 between the task and the walls, floors, etc. It is usually suggested that the task should be the brightest area in the visual field to take advantage of the phototropic response. This precludes the use of materials such as white laminated plastic for desktop materials since the luminance ratio between white paper and the desktop will be too low. Wood or pastel-colored finishes are preferable from this point of view.

Interior decoration should be chosen to achieve a balance of surface luminances gradually diminishing from the task to the surroundings. Brightly colored walls, carpets, furniture, and fittings are contraindicated for this reason.

Glare occurs when there is an imbalance of surface or object luminances in the visual field—the brighter sources exceeding the level to which the eye is adapted. Sources of glare include the sun, bright or naked lamps, or reflections off shiny objects. Although the retina is able to adapt to different levels of luminance so as to operate over a wide range of conditions, it is not able to adapt selectively to large, simultaneous discontinuities in luminance in the visual field. For example, if the ambient luminance is high compared with the task luminance, the retina will adapt to the former rather than the latter and the task will appear dim and will be more visually demanding. This situation can happen when VDT screens are placed against a window such that users face the window. If the illuminance in a room is very low, the retina will dark-adapt and will be more vulnerable to the effects of glare from task elements or extraneous sources.

Glare is categorized in several ways. Disability glare increases task demands, whereas discomfort glare does not. Discomfort glare may occur in offices, for example, when one or more bright objects are seen peripherally. Disability glare occurs when objects brighter than the task interfere with the detection and transmission of visual task data; extraneous light sources may increase the adaptation level of the retina, making the task appear dimmer than it really is. Intense light may be interreflected by structures within the eye itself and reduce the contrast between the retinal image and the background. Older workers may be particularly prone to this type of glare because of age-related changes in the refractive media of the eye. Finally, bright lights may cause retinal afterimages which have a veiling effect on the main task.

Glare may be direct or indirect; i.e., it may be emitted by a source or reflected off an object. If reflected, the glare may be either **diffuse or specular** (mirrorlike). White walls on the inside of a room may reflect sunlight into an operator's eyes and would be classed as a diffuse source. Chrome-plated controls or components such as bevels on dials and gauges may cause specular reflections. More detailed discussions of glare can be found in Cushman and Crist (1987) and Howarth (1990).

LIGHTING DESIGN CONSIDERATIONS

For visual comfort and to meet visual demands, the following should be considered (Grandjean, 1980):

1 A suitable level of illumination
2 A balance of surface luminances
3 Avoidance of glare
4 Temporal uniformity of lighting

The color-rendering properties of light sources might also be taken into account.

Illumination Levels

Although early research indicated that improvements in productivity were possible when illumination levels were increased, "more" does not necessarily equal "better." In fact, high levels of illumination may increase glare and may wash out important visual details. If illumination levels are inadequate, increasing them may improve performance, but a region of diminishing returns will be reached as **nonvisual limits to performance** (such as motivation, fatigue, or manual dexterity) become increasingly important. Older workers generally require higher levels of illumination than younger workers as a result of the loss of refractive power and changes in the light-transmitting media of the internal structures of their eyes.

The Illumination Engineer and the Ergonomist

It is difficult to specify the skill level of the ergonomist with regard to the design and evaluation of lighting systems. The ergonomist is not an illumination engineer, nor should he or she try to be one. However, the ability to use a light meter to measure illumination levels in a workplace must be regarded as a minimum technical skill. Commercially available light meters are shown in Figure 10-8. These meters can be used to measure illuminance levels on the work surfaces in offices, shops, factories, etc., and the readings can be compared with the levels recommended in published standards or in Table 10-3.

Balance of Surface Luminances

In practice, a balance of surface luminances is achieved by specifying appropriate illuminances and corresponding reflectances of the surfaces in a room. Detailed specifica-

FIGURE 10-8   Light meters suitable for measuring illuminances in ergonomics. (Photographs supplied by Gossen-Metrawatt, Nürnberg, Germany.)

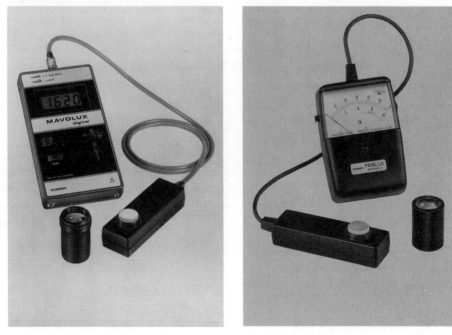

tions can be found in the publications of the IES and ANSI, and examples of this information are included here. Some additional considerations are described below.

Direct lighting is often used to illuminate sculpture because the shadows produced have a "modeling effect." The areas of light and dark enhance the three-dimensional appearance of the works by emphasizing differences in depth. Angled direct lighting is useful for enhancing the surface texture of materials such as cloth or wood and can be used in the production of these materials to aid inspection and quality control. However, direct light can reflect off desktops or surfaces and cause indirect glare and sharp luminance discontinuities between lit and unlit surfaces.

When indirect lighting is used, most of the light is directed onto the ceiling and walls, which reflect it back. Thus, objects are illuminated from many different directions at once, a smoother distribution of luminances is obtained, and shadows are reduced. Direct and indirect lighting can be combined to achieve a balance of surface luminances and minimum glare. In a well-illuminated office for example, all large objects and major surfaces should have a similar luminance, surfaces in the middle of the visual field should have a contrast ratio of no more than 3:1, and contrasts at the sides of the visual field should be avoided. A balance of surface luminances can best be achieved in practice by using materials with different reflectances in a room. The ANSI recommendations for the reflectances of surfaces in offices are given in Table 10-4.

A balance of surface luminances is also an important design requirement for lighting systems in corridors, stairwells, and outdoor facilities such as railway platforms which are used at night. Since **shadow is a cue to depth,** patterns of shadow of the floor can be misinterpreted as changes in ground level and increase the risk of people's slipping, tripping, or falling.

Avoidance of Glare

Glare can be reduced by choosing a suitable combination of direct and indirect lighting. With direct lighting, most of the light is directed toward the target in the form of a cone (Figure 10-9), which produces hard shadows and sharp contrasts between illuminated

TABLE 10-4   ANSI-RECOMMENDED REFLECTANCES FOR OFFICES

Surface	Reflectance	Suitable material or finish
Ceiling	80–90%	White paint (matte)
Furniture	25–50%	Wood (matte, unpolished)
Upper walls (wall-ceiling border)	80–90%	White paint (matte)
Walls	40–60%	Neutral, unsaturated hues (e.g., pastel shades)
Business machines	25–50%	Matte gray finish
Curtains or blinds	40–60%	Cloth
Floor	20–40%	Carpet (slightly darker colors—beige, brown, gray, etc.)

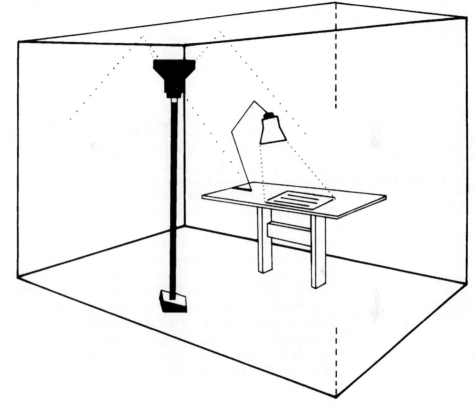

FIGURE 10-9   Direct and indirect lighting.

and nonilluminated areas. Indirect lighting is reflected off other surfaces in a room and produces a smoother transition between surface luminances and reduces shadows.

No light sources should appear within the visual field during work activities. Glare control thus depends on both the design of the lighting system and the general workspace and task design. All lamps should have glare shields or shades and the **line of sight from the eye to the lamp should have an angle greater than 30 degrees to the horizontal.** Overhead lamps should not reflect off desktops or work surfaces into the operator's eyes. Conversely, reflective desktops and work surfaces should not be used. Wood or matte finishes are preferable. Fluorescent lamps arranged in rows should be arranged parallel to the line of sight to present the minimum luminous surface area possible. Generally, use of a large number of low-powered lamps rather than a small number of high-powered lamps will result in less glare (Figure 10-10).

Glare and VDTs

Good lighting is particularly important in offices where visual display terminals (VDTs) are being used. Traditional offices designed for pencil-and-paper work usually have

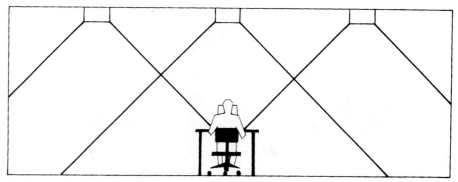

FIGURE 10-10   Place rows of luminaires parallel to the line of sight and to the operator's side. Overlapping cones of light ensure uniform distribution of surface illuminances.

rows of fluorescent lamps designed to provide an appropriate level of illuminance on the horizontal desktops below. Desks can usually be placed anywhere in the room since the spacing of lamps is such that the "cones" of light from each lamp overlap to provide a uniform distribution of illuminance. Because the direction of gaze is downward—toward the horizontal desktop—lamps on the ceiling are more than 30 degrees above the line of sight of a seated worker and do not cause glare.

Because a VDT user looks directly ahead at the screen, light from overhead lamps may cause direct glare or reflect off the screen, having a veiling effect—reducing the contrast between the characters and background (Figure 10-11). Correct positioning of

FIGURE 10-11   Avoid direct and reflected glare at VDT workstations.

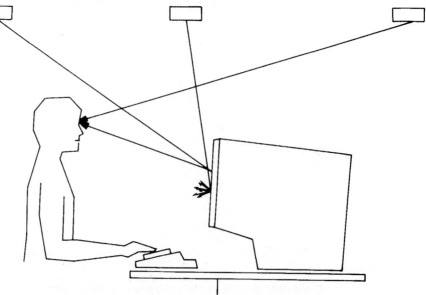

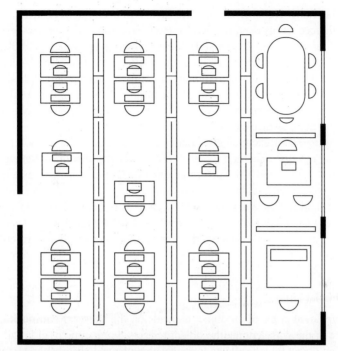

FIGURE 10-12   Layout for VDT and non-VDT work in an open office.

workstations with respect to windows, lamps, and bright surfaces is therefore very important (Figure 10-12). Rows of fluorescent lights should be parallel to and on either side of the operator-screen axis to reduce direct and indirect glare, VDU workstations should be at right angles to windows, and more use should be made of indirect lighting.

At the workstation itself, sharp contrasts due to the dark screen (i.e., light characters against a dark background) and bright source documents can be problematic. Since the VDU screen emits its own light, general illumination can be lower in VDU than in traditional offices. Low-reflectance surfaces can also be used (particularly behind the screen) and task lighting (e.g., desk lamps) can be provided at each workstation to support the visually demanding aspects of the work.

Temporal Uniformity of Lighting

Fluctuating luminances can be more disturbing than static contrasts. The temporal uniformity of lighting can be influenced by both the characteristics of the light source or the requirements of the task. Incandescent bulbs radiate light fairly uniformly over time, whereas fluorescent lamps are known to flicker. Fluorescent lamps work by passing an electric current through a gas in a glass tube. The gas emits visible (almost monochromatic) light and ultraviolet radiation periodically in accordance with the mains frequency. The ultraviolet radiation excites a phosphor on the inside of the glass tube, which emits visible light. Different phosphors are used to produce the various shades of commercially available lamps. The flicker frequency of correctly functioning fluorescent lights exceeds the threshold for the perception of flicker. However, malfunctioning or old lamps can produce visible flicker which may cause visual discomfort. In factories, flicker is a hazard since it can have a stroboscopic effect—rotating or oscillating machine parts may appear stationary or to move more slowly if their frequency is similar to that of the flickering source which illuminates them. Phase shifting of fluorescent tubes (using banks of two tubes and alternating the phase of each) and regular inspection and maintenance can prevent flicker problems.

Task and equipment characteristics can also reduce the temporal uniformity of light entering the eye. In certain types of VDT tasks, the user may alternate between looking at a dark screen and a bright source document up to seven times a minute. Interactive VDT work often involves scrolling, in which screen characters move against a dark background and the eye is presented with a continuously changing array of luminances. For these and other reasons, VDT work is generally regarded as visually demanding. A number of additional considerations concerning its design are presented in a following section.

Color Rendering and Artificial Light

The apparent color of an object depends on the **spectral composition** of the incident light and the wavelengths the object reflects. For example, in broad-spectrum (white) light an object which reflects long wavelengths and absorbs the rest would appear red. The same object, when viewed under artificial light, deficient in long wavelengths, would appear reddish grey or brown. An object which appeared bright red under white light would appear black when illuminated by monochromatic blue light.

The light emitted by a source can be described in terms of the relative amounts of different wavelengths of which it is composed. Monochromatic sources emit a narrow band of wavelengths. Objects which reflect this wavelength will appear very bright when illuminated by the monochromatic source. Objects which do not reflect the wavelength at all will appear black. For example, sodium street lights emit a narrow band of wavelengths and appear yellow. Most objects illuminated by them appear various shades of yellow-grey and can be distinguished only by the contrast or brightness differences between them. A monochromatic light source can be said to have poor color-rendering ability because it does not reveal the way that objects differentially reflect incident light according to its wavelength composition.

In order to have good color-rendering properties, a light source must contain sufficient amounts of light across the visible spectrum. The term " spectral energy distribution" is used to describe relative amounts of different wavelengths which a light source emits. Figure 10-13 depicts spectrums for a variety of light sources.

Daylight, incandescent light, and some fluorescent tubes have broad emission spectrums and therefore good color-rendering properties. The color-rendering properties of a light source cannot be determined subjectively by the appearance of the light itself. Although green and red lights can be mixed to give a yellow light of similar appearance to a monochromatic source, the color-rendering properties of the two sources are com-

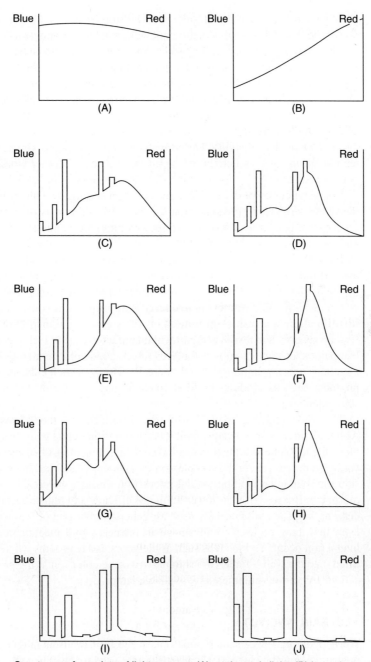

FIGURE 10-13   Spectrums of a variety of light sources. (A) northern skylight. (B) incandescent lamp. (C) deluxe cool white. (D) cool white. (E) deluxe warm white. (F) warm white. (G) daylight fluorescent. (H) white. (I) mercury lamp.* (J) mercury lamp.* (Photographs courtesy of Osram SA, PTY Ltd.) * Mercury lamps look white but have poor color rendering and are deficient in long (red) wavelengths.

pletely different. The red-and-green mix will render the colors of ripening apples very well, whereas under the monochromatic yellow they will appear a dull, greyish yellow.

Selection of light sources for good color rendering should be based on objective data about the emission spectrums of the source. In practice, an index—the **color-rendering index**—has been devised to describe the color-rendering properties of light sources. Daylight is used as a standard and given a value of 100. The color-rendering properties of other sources are expressed with respect to that standard (Boyce, 1982; Howarth, 1990). A high value of the color-rendering index usually indicates that a light source has good color-rendering ability. Unfortunately, there is an approximate negative relationship between color rendering and luminous efficacy—in other words, good color rendering costs more.

The apparent color of an object depends very strongly on the properties of the incident light. Woodson (1981) summarizes some of the effects of using "colored" light. Under red light, for example, white appears pink, red colors are greatly strengthened, and blue appears purple.

Incandescent lamps have good color-rendering properties and strengthen the appearance of red, pink, orange, and yellow objects, but bring out blue and green less well. Standard cool white fluorescent lamps bring out the latter colors but not red, pink, and orange. **Lamp color-rendering property is an important consideration where color identification is a task requirement,** as in assembly and inspection of wiring looms, fruit inspection, and textile and garment manufacture. Dental laboratories, florist shops, and paint shops are examples of places where good color-rendering light is needed. In the retail industry, lighting is used to create displays which enhance the appearance of products such as clothing or food (meat is usually illuminated using lights which strengthen red).

The color-rendering properties of lights also influence the "atmosphere" in a room. Although there is still debate about whether environmental color has psychological effects, the ambient illumination can have social implications because of the way different light sources render the complexion of a person's face. Incandescent lights enhance skin tones and can facilitate social interaction through the creation of a "warmer" atmosphere. Because of their lower luminous efficacy and higher heat production, incandescent lamps are not normally used for general lighting in offices and factories. However, they may be used to advantage in interview and meeting rooms where social interaction occurs and in areas such as elevators and restrooms where mirrors are often fitted onto the walls. Table 10-5 summarizes some color-rendering properties and applications for various commercially available lamps.

VISUAL FATIGUE, EYESTRAIN, AND VDTs

Work involving VDTs is often said to cause eyestrain or visual fatigue. Although not recognized diagnostic categories, these terms are used to describe symptoms such as ocular discomfort or dryness (the feeling of sand underneath the eyelids), viewing difficulties, and more generalized symptoms such as headaches. Although these problems do occur in other types of close visual work, they appear to be more prevalent among users of VDTs than among similar occupational groups who do not use VDTs.

TABLE 10-5 COLOR RENDERING AND LUMINOUS EFFICACY OF LAMPS

Lamp	CRI	Luminous efficacy	Applications and comments
Incandescent	90–99	Low	Home, hotels, rooms dedicated to social interaction.
Fluorescent*			
Color-matching	85–100	Medium	Florist shops, paint shops, etc.
White	50–70	Medium	General purpose. Factories, shops, etc. Reds appear orange; violet-blue is dulled.
Deluxe warm white	70–85	Low-medium	Violet-blue is weakened. Creates relaxed, informal atmosphere similar to that with incandescent lamp.
Tungsten-halogen	85–95	Low	Have many applications where good color rendering is needed
Mercury fluorescent	50–60	Medium	General illumination. Long life. Weakens red.
High-pressure sodium	25–40	Medium-high	Background and outdoor illumination
Low-pressure sodium	Very low	High	Street lighting. Almost no color rendering.

*Fluorescent lamps last 5 or more times longer than incandescent lamps.

Visual Demands of VDT Work

Work with display units may be visually demanding because of the quality of the image on the screen, the design of the workstation, or the design of the operator's job. Although screen design has improved over the years, the characters on a CRT always have blurred edges because the boundary between the excited and nonexcited regions of the CRT phosphor is not sharp. Since blur is one of the stimuluses for accommodation, unresolvable blur may make accommodation unstable. There is evidence for a temporary recession in the near point of vision after prolonged performance of close tasks. This unresolvable blur has also been reported in microscope operators (e.g., Soderberg et al., 1983). Recession of the near point is suggestive of a loss of refractive power and is sometimes referred to as accommodation strain. It has been reported in VDT workers (as has the time to refocus between near and far points) and is exacerbated by organizational factors such as the rigidity of work routines and the duration of work periods (Gunnarson and Soderberg, 1983).

Binocular fixation is controlled by the vergence system and is essential for viewing near objects and for fusing the two retinal images into a single three-dimensional percept. Eyestrain as a result of close visual work may often occur as a result of an imbalance in the muscles controlling the position of the eyes. Tasks which require frequent changing of the direction of gaze (e.g., from the screen to a source document) at different task distances may increase the likelihood of these problems. Operators may complain of diplopia or may feel that they have to work harder to view the screen.

The visual problems associated with VDT operation are not unique to it; they can occur in many other tasks where close visual work is required. Some useful preventive measures are summarized in Table 10-6 (Grandjean, 1987, provides further information). Finally, it appears that stressing of the accommodation and vergence systems by requiring users to carry out close visual work for long periods increases the tension in the muscles of the upper body (Lie and Watten, 1987, 1993). This connection between the ocular muscles and those which control the position of the head is not surprising when it is remembered how the eyes and head move together to locate a stimulus in space. In fact, the two systems share common neural pathways.

Experiments have shown that the wearing of lenses to distort vision also increases the upper body EMG activity. Alternatively, providing subjects with appropriate spectacles has been shown to reduce upper body EMG. These findings all serve to demonstrate the close connection between visual workload, visual fatigue, and pain in the upper body, including headaches, and support the practice of providing VDT users and others who engage in visually demanding work with regular eye tests and well-designed workstations. Users must be able to select their own preferred viewing distance when using a VDT. Jaschinski-Kruza (1991) has shown that healthy subjects, when free to set their own viewing distance, choose distances between 51 and 99 cm for characters 5 mm high. Since reading glasses are normally prescribed for a distance of 30 cm, users may need another set of glasses for VDT work.

TABLE 10-6   GUIDELINES FOR THE VISUAL DESIGN OF VDT TASKS

1   Workstation and work environment:

Provide visual relief.

Never position the VDT against walls or screens.

Design workstations so that users can periodically change focal length and vergence by glancing out a window or along the length of an open-plan office.

Provide potted plants or other visually complex objects as indoor visual relief—to facilitate automatic recovery from visual fatigue when operators are not looking at the screen.

Align VDT screens in correct relation to light sources.

2   Terminal and screen:

Use reverse video to minimize screen glare (i.e., make screen brighter than glare).

Reduce remaining glare with filters.

3   Job design:

Design-in natural rest breaks.

Increase task variety.

4   Operator selection and assessment:

Require eye tests for new operators to determine existing visual defects.

Provide regular eye tests for frequent VDT users.*

*See Stone (1986).

It is usually recommended that dedicated VDT workers take periodic rest periods throughout the day (5 minutes per hour). Jobs can be designed to provide natural rest periods if some non-VDT work is included. Even short breaks (2 to 60 seconds) every 10 minutes may help prevent fatigue (Pheasant, 1991).

More detailed technical discussions of visual factors in VDT operation can be found in the report of the Panel on Impact of Video Viewing on Vision of Workers (National Research Council, 1983) and in Cakir et al. (1980). A broad review of the health issues in VDT operation can be found in Marriot and Stuchley (1986).

PSYCHOLOGICAL ASPECTS OF INDOOR LIGHTING

Sundstrom (1986) has reviewed the research on lighting and satisfaction. The findings of both laboratory and field studies indicate that people are satisfied with illumination levels of about 400 lx. Further increases in illuminance only bring modest increases in satisfaction. Glare is associated with dissatisfaction.

There is also evidence to suggest that people dislike uniform lighting. Light from windows varies throughout the day. Not only the intensity changes but also the spectral composition—sunlight differs from blue skylight which differs from light from an overcast sky. Reflections from objects and buildings also add variety. It is often suggested that uniform lighting leads to boredom, whereas changes in light have a stimulating effect.

The contribution of daylight from windows to the illuminance on indoor surfaces is usually much less than it appears to be (although direct sunshine can cause severe glare on surfaces such as VDT screens, chromed surfaces, etc.). However, anecdotal evidence suggests that windowless workspaces are associated with dissatisfaction. Many indoor workers appear to believe that daylight is superior to artificial light or that working under artificial light is unhealthy.

Windows are a valuable source of visual relief and seem to have psychological value. By conveying information about changes in the weather and change from day to night, they provide continuity and contact with the outside world.

LIGHT, CIRCADIAN RHYTHMS, AND SHIFT WORK

Light is an important behavioral cue in both animals and humans. People who live in countries with high light intensity avoid working in the middle of the day, whereas in countries with low intensity, this time is one of high activity.

It has long been known that many physiological processes exhibit a 24-hour rhythm, or circadian rhythm. Body temperature is a representative example. It is low in the early morning and rises throughout the day before peaking in the late afternoon and becomes lower again at night. Just after midday, many physiological variables, which have been rising steadily since early morning, suddenly dip to a lower level. This phenomenon is known as the postlunch dip. Subjectively, we often feel overcome by feelings of lassitude in the early afternoon, which can be mistakenly attributed to lunchtime overindulgence. In fact, the postlunch dip occurs even if lunch is missed.

Research on Circadian Rhythms

Experimental subjects have been put into bunkers in which all time-of-day cues as well as all social contact have been removed. Under these circumstances, body rhythms change to a slightly longer cycle. Physiological rhythms are thought to be genetically programmed but become entrained to the environment by external cues ("zeitgeber"). People living on the equator, where there is a 12-hour day-night cycle throughout the year, have pronounced circadian rhythms. Social cues are also important in entraining circadian rhythms in humans—probably more so than light. People living in the Arctic under total daylight in summer and total night in winter also have identifiable rhythms due to the powerful effect of social cues.

There is evidence that task performance varies over the course of the day; errors and accidents in industry are more common at certain times than others. Time-of-day effects have been found in the speed of reading gas meters or answering a switchboard, the frequency of car drivers' falling asleep behind the wheel and of train drivers' missing warning signals. These effects roughly track body rhythms—they are more common at night and during the postlunch dip. Problems can occur when workers adapted to a particular cycle of light and dark and sleep and wakefulness have to adapt to a new cycle, as when flying to a different time zone, particularly when the difference is greater than 4 hours.

Adaptation to a new time zone seems to be more rapid if the new schedule is adhered to immediately on arrival with plenty of exposure to the new conditions (such as taking walks early in the morning).

Factors Influencing Adaptation to Shift Work

Social factors may inhibit physiological adaptation to shift work—many individuals prefer to revert to the normal cycle on rest days and at weekends. Incomplete adaptation to shift work may be a cause of increased sickness absence. Colligan et al. (1979) examined the records of 1219 nurses on permanent day, afternoon, or night shifts, and on rotating shifts. They found that those on permanent shifts had a lower rate of clinic visits due to ill health than those on rotating shifts. Although sickness absenteeism was similar for the two groups, those on rotating shifts had more sick days for serious illnesses; that is, rotators were in poorer health. Smith et al. (1982) compared day workers with night workers and people on rotating shifts. The latter two groups had poorer sleep, altered eating habits, greater alcohol consumption, greater sickness absence, and more on-the-job injuries. It appears from the work of Smith et al. that working a permanent day shift is least stressful and working a rotating shift system is most stressful.

Some Considerations in the Design of Shift Systems

One solution to the shift-work design problem may be to identify individuals who prefer permanent night work, thereby avoiding the need to rotate workers between the night and day shifts, and thus reduce the need for adaptation. Some individuals seem to prefer night work and choose jobs which can be done only at night (such as night guard), whereas others may have a clinical intolerance to shift work (Motohashi, 1992). There is still considerable debate about how best to design shift systems to minimize unpleasant health and social effects. Wilkinson (1992) reviewed the literature on the rotation of shift systems and concluded that fixed night-shift systems were superior on most counts. For 24-hour operation of a worksystem, the permanent night shift can be supplemented by two rapidly rotating day shifts. Wedderburn (1992) disagreed with this conclusion and proposed the opposite view—that rapidly rotating shifts provide the best compromise solution, providing for both the physical and the social needs of workers. Finally, Folkard (1992) has argued that Wilkinson probably overestimated the problems associated with rapidly rotating shift systems and underestimated the problems associated with implementing permanent night-shift systems. Folkard has concluded that there is no such thing as a good shift system and that in the design of any such system there is a trade-off between accommodating workers' social needs (using a rapidly rotating system) and health and safety considerations (with workers adapted to permanent night work).

The debate about the design of shift systems illustrates an important point in ergonomics: Fundamental knowledge from core sciences can provide guidelines for design, but implementation requires information about other aspects of system functioning.

Although there is still controversy about how best to design a shift system for a particular industry, the literature suggests the following:

1 Adaptation to a new shift system takes place over at least a week and must be allowed for. Workers may never adapt properly to rapidly rotating systems. The naval shift system in Table 10-7 is an example of a particularly difficult system to adapt to, in which circadian rhythms gradually disintegrate. Two out of three nights are disturbed. It was originally designed because seamen had to do heavy physical work on deck and needed frequent rest.

2 Where possible, minimize the change in working hours. If it is 4 hours or less, workers will be able to accommodate to the change with few problems.

3 If a voluntary, permanent night shift cannot be found, a rapidly rotating three-shift system may be the best compromise. It is pointless to try to adapt workers to night work if they revert to daytime wakefulness in their time off, and a rapidly rotating system may cause less disruption to circadian rhythms under these circumstances. Folkard (1987) suggests rotating shifts in the order: morning shift, then evening shift, then night shift, back to morning shift, and so on.

4 The permanent night shift with rapidly rotating day shifts may be the best compromise if enough people can be found to work the permanent night shift. Workers on

TABLE 10-7   AN EXAMPLE OF A STRENUOUS 4-HOUR SHIFT SYSTEM

	Hours of work		
Day 1	0:00–4:00	08:00–12:00	16:00–20:00
Day 2	12:00–16:00		
Day 3	0:00–4:00	12:00–16:00	

*Work hours are expressed in 24-hour time.

the rotating day shift will be able to accommodate the change since their hours of sleep and wakefulness are unlikely to be disrupted more than 4 hours.

5 Permanent night-shift systems may not work, even with a willing workforce, if night workers live in noisy inner-city areas or in crowded shanty towns in third world countries. Disturbances of daytime sleep may result in chronic sleep deprivation of night workers, which cancels out any health or performance benefits of their being adapted to the night shift.

6 Shift designers must also consider the social customs and circumstances of their employees in order to arrive at the best compromise system. If night workers do not have personal means of transport, special arrangements may have to be made. Social, recreational, and entertainment facilities geared to the needs of night workers may assist them in maintaining their adaptation to night work during their free time.

Pros and Cons of 12-Hour Shifts

It is sometimes suggested that the working week be compressed from 5, 8-hour days out of every seven to two 12-hour days out of every four. Rosa and Bonnet (1993) investigated performance and alertness on 8- and 12-hour rotating shifts. Performance and alertness decrements were found on the 12-hour system as were reductions in sleep. Greater fatigue was apparent over the course of the 12-hour shift than the 8-hour shift, and the investigators suggested that critical activities be scheduled for the early part of 12-hour shifts, particularly at night. However, not all investigations have found there to be more problems with 12-hour rather than 8-hour shifts, and the former seem to be popular among workers because of the increased number of days off.

Williamson et al. (1994) evaluated a 12-hour shift system which involved two 12-hour day shifts followed by two 12-hour night shifts followed by 4 days off work. The workers were computer operators and the 12-hour system replaced an irregular 8-hour system. Data on mood state, sleeping and eating patterns, work quality, productivity, staff turnover, and sickness were collected when the 8-hour system was in place and when the 12-hour system had been worked for 7 months. The 12-hour system produced positive changes in mental state and physical symptoms and improved sleep quality and quantity. There were no significant effects on absenteeism, staff turnover, or productivity. The 12-hour system may be appropriate in certain jobs, particularly when arranged so as to provide 4 days off for every 4 days worked. Rosa and Bonnet caution that although workers may gladly tolerate increased fatigue in exchange for more free time, further research into both work-related and non-work-related accidents and injuries as a function of extended work shifts is needed.

LIGHT, TIME-OF-YEAR EFFECTS, AND BEHAVIOR

Light is known to have biochemical and behavioral effects on humans. **Vitamin D** is produced in the skin by the action of ultraviolet (UV) light. Vitamin D is required for normal calcium metabolism in humans. Research has shown that people living in temperate latitudes exhibit seasonal variation in intestinal calcium absorption, with the lowest levels occurring in winter. This finding was one of the reasons for the fortification of dairy products with vitamin D, which was introduced to reduce the incidence of vitamin D–deficiency bone disease (rickets) in children (Hughes and Meer, 1981).

People living in temperate latitudes are subject to wide variation in daily exposure to natural light over the course of the year. In animals, the level of physical activity and behaviors such as reproduction and migration are thought to be triggered by the daily duration of sunlight. Changes in human mood and behavior may also exhibit seasonal variation. Depression is more common in winter than in summer.

Jacobsen et al. (1987) describe the characteristic features of **seasonal affective disorders** (SAD). Patients suffering from SAD typically exhibit decreased activity, anxiety, increased appetite and sleep, and work and interpersonal difficulties. Autumn or winter usually marks the onset of signs and symptoms of SAD, which persist until spring. In the United States, the incidence of SAD is greater in the northern than southern states.

SAD patients have recently been successfully treated using phototherapy, which consists of daily exposure to bright light (2500 lx) in the morning and evening to extend the "day" by up to 6 hours. It should be remembered that even in winter in the temperate latitudes, illuminance levels in a building can be an order of magnitude lower than those outdoors. Because of the short period of winter daylight, many indoor workers may spend all day indoors exposed to very low levels of illumination, only emerging from their workplaces at dusk. The daily exposure to light of indoor workers may therefore be extremely low. Jacobsen et al. suggests that more research on how ambient light levels affect mood and productivity would be worthwhile.

CURRENT AND FUTURE RESEARCH DIRECTIONS

One of the most controversial research issues at present is whether operating a VDT has detrimental effects on vision. When this issue first appeared in the late 1970s, it was dismissed on the basis of the then-current knowledge. Not enough people had been using VDTs for long enough to enable epidemiological surveys to be carried out.

The current view is that no permanent degradation in vision occurs as a result of occupational VDT viewing. However, the author knows of no definitive study which has proved conclusively that occupational VDT viewing is or is not detrimental to vision. When the seriousness of the legal implications of this is remembered, it is clear that such a study is needed.

A major trend in VDT design is to improve the resolution and quality of screen characters. The change from dot matrix to bit-mapped character generation has been an advance. Although the CRT is likely to be the standard display device in coming years, alternatives have been developed. These include liquid crystal, thin film electroluminescent, and gas plasma displays. More recently, active liquid crystal displays have been developed which promise to deliver at least the clarity and color quality of CRTs. Somewhat surprisingly, little work seems to have been done on the optimum size and shape for the screen of a standard VDT. CRT technology clearly limits the range of sizes and shapes that can be manufactured cost-effectively, but other display technologies may not be so limiting.

The introduction of flat screens will remove one of the biggest constraints on VDT workstation design—the footprint and viewing requirements of the CRT. Flat screens can be mounted on a wall or dividing panel or placed on a document holder. They will make possible new workstation designs and increase the flexibility available to users and facilities managers.

SUMMARY

In the design of appropriate workspace lighting, the following variables need to be considered. The amount and type of light emitted by the light sources themselves determines the illuminance delivered to work and other surfaces. The furniture and materials, by virtue of their reflectances, determine the balance of luminances seen by workers and the amount of indirect lighting and glare. The visual demands of tasks should be analyzed in the evaluation of lighting, particularly with respect to visual acuity and the demands on accommodation and adaptation of the eye and the avoidance of visual fatigue.

An ergonomic approach to lighting must, however, go beyond the purely functional specification of lighting systems and consider the psychological and biological effects of light on people.

ESSAYS AND EXERCISES

1 Measure illuminances on a horizontal plane in the following places:
   a Outdoors
      i Under a shady tree
      ii On an overcast day
      iii On a clear night with no moon
      iv On a clear night with a full moon
   b Under urban streetlights at night
   c In a fast-food outlet
      i During the day
      ii At night
   d In a modern office at noon with the blinds open
      i With the lights on
      ii With the lights off
   e On workbenches in a workshop or factory
   How do the measured values compare with your subjective impressions of light and dark?
2 Carry out a lighting survey in an office. Obtain or draw a floor plan of the office showing the location of desks, luminaires, and windows. Measure desktop illuminances (at least two measurements per desk) and write them on the plan. Compare your measurements with published guidelines. At each workstation, take note of any glare sources. Ask employees whether they have any comments about the present lighting arrangements. Combine this information into a report and make suggestions for improvement.
3 Describe the structure and function of the eye. How can this information be used to analyze practical visual problems in the workplace?

FOOTNOTE: Other Forms of Electromagnetic Radiation and Their Effects

As stated at the beginning of the chapter, visible light constitutes only a minute portion of the electromagnetic radiation to which humans are exposed. Ionizing radiation such as x-rays is known to be hazardous and standards for exposure have long been in existence. Ultraviolet (UV) and infrared (IR) radiation can be hazardous, the latter by its heating effect on body tissue. UV radiation, from the sun, tanning booths, or industrial processes such as welding, may cause skin cancer or cataracts.

Midday sunlight has the greatest proportion of ultraviolet and blue (short-wavelength) light, but the eyebrows shield the eye from the incident ultraviolet rays coming from above. UV intensity drops by a factor of 10 by 3 P.M. Toward late afternoon, the longer atmospheric path through which the sun's rays have to travel filters out UV and the shorter visible wavelengths, giving the sun its characteristic red color at dusk (Sliney, 1983). Working out-of-doors at midday should be avoided in hot countries because of the intensity of visible, IR, and UV radiation.

Interest in the biological effects of nonionizing radiation has been stimulated by a number of recent epidemiological studies. Wertheimer and Leeper (1979) reported that the homes of children who subsequently developed cancer were found unduly often to be near electrical power lines carrying high currents. They suggested that the extremely low frequency magnetic fields produced by the currents might directly or indirectly cause cancer. Several researchers attempted to replicate these findings. Fulton et al. (1980) failed to replicate the Wertheimer findings, although a Swedish study by Tomenius (1986) did find a relationship between the number of cases of childhood tumors and the 50-Hz magnetic field strength outside their dwellings. Higher field strengths were measured among dwellings of cases than around dwellings of controls. Savitz et al. (1988) also found evidence for an association between low-power magnetic fields and childhood cancer. Although the association was modest, they concluded that further examination of the carcinogenic potential of nonionizing radiation was warranted.

The introduction of VDTs on a large scale gave rise to a debate about the possibility of harmful effects of operator exposure to ionizing radiation (it is interesting to note that similar concerns were voiced about television sets many years previously). Although the interaction of electrons with phosphor materials in a cathode-ray tube produces x-rays and sometimes UV radiation, they are absorbed by the glass of the tube. Emissions from VDTs have been measured and found to be negligible (Cakir et al., 1980).

All electrical appliances and cabling emit very low frequency (VLF) and extremely low frequency (ELF) electromagnetic radiation. The field strengths are usually weak (orders of magnitude lower than those required to physically affect living tissue), exposure times are short, and field strengths diminish rapidly with distance from the source. Thus, only devices to which people are continuously exposed for long periods have been suspected of having adverse effects and have been investigated. Wertheimer and Leeper (1986) reported seasonal patterns in low birth weight and spontaneous abortion among users of electric blankets and heated waterbeds. Nonusers did not exhibit these patterns. They attributed the difference to either excessive heat or exposure to low-frequency electromagnetic fields.

Concern about exposure to VLF and ELF electromagnetic fields from VDT control circuitry has been expressed recently. Nurminen and Kurppa (1988) investigated work-related reproductive problems in office work. They found no indication of problems due to office work in general or VDT use in particular. Goldhaber et al. (1988) found that the risk of miscarriage was greater for women using VDTs for more than 20 hours per week in the first trimester of pregnancy. However, whether these findings were due to intrinsic properties of VDTs or other factors such as poor ergonomics or job stress could not be determined. Subsequent research (Brandt and Nielsen, 1990; Schnorr et al., 1991) concluded that VDT use was not associated with increased risk of spontaneous abortion.

Harvey (1984) measured the electric (but not the magnetic) field strength over the frequency range DC to 1 MHz around a selection of VDTs. The estimated user exposure levels were 2 to 3 orders of magnitude below existing exposure guidelines. Field strengths were greater at the back, sides, and top of the units than at the front (as would be expected if the control circuitry rather than the CRT was the source of the electromagnetic radiation at these frequencies). Marha and Charron (1985) came to essentially similar conclusions after measuring the electrical field strengths of VDTs. They concluded, first, that the field strength more than 1 m from the VDT was negligible and, second that operator exposure from adjacent VDTs should be accounted for in the investigation of complaints or in the process of carrying out research into the health effects of using VDTs.

It is not known whether ELF or VLF electromagmentic radiation at the field strengths found at home or in the workplace is a health hazard, although exposure definitely produces biological effects (Anderson, 1993). Included are changes in DNA, RNA, and protein synthesis as well as changes in cell proliferation, membrane permeability, and immune response (Cleary, 1993). However, these changes occur only at particular frequencies or modulations of electromagnetic energy and are subtle. To date, no clear dose-response relationship has been demonstrated for the biological effects of electromagnetism (EM) at low field strengths. Neither does there seem to be a simple threshold field strength at which changes do occur and which would be needed to specify exposure levels or daily exposure doses in practice. Savitz (1993), in a review of the epidemiological evidence, concludes that relationships between certain childhood cancers and residential exposure to EM do exist. Also, the risk of brain cancer and leukemia is now known to be higher in electrical utility workers than in other groups. However, **it is not known whether EM is the causal factor.** Further research and development of better methods for assessing occupational exposure to EM is needed (Bracken, 1993).

Future research may indicate a relationship between exposure to EM and disease. If the relationship is verified, it will have a great impact on the design of the physical environment. At present though, other forms of electromagnetic radiation (such as strong sunlight) and other ergonomic stressors (such as those encountered in VDT operation or manual handling) seem to present much more direct and tangible threats to the well-being of people at work.

11

HEARING, SOUND, AND NOISE

Acoustic waves can be defined as pressure fluctuations in an elastic medium. Sound may be defined as the auditory sensation produced by these pressure oscillations.

In air, sound consists of oscillations about the ambient atmospheric pressure (Figure 11-1). Vibrating surfaces or turbulent fluid flow can act as the source propagating alternately high- and low-pressure areas to the surroundings. The amplitude of the acoustic wave is expressed in units of force per unit area (newtons per square meter) or in pascals. The threshold of hearing (lowest amplitude of pressure oscillations in air detectable by the human ear) is taken to be 0.00002 N/m^2 at a frequency of 1000 Hz.

TERMINOLOGY

Frequency and **amplitude** define a pure tone in physics. Their subjective counterparts are **pitch** and **loudness.** However, the perception of pitch also depends on the waveform and that of loudness depends on frequency. A healthy, young person can hear sounds in the range of 16 to 20,000 Hz. **Noise** is usually defined as a sound or sounds at such amplitude as to cause annoyance or to interfere with communication. Sound can be measured objectively, but noise is a subjective phenomenon.

The amplitude of sound is objectively evaluated by measuring the sound pressure level (SPL). The range of SPLs to which the human ear is sensitive is so wide (0.00002 N/m^2 to 20 N/m^2) that linear scaling would present a problem. For this reason a logarithmic scale—the decibel scale—is used to express the intensity of sound. The decibel (dB) is a dimensionless unit related to the logarithm of the ratio of the measured sound pressure level to a reference level (usually taken to be the threshold of hearing). The smallest noticeable difference in intensity between two sounds is about 1 dB. Commercial sound-level meters measure and display a root mean square (rms) SPL, Lp,

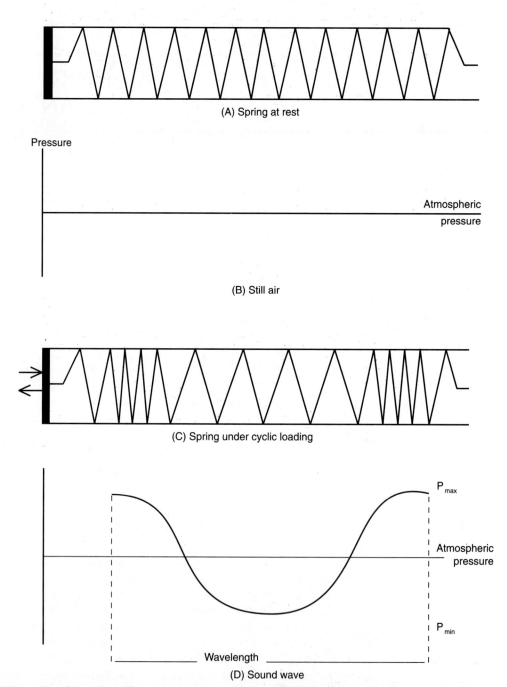

(A) Spring at rest

(B) Still air

(C) Spring under cyclic loading

(D) Sound wave

FIGURE 11-1   Sound is propagated in air as pressure oscillations above and below the ambient pressure. The analogy of a spring at rest and under cyclic loading assists visualization.

$$Lp = 20 \log_{10} (p/pr) \text{ dB}$$

where Lp = sound pressure level, dB
         p = sound pressure, N/m²
         pr = reference sound pressure level (0.00002 N/m²)

For example, if the rms SPL is 2 N/m², then

$$\log_{10} (p/pr) = \log_{10} 2/0.00002$$
$$= \log_{10} 100{,}000$$
$$= 5$$

and Lp = 100 dB

Table 11-1 depicts the range of human hearing in decibels.

THE EAR

The ear converts sound waves in the atmosphere into nerve impulses which travel along the auditory nerve to the brain. Further processing of the nerve impulses in the brain re-

TABLE 11-1   RANGE OF HUMAN HEARING

Source	SPL, dB
	140 (ear damage)
Jet engine (at 30 m)	130 (onset of pain)
Pneumatic chipper	120
Pneumatic jackhammer	
Punch press	115
Hydraulic jackhammer	
Textile loom	105
Power lawnmower	100
Heavy traffic	
Newspaper press	95
Milling machine	90
Diesel truck	85
Very noisy street corner	80
Crowded room	75
Vacuum cleaner	
Passenger car (at 15 m)	70
Conversation	60
Air-conditioning system	
Sound inside car (at 50 kph)	
Private office	50
Quiet room	40
Library	35
Whisper (1 m from ear)	20

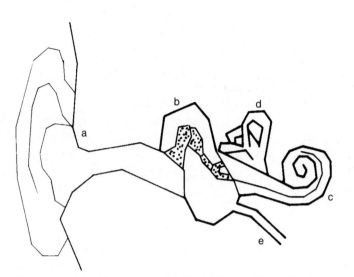

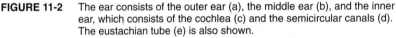

FIGURE 11-2   The ear consists of the outer ear (a), the middle ear (b), and the inner
ear, which consists of the cochlea (c) and the semicircular canals (d).
The eustachian tube (e) is also shown.

sults in the perception of sound and the recognition of auditory patterns. A simplified diagram of the ear is presented in Figure 11-2.

Three main processes are required to stimulate an auditory nerve impulse. In each, energy is converted from one form to another:

1 Pressure variations due to longitudinal waves in the environment are guided into the external auditory meatus where they cause a mechanical vibration of the tympanic membrane (eardrum). The tympanic membrane is connected to three small bones (ossicles) in the middle ear which, in turn, are caused to vibrate.

2 Mechanical vibration of the auditory ossicles is converted to wave motion in the cochlear fluid at the oval window.

3 Wave motion in the cochlear fluid is converted to nerve impulses in the auditory nerve by hair cells in the cochlea.

Anatomically, the ear is divided into three parts at which these energy transformations take place.

The Outer Ear

Most mammals have two symmetrically placed ears to assist in locating sound sources in the horizontal plane (except owls, which hunt at night and by ear and have asymmetrically placed ears to enable them to locate sounds in both horizontal and vertical planes without tilting the head in the way that dogs do when trying to locate a sound source).

The outer ear consists of the ear flap (pinna), the external auditory canal, and the ear drum. The pinna can be likened to an ear trumpet used to funnel sound into the external auditory meatus. Many mammals (and some humans) are able to move their ear flaps in different directions to locate a sound source. The human pinna is relatively small and unimportant. An attempt to listen to a faint sound is best supplemented with a cupped hand. Sound waves travel down the external auditory meatus which terminates in a thin layer of skin known as the eardrum or **tympanic membrane.** The sound causes the membrane to vibrate, and these vibrations are transmitted to the middle ear.

The Middle Ear

The middle ear is an air-filled, bony chamber connected to the outside atmosphere by the eustachian tube, which terminates at the back of the throat. Discrepancies in air pressure on either side of the eardrum are potentially harmful. Swallowing momentarily connects the middle ear to the outside atmosphere, enabling any pressure differences to be equalized. The popping sensation of the ears which is experienced when one is changing altitude is caused by this sudden equalization of pressure.

Vibrations of the eardrum are transmitted to the inner ear via **three connected bones** which together constitute the **ossicular system.** The malleus is connected to the center of the tympanic membrane at one end and to the incus at the other end. Movements of the eardrum cause movements of the malleus which are directly transferred to the incus. The opposite end of the incus is connected to the third auditory ossicle—the stapes. The far end of the stapes is known as the faceplate. It lies against the opening of the inner ear—the **oval window.**

The auditory ossicles are supported by ligaments. A muscle (the **tensor tympani muscle**) keeps the tympanic membrane under tension, enabling it to act like the diaphragm of a microphone. Movements of the tympanic membrane are transmitted by the bony chain of auditory ossicles to the oval window of the inner ear. The auditory ossicles play an important role in increasing the efficiency with which sound is transmitted to the sensitive hair cells of the inner ear. **Since the inner ear is filled with fluid, sound energy has to be transmitted across a physical interface** (i.e., from pressure oscillations in air to oscillations in a fluid). The transmission of sound in air is different from its transmission in water because of differences in the inertia and elasticity of the two substances. Much of the sound energy in air which meets the surface of water is reflected back (as becomes evident when one is swimming underwater and attempting to listen to sounds on the surface). The auditory ossicles provide an elegant solution to this problem.

The auditory ossicles have been described as an **impedance-matching transformer with automatic volume control.** Their function is to transmit sound energy across the media of air and water. They act as a **lever system** in which displacements of the malleus at the tympanic membrane cause corresponding but reduced displacements of the faceplate of the stapes. This **reduction in amplitude** is accompanied by a corresponding **increase in force** by a factor of 1.3. Furthermore, the surface area at the faceplate is approximately 17 times smaller than that at the tympanic membrane. The pressure of the air against the tympanic membrane is thus increased approximately 22-fold at the stapes against the oval window. Without the auditory ossicles, the threshold of hearing would rise by about 30 dB (Guyton, 1991).

Volume control is achieved by the tensor tympani and stapedius muscles. Intense sound (particularly at frequencies less than 1000 Hz) causes these muscles to contract to pull the malleus and stapes away from the interfaces of the outer and inner ear. This pulling away attenuates loud noise and protects the inner ear from damage. Differences in the effectiveness of this attenuation reflex may help explain why some individuals are more susceptible to noise-induced hearing loss than others.

The Inner Ear

The inner ear is known as the **cochlea.** It can be modeled as a fluid-filled tube which is separated into an upper and lower half by a membrane (Figure 11-3). At the end of the tube are two windows—the oval and round windows—situated on either side of the membrane. The structure of the cochlea is apparent if we imagine a tube folded back on itself and coiled to give it its external, shell-like appearance. The central membrane, known as the **basilar membrane,** ascends along the length of the coiled tube from its base to its apex where it is perforated by a hole (the helicotrema). The tube is tapered— it is widest at the helicotrema end.

Since the cochlea is sealed, inward movement of the stapes pushes the oval window inward and causes a column of fluid to move inward into the cochlea. At the other end of the cochlea, the round window moves outward. Fibers in the basilar membrane run along its length, projecting outward from the center of the cochlea. At the base of the cochlea, close to the oval and round windows, these fibers are short and thick. Toward the apex, they become longer and thinner. Sound vibrations at a particular frequency are transmitted by the stapes and cause motion in the fluid column in the cochlea. This sets up a traveling wave in the cochlea. At a certain point along the cochlea, the basilar membrane is just the right thickness to resonate at the particular frequency of the cochlear wave. At this point along the length of the cochlea, maximum vibratory amplitude will occur.

The basilar membrane is connected to an organ on its surface (the organ of Corti). This organ contains hair cells connected to afferent fibers of the auditory nerve. Movement of the hair cells causes depolarization of nerve endings and the transmission of a nerve impulse to the brain.

When the stapes transmits sound vibrations to the cochlea, they pass along its length until a point is reached where the frequency of the vibrations matches the resonant fre-

FIGURE 11-3   Simplified view of the cochlea "unwound" to reveal its tubular structure. (a) = the stapes, (b) = the oval window, (c) = the basilar membrane, (d) = the helicotrema, (e) = the round window. (From Guyton, 1991.)

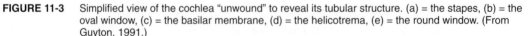
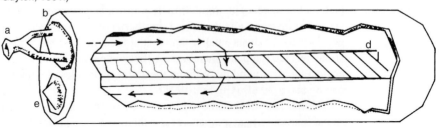

quency of the basilar membrane. At this point the movement of the basilar fibers "short-circuits" the incoming sound wave which is converted to motion of the fibers. This motion causes the hair cells at that point to depolarize. Since the length and thickness of the basilar fibers vary along the length of the cochlea, different frequencies are detected at different parts of the cochlea. High frequencies cause resonance at the base of the cochlea and lower frequencies are detected higher up. This theory, known as the **place-pitch theory,** explains how the **cochlea separates a complex sound into its spectral components.** In this sense, the basilar membrane and fibers are not unlike a harp lying along the length of the cochlea. Detection of sound energy at a particular frequency depends on there being a "string" available which will resonate at that frequency.

Discrimination between sounds of different frequencies takes place in the inner ear rather than in the brain. Supporting evidence for this finding comes from the observation that the cochlea is the only part of the human body that is the same size in an infant as in an adult. If the infant cochlea were a scaled-down version of the adult cochlea, infants would be deaf to low frequencies and sensitive to very high frequencies above the adult range and would probably experience difficulties in learning to speak (because they wouldn't be able to hear their parents' voices properly). Although children are sensitive to higher frequencies than adults, the frequency range to which their ears are *most* sensitive is the same as in the adult because the cochlea is the same size.

Sensitivity of the Ear

The sensitivity of the ear to different frequencies depends on the efficiency of impedance matching by the auditory ossicles and on the response of the cochlear structures. Auditory sensitivity is greatest between 1000 and 4000 Hz—the frequency band in which speech is transmitted (in perspective, the notes on a piano range from 26 to 4096 Hz). Figure 11-4 depicts the sensitivity of the ear to different audible frequencies.

The selective sensitivity of the ear means that the **loudness of a noise** depends on its frequency as well as its sound pressure level. Contours of equal loudness have been worked out using large numbers of subjects. A reference tone of 1000 Hz and, say, 60 dB can be presented to subjects. This tone is followed by a second tone at a higher or lower frequency. The subject then adjusts the intensity of the second tone until its loudness matches that of the reference. In this way, contours of equal loudness can be plotted for all frequencies with respect to a series of 1000-Hz reference tones from 10 to 120 dB. For example, a tone of 100 Hz of intensity approximately 70 dB would sound as loud as a 60-dB, 1000-Hz reference tone.

The unit of loudness is the phon and is derived from the equal loudness contour data. In the above example, the 70-dB, 100-Hz tone would have a loudness of 60 phons. A second loudness measure is the sone. One sone is defined as the loudness of a 1000-Hz tone at 40 dB. Because loudness perception is a logarithmic rather than a linear process, it can be said that a 10-dB increase in intensity represents a doubling of loudness. Thus a 50-dB, 1000-Hz tone would have a loudness of 2 sones.

Old age is marked by compression of the audible frequency range from 16 to 20,000 Hz to 50 to 8000 Hz, a condition known as presbycusis. The higher frequencies are usually lost first and most aged people cannot hear sounds above 10,000 Hz. The mecha-

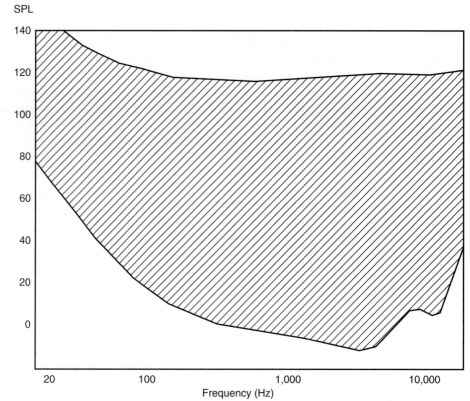

FIGURE 11-4   Sensitivity of the human ear. Sounds below the hatched area are inaudible. Sounds above cause damage to the ear.

nism of hearing loss is at the neural-cochlear rather than the ossicular level and is irreversible.

Noise-Induced Pathology of the Ear

Sudden noises of sufficient intensity may produce trauma, resulting in a reduction in the efficiency of sound transmission to the cochlea or in cochlear response itself. The tympanic membrane may be ruptured or the ossicular chain mechanically damaged. Surgical repair may be necessary.

Sensory neural hearing loss can also occur, particularly after chronic exposure to loud noises. As with age effects, sensitivity to the high frequencies is lost first and the loss is irreversible. In audiometry, such loss is described as a permanent threshold shift.

Audiometric testing consists of determination of the minimum intensity (the threshold) at which a person can detect sound at a particular frequency. As sensitivity to particular frequencies is lost as a result of age or damage, the intensity at which a stimulus can be detected increases. It is in this sense that hearing loss can be described as a

threshold shift. Temporary threshold shifts can occur after exposure to loud noise (e.g., at rock concerts) and are often accompanied by symptoms such as **tinnitis** (ringing in the ears). Repeated exposure leads to permanent threshold shifts (noise-induced deafness). Figure 11-5 shows examples of audiograms of noise-induced threshold shifts. Although permanent threshold shifts are known to occur with age, chronic exposure to noise hastens their onset and severity.

MEASUREMENT OF SOUND

Sound level meters provide several different measures of sound intensity. The **dB (linear) scale** is used to give the **sound pressure level (SPL)**, which is the rms sound pressure at the microphone. If the effect on a listener is the main interest, the frequency of the sound must be taken into consideration. The most commonly used scale is the **dB(A) scale.** When sound is measured with the dB(A) scale, a weighting network **selectively amplifies or attenuates** different frequencies. This process may be likened to adjusting the tone control on a modern hi-fi amplifier. The sum of the pressure levels of the various frequencies is called the **sound level (LS).**

Many devices (such as electric fans or motors) produce intense sound energy at a particular frequency (centered on the 60-Hz mains frequency or harmonics of this frequency). Although the sound pressure levels might be high, the weighted sound level is

FIGURE 11-5 Noise-induced permanent threshold shifts. After 5 years in a noisy job (A). After 20 years in a noisy job (B). Sensitivity to the higher frequencies is lost first (center line gives the patient's average sensitivity). The SPLs are the lowest that the person can hear at a given frequency.

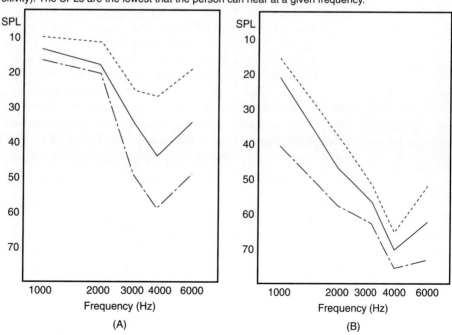

20 or more dB lower since the ear is not very sensitive to sound below 500 Hz. Apart from the A scale, some sound level meters have C and D scales. These specialized scales are used to evaluate noise containing intense pure tone components (such as noise from jet engines).

Frequency Analysis

Frequency analysis is sometimes carried out to obtain further information about the constituent frequencies of a sound source. **Octave analysis** involves dividing the frequency spectrum into bandwidths such that the lower boundary ($f1$) is one-half of the upper ($f2$) where

$$f2 = 2f1$$

The center frequencies of the octave bandwidths are

$$31.5, 63, 125, 250 \ldots \text{Hz}$$

Further resolution can be obtained by using **half- or third-octave bandwidths** where

$$f2 = \sqrt{2} \ f1 \quad \text{(half octave)}$$
$$f2 = \sqrt[3]{2} \ f1 \quad \text{(half octave)}$$

The center frequency of a bandwidth is given by the geometric mean of the upper and lower frequencies.

Octave analysis and its variations are logarithmic scales and are used to take account of the ear's approximate logarithmic frequency selectivity. They have disadvantages when the goal is to analyze the sound source because resolution drops as frequency increases (i.e., the bandwidths get wider and all sounds within a bandwidth are added together). Narrow-band constant bandwidth analysis may then be used to determine intense sound at a particular frequency. The moving parts of many industrial machines, such as pistons, fans, or gears, are often the primary sources of noise. Fortunately, many of these rotate or cycle at known frequencies below 1000 Hz, so their contribution to the total noise can be determined fairly easily. Figure 11-6 shows a third-octave analysis of a pneumatic hammer.

Several Sound Sources

In practice, there are usually several sources of sound and it is the combined effects of these on the worker that is of interest. The combined SPL from several sound sources can be calculated from first principles, although it is usually estimated by calculating the difference between the two loudest SPLs and adding the amount shown in Table 11-2 to the louder of the two.

If there are more than two sources of noise, the exercise can be continued by taking

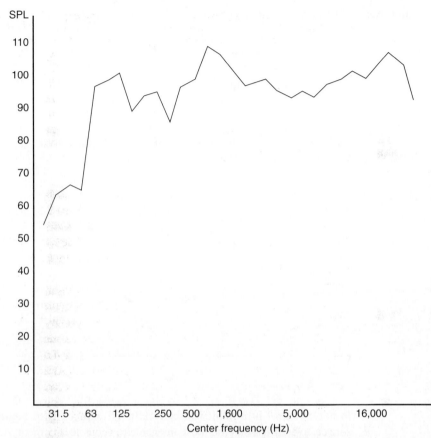

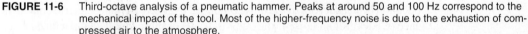

FIGURE 11-6   Third-octave analysis of a pneumatic hammer. Peaks at around 50 and 100 Hz correspond to the mechanical impact of the tool. Most of the higher-frequency noise is due to the exhaustion of compressed air to the atmosphere.

the combined SPL of the two loudest sounds and comparing this value with the third loudest sound. The difference between these is then added to the estimate of the two loudest sounds to obtain a new estimate of the overall noise. This process can be repeated for other noise sources until the difference between the combined estimate and any remaining sources is 10 dB, at which point the addition of more noise sources will, in practice, have negligible impact on the overall noise.

That the addition of two equal sources increases the total sound level by 3 dB (see Table 11-2) is due to the logarithmic nature of the decibel scale. It also applies to the subtraction of noise levels. If three adjacent machines have noise levels of 90, 95, and 101 dB, the combined noise will be 102 dB. Removal of the 90-dB machine will still leave a noise level of 102 dB, whereas removal of the 101-dB source leaves a noise level of 96.5 dB. This principle of subtraction has many practical implications for noise control, as will be seen below. In particular, it emphasizes, for example, that the best approach

TABLE 11-2   SUMMATION OF SPLs $L1$ AND $L2$

Difference between $L1$ and $L2$	Amount to increase louder of $L1$ and $L2$
0.0	3.0
0.5	2.8
1.0	2.5
1.5	2.3
2.0	2.1
2.5	1.9
3.0	1.8
3.5	1.6
4.0	1.5
4.5	1.3
5.0	1.2
5.5	1.1
6.0	1.0
6.5	0.9
7.0	0.8
7.5	0.7
8.0	0.6
8.5	0.5
9.0	0.4
9.5	0.4
10.0	0.3

to noise control in a room with several noise sources is to **begin with the noisiest source,** even if it is not the most amenable to acoustic treatment. Another practical implication is that surveys of the noise produced by machines can be carried out in the presence of background noise if the latter is 10 dB or more below the machine noise.

Measuring Noise Exposure

A number of measures have been developed to enable employees' daily noise exposure to be measured. The equivalent A-weighted noise level, or L_{Aeq}, has been developed for those situations where noise levels fluctuate over the course of a day. It is an integrated value—the average level of sound energy over the measuring period. The L_{Aeq} is defined as follows:

That steady-state sound level which would have the equivalent sound energy as the actual noise over the same period of time.

The L_{Aeq} can be thought of as "smoothing out" the peaks and the troughs of the daily noise exposure to give an average. If a worker's exposure times to different noise levels over the course of an 8-hour shift are known, the L_{Aeq} can be calculated as follows:

$$L_{Aeq} = 10 \, Log_{10} \, (T1/8 \text{ antilog } L1/10 + T2/8 \text{ antilog } L2/10 + \cdots$$
$$+ \, Tn/8 \text{ antilog } Ln/10)$$

where $T1 \ldots Tn$ = exposure time (in hours) at sound levels $L1 \ldots Ln$

Note that the time T spent at a given sound level is expressed as a percentage of the total 8-hour shift.

In practice, employees wear a small recorder or sound level meter with the L_{Aeq} facility during a shift. It continuously measures the noise to which they are exposed. Because the L_{Aeq} takes into account infrequent, but very loud, noises, it can sometimes give a better indication of an employee's total noise exposure than a noise survey consisting of one or more short visits to measure the noise in an area during which loud, but infrequent, noises may not occur.

Safe Exposure Levels

Much research has been carried out to determine cutoff noise levels below which operators can be exposed to for an 8-hour day without increased risk of hearing loss. The concept of a **maximum daily noise "dose"** is essential if measurements made in the workplace are to be interpreted correctly and appropriate action taken. In the United States, **OSHA has specified 90 dB(A)** as the **maximum permissible exposure to continuous noise for an 8-hour shift.** Many other countries, including those in the European Community, also regard 90 dB(A) as the maximum permissible level. However, there is a modern trend worldwide to reduce daily exposures to below 90 dB(A). An exposure level of 85 dB(A) is often regarded, informally, as preferable.

Noise Dosimeters

A noise dosimeter integrates the noise measured at the microphone over a period of time and expresses it as a percentage of the daily allowable noise dose (e.g., as a percentage of 90 dB(A) for 8 hours). Dosimeters can be used to decide whether workers are being exposed to excessive noise and whether they require ear protection, and also to specify for how long per day a worker may be exposed to a particular noisy work situation (Figure 11-7).

L_{Aeq} and noise dose measurements are particularly useful in hazard surveillance of occupations where the worker is mobile because they integrate the different intensities of noise which can vary considerably over the course of the working day. For example, garbage collectors, factory supervisors, and agricultural workers are all exposed to varying noise levels during the course of their work; thus an integrated value of their total noise exposure is often more meaningful then isolated measurements made during particular work activities.

The setting of standards for noise exposure has involved a compromise between the desire to provide maximum protection of the hearing of workers and the cost benefits of noise reduction. Many industrial processes are intrinsically noisy and although practical steps can usually be taken to reduce the noise, a point of diminishing returns is often

FIGURE 11-7   A noise dosimeter. (Photograph supplied by Bruel and Kjaer, Naerum, Denmark.)

reached, at around 90 dB(A), where further reductions in noise become increasingly modest but at greatly increased cost. Also, individuals differ in their susceptibility to noise-induced hearing loss—it is often impossible to lower noise levels to ensure complete protection of all workers. A level of 90 dB(A) for 8 hours is OSHA's maximum permissible exposure level. An **equal energy rule** is used to specify maximum exposure times for louder noises. For example, with a halving rate of 5 dB, a noise dose of 90 dB(A) for 8 hours is held to be equivalent to 95 dB(A) for 4 hours or 100 dB(A) for 2 hours (a halving rate of 3 dB is used in some countries). Standards usually specify a maximum steady noise level (overriding limit) which operators must not be exposed to at all. The OSHA limit is 115 dB(A). It can be seen that even very brief daily exposures to very high noise levels can render an otherwise innocuous work area a noise hazard. Once again, it must be emphasized that very loud noises, those over 115 dB(A), can make an inordinately large contribution to a worker's daily noise exposure even if they are of short duration. Eliminating exposure to noises at this level is usually one of the simplest and most straightforward ways of reducing daily noise exposure.

Integrating Sound Level Meters

To cope with the widely varying types of noise encountered in industry, commercial meters usually have several response modes—normally, slow and fast, and sometimes, impulse response modes. Some modern sound level meters are shown in Figure 11-8.

The slow mode is used to obtain an average reading when rapid fluctuations in intensity measured on the fast mode are greater than 3 or 4 dB. In fairly steady noise with

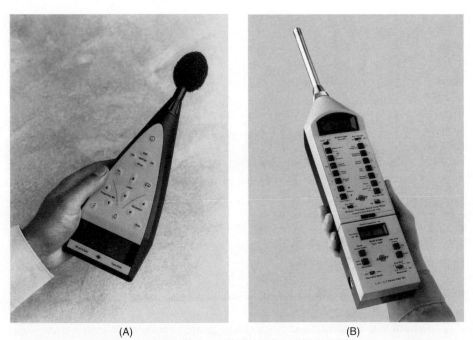

(A)   (B)

FIGURE 11-8   Modern sound level meters. (A) Measures simultaneous Leq and peak levels (a windshield is fitted for outdoor measurements). (B) Sound level meter with octave filter set attached. (Photographs supplied by Bruel and Kjaer, Naerum, Denmark.)

fluctuations of only a few dB, the fast response may be used and a simple average can be taken by the person carrying out the measurements. For example, fluctuations from 85 to 91 dB could be interpreted as 88 ± 3 dB. If the fluctuations are greater than 6 dB, the average sound level can be estimated as 3 dB below the maximum value. Impulse noise cannot be measured accurately with the simple (less expensive) meters since their response time is too slow, but more sophisticated equipment is now available. As a rule of thumb, the impact noise level measured with a standard meter should be regarded as an underestimate of at least 5 dB.

Before a sound source is measured, the sound level meter should be calibrated. Meters are commonly sold with a portable calibrator (which itself should be periodically checked). When measurements are taken out-of-doors, it is advisable to attach a spherical windscreen (made of foam and of about 5-cm radius) over the microphone to prevent wind from distorting the measurements. When operator exposure to noise is measured, the microphone should be situated close to the operator's ear. Octave, half-octave, and third-octave sets are available which can be used in conjunction with standard meters. Figure 11-9 shows a portable sound level meter, fitted with a spherical windscreen, being used to measure noise levels in a mine.

In investigations of the noise produced by machines in offices and factories, a certain amount of supplementary information should be obtained to complement the sound level measurements (Table 11-3).

FIGURE 11-9   Handheld sound meter being used to measure sound levels in a mine (windshield protects microphone in harsh environments).

Noise Surveys

Noise surveys are a useful method of evaluating the distribution of noise in a working area and the exposure of workers. The two main approaches used in occupational hygiene are known as **area sampling** and **personal sampling.** In area sampling the approach is to evaluate the workplace. This approach works well if the noise is fairly constant and workplaces are fixed. If people move around at work or if noise levels are closely tied to a worker's activity, personal sampling may be necessary. In some industries, such as construction and mining, a combination of both approaches is needed.

In area sampling, a floor plan of the workplace is usually obtained. Sound levels at various locations are recorded directly on the map. In this way, areas of high and low noise can be identified so that appropriate remedial action can be taken on the factory floor. It is sometimes useful to plot noise contours—regions of equal sound level superimposed on the floor plan. In conjunction with data about the positioning and movement of personnel, a number of steps can be taken to minimize excessive noise exposure (Table 11-4).

An example of a noise survey which was carried out in a colliery can be found in Bowler (1982). All areas where noise levels exceeded 90 dB(A) were identified and the duration of exposure of persons working therein was determined. Special attention was paid to areas with noise levels greater than 105 dB(A). Within the previously determined 90-dB(A) contour, attenuation of noise was measured by taking repeated measurements up to 1 m from the source (intensity halved approximately as distance from the source

TABLE 11-3   SUPPLEMENTARY INFORMATION FOR NOISE MEASUREMENT

1 A description of the space in which the measurements were made, its dimensions, background noise, and the presence of other noise sources

2 A description of the source itself (for example, data from the nameplate, such as the model number, operating speed, and power; the percentage of maximum load when the measurement was made; and, if different from this, the normal load)

3 Calibration weighting network, and response mode of the sound level meter

4 Background noise level

5 Number and location of personnel in the area

6 Position of microphone with respect to the source

7 Extent of fluctuation of noise levels

8 A-weighted measurements at operator's ear level and at positions of other personnel

9 Time spent at machine by operator each day

10 Results of any previous audiometric testing of workers

11 Previous attempts at noise control

12 Whether ear protection is available

TABLE 11-4   SOME BASIC STEPS IN THE MANAGEMENT OF INDUSTRIAL NOISE EXPOSURE

Short-term measure	Issue earplugs or earmuffs.
Medium-term measures	Reposition noisy machines.
	Soundproof noisy machines.
	Demarcate noisy areas with warning signs.
	Rotate workers between noisy and quiet jobs.
Long-term measure	Institute a comprehensive noise-reduction program:
	Soundproof machines.
	Replace with less noisy machines.
	Change the process.
	Build acoustic refuges.
	Conduct audiometric testing.
	Implement rules and procedures for the wearing of ear protection.
	Implement an audiometric testing program.

doubled). Maximum and minimum noise levels were also measured at the operator's ear level at all fixed workstations. Together with data on exposure times, representative L_{Aeq}'s were calculated. Operators suspected of being exposed to noise over 105 dB(A) were issued dosimeters so that more information on their exposure could be obtained. Noise exposure when traveling to and from working areas was also measured. In this way, both high-risk areas and high-risk jobs were identified.

EAR PROTECTION

Earmuffs or earplugs, either singly or in combination, can protect the ear from excessive noise by providing an airtight (acoustic) seal between the atmosphere and the external auditory canal. Reductions in SPL at the eardrum of up to 40 dB are possible. For very high noise levels (above 140 dB) simple ear protection is inadequate since excessive vibration can be conducted to the cochlea by bone or tissue. Figure 11-10 illustrates the effects of ear protection on the attenuation of different frequencies of sound from a pneumatic hammer.

Even if ear protection is worn, sound energy may still find its way into the auditory system. The protector may vibrate or energy may pass through leaks in or around the protector. Homemade earplugs are often inadequate for this reason. Some characteristics of effective ear protectors are given in Table 11-5.

Sized earplugs are reusable and are usually made of soft, flexible materials such as

FIGURE 11-10 Attenuation of noise from a pneumatic hammer by ear protectors. (a) = third-octave analysis, (b) = attenuation as a result of ear protectors, (c) = A-weighting of attenuated noise.

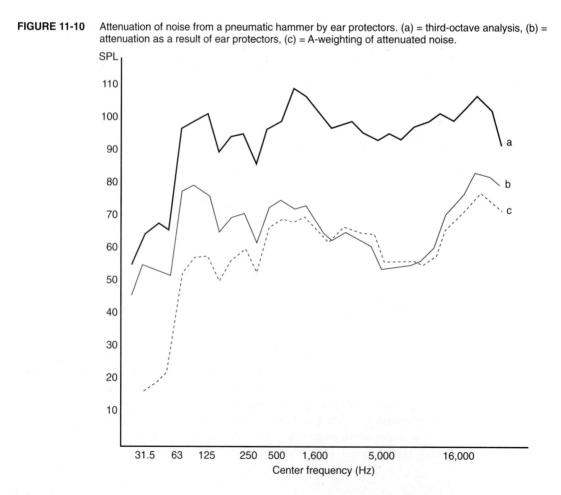

TABLE 11-5   CHARACTERISTICS OF EFFECTIVE EAR PROTECTORS

1 Impervious to air

2 Adaptable to the shape of the user's head or external auditory canal—to form an airtight seal and to ensure comfort by avoiding pressure "hot spots"

3 Remain firmly in place without causing pressure ischemia

molded plastic. They come in different sizes and should be cleaned with soap and water after use. Formable earplugs are usually made of materials such as cotton or wax. The user molds the plug into a cone shape to conform to the dimensions of the ear canal. Protection depends on the material used—cotton provides less protection than wax and certain polymers. Earplugs have the advantages of being small and easy to carry, convenient to wear in confined spaces, and cheaper than earmuffs. They are also compatible with long hair or protective headgear. Some of the main disadvantages are that they take longer to fit than earmuffs and are not always suitable for periodic use because of hygiene problems. Unclean hands may transfer dirt to the ear canal via the plug. Earplugs cannot be worn by people with ear infections.

Earmuffs should provide a good seal between the inside of the cup and the skin surrounding the pinna. They should be adjustable for a range of head sizes. They usually provide better and more consistent protection than earplugs and are usually perceived as being more comfortable. Normally, a single-size, adjustable muff can fit most head sizes. Earmuffs can also be incorporated into other protective headgear such as hard hats and are less likely to be forgotten. However, they can cause skin rashes in hot environments because sweat accumulates at the muff-skin interface. In confined spaces such as mine stopes, they may be dislodged too easily.

Manufacturers of ear protection sold in the United States are required to provide data on the noise attenuation provided by their products (known as the noise reduction rating, or NRR). Because these data are obtained under ideal testing conditions, less protection may be provided in practical settings. Casali and Park (1990) investigated ear protection attenuation performance under different fitting conditions and physical activities. When users fitted the ear protectors using only the manufacturers' instructions, protection was 4 to 14 dB less at 1000 Hz than when they were trained in correct fitting techniques. Muffs were less susceptible to fitting effects than plugs. When subjects carried out tasks involving much upper torso movement and acceleration of the head, losses in protection up to 6 dB occurred. Premolded plugs, muffs, and muff-plug combinations were most susceptible to these losses and compliant foam earplugs least susceptible. It seems that workers should be trained in correct fitting techniques if proper protection is to be obtained. This finding applies to all kinds of ear protection, particularly compliant foam plugs. For tasks requiring a great deal of movement, foam plugs may offer the best protection over the workday if workers are trained to fit them properly. If wearers cannot be trained or for occasional users, muffs may be the best choice.

Acceptance of ear protection is often a major problem in industry. Workers exposed to excessive noise frequently refuse to wear any ear protection, complaining that it causes discomfort, interferes with speech communication, and, in jobs where the machine is the source of noise, degrades task feedback. While these complaints are sometimes justified, ear protectors do not necessarily interfere with speech communication because they lower the intensity of noise as well as of speech (the signal-to-noise ratio remains the same). The lowered sound levels may reduce mechanical distortion of structures in the ear and thus improve speech recognition.

Lhuede (1980) investigated earmuff acceptance among workers in a sawmill. He developed an index for rating earmuffs whereby earmuff attenuation (in decibels) was divided by the product of muff clamping force and weight (in newtons). Out of a sample of commercial earmuffs, the two most highly rated were also found to be the most acceptable to workers. However, it was concluded that the concept of an objective index for evaluating earmuff acceptability required further investigation. Weight was not regarded as an important determinant of comfort by many workers, although the hotness of the muffs in warm conditions was found to be extremely important. It can be seen that ergonomic factors are important determinants of operator compliance in the use of hearing protection. However, because noise-induced hearing loss is a pernicious, long-term phenomenon, many workers may be unaware or unconvinced of the potential threat to their hearing. Ear protection in the workplace should be part of a wider industrial hearing conservation program containing appropriate educational and safety information.

DESIGN OF THE ACOUSTIC ENVIRONMENT

For many animals, hearing is essential to survival, alerting them, as it does, to the presence of predators or prey. The alerting quality of noise makes it an ideal warning signal for humans. By the same token, noise can distract workers and is a major source of dissatisfaction with the environment. Individuals differ widely in their attitudes to and tolerance of noise. Clearly, many factors need to be considered (in addition to the effects of noise on hearing) if an optimal auditory environment is to be designed.

Woodson (1981) has compiled the findings of many studies of the effects of noise on human performance. Although noise levels below 90 dB(A) are not a serious threat to hearing, they can degrade task performance and cause annoyance. Table 11-6 summarizes some of these effects.

Reverberation

The noise level in a room depends not only on the intensity of any noise sources but also on the **reverberation** characteristics of the room. The sound emanating from a source reaches the listener both directly and after being reflected off the walls and other objects (Figure 11-11).

Reverberation is due to the interreflection of sound waves within a room. The more reflection, the longer the reverberation time. If there is continuous sound, background noise caused by interreflection builds up over time to a steady level. This noise is called the reverberant field. Reverberation is a very important acoustic characteristic which influences an occupant's perception of a room. Some pertinent aspects of reverberation are summarized in Table 11-7.

The reverberation time of a room is measured by the time taken for the reverberant

TABLE 11-6   EFFECTS OF NOISE BELOW 85 dB(A)

Noise level, dB(A)	Effect
80	Conversation is difficult.
75	Telephone conversation is difficult. Raised voice is needed for face-to-face conversation.
70	Upper level for normal conversation. Telephone conversation is difficult. Unsuitable for office work.
65	Upper acceptance level when people expect a noisy environment.
60	Acceptable level for daytime living conditions.
55	Upper acceptable level when people expect quiet.
50	Acceptable by people who expect quiet. About one-fourth of the population will experience difficulty in falling asleep or, if asleep, will be awakened.
40	Very acceptable for concentration. Few people will have sleep problems.
<30	Low-level intermittent sounds become disturbing.

level to fall by 60 dB. Reverberation time is a function of room volume and total sound absorption. Excessive reverberation causes problems, because as the distance of the listener from the target sound source increases, the intensity of direct sound decreases (by about 6 dB per doubling of distance). A point is reached where the intensity of the direct field is the same as the intensity of the reverberant field. This point is known as the critical distance. The total SPL at this point will therefore be 3 dB greater than the direct sound and the reverberant sound alone.

Reverberation may be minimized by increasing the amount of sound-absorbing material in a room. It is important to remember that the occupants themselves are sound absorbers (or rather, their clothes are), so the reverberation characteristics of meeting halls and conference rooms will differ when they are empty compared with when they are full, unless the seats are upholstered.

Reverberation can mask speech. Reverberation times of less than 1 second are recommended for rooms where speaking is a main activity. Longer reverberation times are needed for listening to music because reverberation gives depth to music. Rooms suitable for listening to music may not be suitable for conferences and vice versa.

Building Design and the Auditory Environment

The design and construction of buildings have a profound impact on their acoustic characteristics. Old churches, for example, have thick stone walls and small windows, so little outside noise can enter. Noise within the building reverberates, which adds depth and resonance to speech and singing. In open-plan offices and factories, the transmission and reverberation of noise usually has to be reduced by using barriers such as upholstered screens to absorb sound energy and by using acoustic ceiling tiles.

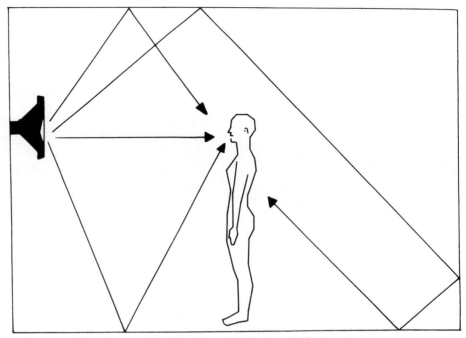

FIGURE 11-11   Reverberation. Sound in a room reaches the ear by direct and indirect routes.

INDUSTRIAL NOISE CONTROL

Several approaches to noise control can be identified:

1 Eliminate the threat to hearing by redesigning the machine or using a less noisy machine.

2 Remove personnel from the noisy environment.

3 Protect personnel by issuing earplugs or earmuffs, or by building an acoustic refuge.

Many machines have both primary noise sources (such as the power unit) and secondary noise sources (parts of the machine which resonate). In tractors, for example, the

TABLE 11-7   SOME CHARACTERISTICS OF REVERBERATION IN ROOMS

1 Reverberation time in a room does not depend on the position of the source or the position of the listener.

2 Excessive reverberation is a major cause of the blurring of acoustic signals such as speech.

3 Room shape has little effect on reverberation time.

4 The intensity of the reverberant field in a room depends on several factors:

The sound level of the source

The volume of the enclosed space

The amount of sound-absorbing material in a room

primary sources are the air intake, fan, engine walls, injector pump, and transmission. The fuel tank, cab structure, paneling, and fenders are examples of secondary sources (Tomlinson, 1971).

The relative contribution of the primary and secondary sources to the total noise depends on the particular machine. For example, pneumatic machines are inherently noisy because of the exhaust gases. The energy loss, due to noise, of machines such as pneumatic jackhammers is only about 0.08 percent (Holdo, 1958) because 87.5 percent of the noise is due to exhaust gases rather than to the impact of the drill steel against the rock or to the pistons and gears of the machine (Weber, 1970). Hydraulic pumps can be very quiet, yet the hydraulic system itself can be a source of considerable noise due to resonance if the pump is not securely mounted and mechanically isolated from fluid lines and other structures.

Noise from the primary sources is often periodic and below 1000 Hz, whereas secondary noise is often of a higher frequency. For example, velocity gradients in the exhaustion to atmosphere of compressed air create random vortexes which move and are dissipated in a chaotic manner, giving rise to wide-band random noise (Beiers, 1966). Frequency analysis with the machine operating under representative conditions of loading is an important first step in machine redesign. The most intense frequencies can be identified and redesign can proceed in a systematic manner. Bartholomae and Kovac (1979) provide an interesting case study of the redesign of a pneumatic rotary-percussive jackhammer. These machines emit noise of 115 to 120 dB(A) at the operator's ear level. By muffling the exhaust, redesigning the drilling mechanism using nylon instead of metal components, and acoustically treating the drill steel, noise was reduced to 95 dB(A).

Some common problem areas and control methods are described below.

Fans   Fans or blowers are much noisier when running at high rather than low speed. The sound level varies as the 5th power of the speed. Use a larger fan running at a lower speed.

Muffling   Pneumatic tools such as paving breakers, screwdrivers, and dentists' drills produce noise due to the exhaustion of compressed air to the atmosphere. This noise can be reduced by piping the air away from the operator to an unoccupied area or by using mufflers which reduce the escape velocity of the compressed air. Reductions of up to 6 dB are possible, but efficiency can drop by up to 15 percent. A third-octave analysis of a muffled and unmuffled pneumatic machine is presented in Figure 11-12.

Part Ejection   Pneumatic ejectors are sometimes used to remove parts from presses. Mechanical ejectors are usually quieter.

Pneumatic Tools   Hydraulic or electric equivalents are usually quieter.

Impact Tools   Large electric solenoids for tripping clutches can be sources of impact noise. Resilient bumpers fitted to the point of impact can be beneficial.

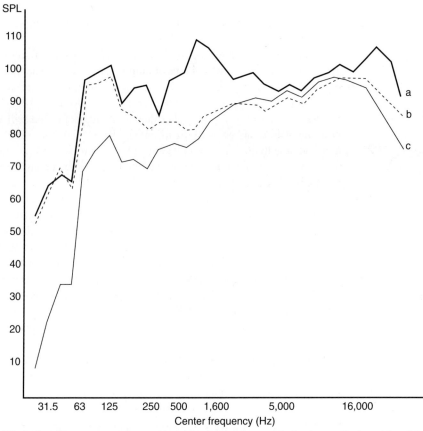

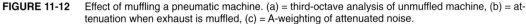

FIGURE 11-12   Effect of muffling a pneumatic machine. (a) = third-octave analysis of unmuffled machine, (b) = attenuation when exhaust is muffled, (c) = A-weighting of attenuated noise.

Hydraulic Reticulation   Cavitation (the release of dissolved air in a fluid due to sudden pressure drops) and turbulence can increase noise from pumps considerably. Cavitation can be reduced by decreasing fluid flow velocity, smoothing the system by removing sudden bends or junctions, and avoiding sudden drops in pressure.

Vibration   Machine vibration can be exacerbated by imbalance or eccentricity in rotating members, inadequate mountings, and wear. Resilient couplings can be used to isolate a source from surrounding objects. Regular maintenance and replacement of worn parts can prevent vibration. In offices, typewriters and daisy wheel printers can be placed on felt or rubber mats to block the transmission of vibration to the supporting table which acts as to amplify the sound. Printer noise can interfere with speech.

A change in process can also reduce noise. Riveting can be replaced with welding, for example, and dot matrix printers can be replaced with laser or ink jet printers. "Dead" materials such as rubber, wood, and some types of plastic are preferable in the construction of flooring, pipes, etc., have high internal damping and do not transmit sound as readily as "live" materials such as metals.

Since sound attenuates with distance, attention should be paid to the location of machines or components such as fans or pumps. Locating these at floor level rather than ear or waist height can significantly lower worker exposure to noise.

Noise Insulation

If machine redesign is not possible or produces inadequate reductions in noise, an acoustic enclosure can be built around the source or an acoustic refuge can be built for the operator.

If the source is enclosed, noise is interreflected by the walls of the enclosure, losing energy as it does so. The inside of the enclosure may be lined with sound-absorbing materials to increase attenuation as the noise interreflects off the enclosure walls. The interreflection increases the noise level in the enclosure by up to 10 dB without absorbent lining.

Sound-absorbing materials are usually porous and lightweight. As the sound waves travel back and forth within the tiny interstices of the material, their energy is converted to heat by friction. The sound-absorbing characteristics of a material are described by its absorption coefficient—the ratio of the energy absorbed to the energy striking the material. A perfect absorber has a coefficient of 1, whereas a perfect reflector has a coefficient of 0. The absorption coefficient of a material varies with frequency; therefore frequency analysis of the problem noise is needed if an optimal choice of material is to be made.

Sound energy does not pass efficiently across the interface of two materials which differ greatly in elasticity. This principle is used in the construction of acoustic refuges. Noise striking the walls of the refuge is reflected away. The best barrier materials have a high density and are nonporous. Rubber seals can be used on windows and door frames to prevent leakage of sound into the refuge.

Concrete, which has absorption coefficients of 0.01 to 0.02 at frequencies from 125 to 4000 Hz is a good reflector and therefore a **good noise insulator.** Fiberglass has a coefficient of 0.48 at 125 Hz, rising to 0.99 at 1000 Hz and is a **good noise absorber.**

Partial enclosures can also be effective provided that the operator's ear remains in the "shadow" provided by the enclosure. Partial enclosures are suitable for protection against higher-frequency noise. The dimensions of the enclosure have to be several times the wavelength of the sound. Noise from daisy wheel and dot matrix printers can be reduced significantly by installing correctly designed hoods over the platen.

NOISE AND COMMUNICATION

Most human work depends on verbal communication and concentration. Noise can interfere with communication in a manner which is analogous to the way glare interferes with vision. According to the place-pitch theory of cochlear function, it would seem that registration of target sounds such as speech will be disrupted by background noise and that the disruption will be greater if the background noise shares frequencies that are similar to those of the target signal. Such disruption is known as masking, and there is evidence that masking is greater when the signal and the noise frequencies are similar (Davies and Jones, 1982).

Several indexes have been developed to predict whether noise will interfere with speech communication. Much information in spoken language is conveyed by the consonants rather than the vowels, and one index of speech intelligibility is the percentage articulation loss (%AL) of consonants. A %AL less than 12 signifies excellent communication for untrained listeners and speakers, whereas speech with a %AL greater than 30 is virtually unintelligible. Bailey (1982) provides a useful introduction to many of the factors influencing speech intelligibility and communication. A straightforward method is to calculate the speech interference level (SIL) of the background noise, bearing in mind that ordinary speech is of approximately 65-dB intensity. SILs are calculated from the arithmetic mean of the SPLs of the noise in the frequency bands 500 to 1000, 1000 to 2000, and 2000 to 4000 Hz. By reference to tables, one can specify the minimum distance between two people for adequate communication (speaking normally or by raising the voice) for a particular SIL. There are several other methods for determining the effects of noise on speech communication. However, it must be remembered that speech is highly redundant and comprehension depends on the hearer's knowledge of the subject matter as well as on the level of noise (Haslegrave, 1990). Factors such as the subject matter, the knowledge of the people involved, and the content of the messages to be communicated must also be taken into account when the effects of noise on speech communication are being investigated.

Auditory Warnings and Cues

Although vision is the dominant sense in humans, auditory warning signals and displays are often extremely valuable. An effective auditory warning should be of sufficient intensity to stand out above background noise and differ in pitch, waveform, etc. so as to be discriminable. The ear is most sensitive to the frequency range 500 to 3000 Hz, so warnings ideally should be in this range, with high intensities at the lower frequencies because lower frequencies are not deflected by objects and travel farther than high frequencies. Warnings should not be excessively loud so as to increase workers' daily noise dose above acceptable levels (louder is not better beyond a certain point). Certain types of warning signals may be more appropriate for some types of hazards than others. Lazarus and Hoge (1986) found that different types of danger signals were judged by workers to be more compatible with certain situations than others. Sirens were judged to be compatible with danger arising from "rays" (radioactivity, lasers) and danger areas. Horns were compatible with danger from machines. An alarm similar to that used on police cars was associated with fire or vehicle danger. Impulsed sounds such as bells had low compatibility with both specific and general danger situations and appeared to be more suitable for signifying rest pauses or the end of a shift.

For a sound to serve as an auditory cue, it must first be heard above the background noise (i.e., be sensed) and, second, it must attract the person's attention. These two requirements, along with the theoretical discussion of cochlear function, would suggest that the two key design parameters are cue intensity and frequency. If the frequency of the cue is very different from that of the noise, the cue can be more easily sensed. Murrell (1971, quoted in Oborne, 1982) suggested that auditory cues should be at least 10 dB louder than the background noise. Miller and Beaton (1994) recommend a signal-to-noise ratio of 8 to 12 dB(A). Sound attenuates according to an inverse square law, so the farther away the listener is from the source, the greater the attenuation. Less attenuation takes place in confined spaces than in open spaces because of interreflection of sound waves. In practical situations then, the following considerations are essential in the specification of auditory cues:

1 *Intensity of background noise at the point(s) where the cue is to be heard.* A noise survey of the premises at times when the background noise is at its loudest is required to determine background sound levels.

2 *Frequency of background noise.* A frequency analysis of the background noise can be carried out to assist in the selection of an auditory cue of a different frequency. Much machinery produces complex sounds with many different frequencies (Sorkin, 1987). In this case, the cue must be selected so that it is more intense than the higher-intensity frequencies, particularly if these are of a frequency similar to that of the cue. Sorkin suggests that in the case of very high intensity background noise [pneumatic machines produce sound levels of over 115 dB(A)] auditory cues not be used at all.

3 *Attenuation of cue intensity from source.* If the cue is produced at a place different from where it is intended to be heard, it will be necessary to determine by how much the source intensity of the cue is attenuated at the points where it is to be heard. It is at the listening points that the cue must be distinguished from the noise.

Apart from intensity and frequency, the cue must have high discriminability; it must be different from any other sounds the operator is likely to hear at work. Sirens, for example, can be made to warble or fluctuate in intensity or frequency. A fast onset time can be very attention-getting, but Sorkin (1987) cautions that onset rates of more than 10 dB per millisecond can cause a startle response which can have unanticipated consequences. Onset rates of 1 dB per millisecond are recommended. Auditory cues should persist for at least 100 milliseconds.

Some of these design considerations are illustrated by Miller and Beaton's (1994) analysis of the problems in detection of emergency vehicle sirens by motorists. In the United States, emergency vehicle sirens have an intensity of 118 dB(A) at a distance of 10 m from the source. This is a maximum limit which was set to prevent ear damage and limit noise pollution. Although pedestrians will usually hear an oncoming emergency vehicle at a great distance, drivers of other vehicles will not because the interiors of many modern cars attenuate sound by 20 to 30 dB(A) if the windows are closed. Thus, at 10 m from an emergency vehicle, the intensity of the sound at the driver's ear will be about 98 dB(A). At 50 m, it will be about 74 dB(A). Since the intensity of sound in a car with the radio turned off is about 70 dB(A), the driver is unlikely to hear the siren of an oncoming emergency vehicle until it is about 25 m away. If the emergency vehicle is traveling at a speed 50 km per hour faster than the car, the driver will have 1.8 seconds to take evasive action. Clearly, this is not long enough and explains why sirens alone are inadequate to prevent emergency vehicles from being impeded by traffic (alternative methods of alerting drivers to the presence of emergency vehicles using intelligent vehicle highway system, IVHS, technology are discussed in Miller and Beaton, 1994).

A recent development in the design of auditory warnings is the use of stereo headsets to provide a three-dimensional "head-up" audio display. In research with pilots, Begault (1993) compared a monaural warning with a binaural warning presented over stereo headphones. Pilots had to visually locate a target aircraft while flying a simulator. Auditory collision-avoidance warnings were provided under the two conditions of presentation. Crew members using the three-dimensional display acquired targets approximately 2 seconds faster than those using the monaural warning. The use of the auditory sense to provide spatial information may have promising application in many information-intensive tasks.

Voice Warnings

Wolgalter and Young (1991) investigated behavioral compliance to voice and print warnings. They found that for simple warnings, compliance was greater to voice then to print and greater still when the two were combined. They argue that voice warnings have been relatively unexploited and offer many advantages over other types of warning. Voice is attention-getting and omnidirectional and can orient people to hazards such as slippery floors in complex and visually distracting environments such as malls. Compared with other auditory warnings, voice conveys information directly. However, voice is less useful for conveying complex warnings because of its serial nature, transmission load, and fragility. Wolgalter and Young suggest that it may be a useful supplement to complex visual warnings, capturing attention and providing effective contextual orientation.

Representational Warnings and Displays

Representational auditory warnings have been termed "earcons" on the basis of an analogy with icons, their visual counterparts (Blattner et al., 1989). Stanton and Edworthy (1994) distinguish between representational and abstract earcons. The former sound, in some way, like the object they represent (which itself must be associated with some kind of sound). For example, applause can represent approval, the sound of a closing drawer can represent closing a file on a computer system, and a tone of steadily falling frequency can represent a reduction in the value of a variable. Abstract earcons have no direct linkage with their referents and may be more confusable except when there is some well-known stereotype associated with their use (e.g., a ringing bell signifies alarm). Representational alarms can potentially increase the amount of information that can be communicated, according to Stanton and Edworthy, and may be advantageous in systems or products used by different language speakers. New population stereotypes (see Chapter 13) may emerge among user populations if earcons are implemented in a consistent way in future systems.

A related consideration is the use of sound effects in the design of interactive systems. Nielsen and Schaefer (1993) investigated the usability of a paint program which incorporated sound effects. Paint programs provide users with several "brushes" for drawing on the computer screen (e.g., thin or thick straight lines, airbrushes, jagged lines, lines which "drip" paint, etc.). In one condition, different brush options were accompanied by representational sounds when used—the jagged line "crackled," the dripping line "dripped," etc. Although younger users enjoyed the sound effects, older users did not and found the interface more difficult to use with sound rather than without sound. Nor did older users find that the sound effects helped them remember the different paint functions. The results do not suggest that sound effects in themselves are ineffective because in this study, they were redundant in that paint functions were represented with visual icons as well. For relatively simple interfaces then, the addition of another modality to represent system functions may be unnecessary. For more complex interfaces and when not all system components can be represented visually, sound effects may have greater value.

Bailey (1982) summarizes some of the design considerations in using auditory displays: Auditory displays are more alerting than visual displays. They are suitable when the message is short and need not be referred to later and when the user's job requires movement (for example, among banks of complex control panels). They are also useful to communicate with people working in dark places (such as miners) or at night.

THE AUDITORY ENVIRONMENT OUTDOORS

Several factors influence the transmission of sound out-of-doors. Climatic factors such as temperature gradients, wind, and relative humidity can attenuate and refract sound. Buildings can have a major effect by reflecting sound in a particular direction or by causing echoes (if the reflected sound reaches a listener more than 50 milliseconds after the direct sound, an echo will be perceived). Such factors should be considered when, for example, positioning loudspeaker systems out-of-doors.

Some of the major sources of outdoor noise are traffic, aircraft, factories, and outdoor machinery such as paving breakers and lawnmowers. The methods for characterizing outdoor noise are similar to those used indoors. Measurements such as the L_{Aeq} may be weighted at night to account for people's lower tolerance to noise when trying to sleep. The L_{N} is an index which gives the percentage of the time a criterion noise level is exceeded.

Fisher (1973) found that as a result of the increase in traffic, urban noise inside and outside buildings had increased in intensity to a point which is beyond the physiological and psychological level of adaptation of urban dwellers. Two common approaches to the reduction of traffic noise are the use of barriers or embankments to shield residential areas from busy roadways and the installation of windows with double glazing in residential and other buildings. Measurement of the annoyance level of community noise is a complex area beyond the present scope. As well as objective measurements of ambient and peak noise, survey and questionnaire data are needed to evaluate people's attitudes and reactions to noise.

EFFECTS OF NOISE ON TASK PERFORMANCE

It would seem obvious that noise can affect the performance of auditory tasks by masking important sounds, making the task more difficult to perform. How much more difficult would depend on factors such as the intensity and nature of the noise and the contextual importance of the information conveyed acoustically. In addition, the magnitude of decrements in task performance depends on nonauditory factors such as the difficulty of the task without noise and the motivation of the person doing the task. It has been hypothesized (Poulton, 1976a) that noise may improve task performance under some circumstances of increasing arousal. Arousal is the central state of activation of the nervous system which underpins preparedness for action and is controlled by the reticular activating system (RAS) of the brain stem. Sleep is accompanied by decreased activity in the RAS. The RAS receives nerve impulses from the senses which can cause an arousal reaction—nerve impulses from the RAS are sent to the higher centers of the brain, including the cortex. Noise is known to be a particularly arousing stimulus.

From a neuropsychological point of view, terms such as "mental fatigue" and "boredom" really refer to a state of deactivation of the RAS which leads to decrements in the performance of tasks or in the readiness to perform tasks. Thus, it seems possible that noise can prevent boredom or mental fatigue.

On the other hand, noise may degrade performance by interfering with the mental processes involved in doing a task. Poulton (1976b, 1977) has suggested that continuous noise can interfere with work by masking auditory feedback and inner speech. It is known that human short-term memory typically utilizes a verbal or articulatory method of coding information. In this sense, it is easy to imagine how noise, particularly conversation or singing, might interfere with task performance by masking inner speech, with workers not being able to "hear themselves think."

Arousal has been used as a hypothesis to explain the effects of combined stressors on task performance. Poulton (1976a) has suggested that under certain conditions, stressful levels of noise, vibration, or heat can improve task performance by increasing arousal and that the ideal working environment for a task may not necessarily be stress-free. The facilitating effects of noise-induced arousal may cancel out the masking effects of noise, and no reliable effect of noise on performance will be observed. An alternative view that noise can degrade task performance by causing "overarousal" has been discredited by Poulton (1976b), who argued that the concept of stress-induced overarousal was unnecessary to explain the experimental findings concerning the effects of noise on performance.

Sundstrom (1986) has summarized the complex literature on noise and performance. Predictable noise, such as the background noise in a room, can reduce accuracy in clerical tasks, in highly demanding motor tasks, and during vigilance tasks if the noise is very loud and when two tasks are being carried out simultaneously. Unpredictable noise can also degrade performance of the above tasks as well as tasks involving mental calculation or short-term recall.

There does not appear to be one simple effect of noise on performance. Haslegrave (1990) concluded her review by stating that effects are difficult to determine and are task-specific. Sundstrom concluded that laboratory findings on noise and performance have limited applicability to offices and factories.

Industrial Music

In developing countries it is still common for people to sing while carrying out work. In developed countries, music is often played to workers in factories in the belief that it will improve task performance. There has been very little recent research on the effects of industrial music on performance. It does seem clear that factory floor workers enjoy listening to music while they work, but there is no theoretical reason to think that this enjoyment is related to improved output—it might equally be related to lower output.

On the basis of the above theoretical discussion, it might be hypothesized that "music while you work" would be appropriate for those engaged in nonverbal tasks at risk of falling asleep through boredom. In this case, a better approach would be to redesign the task or the operator's role to make it more meaningful. Improvements in the performance of tasks high in verbal mental activity or requiring a rapid rate of communication would probably not be expected from introducing music. Nonverbal tasks requiring a high level of arousal, such as industrial inspection, would seem suitable for industrial music, though. Attempts to demonstrate the efficacy of industrial music will be complicated if only limited gains in production are possible as a result of the design of the system being investigated.

NOISE AND SATISFACTION

Whether sound is perceived as noise depends as much on the listener and the context as it does on the sound itself. Sound levels of 100 dB(A) are commonly found in factories and rightly termed "noise." Fisher (1973) measured sound levels of over 100 dB(A) in a busy hotel bar—whether this level was perceived as "noise," and by whom, is much more debatable.

Noise is often associated with dissatisfaction with the environment. Nemecek and Grandjean (1973) studied noise levels in 15 landscaped (open-plan) offices. Mean noise levels were low—from 48 to 53 dB(A) with peaks 8 to 9 dB(A) higher. However, 35 percent of employees complained of being severely disturbed by noise and 61 percent said their concentration was worse in the landscaped than in a conventional cellular office. Conversation seemed to be the most common cause of annoyance. Its content, rather than its decibel value appeared to be the biggest problem. Other sources of complaint in open-plan offices are sudden noises, such as ringing telephones.

Office work usually involves conscious mental activities which are prey to the intrusive nature of noise. Several factors determine whether noise in offices will be a source of annoyance. First, if the noise level itself is too high, telephone or speech communication will be rendered more difficult and satisfaction with the environment reduced. Second, if the noise level is too low ("quiet enough to hear a pin drop"), intermittent noises, such as conversation, ringing telephones, loud traffic noise, will be more likely to distract people. Sundstrom notes that the intensity of sudden noise above the background (the signal-to-noise ratio) is more important than the noise level itself in causing annoyance. White noise (noise in which amplitude is equal at all frequencies) is sometimes deliberately introduced into open-plan offices to reduce the intensity of noise with respect to the background (to mask conversation, telephones, etc.).

Unexpected noise is more distracting than regular noise. By a process known as **habituation,** people can block out regular noise (the experience of becoming aware of the presence of a clock only when it stops ticking indicates habituation). Auditory perception is a complex topic beyond the scope of the present discussion. However, it is known that inhibitory fibers exist in the neural pathway from ear to brain which inhibit specific parts of the organ of Corti (Guyton, 1991). These are thought to enable people to direct attention to particular sounds by blocking out others. Guyton uses the example of a person listening to one instrument in an orchestra to illustrate the point.

Attitudes toward a noise source play a major role in determining its disturbing effect. Approval of the source minimizes disturbance (e.g., noise from one's own or one's subordinate's machine), whereas disapproval increases disturbance. The listener's control over the noise and the social distance between listener and noisemaker have a similar mediating effect on disturbance.

VIBRATION

A detailed treatment of vibration is beyond the present scope, although a few of the most important aspects in ergonomics can be briefly introduced. Vibration is defined as the oscillation of a body about a reference position and can be described, like noise, in terms of amplitude, frequency, and phase (Cole, 1982). Of most interest in ergonomics are the effects of vibration on health, task performance, and communication. Vibration from railways and heavy vehicles can be a cause of annoyance in residential areas (Howarth and Griffin, 1990).

Vibration and Health

Troup (1978) reviewed driver's back pain and its prevention. Vibration is a major risk factor in the development of spinal problems in certain occupations. In addition to the postural stress of sitting, drivers are often exposed to vibration in the vertical plane. Vibrations in the frequency range 4 to 8 Hz (the natural frequency of the trunk) are particularly hazardous. Seat design to minimize the transmission of vibration at these frequencies can reduce the risk of back injury.

Raynaud's disease affects the blood vessels and nerves of the hands or feet. When associated with the use of vibrating tools it is called **vibration white finger (VWF).** A more modern term for the disorder is **"hand-arm vibration syndrome."** It is characterized by local ischemia, pain, or numbness and is exacerbated by working in cold conditions. It is caused by operating hand tools with low-amplitude vibration below 500 Hz. Paving breakers, power sanders, grinders, chipping hammers, and power wrenches are examples of vibrating hand tools which can cause VWF. If detected early, the disease is reversible. If not, it can cause permanent disability in the use of the hands. Exposure to vibration from hand tools can be reduced by wearing gloves (which also reduce cold) or by using tools with specially designed vibration-damping handles. Hansson et al. (1987) report that the level of vibration from different makes of hand tools can vary considerably and that rubber dampers placed between the tool and handle can reduce vibration by about 65 percent. They suggest that products be tested and that this information be made available to industry to assist in making equipment purchasing decisions.

Vibration and Task Performance

Vibration can affect performance by causing parts of the body essential to the task to resonate. At 5 Hz, resonance of the shoulder girdle, occurs which can degrade manual actions. Between 20 and 30 Hz, the head resonates between the shoulders, and at higher frequencies, the eyes themselves may resonate. Thus, both vision and manual control can be degraded by vibration. Communication problems caused by speech difficulties can occur at frequencies from about 7 to 20 Hz.

Vibration and Public Transport

Mention must also be made of how people respond subjectively to vibration of different frequencies and intensities. This information is particularly important in the design of public transport systems if passenger comfort is to be optimized in relation to engineering constraints. Oborne has carried out several studies on the subjective response to vibration. Oborne and Boarer (1982), for example, investigated the effects of posture on subjective response to whole-body vibration. They found that vibration affected standing subjects significantly less than seated subjects, but sitting posture itself did not have a significant effect (no one sitting posture was better than the others).

A final consideration in the design of transportation systems is motion sickness and its causes and cures. Individuals differ considerably in their susceptibility to motion sickness, but it seems to be caused by low-frequency (0.1 to 0.8 Hz) vibration in the vertical plane. It is not surprising that vibrations of this nature are commonly encountered at sea. Further information may be found in Griffin (1992).

CURRENT AND FUTURE RESEARCH DIRECTIONS

The technology for evaluating the auditory environment and for dealing with troublesome noise sources is well established. There is an international trend to lower industrial noise exposures beyond their present levels in order to further safeguard the hearing of the workforce and improve working conditions. Applied research to investigate cost-effective ways of achieving these reductions is appropriate. As described in the Footnote to this chapter, there are new technologies which can be used to control industrial noise.

SUMMARY

The most important consideration in the design of the acoustic environment is to protect the hearing of people exposed to noise. Noise can be evaluated by measuring sound levels, noise doses, and the L_{Aeq}, and by carrying out frequency analysis. Unacceptably noisy machines should be quietened, replaced, or isolated from occupied areas. If none of these options are possible, operator exposure should be reduced or ear protection provided. Further information on the physical aspects of noise and noise control can be found in the publications of the International Organization for Standardization and the American Industrial Hygiene Association, the *Handbook of Noise and Vibration Control*, and in texts such as Diehl (1973).

An ergonomic approach to the auditory environment should go beyond the straightforward measurement of sound levels and consider the satisfactoriness of the acoustic environment as a whole. Creative, intellectual work may suffer from noise more than routine work in which operators have a high level of skill. A knowledge of people's attitudes to noise is essential in the evaluation of physical measurements and in designing hearing conservation programs.

ESSAYS AND EXERCISES

1 With a meter suitable for the measuring of sound levels, measure sound levels in the following environments:
   a A busy urban street during rush hour
   b The same street at night
   c A library reading room
   d An empty church
   e A canning or bottling factory
   f An automated office

Comment on the magnitude of differences in sound levels in the various environments.

2 Carry out a noise survey in a small factory. Obtain or draw a floor plan of the factory floor, indicating the locations of machines and other noise sources. Indicate all fixed work positions. Write the measured sound levels on the map. Identify demarcated areas beyond which hearing protection should be worn. What other recommendations can you make?
3 Describe the structure and function of the human ear. How can the ear be damaged by noise at work, and what can be done to prevent this damage from happening?

FOOTNOTE: The Active Control of Noise

Recent advances in acoustic engineering have led to the development of so-called active noise cancellation technology. Noise cancellation works on the principle whereby a noise source is analyzed and an antiphase version of equal amplitude is propagated from a loudspeaker back toward the source. If the propagation of "antinoise" is correctly synchronized, noise cancellation takes place. The method is only really effective at lower frequencies but has many applications since much machine noise is of a periodic-cyclic nature. Fast processing of noise from a cycle can be used to predict the noise from the next cycle. An antinoise waveform is then generated and directed back to the noise source.

The method has been used to deal with noise from a variety of sources, including diesel engines (where reductions of up to 20 dB are possible) and exhaust noise from ships' engines and ventilation ducts. Further information can be found in Chaplin (1983).

12

HUMAN INFORMATION PROCESSING, SKILL, AND PERFORMANCE

Human-machine interaction depends on a **two-way exchange** of information between the operator and the system. Although we are accustomed to talking about an operator's control actions or the commands entered into a computer by a user, it is also necessary to consider the information that is "entered" into the user by the machine from displays or as direct feedback from the controlled process. Designers usually have detailed, explicit models of machines and machine behavior which can be used to improve human-machine interaction. However, equivalent models of humans and human behavior are usually lacking. In ergonomics, it is the task of those with a background in psychology or cognitive science to devise applicable models of the user which can be used in the system design process to optimize information exchange in human-machine interaction.

AN INFORMATION-PROCESSING MODEL OF THE USER

The idea that the human brain is an information-processing system has had a profound impact on the study of human behavior since it first appeared in the 1950s. Rather than investigate human psychology from a neuroanatomical or psychoanalytic viewpoint, psychologists have attempted to understand how information is processed by the brain. To use an analogy with computing, one could say that the emphasis is on the software rather than the hardware of the system. Computer models intended to serve as psychological theories were first developed in the 1950s, as were the first computer realizations of artificial intelligence (Simon, 1979).

In the information-processing approach, flow diagrams rather than anatomical drawings are used to represent how the brain processes information. The as yet unsolved problem has been to determine the nature of the programs installed in the information processor, the system architecture, and the control processes for handling information flow.

Wickens (1992) presented a composite, general model of human information processing (Figure 12-1). The model emphasizes that information flows through the system and is processed in stages. Different programs or processors are said to operate on the information at the different stages.

Information enters through the senses as a result of some physical stimulus impinging on the sense organs. A representation of the stimulus is created by the central nervous system (CNS). This representation persists for a short period but decays rapidly once stimulation has ceased (in several hundred milliseconds). In visual perception, this persistence is attributed to an iconic store (Sperling, 1960). Iconic storage is automatic, in that all incoming information is stored in this way, and **analog,** in that the CNS representation has the same structure as the stimulus which gave rise to it. The function of the sensory store appears to be to prolong the availability of stimuli which have been detected by the sense organs.

From Sensation to Perception

Sensation occurs when environmental stimuli impinging on an organism give rise to neural events. These events may or may not have further consequences. Perception, however, involves the identification of the stimulus which gave rise to the initial sensation. Perception is followed by higher mental processes such as recognition of the stimulus and its comparison with previous experience. Table 12-1 depicts the hierarchy of stages from sensation to perception and the operation of high-level mental processes.

We can further distinguish between sensation and perception by noting that sensation

FIGURE 12-1 Wickens's general model of human information processing. (From *Engineering Psychology and Human Performance*, 2nd edition, by Christopher D. Wickens. Copyright © 1992 by HarperCollins Publishers. Reprinted by permission of HarperCollins College Publishers.)

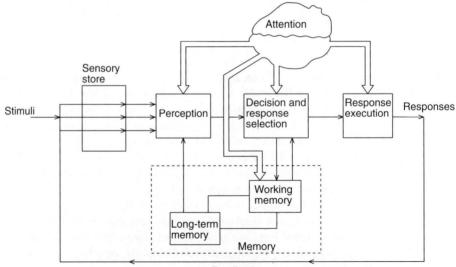

TABLE 12-1   HIERARCHY OF SENSORY AND PERCEPTUAL THRESHOLDS

Conceptual categorization
  Naming

---------------------- RECOGNITION THRESHOLD ----------------------

Features of stimulus situation
  Shape
  Color
  Contour

----------------------IDENTIFICATION THRESHOLD ----------------------

"Something is there"
  Boundary information
  Brightness differences
  Movement

---------------------- DETECTION THRESHOLD ----------------------

Sensory nerve impulses

----------------------PHYSIOLOGICAL THRESHOLD ----------------------

Environmental stimuli
  Impingement on sense organs
  Transduction and transmission of energy
  Stimulation of receptor cells
  Depolarization of neurons

----------------------ENVIRONMENTAL ENERGY ----------------------

is characterized by a direct relationship between the intensity of the stimulus and the evoked neural discharge or inner experience. Perception, however, is mediated by individual factors, such as personality and mood, and by the context in which sensation takes place. For example, a pilot landing an aircraft in difficult conditions might see a series of lights below and ahead of the aircraft. Whether these are perceived as lights on the runway or streetlights will depend on internal factors, such as the pilot's expectations, level of training and its retention, and external factors, such as visibility and the visual cues which distinguish between the two types of lights.

For perceptual processing to be initiated, impinging stimulus energy must exceed physiological thresholds. Impulse conduction leads to the detection of something in the environment which is accompanied by orientation of the senses toward the source of stimulation. The identification threshold is passed when the "something" that is detected is refined into a specific object by **detection of contours and main features.** Comparing the emergent percept with information stored in memory makes recognition possible. Further processing of the incoming information may take place at a number of levels, for example:

1 *Physical level:* This is the letter "A" written in italics.
2 *Phonetic level:* This word sounds like "cheese."
3 *Semantic level:* This flashing red light means danger.
4 *Other levels:* This is a nice photograph of my grandmother.

The distinction between sensation and perception is by no means an academic one. It is well known, for example, that motorcycles have a greater involvement in certain types of traffic accidents than other road vehicles. It has been suggested that such accidents occur because an approaching motorcycle is more difficult to see than an approaching car or truck, making other vehicles more likely to pull out in front of oncoming motorcycles. According to this hypothesis, collisions can be reduced by making motorcycles more conspicuous. In some U.S. states and various other countries, legislation had been introduced compelling motorcyclists to use their headlights during the day as well as at night. Olsen (1989) has pointed out that although there is evidence that the use of daytime headlights on motorcycles reduces accidents, this finding also applies to cars and there may be other factors which contribute to the motorcycle accident figures. Two perceptual factors have been suggested by Olsen: Because motorcycles viewed from the front are much smaller than cars, drivers may overestimate the distance of an oncoming motorcycle and be more likely to pull out in front of it. They may also have more problems judging the speed of oncoming motorcycles compared with that of cars.

The motorcycle accident problem illustrates why correct analysis of the processes is important. If there is more to the problem than just motorcycle conspicuity, then further measures, which take account of drivers' perceptions of road events and decision making, will be needed to improve road safety.

Coding

Perceptual processing is said to be **stagewise and hierarchical**—the inputs to higher-level processes are the outputs of lower-level ones. Only a fraction of the information impinging on the senses reaches conscious awareness. As soon as environmental energy impinges on the senses, it is transduced, distorted, and **coded** by the very processes which transmit it to the higher centers of the brain. The concept of coding is central to the information-processing approach. In a similar process in computing, data are often coded in some way before being entered into the machine (names might be abbreviated, for example) according to the predetermined requirements of the software designer. Having been entered and stored in some way, the coded data can then be operated on by programs. Thus, what is operated on is a coded version, or **representation,** of the external (real-world) data.

Coding and Cognition

The concepts of stagewise, hierarchical information processing; coding; and internal representation of stimuli are central to the model of the human as an information processor. The perceptual system is a first stage in the coding of environmental stimuli. Since cognitive processes act on percepts, not on stimuli, the way the system codes a stimulus determines what can or will follow. The stimulus configuration and the coding mechanisms of the perceptual system determine what will be perceived. Memory also plays an important role.

Coding of stimuli by the perceptual system has major implications for ergonomics. **The ergonomist must ensure that the information displayed to an operator is perceived in an appropriate way for the requirements of the task.** In order to design effective displays, the task requirements must be analyzed early in the design process and the skills and knowledge of the operator taken into account.

Two Kinds of Memory?

Psychologists normally distinguish between short-term and long-term storage of information. Short-term memory (STM) can be likened to a temporary store (or buffer) in which small amounts of information are briefly retained while a particular mental or physical operation is carried out. Remembering a phone number while writing it down is an example. STM can contain symbols related to current processing but has limited storage capacity. STM storage limitations can cause errors in task performance, for example, forgetting important data before it can be consolidated in long-term memory (LTM) or used in decision making or acted on during a sequence of physical operations. Forgetting of intermediate results when doing mental arithmetic or omitting one of the ingredients of a recipe are examples of STM errors. STM is sometimes referred to as **working memory.** The relationship between perception, working memory, LTM, and the operation of mental "programs" is shown schematically in Figure 12-2.

LTM contains symbolic structures built up through learning in which new data can be embedded as a single symbol after being suitably coded. Our general knowledge of the world and of events in our lives are stored in LTM. We may sometimes be unable to

FIGURE 12-2   New information is kept in memory by the action of mental operations such as rehearsal, and is stored in LTM. Information in LTM can be written to working memory to be operated on.

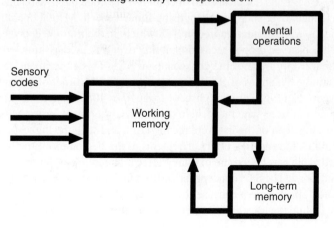

retrieve this information but can be reminded of it by others, which indicates that this mode of storage is more or less permanent. LTM is associative in nature; new data can be represented in the context of past behavior, but this takes time.

In computer analogy, STM can be likened to the assignation of a particular value to a variable in a computer program for the duration of an operation. After the operation is completed, the value is no longer needed and is lost. The variable itself, however, remains in the program (in LTM) and new values can be assigned to it the next time the operation is invoked. Unlike in a computer, however, storing information in STM requires considerable mental resources, particularly conscious attention. Storage is fragile and the information is easily lost if the operator is distracted by unexpected events or overloaded by new information. For example, when driving home in busy traffic, a driver may forget details of traffic conditions just heard on the car radio when distracted by a call on the car phone. The STM trace of these items is displaced by new information, but the route information, which is stored in LTM, is not forgotten.

Two Memory Stores?

Memory has often been modeled in terms of **stores**—a short-term store (STS) and a long-term store (LTS). Much research has been carried out in an attempt to characterize the nature of information handling within these stores. A number of classic experiments are described here.

Peterson and Peterson (1959) presented subjects with three consonants (e.g., CMJ, PQF) and a number. After hearing the three consonants, subjects were required to count backward from the number in three's until a light was switched on, at which time they were asked to recall the trigrams heard initially. The researchers repeated this procedure many times, varying the length of time between presentation and recall (the retention interval).

The probability of correct recall was found to diminish as the retention interval increased—from about 80 percent after a delay of 3 seconds to 10 percent after a delay of 18 seconds. By asking people to count backward in three's, the investigators hoped to prevent their using **mental operations** to keep the stored trigrams in memory (Figure 12-3). One of the simplest such operations is to code the consonants verbally and to rehearse them over and over again using **inner speech.** Mental operations appear to reactivate the trace of items stored in STS which, in their absence, decays rapidly.

The observed drop in recall probability over time was assumed to be due to the decay of the memory trace of the consonants in STS. This experiment demonstrates some of the storage dynamics of STS—storage time is of the order of seconds.

Murdock (1962) drew up lists of from 10 to 40 unrelated words. The lists were presented to subjects who then had to recall the words in any order. Murdock found that the items at the end of the list were recalled first and with the highest recall probability. He called this the **recency effect.** The items at the beginning of the list were next best recalled. This was called the **primacy effect.** The items in the middle of the list were recalled worst. The STS concept can be used to explain these results (see Figure 12.3). The recency effect occurs because the last items have only just arrived in STS and are recalled while the trace is still "fresh" in the subjects' STS. The primacy effect occurs be-

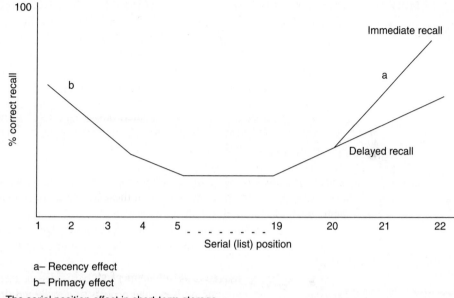

a– Recency effect
b– Primacy effect

FIGURE 12-3   The serial position effect in short-term storage.

cause the first items to enter STS get rehearsed the most and are thus more likely to enter permanent storage. As more items enter STS, it "fills up" and less time is available to rehearse them (this situation happens after about the first five items). Thus, these items are less likely to enter long-term storage. Increasing the presentation time of each item was found to improve recall of even the middle items. Increasing list length lowered the recall probability of all words, as might be expected.

In addition to a rapidly decaying trace, the STS seems to have limited capacity. Longer-term storage of information depends on items in STS being subjected to mental operations. Miller (1956) concluded that the **"magical number 7, plus or minus two"** was the limit on our capacity to consciously process this type of information (i.e., the capacity of STS). Some clear implications for the **design of codes** can be derived from this finding and are discussed in the next chapter.

Table 12-2 summarizes the differences between the long- and short-term memory stores. It can readily be seen that short-term memory is very fragile. **The performance of people who have to use short-term memory in their jobs can be degraded by interruptions of even a few seconds—particularly by conversation—or by having to carry out verbal secondary tasks (such as answering the phone while doing mental arithmetic).**

Attention

Attention is presented in Wickens's model as a **limited resource** which can be channeled to "drive" processes such as working memory, response execution, etc. It can be

TABLE 12-2   DIFFERENCES BETWEEN SHORT- AND LONG-TERM MEMORY STORES

	STS	LTS
Storage capacity	7 items ± 2	Extremely large
Retention interval	5–30 seconds	Many years
Mechanism of information loss	Trace decay. Displacement by new items	Inability to retrieve item
Process for coding information	Phonetic Articulatory	Semantic

thought of as the spotlight which illuminates the information world or, in a computer analogy, as the central processing unit (CPU) responsible for scheduling tasks and allocating them to subroutines. Industrial inspection is an example of a task which requires sustained, focused attention for long periods. A knowledge of factors which influence our ability to attend to incoming stimuli can be used to improve the performance of such tasks.

Attention can be **selective,** as in listening to a particular instrument in an orchestra; **focused,** as in concentrating on particular aspects of a task; and **divided,** as in reading a newspaper while watching television. In divided attention, two or more tasks are not necessarily carried out simultaneously. A time-swapping mode can be used (in which attention switches rapidly between each task so that it appears that both are being carried out simultaneously). Car driving often requires attention to be divided between different tasks such as lane tracking (keeping the car on course in a lane while negotiating an intersection) and visual analysis (scanning the traffic ahead and detecting and interpreting road signs, etc.). It is known, for example, that older drivers are overrepresented in road accidents when turning left—a time of high mental workload. Older drivers seem to be less able to effectively divide their attention between tasks, which suggests that intersections should be designed so as not to require concurrent activities (see, for example, Brouwer et al., 1991).

The information bottleneck in the processing system is well documented and performs a valuable role in protecting the system from information overload. Broadbent (1958) was one of the first investigators to propose a theory of attention. He proposed that information enters the system through a single channel of limited capacity. A filter, situated between the sensory (iconic-echoic stores) and the perceptual systems protects the limited-capacity processor from overload by filtering out unwanted information. Selection of information was said to be on the basis of the physical characteristics of the stimulus. That the reader may, until this point, have been unaware of the pressure at the surface of his or her seat illustrates the apparent validity of the concept of physical selection of stimuli in human information processing.

Subsequent research indicated that incoming information is automatically processed beyond the purely physical level, and therefore selection is unlikely to be solely on the basis of the physical nature of the stimulus (the filter must operate at a higher level of information processing). This concept is partly the rationale underlying subliminal advertising (we cannot help processing the information even if we don't attend to it) and Neisser's (1976) view that **we see through our cognitions** (that there are top-down processes, beyond conscious control, which utilize higher-level information to process incoming information).

Several other single-channel models of attention were developed over the years as researchers attempted to specify the location of the bottleneck in the system. Poulton (1970) developed a model for skilled performance and noted that the input selector (similar to a filter) can select information from LTM as well as information from the sensory stores.

How Many Tasks Can We Do Simultaneously?

Some of the important issues concern **the number of inputs an operator can attend to at once and how many operations can be carried out simultaneously.** This capacity is of importance if information overload is to be avoided, particularly in emergencies. New methods of interacting with machines, particularly the use of voice, make this an important issue for ergonomics. Some of the research is introduced here and described in subsequent chapters.

The early single-channel theories of attention seem unable to clarify many of the important issues. An alternative **multichannel theory** was proposed by Allport et al. (1972). Humans are multimodal in the sense of being able to process and represent information in many modalities (e.g., visual, auditory, semantic), and many purposeful activities, such as walking and standing, can be carried out without being consciously attended to. Sitting erect on a badly designed seat does require conscious attention, however. The single-channel view of the operator has been expanded to account for this multidimensional aspect of human behavior. Attention is now seen more as a **problem of allocating processing resources to tasks** rather than as a bottleneck in a passive information channel. A multichannel model of attention is presented in Figure 12-4. According to the model, a limited-capacity processor handles incoming data, but specialized subroutines requiring no processing capacity can be utilized to free the processor for additional tasks. The main features of the multichannel view are summarized in Table 12-3.

The model is intuitively attractive in that it can account for how we can walk and talk at the same time but might have difficulty in walking and juggling or in talking and reading simultaneously. Activities which occupy different modalities are fed by different channels to **dedicated subroutines** and can be carried out simultaneously. Activities

FIGURE 12-4   A multichannel model of attention.

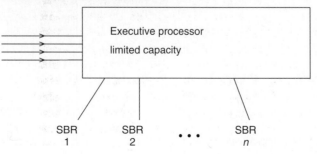

TABLE 12-3   CHARACTERISTICS OF THE MULTICHANNEL VIEW OF ATTENTION

1 The processor and subroutines are fed by limited-capacity, modality-specific channels.

2 The processor handles data in a way that roughly equates with consciousness.

3 The subroutines are built up through practice.

4 The system can break down under a number of conditions:

    a No subroutines are available. Multiple inputs then compete for executive processes which may be insufficient.

    b There is competition for subroutines. More than one task is being carried out simultaneously and the tasks are similar in nature.

    c The existence of subroutines implies skill. If the processor tries to take over the operation, the subroutine is disrupted.

that occupy a single modality can be carried out simultaneously only if channel capacity is not exceeded.

Some evidence for the notion that tasks occupying the same channel can interfere with each other comes from Brooks (1968). Subjects were shown figures (Figure 12-5) and, *from memory,* had to identify whether each corner of the figure was made up from either the bottom or top lines. In one condition, they responded verbally, and in the other, by pointing at the words "yes" and "no." Performance on this visual-spatial decision task was better when the response was in verbal rather than visual-spatial mode. Similarly,

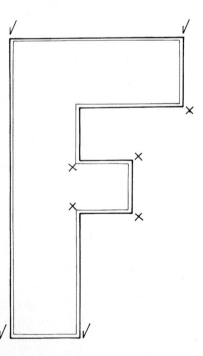

FIGURE 12-5   Example of Brooks's visual-spatial checking task.

subjects given a verbal decision task (identifying which words in a sentence were nouns) did better when they could respond "yes" or "no" by pointing rather than by speaking.

Further experimental support for the multichannel view of attention comes from All- port et al. (1972) and Schaffer (1975). Allport et al. gave subjects a **speech shadowing task** to do and a memory task. Listening to a passage of prose and detecting embedded target words is an example of a speech shadowing task. If the memory task was pre- sented in verbal form, performance was almost completely disrupted, suggesting that the shadowing was taking up all the processing capacity. However, if the memory task was presented in pictorial form, no disruption occurred. In terms of a single-channel view, it is difficult to explain where the extra processing capacity came from. A multichannel view would say that the shadowing task and the memory-for-pictures task **utilize dif- ferent limited-capacity channels** which can operate independently of each other.

In another experiment, Allport et al. found that pianists could sight read and carry out a verbal shadowing task equally well simultaneously or individually. Schaffer (1975) found that typists *could* shadow an auditory message while copy typing visually pre- sented text but *could not* shadow a visual message and audio type simultaneously. Copy typing from visually presented text does not involve the processing of verbal informa- tion (skilled typists do not read text when typing; they just type in the letters), so the copy typists had spare verbal capacity to process the auditory message. However, audio typing does require verbal processing capacity, as does shadowing a visual message, and insufficient capacity was available to do both tasks together.

Wickens (1980) has described how human attentional resources are subdivided along spatial-verbal and other lines. Vidulich (1988) has argued that the traditional all-manual human-machine interface leaves a pool of human resources untapped. New ways of en- abling humans to interact with machines promise to make these resources available and lessen the conflicting demands on operators in situations where more than one task has to be carried out at a time. As **speech-recognition technology** develops, it is becoming possible to implement **voice control** of a variety of machines. Voice control will be of benefit in tasks requiring **a large amount of visual-spatial processing** (such as com- puter-aided drafting and graphic design) but might interfere where the primary task is verbal (such as text creation).

Response Selection and Execution

Response selection and execution may occur automatically, as the output of a specific subroutine, or they may be initiated by the executive processor. This corresponds to the automatic, or reflex, response of the skilled performer or to the rapid avoidance behav- ior which overrides all other outputs to avoid impending danger. When initiated by the executive processor, response selection and execution are more akin to a measured or calculated decision to act.

Skilled vs. Unskilled Operators

The multichannel view of attention highlights the importance of task modality and skill. In Schaffer (1975), the copy typing (visual-manual) task did not interfere with the ver- bal shadowing task. Since the typists were skilled, it can be assumed that the typing was done automatically by means of a subroutine. However, the audio typing (verbal-manual) task did interfere with the visual shadowing (visual-verbal) task. Even though the typists had subroutines for typing and for reading the visual message, the two tasks shared a verbal component which overloaded the available processing resources. It is in this sense that the requirements of tasks can be analyzed to determine whether the performance of one of them will interfere with the performance of the other. Tasks which are unlikely to interfere can be said to be compatible.

A skilled driver has subroutines, built up through practice, which handle the routine tasks of steering, changing gears, scanning the road ahead, etc. Thus, executive capacity is left to make plans, carry out a conversation, or listen to the car radio. At times of high load or in nonroutine situations (such as having to negotiate a busy intersection) the subroutines may be unable to cope and executive processes will have to be used. Even a skilled driver may suddenly break off a conversation with a passenger under such circumstances as the executive processes dedicated to carrying out the conversation are diverted to the car control task. Alternatively, a novice driver may be unable to carry out even the simplest conversation on the first driving lesson—in the absence of any subroutines for driving, the executive processor has to handle the entire burden.

A skilled operator has subroutines to handle the low-level aspects of task performance which spare the executive processes. The subroutines which handle motor actions (sometimes referred to as motor programs) are of particular importance in integrating different muscular actions into a unified whole. The skilled golfer or tennis player conceives of and executes a swing or a serve as one smooth action, whereas the novice has to assemble the action out of a set of more primitive movements. For the skilled operator, detailed actions are derived in a top-down fashion from higher-level representations. For the novice, purposeful actions are assembled from smaller components in a bottom-up fashion.

The unskilled operator is said to be **tied to the present** through lack of automatic processes—the mechanics of the task require all her or his attention, leaving none for other activities. The skilled operator has subroutines built up to deal with the mechanics of the task and therefore does not have to monitor his or her own control actions and can predict and continually look ahead. Thus, the difference between skilled and unskilled operators is more than quantitative. The skilled operator exhibits more **parallel processing** of inputs and an increased ability to separate out doing the task from thinking about doing it. A critical discussion of skill and the possibility of producing "skilled" or "expert" machines can be found in Dreyfus and Dreyfus (1987).

The distinction between skilled and unskilled users is made explicit in the multichannel view of attention and can be seen to be profound. Empirical studies have demonstrated **qualitative differences** between the way skilled and unskilled operators carry out work tasks. Prumper et al. (1992), for example, measured the number of computer errors and the time taken to rectify them of 260 clerical workers in 12 German companies. Before collecting the data, they classified the workers as skilled or unskilled. Skilled computer users made fewer errors due to a lack of knowledge than unskilled users but made more low-level errors. Experts are usually held to have knowledge which is represented in a **high-level, abstract way** and may thus be more error-prone when the situation calls for **precise, concrete actions** (e.g., switching from one program to another with a different interaction style or command language notation). Unskilled users, because they lack high-level or abstract knowledge may be more prone to error when confronted with **conceptual problems** in the operation of a system. Surprisingly, in Prumper et al.'s study, the skilled users made *more* errors overall than novices, which suggests that simple error counts are not sufficient to separate skilled from unskilled performance. However, the big difference between experts and novices was found in the time taken to deal with the error—experts were consistently faster. One of the interesting and important implications of this finding is that perhaps less attention should be paid to minimizing the number of errors in the operation of a system and more should be paid to teaching users the types of errors that can occur and effective **error-handling skills.** In other words, design effort should be spent not only in minimizing errors, since even experts are error-prone, but also on minimizing the **consequences** of an error. This seems to be one of the key requirements in the operation of interactive systems. In the evaluation of existing systems, data are needed not only on the number of errors, accidents, or unexpected events but also on their effects on system operation and the resources required to restore the system to its previous level of functioning.

A first step in the design of any human-machine system, then, is to **characterize the expected skill level of the user population** since the interface design requirements may be completely different for different groups. Of particular importance are the constraints on display complexity and layout and the requirements for on-line help, training manuals and operating instructions, warning signs and labels, and feedback.

Feedback

Most human behavior gives rise to feedback of some type or other. Feedback may be **intrinsic** (such as proprioceptive feedback from the muscles and joints or a mental outcome such as a decision) or **extrinsic** (from the environment). **Feedback is essential for most purposeful activity and is the basis for control, learning, and improvement of performance.**

Much research has been carried out on the factors affecting the ability of an operator to detect faults in a system. The industrial inspection task in which an operator scans newly made items for defects is a classic example. Frequently, the operator's task is to classify items as good or faulty according to preestablished criteria. Drury and Addison (1973) investigated the effects of feedback on the performance of inspectors at a glass-making factory. They found that increasing the speed of feedback about fault-finding performance improved the inspectors' ability to detect faulty items.

The **lag between executing an action and receiving feedback** is one of the most important variables affecting our ability to carry out continuous tasks. The classic experiment in which a subject is given delayed feedback about his or her own voice illustrates the importance of feedback in the production of speech. Even slight delays severely disrupt the ability to speak. In the field of human-computer interaction, the lag between user action and feedback is often due to the computer system's response time (the time taken to process the entry and display the result). Excessive response time can be a cause of user dissatisfaction. Goodman and Spence (1981) found that it was response time **variability** which particularly degraded performance in an interactive graphic problem-solving task.

Modern approaches to user-interface design stress the importance of appropriate feedback (of the system talking to the user). For example, when a command has been executed, some immediate feedback can be given to indicate that the command has been received and is now being executed, such as the display of the response "working." A flashing cursor on a VDT screen or line of periods growing in length can be used as feedback to fill up the time lag between acceptance of the command and display of the end result.

Bank statements, bills, and receipts are examples of the complex array of feedback individuals receive in their everyday lives. An interesting and potentially important area, in the context of the increasingly integrated modern communication and data processing systems, is the design of feedback about everyday behavior. Feedback, in this context, is useful only if users **recognize** it as such and are motivated to use it, and **if it is in a form that they can easily map onto their own behavior.** McClelland and Cook (1979), in an early study, investigated the effects on domestic energy consumption of providing continuous feedback (in cents per hour) in 101 all-electric homes. They found an average energy saving of 12 percent, with the greatest improvement in the more temperate months, suggesting that the savings were due to changes in energy use other than energy for heating and cooling.

The **design of feedback** is an area that requires careful consideration because people have memory and other cognitive limitations. If the **grain size** of feedback is too small, major trends may be obscured by minor variations. Alternatively, the patterning of feedback over time may be just too complex for the user to extract meaning from it. If the grain size is too large, people may not be able to map the information in the feedback onto specific behaviors amenable to self-modification. Thus, there is a trade-off between two extremes. At one extreme, detailed feedback which can be related to particular behaviors must not cause information overload. At the other extreme, general feedback may be easily grasped but, if too general, cannot be related to specific actions.

We now return to the domestic energy-consumption example. Most people do not regularly monitor how much electricity they are using. Often monthly or even quarterly bills are the only feedback they receive. Alternative energy purchasing and payment schemes have been developed, including the pay-as-you-use system in which users buy electricity credits, which are entered into their domestic energy meters using a plastic card similar to a credit card. The meter displays the amount of electricity remaining before more credits have to be purchased. It seems likely that users of this type of system would be much more able to modify their energy usage because of the much closer mapping between consumption behavior and feedback.

Feedback is also important in the development of social skills such as interviewing (Randall, 1980) and in job design (Wall, 1980) to enable employees to assess their own performance. The term **"knowledge of results"** is sometimes used in this context. More general feedback about job performance (including recognition from superiors and social support from peers) is one of the key principles of the sociotechnical design of jobs (Cherns, 1976).

Long-Term Memory

The information-processing approach uses an analogy in which information flows through the system and at some stage becomes amenable to the action of higher processes (or programs) stored in LTM. LTM contains symbolic structures, built up through learning, in which new data can be embedded. Of particular interest in ergonomics is the set of cognitive processes—those which have to do with knowing (as opposed to conative processes, which have to do with desire and volition). Some examples of cognitive processes in the context of ergonomics are the following:

   1 Learning what facilities a new word processor offers and how to use them to create documents
   2 Deciding which of a number of packages to buy on the basis of an analysis of the usefulness of their attributes and of the cost
   3 Solving comprehension problems which arise when learning to use the system

There is much anecdotal and experimental evidence to suggest that LTM is organized on the basis of meaning or **semantics.** If people are given lists of random words to remember, they tend to recall them in semantically similar clusters, which suggests that they have actively organized the information in this way when storing it.

Bartlett (1932) was one of the first researchers to systematically investigate LTM. He showed people pictures or stories which they repeatedly recalled up to 6 years later. He found systematic differences between the original stimuli and the recalled material and noted that people abstracted out the details, integrated them with other information, and omitted "irrelevant" information, producing a condensed form of the original material. People also made additions to the original material in an "effort after meaning" in which "we fill in the lowlands of our memory with the highlands of our imagination." This finding has important implications for the assessment of eyewitness testimony in court cases and the need to avoid leading questions when interrogating witnesses (a leading question is one which implies its own answer).

A fundamental distinction can be made between episodic and semantic memory (Tulving, 1972). Episodic memory is memory about personal experiences and events, whereas semantic memory contains world (or general) knowledge. The former is more easily forgotten than the latter.

Encoding Factors Influencing the Retrieval of Information from Memory

The way an item is encoded in memory plays an important role in determining how easily it can be recalled. In fact, some theorists regard memory as a by-product of processing rather than a store in which items are placed.

Paradoxically, in order to retrieve something from long-term memory, we first have to know what it is we are trying to remember. A retrieval cue acts like a description of the item to be remembered. If the cue is sufficiently detailed, recall probability is increased. In addition, the desired item has to be discriminated from other similar items in memory in order to be recalled. This depends on how it was **encoded** when it was stored—similar items which were encoded in sensorily or contextually different ways can be discriminated more easily than those which were encoded in a similar way.

Elaborative rehearsal (Craik and Lockhart, 1972) is an encoding method which increases the number of retrieval cues for an item. Items can be encoded multidimensionally (e.g., visually, verbally, and semantically) to increase information redundancy and meaningfulness. According to Bower (1972), multidimensional coding works on the principle that if more features of an item are encoded, it is more likely that at least one will be retrieved later on. Most **mnemonic systems** rely on some method of elaborative rehearsal to code new information.

Mnemonics, Verbal Elaborative Processing, and Visual Imagery

The use of mnemonics has a long history. Ancient Greek orators used a system known as the method of loci to help them remember the details of a long speech. In the rehearsal of the speech, a familiar street would be imagined. At the start of the speech, the orator would imagine himself at the beginning of the street. Each important section of the speech would be associated with a particular part of the street, such as a shop. As the orator "walked down" the imaginary street during the speech, each section of the speech was recalled in the order it was memorized. The method worked by linking new information (the parts of the speech) to an intrinsically sequenced set of previously memorized objects (locations along the street) which acted as retrieval cues.

Verbal elaborative encoding can be used to assist in the initial stages of learning new material. Bailey (1982) describes how the material may either be reduced or expanded. Reduction usually involves taking the first letters of each word to be remembered and using them to create a new word. The original words can be derived from their first letters, which are stored together as one item. For example, in the days before pocket calculators, many students of trigonometry used the acronym "SOHCAHTOA" to remember the relationship between the trigonometric functions and the sides of a right-angled triangle (sine = opposite/hypotenuse, cosine = adjacent/hypotenuse, tangent = opposite/adjacent).

Expanded mnemonics usually involve the addition of meaning to the material to be remembered, as when one is learning the lines of the treble clef in music. E, G, B, D, F can be expanded to "every good boy does fine." At first it may seem paradoxical that one of the best ways to improve the ability to remember something is to increase the amount of information to be remembered. The paradox disappears when it is remembered from memory theory that the process of storing an item in memory involves linking the new item into a previously established associative network of meaningful information. Expansion works by increasing the number of links between the new item and the network.

Bower (1972) described the principles for the use of visual imagery as an aid to learning:

1 Both the cue and the item must be visualized.
2 The images must interact.
3 The cues must be self-generated.
4 The semantic similarity between the cue and the item must be minimized.
5 Bizarre images are no better than obvious relationships.

It is interesting to note that imagery seems to be necessary only during the initial learning of new material. With use, retrieval becomes automatic and the person may eventually forget the image itself even though the word can now be remembered.

Figure 12-6 illustrates the use of a composite mental image generated by the author

FIGURE 12-6   Visual image to assist learning the Spanish word for celery ("apio"). The item is first visualized; then it is combined with an image of a similar-sounding English word ("happy"). The image itself acts as a retrieval cue for the word to be memorized.

to assist him in learning the Spanish word for "celery." First, a piece of celery was imagined. Next, the Spanish word for celery ("apio") was visualized by converting it to something similar in English, in this case, "happy." By making the celery happy, the author could reconstruct the Spanish word at a later date using the general rules that the letter "h" is silent in Spanish and that many Spanish words are very similar to English ones but have an extra "o" on the end.

Information in long-term memory is stored at a semantic level but is linked to words, rules, and other structures. Many concepts can be represented in terms of other symbols or sensory modalities at will. In this way, it is possible to talk about the **imagery** value of words. Words such as "justice" and "philosophy" are linked to abstract concepts, which are difficult to link with information received from the senses. In contrast, words such as "balloons," "birthday cake," "cotton candy" are linked to more concrete concepts. We obtain knowledge about them directly via our senses, and they evoke strong visual and other images when we hear them. There is evidence that **abstract words are more difficult to learn than concrete words** (Baddeley, 1983), which may explain some of the difficulties experienced by novice computer users in learning to interact with a system. Many of the concepts associated with information technology are abstract, as are the command names used to describe them (Rogers and Oborne, 1985).

Chunking

Chunking involves the assembly of several items of information into larger units. This process increases the amount of information which can be held in short-term memory.

A word is a way of chunking letters, for example. Codes such as telephone numbers can be printed so as to explicitly encourage people to store them in a particular way. For example, an 10-digit number such as 2125126044 appears less daunting when it is revealed to be the telephone number 212-512-6044, where 212 is the area code; 512, the exchange; and 6044, the extension. By making the structure explicit, one can **reduce a 10-digit string to six items** (the two chunked codes and the unique 4-digit user number).

Chunking can also take place at a semantic level. Chase and Simon (1973) showed novice and experienced chess players a board containing about 25 chess pieces. Grand masters were able to reconstruct the board after viewing it for a few seconds, using information stored in long-term memory. Novices, however, could replace only about six pieces. The difference in performance can be explained by the notion of semantic chunking. The grand master is able to encode the positions of the 25 chess pieces in terms of a smaller number of higher-level representations of the state of play. The novice, lacking the appropriate concepts, cannot encode the board in this way, so performance is determined by the limitations of short-term memory. As might be expected, when the pieces were arranged in a random manner, rather than as a stage in a real game of chess, the performance difference between the novices and the grand masters disappeared. A knowledge of chess does not contain information which can be used to semantically code the randomly positioned pieces, so the grand masters, like the novices, had to rely on short-term memory.

The ability to organize many small details into higher-level conceptual units is one of the main characteristics of skilled **cognitive** behavior. It is also found among computer programmers—skilled programmers are able to conceptualize a program as a number of interrelated modules from the function of which the individual lines of codes can be inferred (modern structured programming languages actively encourage this manner of conceptualizing a piece of software). Skilled cognitive behavior is characterized by a top-down approach to the storage and retrieval of task data, whereas novices often have to think in terms of fine details from which higher-level structures are built (a bottom-up approach to the storage and retrieval of data).

When semantic chunking begins to take place vertically as well as between categories of items, we can see the development of a hierarchical system of semantic memory organization. For example, after a walk in the country a skilled biologist may be able to recall seeing many more types of plants than a novice because of his or her more organized system for categorization—making distinctions between the various plants seen and organizing the data using the existing hierarchical structure. The ability to categorize often depends on or is facilitated by the acquisition of verbal labels for the different categories (Liublinskaya, 1957), a point that will be returned to in a later chapter.

Network Theories of Memory

Modern theories of semantic memory emphasizes its associative nature—all items stored in memory are ultimately connected to all other items **on the basis of their meaning.** LTM is modeled as a **semantic network** (Figure 12-7). The problem facing these theories of memory is twofold—to describe the organization of information stored in memory and to account for the ways it can be accessed (Simon, 1979).

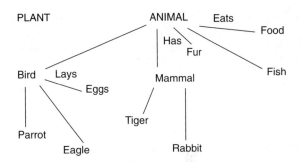

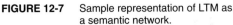

FIGURE 12-7   Sample representation of LTM as a semantic network.

Since all items in semantic memory are ultimately linked (memory is associative), items can be retrieved by searching along the links that connect them. Thus, in trying to remember something, we may search for a cue—some other item known to us that is closely connected to the required item. A good retrieval cue can amount to a full description of the desired item of information which distinguishes it from all other items. **Discriminability** is a major determinant of the ability to remember something.

Memory can be modeled as a series of nodes (concepts) interconnected by relations. The nodes themselves are not words but are connected to words in a mental dictionary. Several words may be linked to the same concept (e.g., car, automobile), or the same word may be linked to different concepts (e.g., plane). Figure 12-8 shows an example of a network theory that was developed by Quillian (1969). In order to understand a sentence, one must decode each word and flag the appropriate concept. A parallel search then takes place in all directions. Thus, understanding the sentence "canaries can fly" would involve searching the network for the existence of interconnections between the different nodes. If the nodes are found to be connected, then the sentence is understood and can be verified. Despite some evidence supporting the theory (Collins and Quillian, 1972), it could not account for certain features of sentence comprehension. For example, the theory predicts longer verification/falsification times for sentences with concepts from distant parts of the network compared to closer concepts. However, subjects could very quickly recognize as false, sentences with a large network distance between their component concepts.

More elaborate theories of LTM have been developed over the years to overcome the limitations of earlier models (e.g., Anderson, 1983). Anderson distinguishes between **declarative** knowledge ("knowing that") and **procedural** knowledge ("knowing how"). Declarative knowledge is stored as a network of linked nodes, a small part of which can be active at any one time (which corresponds roughly to working memory). Activation can spread out from a node to other nodes, for example, thinking about a particular person may remind us of other related people, or when trying to recall the details of a report, we may think of a similar report whose details we do remember. The more nodes through which activation spreads, the more slowly it spreads. Since declarative knowledge is said to be stored in the form of abstract propositions, the theory explains why people remember only the gist of things. Procedural knowledge is stored as a set of **production rules** specifying what should be done if a given set of conditions becomes active (e.g., **if** the light is green, **then** cross the street).

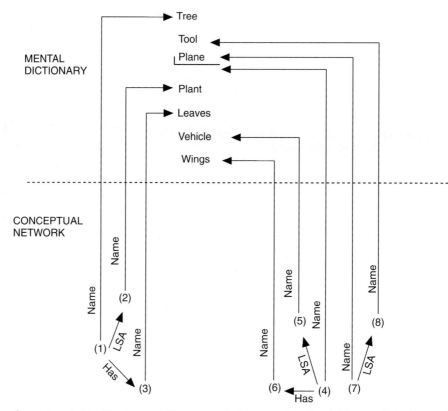

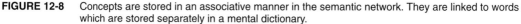

FIGURE 12-8   Concepts are stored in an associative manner in the semantic network. They are linked to words which are stored separately in a mental dictionary.

Some Implications for Training People to Operate New Systems

The general idea of knowledge's being represented as a semantic network is useful in characterizing and analyzing many human-machine interaction issues, such as how to represent complex systems to users and how users learn to operate new systems. For example, existing concepts can be used to assist in the learning of new ones by drawing analogies between the new system and systems users already understand.

Theories such as Anderson's predict that **how well we remember something will depend on the processes which we used to learn it.** Storage of items in long-term memory is slow because new items have to be linked with the existing network. Linking a new item such as a computer command word could be achieved by rehearsing it over and over again or by associating its sound with another known word (finding a rhyme for it). Alternatively, the user could concentrate on the meaning of the word in relation to the functions it involves and to other command words and their functions. **Semantic processing** such as this would be predicted to lead to better retention. There is evidence that this is the case, that deeper or more elaborative rehearsal of new material does lead to better retention (Craik and Lockhart, 1972). This finding implies that when one is trying to learn something new, such as the command words of an interactive system, the act of trying to understand them in relation to the system will itself lead to their being committed to memory. Retrieval of information will also be improved if we try to understand something from all angles, relating it to what we already know in several different ways.

There is evidence that well-learned information really is stored in this kind of multi-dimensional manner. For example, when searching for a particular word, we may experience the **tip-of-the-tongue** phenomenon. Unable to retrieve the word at a lexical level, we access the code which describes how it is articulated or is spoken. When unable to remember how to spell a word, we may write down differently spelled versions of it to see which one looks right.

The network approach can be said to have heuristic value because it provides a general framework and vocabulary for discussing people's knowledge of systems. Some conclusions about the user's knowledge are presented in Table 12-4, using network theory terminology.

COGNITION AND ERGONOMICS

Cognition can be described as a form of information processing where that which is processed tends to be internally stored rather than gleaned from the external environment (C. D. Wickens, personal communication, 1993).

Advances in microelectronics and information technology have produced changes in the task requirements of many jobs. In systems which utilize rudimentary technology, the tasks which people are required to perform are mainly physical—the emphasis is on manual labor and perceptual-motor (craft) skills. Mechanization of these systems reduces the need for manual labor and places more emphasis on manual control skills. Control takes place directly and depends on analog visual and proprioceptive feedback. As technology continues to develop, the tasks required of the user or operator of a system become more abstract and indirect, **with the emphasis on the manipulation of**

TABLE 12-4   CONCLUSIONS ABOUT LTM*

---

1 Information is stored in LTM on the basis of meaning.

2 Concepts are organized in terms of a network of propositions (e.g., "a dog is an animal").

3 Working memory is that part of the network that is active.

4 The spread of activation from a concept depends on the strength of the links.

5 Successful learning depends on the following:

   a Activation of appropriate parts of the network

   b Linkage of new concepts to existing ones

   c Elaboration and explanation to build more links

   d Practice to strengthen new links

6 Successful recall depends on retrieval cues to direct search to appropriate parts of the network.

---

*Adapted from Anderson (1983).

symbols according to a set of **previously prescribed rules.** Many of the most important issues in the design of these systems are to be found at the abstract, symbolic level and can best be dealt with using the concepts and methods of cognitive psychology. It is in this sense that the term "cognitive ergonomics" is used in much modern literature.

Rasmussen (1983) has identified three types of task performance in industry. Skill-based performance depends on the existence of specialized subroutines for the performance of routine tasks and has already been discussed in an earlier part of this chapter. Rule-based performance makes greater demands on conscious processing capacity because explicit rules have to be kept in mind and followed as the operator carries out the task. Knowledge-based performance is essential in complex tasks, particularly where the operator has to engage in problem solving and decision making in novel situations. In complex systems which are largely automatic, such as nuclear power stations, operators operate primarily in the cognitive mode (Rasmussen and Rouse, 1981). They engage in supervisory control which involves goal setting, monitoring system performance, and dealing with nonprogrammed events. Detection of signals and manual control are relegated to a more secondary role for the operator. Car driving is an example of a perceptual-motor control task. Operating a power station is an example of a cognitive control task.

The purpose of this portion of the chapter is to acquaint the noncognizant reader with the science of cognition and to illustrate why a cognitive approach is often the best approach to the solution of certain design problems.

Cognitive Systems

Hollnagel and Woods (1983) argue that human-machine systems should be analyzed, conceived of, and designed at a cognitive level—as **cognitive systems.** They define a cognitive system as follows:

> A cognitive system produces "intelligent action"; that is, its behaviour is goal-oriented, based on symbol manipulation and uses knowledge of the world (heuristic knowledge) for guidance. Furthermore, a cognitive system is adaptive and able to view a problem in more than one way. A cognitive system operates using knowledge about itself and the environment, in the sense that it is able to plan and modify its actions on the basis of knowledge. It is thus not only data driven, but also concept driven. Man is obviously a cognitive system. Machines are potentially, if not actually, cognitive systems. An MMS (man-machine system) regarded as a whole is definitely a cognitive system.

In the first portion of the chapter some psychological aspects of system design have been considered using the human information-processing model of Wickens to structure much of the discussion. Models such as this have been valuable in improving the design of human-machine systems. One of the advantages of using such generic models is that any validity they have is independent of the particular system, and any guidelines or design principles derived from them can have general application. This can also be a disadvantage—a generic model of human information processing may lack the precision necessary to deal with the most important ergonomics issues in a particular system. In many situations designers are presented with **system-specific** problems and here, generic models may be of limited value. A more focused approach which yields system-specific models of the human operator or user can better take into account the worksystem context in which human-machine interaction takes place. Such models may be more powerful in enabling specific design issues to be addressed and solved.

Cognitive Models of the Human Operator

Some of the first **cognitive models** of human operators were developed in the process control industries. In these industries, the operator monitors the system while the system controls the process (e.g., Bainbridge, 1979; Umbers, 1979). The operator's role is supervisory and the task requirements are to monitor the process and diagnose faults. This type of task can be described as a **cognitive control task**—the operator deals with facts and rules rather than responding to stimuli in the perceptual-motor mode of interaction. If we want to analyze the task, the best approach is to discover the facts and rules and how they are used to carry out the task, rather than looking at the design of the workplace.

In process control tasks, the operator must ensure that the system operates in an optimum way with respect to certain criteria. Optimum operating ranges for important system variables are usually specified in advance by the designer, and the operator has to monitor the system to see that the magnitude of key variables remains in the optimum range (e.g., typical variables are the temperature or pressure in boilers, the flow rates of reactants or products, and the consumption of energy). Only if departures from the optimum occur are control actions necessary.

The operator can be modeled as an information-processing system which has to solve a finite set of problems (dependent on the particular system) using information from the system (feedback about system behavior) by the application of programs (usually algorithms or rules which specify how particular events are to be dealt with). The key elements of the model which have to be determined are the following:

1 The rules which the operator uses to control the system
2 The strategies which determine how the rules are used
3 The types of system feedback which influence the operator's control strategy

Table 12-5 shows common uses of models of the operator.

TABLE 12-5   USES OF MODELS OF THE OPERATOR

---

1 Prediction of operator behavior

2 Facilitation of task load evaluation

3 Direction of equipment design evaluation procedures (using other ergonomics approaches)

4 Evaluation of the adequacy of operating procedures

5 Evaluation of training programs

6 Implementation on computer to simulate the behavior of the operator and find answers to what-if questions (questions not easily answered in other ways)

CURRENT AND FUTURE RESEARCH DIRECTIONS

The development of psychology as a body of knowledge applicable to the design of human-machine systems depends on advances in psychology itself and on technological change outside the discipline. Advances in experimental and cognitive psychology have provided ergonomists with a better model of the user as a processor of information. The information-processing analogy and much of the experimental research which it has generated are closer to the problems of ergonomics than much of previous psychology. Technology "push" is forcing ergonomists to use a more formal approach to solve truly novel problems of human-machine interaction. It is likely that the theoretical grounding provided by psychology will be of increasing importance to ergonomics and that the gap between applied experimental psychology and information ergonomics will become even narrower.

SUMMARY

Some of the first formal models of the human operator were developed in the study of manual control of systems, using concepts and principles from control systems engineering. The study of qualitatively and quantitatively different systems demanded new approaches. The operator or user can be considered a processor of information. Information flows from the senses, via the perceptual system, to working memory or to specific subroutines in which automatic processing takes place, possibly leading to a response. Information in working memory decays rapidly unless operated on by mental operations stores in long-term memory. The processing capacity of the system is constrained by channel capacity, the availability of subroutines (level of skill), and the demands made on executive (conscious) resources.

The information-processing approach has heuristic value in ergonomics since it provides a framework and vocabulary for the structured analysis of design issues in human-machine interaction. It has made possible the field of cognitive ergonomics, which is concerned with the symbolic processes involved in human-machine interaction—how users conceive of a system and how the system should be designed so that it will be conceived of in an appropriate way.

ESSAYS AND EXERCISES

1 Describe the elements of the human information-processing model presented in this chapter. Illustrate each part of the model with examples from everyday life.
2 Ask colleagues or friends to memorize the recipe below without taking notes. Allow them 5 minutes of study time. Ask them to recall the recipe immediately, the next day, after 1 week, and after 1 month. Describe the differences and similarities between the recalled and original versions. In particular, look for errors of omission and sequence and for additions and simplifications. Develop a system for categorizing these errors. Does previous culinary experience influence recall?

Pedro Ruiz's Paella Recipe
Obtain 12 large prawns and remove their heads. Simmer the heads in 3 L chicken stock for 2 hours together with 1 large, whole peeled onion. Add 2 tsp. salt to the stock. If you have any fresh fish heads, add them to the stock as well.

Debone, skin, and cut into small, lean pieces 1 large, fresh chicken. Remove the fat from 1 kg tender pork, either chops, leg, or steak, and cut the pork into small pieces. Heat 2 tbp. olive oil in a large, flat-bottomed frying pan (or a "paellera" if you have one). Finally chop 3 large cloves of garlic and add to the olive oil. Fry the garlic for several minutes.

Add the diced chicken and pork to the frying pan and stir, making sure that the meat cooks evenly. When the meat is nearly ready (i.e., when it is almost cooked on the inside), add 2 large red and 2 green peppers that have been previously cut into thin strips. Fry the strips of pepper with the rest of the ingredients until they are soft on the outside, but still crisp.

Next add the contents of 2 tins of peeled plum tomatoes and stir. Heat until the mixture simmers; then reduce for about 5 minutes. Remove the stock pan from the burner and strain the stock to remove the solids. Put 3.5 c. rice (heaped) into the frying pan and cover it with 1.5 L stock. Simmer gently for 10 minutes.

Add more stock as needed to prevent the mixture from sticking to the bottom of the pan. Season to taste and add 10 g saffron. Add 500 g mussels (or clams); 500 g diced, deboned hake; and 300 g calamari rings. Make sure there is sufficient stock to cover the ingredients. Finally, arrange the prawns, nicely, around the circumference of the pan. Simmer gently for 5 minutes or until the prawns are cooked.

3 What is the difference between skilled and unskilled performance? Discuss the ergonomic implications of this difference.

FOOTNOTE: Changing Models of the Human Mind

The study of thought has a long history traceable to the extensive writings of ancient Greek philosophers and beyond. Plato attempted to reconcile the search for true knowledge with the apparent uncertainty of a constantly changing world. All objects of a particular type were believed to be imperfect copies of an unchanging general idea or form. Linguistically, this view implies that a word such as "cat" attains its meaning from the ideal form and not from any particular cat. In more modern terminology, this is rather like saying that we can have a concept of cat (linked to the word "cat") which does not depend on observation of individual cats. This view has been extremely influential, since it implies that true knowledge lies beyond perception and can be attained only by contemplation. It draws attention away from physical matters toward an abstract world of intellectual truth accessible only by thought.

The search for the rules of thought has continued, with interruptions, throughout history. Descartes (who is said to be famous for saying "I think, therefore I am") distinguished between mind, as an immaterial entity, and body, as a material object, and set down certain rules for the direction of the mind. Since the mind was not a machine, according to Descartes, it could not be studied as one.

The first psychologists attempted to study the mind using introspection. Subjects carrying out a specified activity attempted to analyze their own mental processes by thinking about them. A problem with introspection is that the contents of consciousness are produced by processes that are not amenable to conscious analysis. For example, much of visual perception is preconscious—the perception of depth and of illusions is due to processes to which consciousness has no access. Thus, introspection is limited because it has no access to the processes on which its operation depends.

It is often said that psychological theories are products of their times. Technology has a powerful effect on how people perceive the world. This may turn back on itself and affect how people perceive themselves. Freud, at the beginning of the century, stressed the importance of primitive, unconscious drives in motivating behavior. The *id* was the furnace of the psychological "steam engine," which had to be held in check by the *ego* (the driver of the psychological train) and channeled in socially acceptable directions by the *superego* (the psychological equivalent of the railway system operating procedures).

Behaviorism, first developed by John B. Watson and later refined by B. F. Skinner, rejected introspection and the abstract constructs of psychoanalysis. Behavior which was directly observable was said to be the only acceptable data for use in psychology. The study of mental states, since they could not be directly observed, was rejected. This change in approach was better suited to the study of learning than to perception or emotional behavior and led to a change in the classes of problems investigated by psychology. The relationship between stimuli and responses under different patterns of environmental reinforcement received much attention. Learning was seen to consist of building up an ever more complex repertoire of stimulus-response (S-R) links. This had direct application to the study of training and the search for more efficient methods of training people to operate machines.

Behaviorism substituted the telephone switchboard for the steam engine and was itself replaced by the digital computer. The information-processing approach, which began in the 1950s, made an analogy between the brain as a processor of information and the digital computer. It is appropriate to talk about the symbolic processes going on inside a computer when it is carrying out some directly unobservable operation and even to make inferences about the processes. In the same way, inferences can be made about the symbolic processes going on in the human mind, and these inferences can be tested by implementing them in the form of computer programs. Thus, the computer provides both an analogy and a testing ground for psychological theories. The challenge of the information-processing approach is that it forces psychologists to express their theories in an explicit way and with sufficient detail for implementation on a computer.

Johnson-Laird (1988) provides an introduction to the field of cognitive science that has grown out of the information-processing approach and from developments in artificial intelligence. The cognitive approach takes the view that "all psychological explanation must be framed in terms of internal, mental representations and processes (or rules) by which these representations are manipulated and transformed" (Still and Costall, 1987). This strong view, which sees the cognitive approach as *the* paradigm for investigating human behavior, has been criticized by Costal and Still (1987) (although the weaker view that it is the paradigmatic approach to the study of **cognitive** behavior would be less open to criticism).

It is likely that further developments in technology will continue to influence psychological theory-making by providing researchers with new analogies for conceptualizing human behavior.

13

DISPLAYS, CONTROLS, AND HUMAN-MACHINE INTERACTION

Displays and controls are the interface through which human-machine information exchange takes place. The design of the displays and controls of a machine can either facilitate interaction or increase task difficulty and the probability of error. General principles for the design of displays and controls exist, but task analysis, as described in Chapter 1, is normally required before these principles can be applied. As C. D. Wickens (personal communication, 1993) and others have put it, we must understand the task before we attempt to design it.

PRINCIPLES FOR THE DESIGN OF VISUAL DISPLAYS

The study of visual perception is a useful point of departure for discussing the design of visual displays. The German Gestalt psychologists in the first half of this century stressed the importance of structural information in determining what is perceived. To them is attributed the famous saying "the whole is greater than the sum of its parts," which can be interpreted to mean that conscious perceptions result not just from an analysis of objects in the field of view but from a synthesis of the objects themselves and the structural relations between them. Figure 13-1 illustrates that perceptual mechanisms, beyond conscious control, actively construct a meaningful representation of reality from information in the proximal stimulus—that there is more to visual perception than meets the eye.

The Gestalt psychologists identified a number of laws by which the perceptual system was organized. These laws provide a framework for elementary discussion of the design of visual displays.

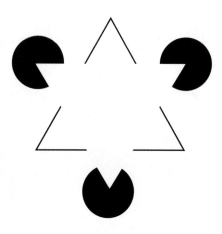

FIGURE 13-1   The Kaniza triangle. There is more to visual perception than meets the eye.

Figure-Ground Differentiation

The stimulus array impinging on the sense organs is transmitted to higher processing centers in the brain, where it is differentiated into a meaningful or perceptually intrusive figure and an unobtrusive background. Neonates can follow a moving object (the figure) against a background. Figure-ground differentiation is a fundamental step in perceptual processing, in which:

1 The perceived figure has form while the background is formless.
2 The figure appears to stand out against the background.

Since the perceptual system is of limited capacity, figure-ground differentiation can be seen as a way of reducing incoming data to manageable proportions. It is not restricted to the visual system—listening to a radio in a noisy room or savoring the smell of roasting meat over a smoky fire are examples from other sensory modalities. Although information in the figure receives preferential processing at the expense of ground information, the latter is not entirely lost. Ground information provides a **context** that influences the way the figure is perceived, which demonstrates that ground information is processed beyond the physical level. In advertising, journalism, and report writing, the use of special typefaces can influence the way a message is interpreted. Italics or bold letters may be used to convey important points. Advertisers use different typefaces to impart connotative dimensions which would be otherwise lacking in the message—sophisticated typefaces used for the labels on bottles of sparkling wine or rugged typefaces on bottles of aftershave are examples.

Contour, closure, and previous experience influence the differentiation of a stimulus array into figure and ground. Contour is a characteristic of a stimulus array which provides cues for figure-ground differentiation (Figure 13-2). In Chapter 10, it was seen that the retina functions as an edge detector rather than a light meter—sharp contrasts enhance contour detection and produce strong figures.

Sudden changes in brightness or color provide natural contours between objects. Contour enhancement is a useful method for helping operators differentiate parts of a display panel or recognize a symbol. It is known that well-designed road signs are more

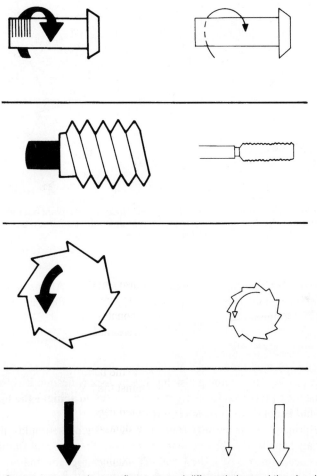

FIGURE 13-2   Strong contour enhances figure-ground differentiation and the visual impact of a figure. (From Easterby, 1970, with permission of Taylor and Francis.)

easily recognized than written messages, giving drivers of all ages almost twice as much time to respond to them (Kline et al., 1990). Kline and Fuchs (1993) have also shown that existing designs of road signs can be improved. They investigated the visibility of symbolic highway signs as a function of driver age, using the signs to indicate divided highway, road narrows, workers ahead, and hill. They identified and altered details of the symbols that were difficult to resolve or susceptible to confusion with adjacent contours and showed that visibility of road signs could be improved for adults of all ages.

Closure can be described as the rendering of form—a tendency to complete, or "close," a figure. Closure is the tendency to produce a meaningful percept from incomplete cues and is the mechanism which prevents us from being aware of the effects of the retinal blind spot on vision. The ability to create meaningful percepts from incomplete cues depends on past experience and is a characteristic which distinguishes the skilled from the unskilled operator. When one first views a map, the initial impression

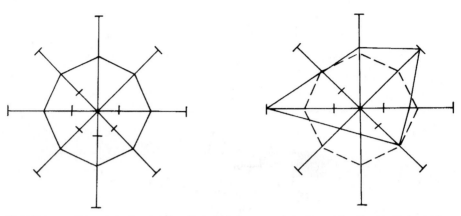

FIGURE 13-3   Eight linear scales arranged as a polar display. Under normal plant operation the separate readings create a closed figure (an octagon). Departures from normal can be readily perceived as a distortion in the octagon. (From Wickens, 1984.)

may be of a bewildering complexity of unrelated parts. After study and use, closure may be achieved and the map seen as a whole. Figure 13-3 illustrates the principle of closure in display design.

Grouping

The Gestalt idea of grouping also has use in display design. It is based on the observation that the **perceptual mechanisms always try to achieve the best possible form so that what is perceived has maximum meaning.**

The principle of **proximity** states that elements of a stimulus array which occur together are perceived to belong to each other, in the sense that flashing lights on an ambulance accompanied by a siren "belong" to an emergency. In Figure 13-4, most people tend to see three rows of 4 dials, rather than 12 unrelated dials or four columns of 3 dials. The dials are closer together horizontally rather than vertically, and this distance information is automatically included in the production of a percept. The principle of proximity is extremely important in panel design since similar display elements that *appear* to belong together because of their proximity *should* belong together at the level of sys-

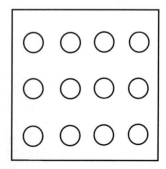

FIGURE 13-4   The principle of proximity.

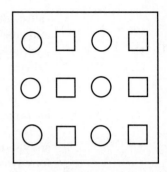

FIGURE 13-5   The principle of similarity.

tem operation. Looking at Figure 13-4, one finds it extremely difficult to consciously reverse the percept and "see" four columns of three. There is experimental evidence that operators are better able to integrate probabilistic information when proximity is used to create more integrated displays (Barnett and Wickens, 1988).

The principle of **similarity** (on the basis of color, shape, or size) states that similar items will be grouped together. This principle can override the proximity principle, as is apparent from Figure 13-5, where the proximity of elements is the same as in the previous figure. However, with the introduction of similarity, the tendency to perceive four columns of three is now dominant. In display design, similarity can be used to partition various parts of a display and strengthen the tendency to perceive related elements as belonging together (by making them appear similar).

People have a very strong tendency to perceive similarly colored objects as belonging together (van Nes, 1986) as long as **no more than three or four colors** are used together. Color can be used to provide conceptual grouping of text in itineraries, timetables, and so on. For example, station names and cities can be presented in one color, destinations in another, and departure and arrival times and on-board facilities in a third. Some of the advantages and disadvantages of the use of color have been summarized by Filley (1982) and are given in Table 13-1.

Generally, red is interpreted to mean "stop" or "danger," green to mean "go" or that the system is running normally, and orange to signify caution. This use of color is unfortunate because approximately 5 to 8 percent of people have color vision defects (so-

TABLE 13-1   ADVANTAGES AND DISADVANTAGES OF THE USE OF COLOR IN DISPLAY DESIGN

Advantages	Disadvantages
Draws attention to specific data	May have little or no benefit for the 8 percent of males who are color blind
Enables faster uptake of information	May cause confusion
Can reduce error	May cause fatigue
Can separate closely spaced items	May cause unwanted groupings
May speed reaction	May cause errors
Adds another dimension	Can cause afterimages
Seems more natural	Appears more frivolous

called color blindness). **The most common deficit is an inability to distinguish red and green!** (They both look brown.) Approximately 7 million drivers in North America have this problem (University of California, 1993) and some studies suggest that color-deficient drivers have problems seeing traffic lights. This finding suggests that **designers should use shape as well as color for coding purposes.** It is also unfortunate that there does not seem to be a convention for the color blue because disorders involving the perception of blue are rare.

The principle of **symmetry** describes how, in the attempt to interpret the stimulus array so as to maximize its meaningfulness, elements of the array will be included to form symmetrical or balanced figures.

Finally, the principle of **continuity** describes how the perceptual system extracts qualitative information from a stimulus to form a unity. All parts of an array having this characteristic will be included in the percept, which will be perceptually separate from all other parts of the display even if the parts physically overlap in places. Figure 13-6 illustrates the use of the continuity concept to enhance graph design.

Resolution of Detail

The ability to resolve detail depends on the accommodation of the lens of the eye, the ambient lighting, and the visual angle—the angle subtended at the eye by the viewed object (Figure 13-7).

Visual angle is a more useful concept than absolute object size since it takes into account both the size of the object and its distance from the viewer. Under good lighting, a minimum visual angle of 15 minutes of arc is recommended. Many instruments such

FIGURE 13-6   The principle of continuity. Even though the two lines on the graph share a common point, they remain perceptually separate.

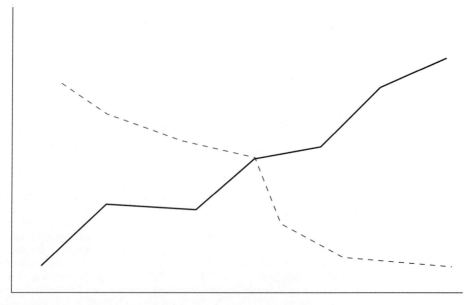

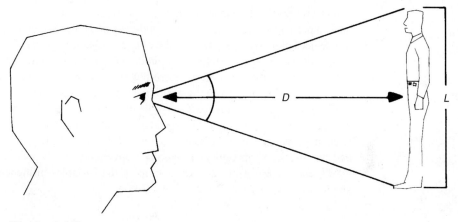

FIGURE 13-7   The visual angle.

as dials and gauges are designed to be read at **a reference distance of about 70 cm.** If the reference distance is known, the required change in object size to equalize the visual angle when the object has to be viewed at a different distance can be calculated by

$$\text{Size at new distance} = \frac{\text{size at reference} \times \text{new distance}}{\text{reference distance}}$$

For example, if the reference distance is 70 cm and the new distance is 140 cm, the size of the object would have to be doubled to maintain the visual angle (see Bailey, 1982, for further information). The ability to resolve detail also depends on human informa-tion-processing limitations. The early stages of information processing can handle large amounts of information at a primitive level, but capacity diminishes as soon as more complex processing is required.

Humans can visually apprehend and make intuitive judgments about small numbers of objects quite easily (Figure 13-8), but this ability is lost as soon as the number of ob-

FIGURE 13-8   Limits to visual apprehension of quantity. In (A) we can "see" that there are 3 objects. Intuitive visual judgments about quantity are accurate for small numbers. In (B), we have to count to know that there are 12 objects.

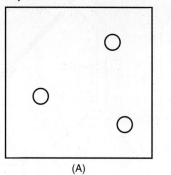
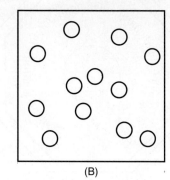

(A)   (B)

jects increases by more than about five. This finding has implications for display design and, in particular, the design of dials and gauges. Dials and gauges are essentially analog devices used to display the change or rate of change of a variable and to allow visual judgments to be made about its magnitude. For example, some of the modern, fashionable analog wristwatches do not have numbers on the dial since the time can be estimated quite accurately from the position of the hands on the face.

To allow more accurate readings to be made, scales are often superimposed onto the circumference of dials (as in the speedometer of a car). This introduces a trade-off between the desire to improve scale resolution by increasing the number of interpolations and the ability of operators to **visually apprehend** the displayed quantities. Murrell (1965) carried out detailed investigations of the variables influencing dial design. Examples of designs are given in Figure 13-9.

Color Coding of Dials

Color coding of parts of a dial face is an example of a qualitative way of facilitating visual judgment. In industrial applications, green is used to indicate the part of the dial associated with safe or normal operation, orange to suggest caution, and red to suggest unsafe or abnormal operation of the displayed parameter. Similarly, in car instrumentation, the temperature gauge is often color coded. Red is used to indicate the high-temperature part of the scale (which signifies overheating of the engine if the pointer moves into the red region) and blue to indicate the cool part of the scale (which indicates underheating and possible thermostat malfunction if the pointer remains in this part of the scale after the engine has warmed up).

Color coding can be combined with quantitative scaling in dial design. Color-coded dials do not require the operator to memorize exact scale values which correspond to normal, unsafe, or suboptimal states, but exact values can be taken when needed. Color coding of dials can reduce the memory load of routine checking tasks.

FIGURE 13-9   Design of scales. The dial on the right has been made easier to read by having fewer, bolder graduation marks. The scale length has been increased by putting the numerals on the inside. The numerals themselves are larger and level. The whole of the pointer is visible. (Adapted from *Applied Ergonomics Handbook*, 1974, with permission from Butterworth Heinemann.)

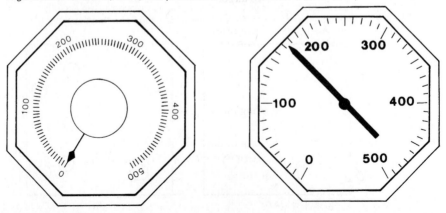

Digital Displays

Digital displays should be used in situations where highly accurate reading of displayed quantities is required. Digital displays are inferior to analog displays for conveying information about rates of change, so in those cases where both types of information are required, a compromise design can be used in which a digital counter is superimposed on the dial face.

Multiple-Display Configurations

Despite the existence of display design guidelines, multiple-display configurations are often neglected by designers. In an investigation of control room design in nuclear power stations, Seminara and Smith (1983, 1984) commented on the complexity of the control rooms and on the massive arrays of control panel elements which were often identical in appearance. There were insufficient visual cues to facilitate the differentiation of subsystems. It was observed that experienced operators became familiar with the panels they operated most frequently—usually through routine operation—and were least familiar with least used panels—those that were most critical in emergencies.

The Three Mile Island nuclear incident drew attention to the need for ergonomics in the design of control rooms and in the training of operators. Malone et al. (1980) evaluated the control room at Three Mile Island. The information required by operators was often found to be nonexistent, poorly located, or difficult to read. Annunciators were not color coded and in one subsystem, **91 percent of the applicable human engineering criteria for display design were not met.** In addition, in the control room, 1900 displays were located on vertical panels, 503 of which could not be read by a short (5th percentile stature) operator. The labeling of panels was also inadequate—it was discovered that the operators themselves had made **800 changes** to the existing panels in an effort to improve them.

An example of an ambiguous panel design is given in Figure 13-10A. The same panel, after some remedial ergonomics, is depicted in Figure 13-10B.

The ability to gather information quickly by shifting focal attention around a complex display depends on several factors. Sanders (1970) demonstrated that as the visual angle between two targets was increased, the reaction time and number of errors needed to make a same-different comparison between them also increased. Sanders identified three **functional attentional fields.** Objects placed within 30 degrees of each other fall into the **stationary** field and can be detected with the stationary eye. The **eye** field encompasses a region from 30 to 80 degrees, in which eye movements are required to compare objects. Beyond 80 degrees, both **head and eye** movements are required to detect targets. Focal attention can be likened to a spotlight which illuminates parts of a complex display, making them accessible to higher cognitive processes. Parts of the display within the spotlight can be processed simultaneously, whereas those outside the spotlight require shifting of attention and search.

Data on the extent of the various functional attentional fields are useful in designing complex control panels. If an operator has to monitor several displays simultaneously and make judgments using the data from all of them, they should be placed in the stationary or eye fields with respect to a predefined point of observation (such as the oper-

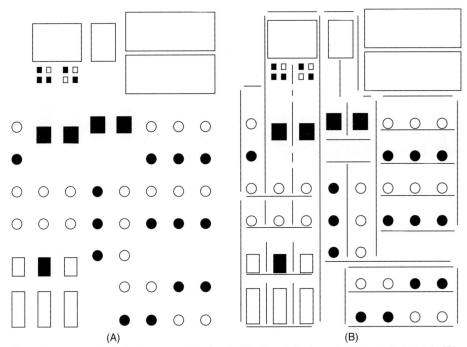

FIGURE 13-10   Remedial ergonomics to improve panel design. In (A), the original panel design is depicted. In (B), black tape has been stuck onto the panel to make clear the true subsystem boundaries. (From J. L. Seminara and D. L. Smith, "Remedial human factors engineering—Part 1," *Applied Ergonomics,* vol. 14, no. 4, December 1983, pp. 253–264. Reproduced by permission of the publisher, Butterworth Heinemann ©.)

ator's desk). Careful analysis of the operator's task and operational priorities, including emergency recovery procedures, can be used to assist in optimizing a panel layout.

Radar screens, switchboards, computer terminals, and so on consist of control-display configurations of varying complexity. Operator interaction with these may be constrained or unconstrained. In the former, the system may permit only a limited number of sequences of specified control actions. In the latter, many more complex sequences may exist as a result of richer task variety (e.g., creating text, programming a computer, or playing computer games). In either case, frequent sequences of operator-system interaction should be identified as part of an early attempt to simulate the operators' tasks. Optimal locations for the placement of controls and displays can be arrived at by considering the sequences of hand and eye movements needed to carry out a task. Some examples of criteria for evaluating prototype layouts are given in Table 13-2.

Guiding Visual Search in Complex Displays

In large, complex displays, special methods for highlighting potentially hazardous situations and guiding visual search to the appropriate part of the display may be required.

TABLE 13-2   COMMON CRITERIA FOR EVALUATING THE LAYOUT OF COMPLEX PANELS

1 Eliminate unnecessary or complex movements.
2 Locate controls and displays to reduce postural load.
3 Reduce movement complexity and encourage muscle memory by consistent choice of movements.
4 Ensure that control actions do not obscure important display areas.
5 Avoid spatial transformations.
6 Provide spatial proximity of displays and controls used frequently in a sequence of tasks.

The process of interrogating a complex display can be decomposed into a number of stages:

1 Alerting operators to the existence of a signal or target data
2 Orienting the perceptual system to the appropriate part of the display
3 Attending to the data so that it can be transmitted to the central processing centers of the brain

Posner et al. (1976) found that **visual stimuli are generally less alerting than nonvisual stimuli,** and for this reason people consciously attend to visual stimuli. The use of nonvisual cues is potentially a better way of alerting operators to hazardous situations, since these cues demand less attention. Alerting cues appear to improve an operator's readiness to respond and to speed up reaction time but may do this at the expense of accuracy. Informative cues (which provide information about the type of event to come) seem to assist in the **encoding** of the signal being attended to (the operator is primed to respond in a particular way) as well as having an alerting quality. Improved encoding of a signal would be expected to result in improved response accuracy. It appears that informative, nonvisual cues can speed up reaction time and response accuracy and are particularly effective when the positional uncertainty of a target is high. For example, Perrot et al. (1991) found that visual search could be aided considerably when the appearance of a visual target was accompanied by a spatially orientated sound (a 10-Hz click train directly behind the part of the display where the target appeared). Improvements in response latency of up to 300 milliseconds were obtainable using this method. The effectiveness of the auditory signal increased as a function of visual load and the distance of the target from the subject's line of gaze. Auditory spatial information seems to offer benefits in situations where operators have to detect multiple targets in spatially extended displays, as are found in aircraft flight decks and some process industry control rooms (see Posner, 1980, and Wickens, 1984, for further discussion of factors influencing visual attention). The information content of auditory cues can be increased by varying pitch, continuity, or waveform.

Computer-Generated Displays

Display configurations based on electromechanical technology can be inflexible and excessively complex. In many tasks, much of the total information displayed is redundant most of the time. In several applications, notably process control, aerospace, and other complex systems, designers have turned to more flexible CRT screen-based methods of display design in an effort to do away with the bulk and complexity of existing hardware. Screen-based displays offer, in addition to increased flexibility, the possibility of using innovative graphic techniques for representing the changing state of a system in a way that is more compatible with operator characteristics. Instead of displaying the values of system variables on dials, scales, and counters, more dynamic, integrated displays such as on-screen histograms, time series, and pie charts can be used. These displays can be continually updated as the system state changes. Research is currently under way to determine criteria for choosing between the various design options. The goal is to ensure that a display supports *direct* perception of a system state. With direct perception the need for interpretive mental operations is minimized, so operator reaction time and mental workload are reduced. Hollands and Spence (1992) compared display options for change and proportion in system variables. They found that the speed and accuracy of detecting change was superior when system data were displayed in the form of time series or histograms (bar charts), whereas perception of proportion was best when in pie chart or divided bar chart presentation mode.

Figure 13-11 presents a composite, screen-based display which illustrates many of the design principles which have been discussed.

FIGURE 13-11 Good application of grouping principles. Proximity, similarity, symmetry, and continuity have been used to create a coherent display which naturally decomposes into subsystems and different functional areas.

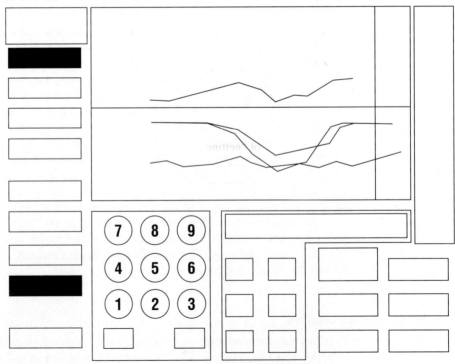

Many of the guidelines for display design can be applied equally well to the design of information on the computer screen as to "knobs and dials" hardware. One of the characteristics of information technology is that it offers increased flexibility in the way information is stored, handled, and displayed, and in the opportunities for interaction it provides. Interactive systems typically display in detail only that information relevant to the current task. Although this approach is very economical and avoids some of the problems of visual complexity discussed previously, potential problems are introduced, particularly that of the user's **losing track of where she or he is with respect to the rest of the system.** A doctor searching a new, computerized medical database for information about the treatment regimes of the patients on the ward can be likened to a person going shopping in a strange town without a map. In the process of executing the sequential task there is always the problem of getting lost in the system.

Several design techniques have been proposed to assist users in finding their way around a system. A common approach is to use **icons** to represent system objects, thereby reducing the need for users to learn system terminology and command names. The icon is a symbolic representation of the object (usually a stylized, abbreviated picture of the real object with strong figure-ground differentiation and closure). A diary can be represented as a book, the calculator facility as a pocket calculator, and so on. Icons can be used to create unambiguous, concise abbreviations of system objects but may be less successful when used to represent more abstract parts of the system (such as a file directory). Figure 13-12 gives examples of icons commonly used in computer systems with graphic user interfaces.

Miniatures (Nielsen, 1990) are similar to icons but are more flexible. They are created by the system from the information to which they refer. Thus, a page of text and graphics or a letter which a user has recently created can be miniaturized (represented in miniature) and displayed in a window on a separate part of the screen. While carrying out a sequence of tasks, a user can be presented with a symbolic representation of the miniaturized information landscape in which he or she is working. Some modern word processing systems present the user with a miniature of a to-be-printed page to indicate how it will appear when printed. This is particularly useful when different sizes of type are being used together or when text and graphics are combined on one page.

The problem of "getting lost" in large databases (not knowing the location of the current position in relation to other parts of the database and other system facilities) is exacerbated if the system structure is complicated or ambiguous or if the user's **mental map** of the system is incorrect or incomplete. For this reason a number of attempts have been made to present the user with an explicit, graphic representation of the system structure. Spence and Apperley (1982) used the concept of database navigation to design an interactive environment for the professional user. In this approach, the VDT screen becomes a window on the system and system objects are presented as an information landscape by means of logos and miniatures. The screen is designed as a **bifocal viewing system.** This idea has been further developed by Spence (1993) (Figure 13-13).

The focal part of the display can be likened to a telescope used to scan the information landscape in detail. In the example given, it can be used to read the contents of messages and other documents. The peripheral areas of the display contain miniatures of the

FIGURE 13-12   Icons used in graphic user interfaces.

other documents in the information landscape. Thus, information can be retrieved by visually scanning the space. **This approach minimizes the need to learn a command language, as most tasks can be accomplished using direct manipulation.**

Beard and Walker (1990) also addressed the problem of user navigation of large data spaces. Such spaces can often be represented as a **hierarchical tree structure.** Because of the small size of the VDT screen, only parts of the tree can be displayed at any one time and the user may lose track of his or her location in the system. Beard and Walker developed a **map window** in which a miniaturized version of the entire data space is presented. Inside the map window, a box (called the wire frame box) is superimposed over a part of the miniature version of the data space. The part of the screen within the box is

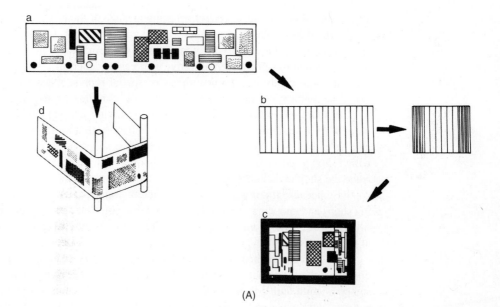

(A)

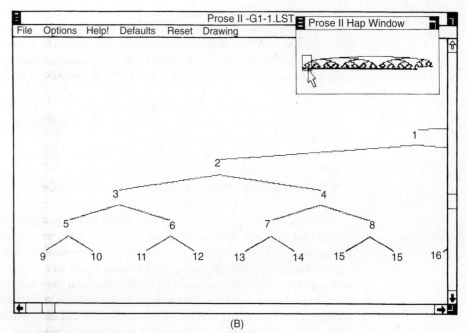

(B)

FIGURE 13-13   Database navigation aids. (A) Bifocal and wraparound displays. Spence's solution to the problem of fitting a large data landscape onto a small screen. The bifocal display provides high resolution of desired items and displays the rest of the data landscape in compressed form using miniatures. (From Spence, 1993; illustrations provided by the author.) (B) Beard and Walker's map window concept. The arrow indicates the location of the wire frame box used to select a part of the database. (Adapted from Beard and Walker, 1990, with permission of Taylor and Francis.)

shown on the rest of the screen. Decreasing the size of the box increases the magnification of the data displayed on the screen (the zoom facility). Users can explore the data space by moving the box to different parts of the miniature data space. The technique has been found to significantly improve user performance.

With large databases, the usefulness of spatial representations may be limited because the spatial representation becomes visually complex. Other ways of assisting users may be more effective. One alternative is to provide an analogy—an explicit, verbal model which compares interrogating the database with a more familiar activity (Webb and Kramer, 1990). For example, "It's a bit like shopping in a mall. You first decide which shop to go in; then you enter and choose among the items in the shop. Then you leave the shop and choose another shop." This analogy is appropriate for a hierarchical database because it conveys the idea of there being separate levels of activity—choosing shops and choosing things in a shop and that you first have to leave a shop (return to a higher level in the hierarchy) before you can choose something in a different shop (part of the database).

The recent emergence of **hypertext** systems—databases in which items of information are linked together nonsequentially—poses challenging problems for display designers (Smith and Wilson, 1993). Hypertext spaces contain information at each node and are "n-dimensional" in the sense that a given node may be the point of departure for many other nodes. A characteristic of user interaction with hypertext systems is that the development of an optimum search strategy is de-emphasized in favor of a more intuitive approach in which, en route to a desired item, other related items are visited. The user is encouraged to acquire a mental map of the space as a consequence of browsing through it. Failure to develop an adequate mental map can lead to disorientation when using hypertext which, according to Foss (1989), may result in users:

1 Reaching a particular node and then forgetting what was to be done there
2 Forgetting to return to the main path after making a digression
3 Forgetting to make planned digressions
4 Not knowing if there are other relevant frames in a document
5 Forgetting which areas were previously visited
6 Having difficulties summarizing information examined during a session

Smith and Wilson (1993) discuss some of the issues in the development of navigation aids for hypertext systems. Aids can be either schematic or spatial, and they can be displayed in either two or three dimensions. Of particular importance seems to be the correspondence between the type of navigational aid chosen and the semantic content of the hypertext system. Two-dimensional schematic aids, such as those described above, are appropriate for smaller systems. With larger systems, two-dimensional schematic displays can be confusing because of the spaghetti-like entanglement of links between nodes. Three-dimensional schematic displays (Figure 13-14) can aid user navigation through complex spaces. Spatial displays are effective for representing complex knowledge domains possessing definite spatial characteristics. Smith and Wilson give the example of a hypertext system for the domain of a university department (see Figure 13-14).

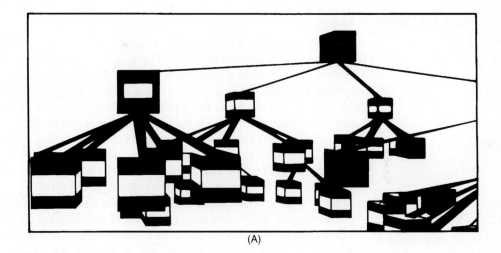

(A)

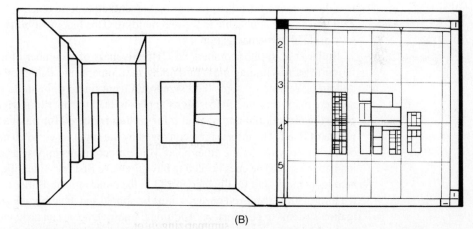

(B)

FIGURE 13-14   Alternative representations for the data space in hypertext systems. (A) Three-dimensional schematic display. Each node has many links which can be better represented in three rather than two dimensions. (B) Use of virtual reality to represent the information space. The user walks through a virtual world, accessing screens of information on the way (this approach shares common features with the method of loci discussed in the previous chapter). (From P. A. Smith and J. R. Wilson, *Applied Ergonomics*, 24(4):271–278, 1993. Reproduced by permission of the publisher, Butterworth Heinemann ©.)

The representation of complex information spaces seems to require a consideration of the nature of the information in the knowledge domain and the achievement of compatibility between this and the user's typical representations of the knowledge when carrying out representative tasks. It is suggested that certain domains will be best represented using the techniques of **virtual reality.**

Maps and Navigation Aids

Some tasks require an operator to hold a spatial representation of the display in short-term memory. This requirement introduces an added component of mental workload. Wickens (1984) provides a general discussion of the issues involved in navigation and map reading. Map reading is a skill which has to be acquired like any other and does not seem to be well developed in the population at large. Problems can occur when members of the public are expected to navigate complex buildings and facilities. For example, modern shopping centers often have complicated floor plans. Maps of the floor plan ("you-are-here" maps) are usually provided at intervals to assist shoppers in finding their way around. The difficulty of using these maps depends not only on the map's intrinsic design, but also on its orientation with respect to the shopping center. We can imagine a compatible situation in which the map is situated horizontally in the middle of the shopping center with its north end facing the north end of the building. A one-to-one correspondence exists between a route to a destination on the map and a route to a destination in the real building. Alternatively, we can imagine the map situated on a vertical pillar with the north end of the map facing the west end of the building. The difference in orientation between the map and the building now has to be taken into account in order to convert the directions of movement required to reach a destination using the map into a route applicable to the real building. This incompatibility increases the mental workload of the navigation task.

Butler et al. (1993) compared the effectiveness of a number of navigation aids (referred to as wayfinding aids) in helping newcomers find their way around a complex building. Wayfinding signs (and signposts) were found to be more effective than you-are-here maps (Figure 13-15). **Signs** appear to lead to superior navigation because they **place the directional and spatial information required for navigation into the environment itself.** The information conveyed by the sign is used when needed and can then be dispensed with. On the other hand, you-are-here maps impose a considerable mental load on the user. The cues needed to navigate have first to be identified on the map, then held in memory, and finally recognized in the environment. Butler et al. emphasize that the information displayed on signs must be simple and unambiguous for sign-based navigation systems to be effective. Additional improvements can be obtained by numbering the objects in the space to be navigated and incorporating the numbers into the sign system. Butler et al. use the example of hotels where bedrooms are often numbered and are easy to find, whereas rooms containing other facilities are only named and are often harder to find. The navigability of many shopping centers could no doubt be improved in a similar way by placing more emphasis on the use of numbers, rather than names, as cues to the location of particular shops.

Three-Dimensional Displays

Our experience of a three-dimensional visual world is spontaneous and is dependent on a number of cues, both internal and external to our senses. A knowledge of these cues is essential in the design and analysis of three-dimensional displays, which are becoming of increasing importance. They are used in the design of simulators, computer systems

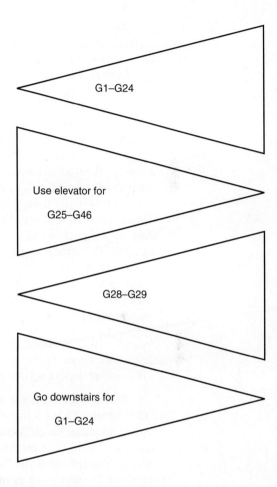

FIGURE 13-15   Wayfinding signs. (From Butler et al., 1993. Reprinted with permission from *Human Factors,* vol. 35, no. 1, 1993. Copyright 1993 by the Human Factors and Ergonomics Society, Inc. All rights reserved.)

for the display of three-dimensional structures such as protein molecules, and in the design of virtual reality systems.

Our visual experience of three-dimensional space is a result of the detection of combinations of the following depth cues.

Monocular depth cues can be detected using one eye only but also operative when both eyes are used:

• *Accommodation.* Proprioceptive feedback from the ciliary muscles provides valuable information about the distance of an object. However, this feedback is of most value in judging the distance of objects up to approximately 6 m.

• *Movement parallax.* Depth perception is strongly influenced by head movements which give rise to the relative movement of close objects against an unmoving background. When one is traveling by car, distant objects appear to travel in the same direction as the car and close objects in the opposite direction.

• *Interposition.* A near object interposed between a more distant one and the viewer provides depth cues when it obscures part of the far object.

- *Relative size of familiar objects.* The size of the retinal image can be used as a depth cue if the size of the object is known. Familiar objects exhibit **size constancy.** A car or a house is perceived as being distant if its image on the retina is small. Young children learn to associate decrease in retinal image size with an increase in viewing distance rather than a change in the size of the object itself (people don't shrink when they walk away from you).
- *Texture gradient.* Seen in perspective, the surfaces of many objects present a texture gradient, with texture changing from rough to smooth as distance increases.
- *Linear perspective.* This is an important monocular depth cue which is used by artists to give the third dimension to a painting. Perhaps the most obvious example of perspective is the way that parallel railway lines appear to converge as distance increases or the way an object's shape appears to change as it is rotated about an axis. Bemis et al. (1988) used perspective to change a two-dimensional circular graphic display to a more three-dimensional format by changing to an oblique rather than a straight-ahead viewing angle. Response time and errors on a detection task were greatly reduced using the perspective display.
- *Casting of shadows.* A light source in the direction of a near object will cause it to cast a shadow on objects farther away. This is often used as a cue to depth on computer systems equipped with graphic interfaces.

Figure 13-16 illustrates the effects of some of the monocular depth cues which can be used to add a third dimension to two-dimensional displays.

Binocular depth cues depend on the use of both eyes:

- *Retinal disparity.* Since the eyes are approximately 6 cm apart, the retinal images differ slightly. The two images are fused to create a single percept characterized by depth. This phenomenon is known as stereopsis. Retinal disparity is a powerful cue sufficient to give the perception of depth in the absence of any other cues. It is the basis of stereophotography and "3-D" films. The most common method of producing three-dimensional displays using the disparity principle is to use two cameras a small distance apart to create a stereo pair of images. Special spectacles (differing in color or plane of polarization) can be used to view the differently colored or polarizing members of the stereo pair such that a different image is presented to each eye in much the same way as when the real object is viewed. As viewing distance increases, the lines of sight of the two eyes tend to the parallel and retinal disparity diminishes (and is negligible after about 20 m).
- *Convergence.* Kinesthetic feedback from the vergence system is a cue to depth which functions up to about 15 m.

Figure 13-17 shows a system for three-dimensional viewing of images on a VDT screen. Two separate views of an object (a stereopair) are displayed alternately on the screen. The viewer wears special glasses with liquid crystals in them. The crystals in each lens darken and lighten alternatively at a frequency and phase corresponding to the presentation of the images on the screen. The left and right eyes are each presented with only one of the stereo images. This system replicates the conditions for retinal disparity which give rise to the perception of depth. Uses of such displays include visualization

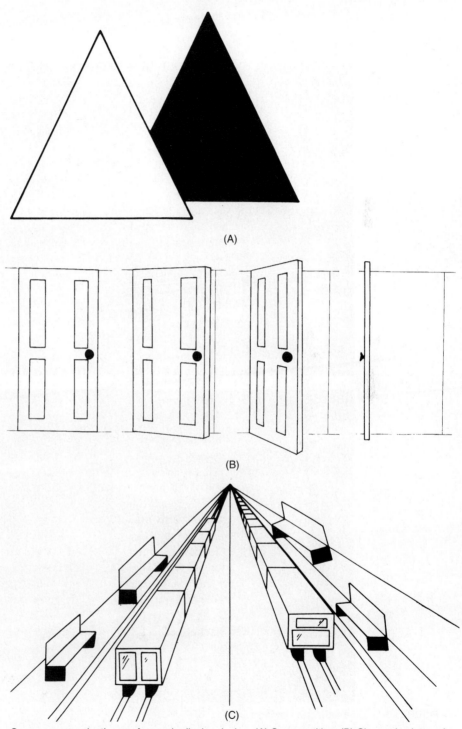

FIGURE 13-16 Some common depth cues for use in display design. (A) Superposition. (B) Change in shape of three-dimensional object. (C) Linear perspective (the upper bench looks bigger because perspective leads us to perceive it as farther away; the object which gave rise to the retinal image must therefore be larger). The cues in (A) and (B) are commonly used to give depth to icon-based CRT displays.

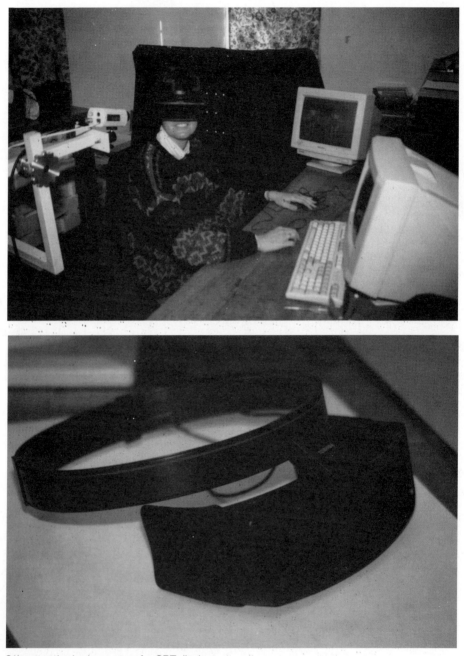

FIGURE 13-17   Stereoscopic viewing system for CRT displays.

of protein molecular structure in organic chemistry and scrutiny of surfaces in surveying and topology.

A review of human stereopsis can be found in Patterson and Martin (1992).

Synthetic Speech

Voice input to humans (by machines) is possible and is a potentially very powerful display modality. Cowley and Jones (1992) have reviewed some of the ergonomic aspects of the use of computer speech and compared the relative advantages and disadvantages of using digitized versus synthesized speech. Synthesized speech is generated using a set of rules, whereas digitized speech consists of real utterances by humans which are stored in digitized form and can be called up and manipulated at will. Synthesized speech is less intelligible than digitized speech but may be more flexible.

Marics and Williges (1988) investigated some of the factors influencing the intelligibility of synthetic speech. Subjects transcribed synthetic speech messages via a computer keyboard. Prior knowledge of context reduced transcription error rates by 50 percent; e.g., if a message such as "rain ending later today" was preceded by the word "weather," subjects were less likely to misunderstand the message. Words at the end of an utterance were less likely to be misunderstood than words at the beginning, which suggests that designers might design phrases which begin using common or easily understood words, with low-frequency words being left to the end. Speaking rates of less than 250 words per minute were found to be preferable for novice users, but the authors suggest that user control of speaking rate might be a desirable feature of synthetic speech interfaces, as would a "would you repeat that?" facility. A final interesting finding which contrasts speech with other types of displays and feedback is that subjects were aware of errors even though the system did not inform them about incorrectly transcribed messages. Speech may therefore be a valuable form of display in situations where it is necessary to block the transmission of errors from one part of a system to another. However, synthetic speech appears to place greater encoding demands on listeners than natural speech and is comprehended more slowly (Ralston et al., 1991). These authors suggest that low-quality synthetic speech not be used for tasks where rapid response to spoken words is necessary or for linguistically demanding secondary tasks where there is competition for conscious processing resources.

DESIGN OF CONTROLS

Some of the basic considerations in the design of controls have already been discussed in the chapter on musculoskeletal ailments and hand tools. Controls should be designed to be operable in low-stress postures and without static loading of body parts, particularly the fingers. Control dimensions should be determined using appropriate hand and foot anthropometry and a knowledge of the mechanical advantage needed to enable the user to actuate the control easily.

Vehicle Controls

Steering wheels, joy sticks, and pedals are commonly used to control vehicles. Control usability can be affected by the resistance of the control, which should be operable using forces which are a fraction of the operator's maximum voluntary contraction. However, the control should offer some resistance to movement so that bumping errors and muscle tremor do not cause control errors (Young, 1973).

With position controls, the displacement of the control is in direct relation to the change in the controlled part. Pressure (sometimes called isometric) controls are almost immovable and the force on the control stick provides the control signal. Pressure controls are of use in controlling higher-order systems which exhibit lag (see below). Some options for control design are control stick friction, preload, control-display ratio, control system hysteresis. Static friction can aid performance in environments in which operators are exposed to vibration. Inertia and viscous damping can be built into controls to assist with the execution of smooth control actions.

Control Distinctiveness

In many instances, numerous controls are grouped together on a panel and the designer's task is to ensure that operators can easily distinguish between different controls. Some of the display design guidelines discussed above are relevant here. In addition, manual controls provide the operator with tactile feedback which can be used to give a distinctive identity to a control or related set of controls.

McCormick and Sanders (1982) recommend using several dimensions to code different controls so as to enhance their distinctiveness. Designers may choose from shape, color, texture, size, location, operational method, position, and labeling. The main tactile cues which may be used to identify controls are texture (e.g., knurled, fluted, or smooth surface), shape (e.g., circular, triangular, square), and size. Size differences seem to have to be fairly large to reliably separate different controls (approximately 1.5-cm diameter difference and 1-cm thickness difference). From the research reviewed by these authors it seems that several combinations each of shape, size, and texture can be used to provide a set of easily distinguishable controls. The selection of controls which provide good tactile cues is particularly important when operators have to work in poor lighting or in emergency situations where multiple cues may assist recognition.

Voice Control: Problems and Prospects

The development of speech-recognition technology introduces the possibility of using voice as a control device. Voice control has the potential to radically alter the design of human-machine interfaces and is therefore discussed in some detail here. Further considerations are discussed in the next chapter in the section on language. Some potential advantages of voice control are the following:

1 It provides an extra communication channel which may take some of the load off more conventional channels.

2 It frees the hands to carry out other activities.

3 Subroutines are already built up for the production of voice commands, so training time should be reduced.

The processing requirements for issuing voice commands would not be expected to conflict with those for manual control; therefore voice is often thought to have the potential to speed up task performance or increase an operator's information-handling capacity. Some evidence for this comes from Martin and Long (1984), who investigated a simulation of a ship's gunfire control task which involved compensatory tracking as a primary task. They found that a simultaneous pointing task degraded tracking performance, whereas a spoken version of the same task did not. It was concluded that the findings were consistent with a multichannel model of performance in which speech and tracking occupied parallel, but independent, channels with no common capacity limitation.

A potential problem with speech as a medium of control is the time taken to produce it. Visick et al. (1984) compared a voice recognizer with a keyboard in the performance of a post office parcel-sorting task. As each of a number of parcels was sorted, the operators either spoke the name of the destination into a simulated voice-recognition machine or pressed a key on a special keyboard where each key was a destination name. It was observed that voice input was slower than keying, despite the bulk of the keyboard which was used, since the average time to produce an utterance was about 600 milliseconds (compared with 50 milliseconds to press a key). In practice, further delays would occur as a result of the time taken for the machine to process the utterance. The authors concluded that **voice input may be unsuitable for tasks requiring little manual component and may offer inadequate timing feedback for tasks requiring precise sequencing of manual and verbal operations.**

Problems the Voice Recognizer Faces

Speech is a complex acoustic signal containing bands of energy centered on 500, 1500, 2000, and 3000 Hz. These are known as **formants** and they consist of an initial transient segment followed by a steady-state segment. The lower two formants are sufficient for the perception of speech, which enables bandwidth compression to be implemented to reduce the information load of speech transmission systems and speech synthesizers.

Many investigations of speech perception by humans have been carried out by researchers in the field of psycholinguistics (see Howard, 1983, for an introduction to this research). Speech perception is thought to depend on the interplay of both top-down and bottom-up processes. Bottom-up (analytic) processes detect particular features in the speech signal as it is received. Combinations of features detected in speech form the characteristic signatures of basic speech sounds such as phonemes, syllables, or words. When matched with a stored representation of known features, the speech can be said to have been recognized. Top-down (synthetic) processes use higher-order information to synthesize possible items which can be matched against the incoming signal (trying to "guess what's coming next"). A knowledge of grammar, syntax, and context is required to drive these higher-order interpretive processes.

Real speech is a "messy" signal. It is mixed in with other noises which humans make (coughs, grunts, sneezes, etc.). The human speech perception system seems to be able to disentangle connected verbal and nonverbal utterances with ease, but it is not clear how this is accomplished or how a machine might best be programmed to do this. People experience **voice fatigue** if they have to speak over the course of a day, which adds to the variability in speech that the voice recognition system has to be able to cope with. Stress, alcohol, colds and flu, and other factors also increase the physical variability of a speaker's utterances.

Understanding speech would appear, at first, to be simply a problem of identifying the basic building blocks of speech and then combining them. However, connected speech is characterized by a lack of **segmentation.** It is as if the building blocks of speech blur into one another in continuous speech. It is not possible to identify a unique set of building blocks at one level.

Many elements of speech are **co-articulated;** that is, the acoustic form of an utterance differs depending on what precedes and follows it. For example, the spoken form of the word "bag" is different from that of the word "beg." It is not possible to record "bag" on audiotape and remove the "a" and replace it with "e"—the whole utterance has to be changed. Similarly, compare the shape of the lips when one pronounces the "st" in "stew" and "stay." When the limitations of the anatomical structure of the human vocal apparatus are considered, it is easy to see why there is a difference between the two versions. The human mouth, lips, and tongue move slowly, have inertia, and exhibit hysteresis when speech is produced. Although continuous speech is *perceived* consistently, it is not *produced* consistently. Parts of an utterance may sound the same to a human listener but differ acoustically depending on the words which surround them. For example, the "y" sounds the same in the phrases

"can you" and "did you"

but is accoustically different (Ballantine, 1980).

Connected speech also shows a lack of segmentation at the word level. Acoustic analysis of spoken phrases does not reveal the gaps between words which listeners perceive. A knowledge of grammar and context is often needed to distinguish between phrases such as

"more ice" and "more rice"
"some more" and "some ore"

The problem is even more apparent when people listen to speakers of foreign languages with which they are unfamiliar. It is often impossible to identify where one word ends and another begins, which often leads to the mistaken view that foreigners speak too fast. This view occurs because of the listener's inability to segment the utterance. It is likely that humans use several types of information to break up continuous speech into linguistically meaningful segments. In particular, it is thought that we use a knowledge of how we, ourselves, produce speech to recognize other people's utterances.

Listeners also adjust their preceptions of speech to take account of the way other people produce speech, which means that speech perception can be conditioned by contextual factors. We automatically take account of accents different from our own when listening to someone. It is not clear how this is done or how similar flexibility could be built into an automatic system.

These examples of the nature of the speech signal and its perception by humans demonstrate some of the difficulties of constructing machines to recognize fully connected speech. More common are voice button systems which recognize isolated words. The ergonomics issues in the design of these systems center on the choice of words for the voice buttons and the recognition accuracy of the speech recognizer. Recognition accuracy is a crucial factor determining user acceptance. Casali et al. (1990) found that improvements in accuracy from 91 percent to 99 percent improved task completion time and acceptance in a task which simulated data entry by voice. Given that words may be recognized, not recognized, or confused for another word, it is often paramount to design the vocabulary set to minimize the confusability of words rather than simply to provide ease of recognition. This should be done at the development stage by carrying out user trials.

Applications of Voice Input

Noyes and Frankish (1989) reviewed the applications of speech-recognition technology in the office. A number of applications have been proposed, including voice messaging (the intelligent answering machine which answers back and takes a message), word processing (using a speechwriter rather than a typewriter), dictation, data entry, and environmental control. The limitations of present voice-recognition technology seem to pose severe limitations on the usability of these systems. Where speech-recognition technology has been successfully implemented, it seems to be because it has clear advantages over other methods and because the task domain is very specific. Voice input in **computer-aided design (CAD)** is a good example. When one is digitizing charts and maps, a cursor has to be held in precise locations on the map while textual and other data are entered by voice. The **eyes busy, hands busy** task seems to be the paradigmatic one for the successful application of voice-input technology (in principle if not always in practice). Over time, more applications have been found for voice-recognition technology. Some banks use it to support telephonic account enquiries. By means of a limited vocabulary and auditory menus, a number of services can be offered using voice recognition and machine-generated speech.

Jones et al. (1992) have reviewed speech-recognition technology in practice. In discussing the criteria for the successful implementation of the technology, they distinguish between voice-driven facilities and services which are truly novel and for which there is a public demand, and voice as an alternative interaction mode for existing tasks (for example, as a replacement for cash registers or scanners at supermarket checkouts). In the former case, voice-driven systems will depend for their success on market forces and other factors extrinsic to the technology itself. In the latter case, success will depend on voice being a superior or preferable way of carrying out a task, and ergonomic considerations will be of prime importance. Norman (1993) uses the telephone as an example of a product which can be improved upon using voice control. Manual dialing can be replaced with programmable voice dialing (instead of keying in the number, you speak the person's name, having previously entered the name and number into the telephone's memory). The success of this application of voice control depends on its being superior to manual dialing (e.g., faster, less error-prone, more convenient because the caller does-n't have to be physically near the phone to dial). New facilities could also be added: When the phone rings, the receiver could answer it or instruct it to return the caller's call in 5 minutes, take a message, or ask the caller to phone back some other time, appropriate feedback being given to the caller. The success of these facilities depend on public demand for a more flexible communication device than existing telephones.

Unnatural Uses of Natural Language?

Although interaction with machines by voice is often said to be more "natural" than other modes, Noyes and Frankish (1989) point out that the limitations of current technology make voice interaction extremely unnatural because the user is restricted to a limited vocabulary and syntax which has to be spoken in an artificial way. Additionally, it can be argued that there is nothing intrinsically natural about talking to a machine.

COMBINING DISPLAYS AND CONTROLS

Controls, Displays, and System Dynamics

Much of the work on human control of systems has concentrated on a class of percep-tual-motor tasks known as **tracking tasks.** Car driving, gunnery, and flying are exam-ples. In a tracking task, the operator has to maintain the output of a system at a target level by means of control actions. Murrell (1965) describes the types of problems that interested researchers at this time. Generally, the operator is regarded as a **system com-ponent,** part of the control loop required to carry out amplification, integration, or dif-ferentiation of system inputs (Figure 13-18). It seems that operators function best as sim-ple amplifiers and that the hardware should carry out the higher-order aspects of the task and handle more complex aspects of system behavior (such as the time lag between con-trol actions and feedback).

In tracking tasks, system dynamics are usually fast and operation is by manual con-trol. If the target is known, the task is called **pursuit tracking.** If the operator has only an error signal (the difference between the current and desired states), the task is called **compensatory tracking** and is usually more difficult. Pointing a gun at a target is an ex-ample of a pursuit tracking task which involves position control. Movement of the gun produces a corresponding displacement of the point of aim with respect to the target. This is known as a zero-order system. In car driving, a first-order system, displacement of the control (the accelerator) produces a change in the velocity of the car. In second-order systems, displacement of a control determines the rate of change of velocity (ac-celeration) of the output, and in third-order systems, it is proportional to the "jerk," or rate of change of acceleration of output. The higher the control order, the more move-ments are required to track the input with a simple control system.

Concepts from **engineering control theory** have been used to characterize the be-

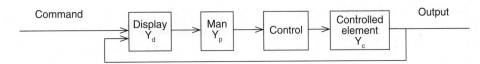

FIGURE 13-18   The human as part of the control loop.

havior of human operators in formal terms. These operator transfer functions describe the ratio of system input to output and are examples of formal, albeit highly specific, models of the human operator (Huchingson, 1981, provides a useful introduction to these operator transfer functions in the discussion of aerospace controls and control systems).

A number of factors can affect human tracking ability and performance. Higher-order systems are normally more difficult to control. Most tracking tasks are self-paced (such as driving a car). Flying an aircraft is a paced task since air speed and glide path restrictions have to be adhered to. Pacing increases tracking difficulty.

Several methods of improving human tracking performance exist. **Aiding** attempts to change the response of the controlled object to control actions. In rate aiding, a single movement of the control affects both the rate of motion and position of the controlled object. **Quickening** involves changing the system feedback. For example, in submarine depth control, a change in control position affects the rate of change of acceleration of depth and it may take a long time for a new stable depth to be reached. Quickening information may be fed back to the operator to assist him or her in deciding what, if anything, to do during the course of a maneuver. Typical information might be depth, change of depth with time, diving angle, rate of change of diving angle, and acceleration of diving angle. **Predictor** displays have also been developed for aircraft and submarines to calculate the future state of the system under present control parameters. This type of display enables the operator to better select control actions which will lead to the desired future state.

The same concepts can be used, for example, to investigate driver behavior. It is known that **preview** (of the track directly ahead) aids tracking performance; performance improves steadily as the preview span approaches 0.5 seconds. This type of information can be applied to the investigation of the problems and difficulties of driving in poor visibility. If the degradation in preview can be measured and related to tracking performance, a safe maximum driving speed can be recommended for a given set of conditions.

Like preview, **anticipation** also improves tracking performance. The ability to anticipate increases with system redundancy. This is the rationale for integrated road holding-steering-braking systems in cars. Cars fitted with these systems appear to behave in the same way irrespective of the driving conditions, thus assisting the driver in correctly anticipating the vehicle's future behavior irrespective of variations in road and weather conditions.

A detailed review of classic human engineering approaches to the design of "human in the loop" systems can be found in Young (1973).

Approaches to Control-Display Integration

Shepherd (1993) identifies a common problem in control-display integration which he calls a "breaking the loop" problem. This happens when designers pay insufficient attention to what operators actually do to carry out their work or when they neglect to carry out a sufficiently detailed task analysis. Shepherd cites as an example a case in which the display was two floors up from the valve where the control action took place. Under these circumstances, it was physically impossible for the operator to control the system on a human-in-the-loop basis even though the designers had deliberately decided not to automate this part of the system.

Controls and displays are the means by which system inputs and outputs are made amendable to operators. For efficient operation, error avoidance, and ease of learning, it is important that the design of the interface has some logical or conceptual correspondence to the input-output relationships of the system and is compatible with operator characteristics such as previous training and knowledge. Three general guidelines can be applied to improve the layout of control-display panels:

1 The Gestalt principles can be used to ensure that the functional relationships between system inputs and outputs are correctly represented in the panel layout.

2 Considerations of spatial memory and of operators' limited ability to process mental images can minimize unnecessary mental workload.

3 Learned associations between displays, displayed variables, and control movements (conventions) should be incorporated to minimize errors.

These guidelines can be combined with information obtained from task analyses to optimize the integration of controls and displays.

Use Grouping Principles in Panel Design

Proximity and similarity can be used in a consistent way so that controls are perceived to belong to their associated displays (Figure 13-19). One problem noted by Seminara and Smith (1993) in their evaluation of a nuclear power plant control room was the lack of correspondence between the proximity and similarity of displays and controls and the actual subsystem boundaries. Under these circumstances, operators may have to work against the tendency to perceptually group together unrelated displays and controls or expend extra effort monitoring groups of apparently unrelated instruments.

Avoid Spatial Transformations

It has been shown that people are capable of retaining a visual image of a real object in memory and even of manipulating the image in a manner analogous to the real object. Shepard and his colleagues (e.g., Shepard and Cooper, 1982) have shown that this **mental rotation** of images takes time, depending on the amount of rotation required. The more one has to rotate a mental image to make some kind of decision about it, the more time it takes. **It seems that correct judgments about displays and their associated controls can be reached more easily when the spatial correspondence between the two is high** (Figure 13-20).

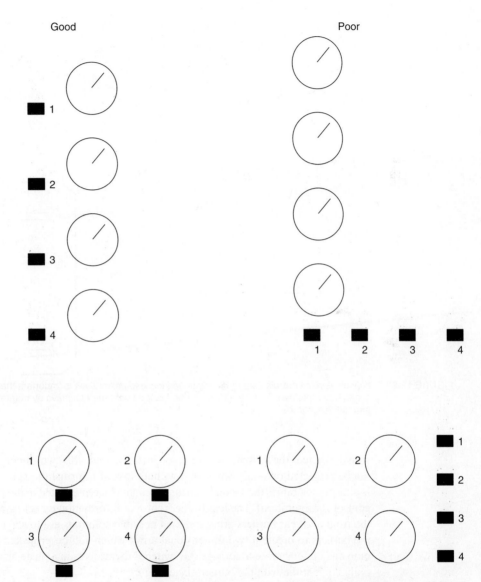

FIGURE 13-19 Proximity can be used to ensure that controls appear to belong to their corresponding displays.

The domestic stove is a good example of a device in which operational problems can arise because of a lack of spatial correspondence between the layout of the switches and the layout of the burners on the range. This lack of correspondence renders the true control-display relationships ambiguous and leads to error. Two typical arrangements for a four-burner range are given in Figure 13-21.

Problems arise because the burners are arranged in a plane (which has two dimen-

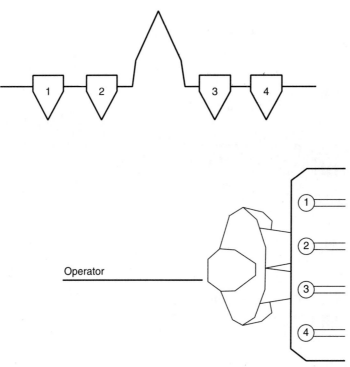

FIGURE 13-20   If displays and controls cannot be positioned in close proximity, it is important that their layouts be spatially compatible. These controls can be mapped onto their displays by mentally rotating either through 90 degrees.

sions), whereas the switches are arranged linearly. Since this two-dimensional array cannot be projected onto one dimension without loss of information, spatial incompability is almost inevitable (the second dimension cannot be represented in the one-dimensional control configuration). The loss of the front-rear information makes switch selection for the front and rear burners arbitrary. Two possible solutions are to stagger the layout of the burners to create a two-dimensional array which can be projected unambiguously onto one dimension or to arrange the switches themselves in a plane arranged to be spatially compatible with the burner plane (Figure 13-22).

Chapanis and Lindenbaum (1959) found the staggered burner arrangement to be freer of operating error than other arrangements. Many different burner-switch arrangements have been developed by manufacturers over the years, and none appear to be optimal. Fisher and Levin (1989) investigated this problem using an unconstrained approach in which subjects were asked which of the switches of a stove burner simulator would turn the burners on. The CABD arrangement was found to be the most consistently chosen and preferred. However, other arrangements were also found to be acceptable, although less so. The fundamental incompatibility between the traditional burner-switch layouts probably explains why so many different designs are commercially available and why designers have resorted to techniques such as the use of logos, overlays, and mimic displays in an effort to make explicit the functional relationships between the controls and

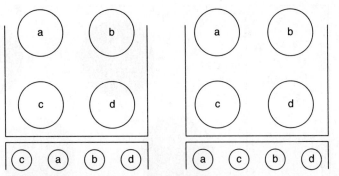

FIGURE 13-21   Two common arrangements for the controls and burners of domestic stoves. (From Fisher and Levin, 1989, with permission from *Ergonomics SA*.)

the burners. The basic ergonomics of the problem is understood and real solutions exist, as depicted in Figure 13-22. These solutions do not appear to have been accepted by the manufacturers of domestic appliances, however. The author offers the following illustration (Figure 13-23) of the layout of the stove in his own kitchen as evidence for the apparent insolubility of this design problem.

FIGURE 13-22   Possible solutions to the cooker control problem.

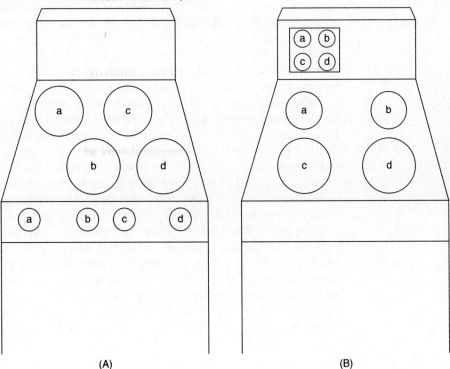

(A)   (B)

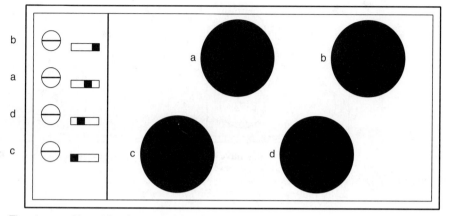

FIGURE 13-23 The staggered layout has been used, but the switches have been put into a spatially incompatible position!

Population Stereotypes

Humans have an inherent tendency to attribute meaning to the movements of objects and a tacit understanding (or naive physics) of mechanical cause-effect relationships. The famous Swiss psychologist, Piaget, described how children develop these notions and how their ideas change with age. For example, very young children often believe inanimate objects (such as clouds or leaves) to be alive because they appear to move by themselves when blown by the wind. Older children use more sophisticated criteria to decide whether something is a living organism or not, for example, whether an object appears to move in a purposeful way.

Adults ascribe meaning to the movement of mechanical objects such as the pointers of dials and scales. Where this is done consistently in a population, it is known as a **population stereotype.** Displays or controls which behave in a manner consistent with the prevailing stereotype are said to be **compatible.** One of the most common stereotypes is the **clockwise-for-anything** stereotype (Barber, 1988). The clockwise movement of a pointer on a dial signifies the passage of time, an increase in the speed of a car, or an increase in pressure or flow. An upward movement of a pointer on a vertical scale indicates increase, as does a rightward movement of a pointer on a linear scale.

Ballantine (1983) investigated population stereotypes in children. Stereotypes are acquired through learning rather than being innate. The development of a repertoire of stereotypes seems to take place over a period of years and is accompanied by an **increasing lack of flexibility and ability to adapt to new situations.** By adulthood, certain stereotypes appear to be well learned and although new relationships can be acquired, they merely overlay the originals rather than replacing them. In stressful situations, more primitive forms of behavior may replace even well-learned skills. If the operator lacks the processing capacity to monitor his or her own performance, old subroutines containing the original learning will emerge, leading to **reversal errors.** It is for this reason that when it is clear that a stereotype exists, designers should ensure that their design is **compatible.**

A number of exceptions to the clockwise-for-anything rule may be noted. Water taps are a good example of an exception to clockwise-for-anything because a right-handed thread is used in the tap mechanism. People appear to be able to learn this exception as a special case. In some circumstances, the ability to correctly interpret the movement of a dial or a pointer will depend on the operator's knowledge of the system or of the displayed variable. For example, the dial on a vacuum gauge moves counterclockwise to indicate an "increase" in vacuum (i.e., a drop in atmospheric pressure). Whether operators will be able to correctly interpret instructions such as "increase the vacuum" will depend on their understanding of what a vacuum is.

People frequently have strong expectations about how a movement of a control will influence the direction in which a display moves. These expectations are referred to as **direction-of-motion** stereotypes. For example, clockwise movements of a rotary knob are associated with upward or rightward movements of display pointers. Some of the more straightforward direction-of-motion stereotypes have been summarized by Sanders and McCormick (1993) and appear to be fairly strong and well represented in the population. Turning a steering wheel to the right should result in the vehicle's moving to the right. When a radio is tuned, a clockwise movement of the knob should cause the dial to move to the right or up, and so on.

That the acquisition of stereotypes depends on previous exposure to technology has a number of implications. First, operators may not have strong stereotypes about the controls and controlled parts of novel machines, for example, where the controlled part moves in a plane different from that of the controls or in a complex way. Alternatively, stereotypes may not exist for novel control devices such as palm wheels or voice. Under these circumstances, the designer may have to undertake empirical research to learn more about people's expectations and preferences, ensuring that if there are any strong expectations, they are not violated. Whatever control logic is then decided upon should be used in a consistent way throughout the design to facilitate the learning of any new stereotypes and avoid interference between mutually incompatible ones.

Second, there may be large individual differences in expectation depending on previous experience. In the example below, there are two mutually inconsistent conceptual models of the relationship between the control and the display (Figure 13-24). In the logical model, an upward movement of the lever brings about an upward movement of the press and vice versa. The movements of the control and controlled object are **logically** interrelated. In the mechanical model, an upward movement of the lever causes a downward movement of the press and vice versa. The mechanical model conceptualizes the control and the controlled object as being **mechanically** interconnected, with the control lever acting like a real lever with its pivot point as a fulcrum. It can be imagined that an operator, accustomed to following rules and procedures, might characteristically adopt the logical model, whereas a mechanical engineer or industrial designer accustomed to dealing with concrete objects might adopt the mechanical model. This example illustrates the importance of a **user-centered approach** and that **designers' assumptions about what constitutes common sense in the operation of a system may not always be valid.**

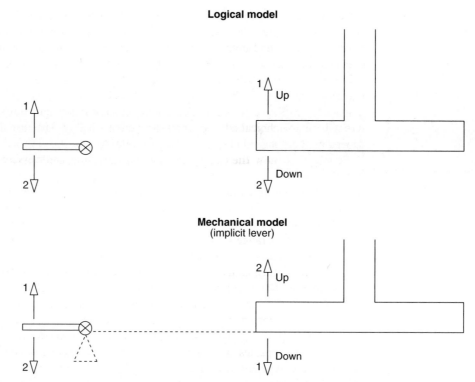

FIGURE 13-24   Logical and mechanical models of control–controlled object relationships. The mechanical model implies that the control and the controlled object are on opposite ends of a lever. Which model would you implement if you were the designer? How would you decide?

Stereotypes in Developing Countries

Assumptions about stereotypes made in industrialized countries may not be appropriate in developing countries. Verhagen et al. (1975) investigated the strength of direction-of-movement stereotypes in north and central Africans. They found that stereotypes were more strongly established in the young, educated people, probably because of their greater exposure to technology. Fisher and Olivier (1992) tested rural and urban south African schoolchildren in order to compare their acquisition of stereotypes in the operation of everyday items. The tasks given to the children were the following:

1 Unscrewing and replacing a bottle top
2 Tuning a clock radio
3 Turning water taps on and off
4 Dialing a telephone number
5 Turning a light switch on and off
6 Operating a four-burner stove

The children's performance was categorized as "correct," "incorrect," or "don't know." Significant differences were observed between the two groups, suggesting that correct performance depended on previous exposure. The rural children's performance was more often classified as trial and error ("don't know") in the operation of clock radios, telephones, and light switches, whereas significantly more urban children failed the tap test (in rural areas of developing countries, a tap, shared by several households, may be a focal point of village life). Significantly more rural children failed to operate the telephone correctly. The authors concluded that there were noteworthy differences in the proportion of rural children unable to manipulate items except on a trial-and-error basis and that technical education was necessary to purposefully uplift them. It would seem that **in the design of technology for implementation in developing countries, it is essential to know the current experience with technology of the intended user population.**

In addition to technology exposure, indigenous culture, alphabets, and symbolism may influence people's expectations about machine behavior. Chinese subjects differ from U.S. subjects in their preferences associated with the control-display relationship of the four-burner range (Hsu and Peng, 1993). Whereas many U.S. subjects map the left-right switch arrangement "ABCD" onto an inverted "U" (i.e., far-left, near-left, near-right, far-right), Chinese subjects prefer the inverted "N" (i.e., far-left, near-left, far-right, near-right). Courtney (1994) has found that although the Chinese share many of the same stereotypes as westerners, there are significant differences; sometimes the Chinese have opposing stereotypes and sometimes they have stereotypes that are not found among westerners. Courtney reports that China has many joint technical projects with western companies, including the construction of a nuclear power plant. Care must be taken in equipment design to account for any differences, even though the Chinese people involved live in a milieu strongly influenced by western culture and technology.

INTERACTING WITH MACHINES, MENTAL WORKLOAD, AND ERROR

Slow Response Systems

Many industrial tasks, particularly those found in the process industries, involve systems which respond slowly to control actions. Research on the control of these systems has identified differences between the behavior of operators and the behavior of automatic controllers. Because the state of the system at any one time is the result of the complex interplay of many variables and the system dynamics are slow, the control task has a large **cognitive** rather than neuromuscular component. For example, Crossman (1974), in an investigation of a simulated slow response system, found that a linear feedback system could not mimic the performance of a practiced subject. Subjects exhibited superior performance by making use of open-loop control based on their **knowledge of system behavior.** Bainbridge (1974) attempted to discover the processes underlying a process control skill by analyzing the verbal protocols (running commentaries) made by controllers. Operator control behavior appeared to differ from that of simple control devices. Operators made use of a flexible **mental picture** of the process. This mental picture (or mental model of the system) enabled them to predict the outcome of potential control actions. Reference to displays and adjustment of controls was intermittent, which suggested that the operators were not tied to the current state of the system. Thus, they relied on an internally represented model of the system (rather than the explicit display data) which constituted their perception of the system's current state. An operator's internal model was said to be modifiable either by prediction, learning, or new data.

Umbers (1979), in a detailed review of research on operator mental models, concluded that the **ability to predict** was an underlying feature of the human controller. In a similar vein, an experiment by Flowers (1978) showed that even in a simple tracking task, subjects made use of predictive control, beyond the explicit data, which enabled them to continue accurately controlling a system despite gaps in the sensory feedback. Subjects developed the ability to predict future system behavior whenever learning was allowed.

In many situations then, it is more appropriate to **consider the operator as a problem solver and decision maker rather than a simple system component** such as an amplifier. Human operators seem to have unique mental qualities which can be contrasted with the qualities of automatic controllers. The challenge to ergonomics is to describe these mental qualities at a sufficient level of detail to enable criteria and guidelines for the design of interactive systems to be specified.

Although the layout of controls and displays can influence the workload of a task, interacting with machines is a dynamic process and the rate at which interaction takes place depends on external constraints and demands beyond the designer's (or operator's) control. It is often useful to determine the mental workload of interacting with a machine under specified, or specifiable, conditions. Some possible reasons for measuring mental workload are the following:

1 To compare alternative methods or designs
2 To evaluate the usability of a prototype
3 To identify the high-stress aspects of a job
4 To evaluate operator performance

Physical workload can be measured objectively at the level of both physics and physiology. The problem with measuring mental workload is that the **operator's thought processes are not directly observable.** The most direct and obvious approach would be to ask an operator or user how difficult it is to carry out a task. Some potential limitations of this approach are that the operator might not want to say, he or she may confuse the important aspects of a task with those that are difficult, or there may be other sources of bias.

To overcome these problems, investigators have developed a number of techniques to measure mental workload (further details can be found in Rohmert, 1987; Ogden et al., 1979; Wierwille, 1979; and Williges and Wierwille, 1979):

1 Physiological measures of mental workload
2 Psychological measures of mental workload
3 Evaluation of the performance of the main task (including error)

Physiological Measures of Mental Workload

These methods typically assume that high workload causes an increase in the state of arousal of the central nervous system. Increased arousal is hypothesized to cause involuntary changes in certain physiological variables which can be objectively measured. Some of the main physiological variables used are described in Table 13-3.

All the physiological methods require specialized equipment and can be influenced by factors other than workload. However, they can be profitably used in conjunction with other methods.

Psychological Measures of Mental Workload

The main psychological methods of measuring mental workload are:

1 Subjective opinions
2 Measures of spare mental capacity by means of secondary tasks

Operator opinions about mental stress can be obtained using rating scales or by questionnaire and interview. In aviation, rating scales such as the Cooper-Harper scale have been developed for assessing the handling quality of aircraft. The validity of these scales depends on the items included and how they were selected and the size of the sample on which the scale was validated.

Interviews and questionnaires are primarily supplementary sources of information on workload. In aviation and astronautics, debriefing of pilots is commonly used, as is self-reported logging of stressful activities. Data from these sources can correlate well with that from other sources.

TABLE 13-3   PHYSIOLOGICAL MEASURES OF MENTAL WORKLOAD

GSR (galvanic skin response). The electrical resistance of the skin changes when sweating increases, because of changes in the concentration of ions in the skin cells. GSR is often measured on the hand, foot, wrist, or forehead. Changes in GSR have been shown to correlate with anxiety. In studies of driving, GSR has been shown to correlate with roadway events which increase workload. Although stressful situations do reduce the GSR, temperature, humidity, and physical workload can have the same effect and need to be controlled when GSR measurements are made.

Heart rate. Heart rate does not necessarily change with mental workload, although it can increase under stress. Tea, coffee, cigarette smoking, and heat can also increase the heart rate. Heart rate variability measured on the basis of the instantaneous heart rate or of the interbeat intervals is sometimes used. Heart rate variability may decrease with mental workload, but it is also affected by changes in respiration rate which can confound the results.

EMG (electromyography). Increases in forearm and forehead EMG have been correlated with high task loading in a variety of air crew tasks.

EEG (electroencephalogram). The electrical activity of the brain can be measured using scalp electrodes. Semiperiodic activity in the EEG correlates with states of alertness (4 to 10 Hz—drowsiness or sleeping; 20 to 30 Hz—alertness). The EEG is an extremely complex signal which requires expert interpretation. Its use as an instrument for mental workload measurement is limited.

Eye movements and blinking. Fixation time, look-away time, and sequencing of eye movements can be used as an index of workload. Alertness and directed attention may reduce the blink rate. Overarousal or emotional stress may increase the blink rate.

Speech pattern analysis. Changes in speech are detectable when a person is under stress (the modulation of speech may change, for example).

Measurement of spare mental capacity is based on the rationale that the operator has a limited capacity to process information. As task difficulty increases, this capacity is used up and less is available to be allocated to other tasks if they have to be carried out at the same time. If the operator is given a secondary task to do concurrently with the main task, the level of performance on the secondary task will reflect the difficulty of the main (primary) task. If the main task is not too difficult, processing capacity will be available for performing the secondary task as well. If the difficulty of the main task increases, less capacity will be left for the secondary task and performance on the secondary task will suffer. Thus, **measures of secondary task performance can be used as an index of primary task load.** For this reason, secondary tasks are designed so that performance can be measured objectively by means of performance time and/or number of errors. Secondary tasks have been used to measure the mental workload of pilots. Some examples of the types of tasks used are given in Table 13-4.

Brown carried out a number of classic experiments on driving using the secondary task technique (Brown and Poulton, 1961; Brown, 1966). Brown (1966) found that a secondary task could predict success and failure in trainee bus drivers. Trainees were given a subsidiary task to do while driving (checking eight-digit strings for changes). The trainees' scores on the subsidiary task while driving were expressed as a percentage of their scores while doing only the subsidiary task. The percentage score was taken as an index of the amount of spare mental capacity while driving. Previous driving experience was found to be a good predictor of whether the trainees were successful in passing a final driving test. Secondary task performance while driving was also found to be a good predictor of future success but only if performance on the secondary task was investigated at the beginning of the training program. These findings were interpreted as follows:

1 Those trainees who initially experienced the smallest decrements in secondary task performance when driving already had driving skill (either naturally or through previous experience). This meant they had spare processing capacity to carry out the secondary task even during the early stages of learning to drive a bus.

2 Those whose secondary task performance suffered while driving had little skill (i.e., few subroutines) and had to use conscious mental capacity to drive at the early stages of training. This left less mental capacity for the subsidiary task.

TABLE 13-4   SOME COMMON SECONDARY TASKS

---

1 Binary monitoring (flicking a switch in response to one of two lights being illuminated).

2 Shadowing (visually presented digits have to be named). This task has been used to measure visual free time in helicopter flying.

3 Mental arithmetic. Pilots have to add 3 to strings of digits presented to them. The time taken depends on the difficulty of the flying task at the time.

4 Classification of digits as odd or even.

5 Reaction time to say whether a presented digit is a member of a previously memorized list.

6 Time estimation. The operator has to estimate when a particular interval of time has passed (based on the idea that time passes more quickly when you are busy).

Some problems with secondary tasks are that they may not be as sensitive as required and they may disrupt performance of the primary task. This, of course, violates the logic of the procedure and renders it invalid. However, the secondary task procedure is intuitively appealing in the sense that at certain times, when the primary task load is high, operators will be too busy to perform the secondary task well. Wetherall (1981) concluded that the technique has merit if the secondary task is selected carefully with respect to the primary task (so that it does not interfere with primary task performance) and is used in conjunction with other measures.

Performance of the Main Task

Performance of the main task is a source of information about the rate of work from which the mental workload of the operator may be inferred. In tasks such as driving, specially instrumented vehicles may be used in which control actions (e.g., steering, braking, accelerating) can be recorded. Although these variables objectively quantify task performance, they do not necessarily have a one-to-one relationship with mental workload. A skilled operator may be able to cope with a wide range of task loadings with minimal demands on conscious processing capacity. Subjectively, these may all be perceived as being equally easy to perform.

Error Data Banks

Establishing data banks on operator **errors and difficulties** is a common way of obtaining information on the mental workload and the adequacy of the design of a system. It is hypothesized that increasing workload raises the probability of error—the more difficult the task, the more likely it is that the person doing it will make mistakes. Several categorization schemes have been developed for handling error data. A distinction is commonly made between errors of **omission, errors of commission,** and **psychomotor,** or bumping, errors. Errors of omission involve an operator's not doing something he or she was supposed to have done (e.g., forgetting to save an updated file on a computer system before calling up a new file to edit). Errors of commission involve executing a correct action at the wrong time or performing an incorrect action. Psychomotor errors involve accidently operating a control or executing a sequence of actions in the wrong order.

Even a simple error categorization scheme such as the one above can point to different aspects of mental workload which can be traced back to particular aspects of the system design. For example, in keyboard operation, people build up **repertoires of keying behavior** through exposure to common sequences. Coding systems which make use of rarely used characters such as "@" or require excessive use of the shift key may increase the probability of psychomotor errors—users may forget to press the shift key or pass it out of sequence as a result of a lack of familiarity with its operation. Bailey (1983) has extensively reviewed sources of human error in computer systems.

In addition to objective measurement of the performance of a task, a technique known as **verbal protocol analysis** can provide data about the mental processes involved in performance. Verbal protocol analysis requires operators to give a running commentary of their thoughts and actions as they carry out the task. The commentary is recorded and subsequently analyzed and can provide information on the **concepts** and **strategies** used by operators. This topic will be returned to in a later chapter.

Finally, Norman (1981, 1984) has carried out extensive investigations of certain types of human error. A sequence of actions is represented in Norman's approach in terms of an **action schema** (a kind of memory-motor program which can be triggered by particular events to enable a person to carry out a sequence of actions without undue concentration). Faulty action sequences can occur as a result of the following:

1 Errors in the formation of the intention to act
2 Faulty activation of the action schema
3 Faulty triggering of the schema

Norman (1981) provides some interesting examples of faulty action sequences. Misclassifying a situation can lead to faulty action sequences, as, for example, trying to enter text into a word processor when the program is in command mode or trying to enter commands in text mode. This misclassification can lead to loss of text or the invocation of unwanted procedures in less robust systems. In this case, the action is correct, but is misapplied. Insufficient information (possible due to a failure to scan the environment fully) can also lead to errors in intention formation. Putting the lid of the sugar container on the coffee cup is an example cited by Norman.

Another group of errors can occur when the intention to act is correct but the wrong action schema is activated. Very well-learned schemes often turn up in unexpected places. On weekends, the author typically finds himself phoning his wife's office when he wishes to contact her, knowing full well that she is at home. Faulty triggering may cause people to make errors such as mispronouncing words or confusing thinking about executing an action with actually doing it.

The intention to act involves a hierarchy of action schemes. A general intention gets broken down into more detailed components which must be executed in order. A particular kind of action slip can occur when part of a sequence gets disassociated by interference or by working memory overload. The higher-level goal being lost, the person no longer knows why she or he is doing something (as in searching through the contents of a drawer but not knowing what you are looking for).

Norman has used these concepts in the analysis of human-machine interaction and presents a framework which categorizes interaction into various stages:

• Forming the intention to act
• Selecting an action
• Executing the action
• Evaluating the outcome

Frameworks such as these make possible the systematic classification of data on errors and difficulties in human-machine interaction. Errors and difficulties experienced by users may be analyzed and placed in one of the above categories to facilitate a structured approach to design, redesign, and the analysis of system usability. For a detailed discussion of human error, see Reason (1990).

CURRENT AND FUTURE RESEARCH DIRECTIONS

Human-machine interaction and display design are areas subject to considerable technology push and only a few examples of research issues will be given. There is currently research on the design of virtual reality systems using novel display modalities such as real-time or near-real-time three-dimensional computer-generated graphics and novel control devices such as the dataglove. Voice input is continuing to receive attention and voice-recognition technology has been implemented in a number of applications. Finally, the increased power and improved graphics of personal computers are leading to the development of display interfaces that are graphically much more sophisticated than current systems. At present, formal methods for the design of these interfaces are lacking, so the methods, as well as the design of the displays themselves, are an issue for future research.

SUMMARY

Although the eyes may be regarded as the windows on the world, what is perceived depends on the operation of processes not accessible to consciousness. These processes organize the stimulus array which is projected from the retina to the occipital cortex of the brain using spatial and other information to maximize its meaningfulness. Visual displays can be designed in accordance with the operation of these processes in order to strengthen the correct interpretation of the data. Lack of consideration of these inherent properties of the perceptual system can result in incompatible designs of less than optimal usability.

Guidelines for the design of control-display combinations have been in existence for many years. Population stereotypes and more general user expectations should be taken into account. Information technology facilitates a more flexible approach as displays can be increasingly implemented via software rather than hardware.

Complementary approaches exist for investigating and evaluating human-machine interaction and, in particular, the mental workload of tasks. In addition to established physiological and psychological techniques, new theoretical approaches to human behavior can provide system designers with powerful frameworks for investigating human-machine interaction and human error.

ESSAYS AND EXERCISES

1 Analyze the design of examples of the following domestic appliances. Identify and comment on the design of the controls, the displays, and the relationship between them.
   a A four-burner range
   b A videocassette recorder
   c A food mixer
   d A microwave oven
   What feedback do the devices provide users to indicate response adequacy?
2 Describe the design options available for human-machine interfaces. Discuss the advantages and disadvantages using examples from everyday life.
3 Use the material in this chapter to write a checklist for evaluating the design of complex displays.

FOOTNOTE: Automation

One of the major trends in the development of technology has been the shift from a reliance on manual labor and primitive tools to a reliance on more sophisticated tools supplemented by animal labor and then by machine power. The role of the operator has also changed—from being a source of muscle power, to being a craftsman, to being a skilled operator of a mechanical system, and finally, to being a supervisory controller of an automated system. Van Cott (1984) draws a parallel between the development of technology and the development of the knowledge requirements of users—a change in emphasis from skills to rules, then to knowledge and understanding has taken place. One of the challenges facing ergonomics in the design of these systems is to develop theories of how users bring together information from varied sources to form concepts appropriate to the abstract level at which human-machine interaction now takes place.

Bainbridge (1982) described some of the ergonomic issues brought about by the automation of systems. Deciding on the role and responsibilities of an operator in relation to a system that should be able to control itself introduces certain **ironies of automation:**

1 The automated system is supposed to perform the task better than the operator. It is ironic that the operator is supposed to determine whether the machine is performing optimally and to take over when the machine fails.

2 If the operator assumes control only when the machine breaks down or in emergencies, the operator will be less skilled than if he or she always controlled the system. In automated systems then, operators are required to assume control at the most difficult times and with suboptimal control skills.

3 Second-generation operators (who have never controlled the system) will have no skills (subroutines), only knowledge.

4 Operators of poorly designed automated systems will have better control and recovery skills than those of well-designed systems.

Some possible solutions to these dilemmas are to build manual control practice into the organization of the work or to use simulators to train operators in the necessary manual control skills.

14

MEMORY, LANGUAGE, AND THE DESIGN OF VERBAL MATERIAL

An analog mode of presentation is usually the most appropriate one for the design of interfaces in perceptual-motor tasks such as car driving. However, many tasks are carried out by **exchanging symbols** rather than executing perceptual-motor skills. This is usually the case with information systems. Interaction with a computer usually requires the user to enter commands and recognize and interpret the computer's verbal (or pseudo-verbal) reply.

Interacting with an information system in this way is fundamentally different from performing a perceptual-motor task such as car driving. In perceptual-motor tasks, performance is **graded;** we can talk about the performance of task elements (such as operating the controls, performing hill starts, and reacting to traffic signals) in terms of the **degree** of success or safety with which they were carried out. Performance can be represented as a continuum, and people can be evaluated according to predetermined criteria and placed at some point along it. Human-machine interaction, when it is based on the exchange of symbols, is much more brittle. Symbols always have a certain arbitrariness about them. Furthermore, the use of a particular symbol is usually either right or wrong—it is either understood, not understood, or misunderstood, with nothing in between. This chapter is being created on a word processor. It is stored in a file called "Chapter14." If, when resuming work on the chapter, I ask the system to retrieve "Chapter 14" instead of "Chapter14," a new file is opened. The system does not recognize the file name even though only an empty space has been added between the "r" of "Chapter" and the "1" of "14." Because of this change, the system assumes that I am requesting it to open a new file called "Chapter 14." The word-processing software recognizes only those file names that are *exactly* the same as those stored in its file directory; a file name is either right or wrong—there are no shades of grey. In contrast, a human librarian would know what I mean and recognize "Chapter 14" as being almost the same as

"Chapter14" and be able to retrieve a copy of it from a filing system. My current word processor never would.

Human-machine interaction which depends on the interchange of symbols poses novel and interesting problems to the designer. These problems arise first from the nature of symbols themselves and second from the differences between the way artificial information systems process symbols and the way humans do. **A knowledge of the nature of symbolic processing in humans is therefore essential in the design of interactive systems.**

DESIGN OF CODES

Symbols can be constructed to represent system elements in many ways. Icons, such as those described in the previous chapter, have some physical resemblance to that which they represent. Symbols may also be constructed in a more abstract way using words, letters, or numbers. Codes are strings of letters or numbers or both which are used to represent system elements.

Bailey (1982, 1983) has summarized many of the psychological considerations in the design of codes—memory is one of the most important.

Code Design Options

There are visual, verbal, and semantic options for code design which can be evaluated with respect to the characteristics of iconic storage, short-term storage, and long-term storage, respectively.

The main considerations in code design are as follows:

1 Length
2 Sequence
3 Content
4 Chunking
5 Errors

System designers often design codes in accordance with the requirements of the information-processing system rather than the characteristic operation of the human memory system. One of the first outcomes of this practice is that codes are often **excessively long,** in that they exceed the capacity of short-term memory (plus or minus seven items). An alternative approach to code design is first to decide the minimum length of code required to discriminate among the members of the class of objects to be coded and to use this length as a basis for development of the coding system. Bailey (1983) points out that every person in the world could be given a unique code number using only a 10-character code, but the New Jersey driver's license number is 15 characters long even though there are only 4 million drivers in New Jersey. Codes such as these are handwritten and read in many situations and invite unnecessary errors because of their excessive length.

The **sequencing of items** in a code is another important consideration. Short-term memory research suggests that the middle items of a code will have the highest probability of being recalled incorrectly (because of the primacy and recency effects). In short codes the penultimate items will be recalled worst, whereas in longer codes the middle items, which are like figures embedded in a background of similar information, will be recalled worst. The problem with long codes is that the discriminability of the middle items is low—they are like figures embedded in a background of similar information. In alphanumeric codes for short-term storage, it may be advantageous to sequence the letters and numbers together to increase the "chunkability" of the code by increasing the discriminability of its component parts (e.g., "GA2" as opposed to "G2A").

Abbreviations

Abbreviations are also a way of coding information. Some rules for the construction of abbreviations are given in Table 14-1.

Reversal Errors

Attention should also be paid to the problem of reversal errors (typing in two characters the wrong way around). Reversal errors are common in data-entry tasks and can be minimized by avoiding the use of character combinations which utilize adjacent keys and, in the design of codes for classes with few members, giving each member a unique code whose reversed forms are illegal (and will not be recognized by the system). On one occasion, the author's housing loan repayments, which were automatically deducted from his salary, were paid to the wrong lending institution. The author, his employer, and the two financial institutions involved spent considerable time trying to find out why these incorrect payments were being made. It turned out that the employer's payroll system used the codes "68" and "86" for the two institutions and a reversal error at data entry had caused money destined for institution number 86 to be sent to number 68—the coding system had reversal errors designed in rather than designed out.

Code Content

There are several important considerations in determining the content of codes. First, an alphanumeric code may be easier to remember than its purely numeric or alphabetic equivalent; people can remember mixed-category lists of words better than single-cate-

TABLE 14-1   GUIDELINES FOR THE DESIGN OF ABBREVIATIONS*

---

1 Determine the required number of letters in the code.
2 Remove suffixes (e.g., -ed, -er, -ing).
3 Select the first letter and the last consonant (e.g., Boulevard—B . . . D).
4 Fill the remaining spaces with consonants (e.g., BLVRD, BLVD).
5 Avoid double consonants.
6 Use vowels only if there are insufficient consonants.

---

* From Bailey (1982, 1983).

gory lists. Special symbols (!, $, @, #) should be avoided, particularly if their meaning is incompatible with the application. The use of many special symbols is contraindicated in most keyboard work because it requires the use of the shift key and therefore increases the likelihood of psychomotor errors.

Code content can also be designed to increase meaningfulness by capitalizing on already learned associations (e.g., "QU" versus "QV") and by designing rehearsable codes ("BORT" versus "TBRO"). The author's student number when he was a student was "BRDRBT003"—a meaningful alphanumeric mix which decays naturally into three easily retrievable chunks: BRD, an abbreviation of the surname Bridger; RBT, an abbreviation of the first name Robert; and 003, a code number distinguishing the authors from others with the same surname. This example also illustrates the concept of familiarity in code design—most people are accustomed to the practice of giving their surname first when interacting with bureaucracies.

Chapanis (1990) has investigated short-term memory for numbers. Repetitive triplets such as 999, 112, or 099 are easily recallable, as are triplets containing two 0s or two 1s (e.g., 001, 100, 101, 900). They can be used for emergency telephone numbers, for example. Triplets containing the doublets 94, 49, 93, or 64 seem to be difficult to recall.

Coding in Errors by Design

The design of a coding system determines the probability of particular types of error. Visual (iconic) errors are more likely if physically similar characters are included in the character set used to generate codes (e.g., F, E; C, G; B, 3; O, 0; and 1, I). The likelihood of errors also depends on the quality of print used to display codes on screen or paper. The quality of the various media used to display codes can be evaluated with particular attention paid to the appearances of easily confusable characters. If handwriting as well as print is involved, the following characters are likely to be confused:

Z–2	G–9, 6
U–V, W	T–J, 7
Y–7, 4	Q–O, 2
S–5, 8	D–0

Confusions can be minimized by removing O, I, Z, Y, S, and G from alphanumeric character sets and U, Y, Q, T, and J from alphabetic character sets.

Text written in uppercase letters is usually less legible than lowercase text (for most commercial typefaces, at least) because the ascenders and descenders in lowercase text provide additional visual cues for letter discrimination which may be beneficial, particularly when viewing conditions are poor.

A separate class of errors can occur when codes have to be held in **short-term memory** for short periods of time. Apart from the forgetting of codes due to excessive code length or trace decay caused by an interruption of rehearsal, **auditory confusions** can also occur. In this case, items which *look* very different but *sound* the same can be confused (e.g., e, D; c, B). Confusion occurs because storage in short-term memory usually takes place in the acoustic or articulatory modalities.

If codes have to be held in short-term memory, while people speak over the phone, for example, **similar-sounding letters and numbers should be excluded from the character set.**

RETRIEVING INFORMATION FROM EXTERNAL MEMORY STORES

One solution to the problem of retrieving information from an external memory (such as a computerized database) is the graphic database navigation approach which attempts to substitute verbal symbols with a pseudo-analog world of graphics and icons which can be explored using simple motor behaviors such as pointing. This approach reduces recognition and retrieval to a level similar to that of recognizing faces (as opposed to re-membering names) and is sometimes referred to as the **direct manipulation** mode of human-computer interaction (e.g., Shneiderman, 1988). Proponents of direct manipula-tion often claim that it is more natural than other modes. Although this is undoubtedly true for certain systems and tasks, it is not necessarily universally true. In large systems, the complexity of the visual display itself may become difficult to manage and verbal models may be more appropriate (Webb and Kramer, 1990). Also, it has long been known (Bartlett, 1932) that people have preferred thinking styles—what is appropriate for a visualizer may be inappropriate for a verbalizer.

Experiments on Database Retrieval

Gomez and Lochbaum (1984) investigated performance on an information-retrieval sys-tem using the **keyword** method. Subjects were required to find one of a number of recipes by entering appropriate keywords (descriptive of the dish). A potential problem with systems such as these is that the system designer and users may choose different words to describe the same things. People often use many words to describe the same object and it may be difficult to know, in advance, what their preferred words will be when they interact with a computer.

The size of the set of keywords was varied in a number of experimental trials (from 1.96 to 31.96 keywords per recipe). The larger (enriched-keyword) vocabulary set was found to increase the number of recipes retrieved by the subjects. It has been hypothe-sized that the enriched-keyword vocabulary might incur a performance cost as a result of increased ambiguity, but that was not the case—subjects were able to identify target recipes using fewer total entries with the enriched-keyword list. The authors concluded that the enriched-keyword technique (known as **aliasing**) had the potential to improve users' interactions with information systems. It would seem that using a larger keyword set can improve retrieval of information from databases, particularly in semantically rich domains where unambiguous classification of items is not possible.

Lansdale (1988) reviewed the ways in which people organize the information stored in their offices and the implications of the methods used for the design of the "paper-less" office of the future. Naturalistic observations of pencil-and-paper offices revealed some of the characteristic features of personal information management which may have profound consequences for office automation and system design.

Neat versus messy offices may reveal more than just the personality of the occupant.

Repetitive jobs often have an information content which lends itself to the construction of a straightforward system for classifying information. New data can be classified in an unambiguous way under a clear heading or category name. For example, a general practitioner may classify each patient's data under the patient's name and have several separate systems for different types of data (for example, a system for patient records that contains medical histories, a system for accounts, and a system for appointments). This type of classification system favors a neat office. However, some jobs have a semantically rich information content and any item of information can be categorized under one of several possible headings. The messy office may really be a symptom of classification problems caused by the information content of an occupant's job rather than a reflection of the occupant's personality.

Similar problems have been found to occur in the design of videotex systems. Users have been found to make mistakes when interrogating these systems because of the lack of an intrinsically unambiguous classification system for accessing the information on the database.

Files vs. Piles in the Modern Office

Filing systems may be replaced by "piling" systems when documents don't fit into any clear category. Users fear that if they file a document, they will never be able to find it again (because of not knowing where to look or looking in the wrong place because the context has changed). In memory theory parlance, the problem does not occur because the item is no longer in storage—rather, **retrieval is impeded by a lack of cues or a loss of context.**

Storing documents in piles is an informal solution to this problem because a pile always reminds the user of its presence. **Physical cues** indicate how long a document has been stored in a pile (by its distance from the top). Physical appearance (typeface, color, etc.) enriches the set of retrieval cues (we might not remember the name of a document, but we might remember that it had a blue cover and the title was written in startling red letters). According to Lansdale (1988), electronic information systems, if they are to be successful, need to support the types of coping strategies that people typically use to overcome the ambiguity and memory problems associated with large databases. An electronic system which used a neat office as the model for its database query system would be suboptimal for messy applications, because of the categorization problems described above. Multiple keywords may be a partial solution, but the task of filing itself (choosing categories) and how this should be implemented electronically are also design issues, as is the specification of other attributes (in addition to a name) to enrich the representation of electronic documents on computer.

Some of these issues were tested by Lansdale et al. (1990) in a document-retrieval task. Enriching information in the form of either words or icons was added to documents either by the subjects themselves or by the system. Recall was found to be higher when enriching information was provided than when it was not provided. Having subjects provide their own enriching information was most effective at enhancing recall. The extra effort required of users to generate their own recall cues would be expected to enhance recall according to the theoretical discussion of Chapter 12. However, in practice, the benefit of the recall aid would depend on users' willingness to make the extra effort accompanying its use. There was no evidence that any particular modality of enriching information influenced recall better than any other. Lansdale et al. interpreted these findings using the **principle of coding diversity.** The principle states that encoding an item using a diverse set of modalities enhances its recallability.

Many other issues remain to be investigated, such as the circumstances in which a particular coding modality will be most effective and the semantic fit or compatibility between an enriching attribute and an item. These issues, however, are beyond the scope of the present discussion.

LANGUAGE, THOUGHT, AND DESIGN FOR EASE OF COMPREHENSION

The classic theory relating language and thought was developed in the 1920s and 1930s. It is sometimes referred to as the linguistic relativity hypothesis. Language is seen as a mold which shapes the way the speaker thinks about the world. Subsequent research has provided some support for the hypothesis. For example, languages differ in the number of color names they employ (usually between 2 and 11), and the ability to distinguish between similar colors has been shown to depend on whether the viewer speaks a language in which the colors have different names. This does not mean that the number of color names in a language determines the colors which can be seen, but that color names make color concepts more available. Anecdotal and other evidence supports a weaker version of the linguistic relativity hypothesis: **Words, such as the names of the different colors, provide linguistic categories which can be used to organize experiences for subsequent recall.** An event containing color information can therefore be registered in a more precise manner by coding it using a unique verbal label. Eskimos, whose language contains many names for different types of snow, and the British, who use many words for different types of rain, can undoubtedly encode the details of these climatic features more readily than a speaker of some other language who lives in the desert. An alternative way of putting this is to say that the words available in a language reflect those things that are important in the life of its native speakers and the types of distinctions that they habitually make in their particular milieu. In this sense, thought is cloaked by language rather than molded by it and to truly understand a foreign language (as some linguists have suggested), one must understand the culture and way of life of the people who speak it.

Communication Problems

Communication problems of interest to ergonomics can arise when different groups of people use different words or use the same words differently when talking about a system. Such differences in the use of language may be found between designers and consumers or users, and between operators as opposed to maintenance or design engineers, and between line versus staff management.

Designers of equipment, manuals, and instructions must understand the vocabulary and preferences of the various groups of people who will use their systems.

Language Comprehension

Language comprehension can be seen as a cognitive process involving both top-down and bottom-up processes. According to Greene and Cromer (1978), language comprehension requires several **levels of knowledge** (Table 14-2). This statement applies to all linguistic transactions between speakers as well as to the comprehension of verbal exchanges between a user and a computer. **Human-computer dialogue is usually limited by the fact that, unlike their users, most interactive systems can communicate only at the lowest levels given in the list.**

Sound and Spelling of Words   A review of the pattern-recognition processes involved in speech perception and reading is beyond the scope of this discussion (an introduction may be found in Howard, 1983, and Barber, 1988). Speech utterances are thought to be recognized by an analysis of their constituent features and by synthetic processes which use contextual and other information to construct a meaningful representation from the acoustic features of the stimulus. In reading, the eyes appear to scan the text in jumps, taking up "slabs" of information which are analyzed as a whole.

Normal Meanings of Words   A semantic network approach suggests that when it is stated that the meaning of a word is understood, the word in question is matched with its stored representation in the mental dictionary (as described in Chapter 12) and that the stored version is linked to a node in the semantic network. When a computer user sees a word or phrase such as "insufficient disk space" displayed on the screen he or she must, in order to understand the message, first, have these words in his or her vocabulary, and second, be able to link them to the appropriate concepts about the functioning of computer storage media.

A plethora of abstract jargon has evolved around information technology. We can distinguish between comprehension difficulties which arise because of a lack of the proper **concepts** and difficulties which arise because of the choice of **vocabulary** or because an excessive number of different terms are being used to describe the same thing.

In the first case, the user's knowledge of the word is "empty" as a result of a lack of the appropriate concepts. This situation corresponds to being familiar with a word but not knowing what it really means. The user may simply learn to respond in an appropriate way when the word is presented without ever developing anything more than a superficial understanding of the system. In some applications, this type of response may be appropriate to the requirements of the task. For example, it is not usually necessary

TABLE 14-2   LEVELS OF KNOWLEDGE IN LANGUAGE COMPREHENSION

1 The sound and spelling of words
2 The normal meanings of words
3 The grammatical rules for acceptable utterances
4 Metaphorical-idiomatic usages
5 Contextual knowledge
6 General (world) knowledge

to understand how an electric motor works or to know the appropriate terminology (what commutators or brushes are or what the field strength of the magnet is) in order to use a vacuum cleaner or a food mixer. However, if the task at hand *does* require a knowledge of the concepts, a user who has only a superficial understanding of the words may experience problems, for example, in knowing what to do when presented with a message such as "mouse driver not installed."

Designers of instructions, warnings, and manuals must understand how users conceptualize a system and be familiar with the vocabulary they use to describe it.

In the second case, the user may understand the appropriate concepts, but the system may use unfamiliar words to describe them and the user may have to infer the meaning of the word from knowledge of first principles and the current system behavior. Novices are unlikely to either possess the appropriate concepts or know the words which refer to them. It is for this reason that novices need to be shielded from technical feedback such as "input buffer overload."

Information technology presents novel ways of enhancing the presentation of text for comprehension. Lachman (1989) had subjects read text on a VDT. A window was made available in which definitions of technical words embedded in the text could be displayed. Subjects were found to access 80 percent of the available definitions and were able to retain a substantial amount of the semantic content of the text. Lachman described his system as a shallow form of hypertext, inasmuch as cues that are associated with items in the database can be called onto the screen. He concluded that even simple hypertext systems can enhance learning and comprehension of text.

Grammatical Rules for Acceptable Utterances   A detailed discussion of grammar is beyond the scope of this book, and the reader is referred to introductory texts such as Green (1975), Howard (1983), or Johnson-Laird (1988). However, no discussion of the design of verbal material would be complete without a brief review of certain key concepts in linguistics and cognitive psychology.

Grammar can be defined as a set of rules for generating all grammatical sentences in a language and no ungrammatical ones (Chomsky, 1957). Speakers of English, for example, can tell which of these sentences follows the rules of English grammar:

Press the red button in an emergency.
In an emergency the red button press.

"Syntax" is the term used to describe the rules for combining words in an order which will be recognized as "grammatical" by a native speaker of a language. In the example above, it is the syntax of the two sentences which is different. It can be said that the first sentence utilizes recognizable English syntax, whereas the second version does not appear to be correct English. Notice that in the discussion of grammar and syntax, we are not necessarily talking about the meaning of these sentences. It is likely that, in an emergency, the second version would indeed be understood despite its incorrect syntax.

Chomsky proposed a theory of grammar which makes a number of important distinctions about word order, syntax, and meaning. In order for one to understand a sentence, the basic syntactic relationships between the words have to be determined. The actual sequence of words (called the surface structure of a sentence) has to be analyzed in order to determine these basic syntactic relationships. Any grammatical sentence can be said to have a deep structure in which the syntactic relationships are represented, and it is from the deep structure that semantic analysis of a sentence takes place.

A set of transformational rules is used to convert surface structures to deep structures and vice versa. According to the theory, when a sentence is being read, transformational rules are applied to its surface structure to reveal the underlying syntactic relationships between the words (the deep structure). It is from the deep structure that semantic analysis proceeds. Alternatively, when one is uttering or writing a sentence, the deep structure, having been determined first, is implemented in a surface-structure form using the transformational rules.

These ideas can be illustrated as follows. Consider the sentence "Jane hit the boy." Its deep structure can be represented as in Figure 14-1. The sentence itself can be broken down into a noun phrase and a verb phrase. The noun phrase can be rewritten as a noun ("Jane") and the verb phrase as a verb ("hit") and another noun phrase ("the boy"). This type of grammar is called a phrase-structure grammar. In such a grammar, the rules for the ordering of phrases in a sentence and for the types of words the phrases can contain are specified. It can be imagined that, given a set of rules for the construction of sentences and a vocabulary of verbs, nouns, etc., a computer could generate strings of grammatical sentences, and this could take place entirely independently of any semantic analysis. Depending on the words in the vocabulary, the system might generate sentences such as (after Chomsky) "colorless green ideas sleep furiously," which speakers of the language would recognize as being grammatically correct although meaningless.

These examples illustrate that **sentence comprehension requires more than just a knowledge of the meaning of words.** In the "Jane hit the boy" example, the sentence has to be decoded (or parsed) using the phrase-structure rules (see Figure 14-1) to dis-

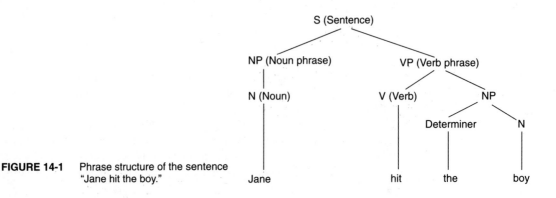

FIGURE 14-1   Phrase structure of the sentence "Jane hit the boy."

close the basic syntactic relationships necessary for understanding it. In the first example, the phrase-structure rules make it clear that the first noun phrase is the subject of the sentence (Jane) and the last noun phrase is the object (the boy)—that it is the boy who gets hit and Jane who does the hitting. In the case of "Jane hit the boy," the surface and the deep structures of the sentence are the same. Transformational rules can be used to create alternative sentences from a particular deep structure and to map incoming sentences onto their basic deep structure. For example, "Jane hit the boy" can be converted to "The boy was hit by Jane." The transformational rule which is applied in this example states that in a passive sentence, the first noun phrase is equivalent to the object of an active sentence.

Thus, in order for a sentence to be understood, a syntactic analysis is required to describe the surface structure (the sequence of nouns and verbs) in terms of verb and noun phrases. Transformational rules are then used to determine the deep structure which reveals the basic syntactic relationships between the words (e.g., subject, verb, object). Semantic processing of the contents of the sentence can then take place. In the design of verbal material or analysis of comprehension problems, it is important to distinguish between the syntactic and semantic aspects of language comprehension.

Comprehension problems can occur if the surface structure of a sentence can be mapped onto more than one deep structure, as in the well-known example "Visiting aunts can be a nuisance." "Aunts" can be interpreted as either the subject (the people doing the visiting) or the object (the people being visited). A similar example concerns the identify of "they" in "They are sailing boats." This brief review of grammar, syntax, and word meaning is of use in the present discussion because it demonstrates why language comprehension requires more than just a knowledge of the meanings of words—a **knowledge of the rules** which govern word order is also required. Some conclusions about the comprehension of sentences are given in Table 14-3.

In ergonomics, language comprehension can be regarded as a **mental task,** which, like any other mental task, has a workload associated with it. Ergonomists should attempt to design verbal material such as warnings and recovery and emergency procedures which can be comprehended with a minimal amount of mental effort.

Metaphorical-Idiomatic Usages   One of the remarkable aspects of the use of language by humans is the tendency to find new ways of using words to convey ideas. Creative writers frequently use these techniques to evoke images in the minds of their read-

TABLE 14-3   SOME CONCLUSIONS ABOUT SENTENCE COMPREHENSION

1   Sentences have a structure over and above their actual word order.
2   Rules have to be applied to disclose the syntax of a sentence.
3   Understanding the syntax of a sentence is not the same as understanding its meaning.
4   Some sentences may have more complex structures than others but mean the same thing. If this is the case, more rules will have to be applied to disclose the syntactic relationships between their constituent words.

402   INTRODUCTION TO ERGONOMICS

ers (the author Tom Robbins once described the flesh of a beetroot as being like "black velvet soaked in red wine"). Everyday speech is also characterized by this type of word usage, as in "John climbed rapidly up the ladder, grasping every opportunity with both hands." In order to comprehend utterances in a language, the listener often has to decide between a literal or metaphorical approach to interpretation.

Because many of the concepts encountered in information technology are abstract and difficult to express using everyday speech, the area is littered with metaphorical-idiomatic uses of language. Performing a "screen dump," for example, involves printing the entire contents of the screen onto the printer. Colorful uses of language such as these have high imagery value and are no doubt **easy to learn and subsequently recall but may cause difficulties on the first encounter because the real meaning of the phrase is nowhere implicit in its constituent words.** Special attention needs to be paid, therefore, to explaining such phrases to novice users either in training or on the on-line help facility.

Contextual Knowledge   Information about the context in which an utterance takes place can be an important determination of comprehension. Specific knowledge about people and events is often essential to disambiguate the usage of particular words. For example, Bransford and Johnson's (1973) sentence "The haystack was important because the cloth ripped" could be understood only when accompanied by the heading "The Parachutist."

Document designers should use headings and subheadings to provide local context for readers of extended documents.

General (World) Knowledge   Communication using natural language is more than just the interchange of verbal symbols ordered according to a set of grammatical rules. The purpose of language is to communicate meaning. To make sense of a speaker's utterances, the listener decodes the sound pattern and performs a grammatical analysis and also uses knowledge stored in memory of the topic under discussion to assist in interpretation.

The **everyday use of a language** by its native speakers **is underpinned by assumptions about shared beliefs and experiences.** Speakers *expect* listeners to share similar beliefs and to keep them in mind during a conversation (we all expect others to use their common sense when listening to us). This is one of the reasons why individuals who speak English (or any other language) as a second language often experience problems when communicating with native speakers. Hill (1972) cites many examples of potentially ambiguous or ridiculous phrases which depend on common sense for their correct interpretation (Table 14-4).

Of particular importance to the construction of user-friendly language are the assumptions made by the writer about what constitutes common sense on the part of the readers. If these assumptions are invalid, comprehension problems and misunderstandings may result.

TABLE 14-4   IS NATURAL LANGUAGE USER FRIENDLY?

---

1 You would scarcely recognize little Johnny now—he has grown another foot.
2 Nothing acts faster than X-brand painkiller (nothing is undoubtedly cheaper as well).
3 The best thing about the match was the shooting of Peter Lorimer.
4 During the present fuel shortage, please take advantage of your secretary between the hours of 12 and 2.

---

* Examples are from Hill (1972).

DESIGN OF VISIBLE LANGUAGE

Applied and theoretical research supports the need to design verbal material for ease of comprehension. Such design will have positive effects both during the reader's initial encounter with the material and on subsequent behavior because **memory for the meaning of a sentence appears to be independent of memory for its wording** (Fillenbaum, 1973). Since, as we have seen, people remember only the gist of things, they may be unaware that their memory of the meaning of a previous instruction is incorrect. If they are unable to retrieve the precise wording of the original material, their initial, incorrect, interpretation will persist.

Most people experience difficulty in answering questions about complex sentences, at least after a first reading. Wright (1978) points out that complex phrases are often found in official and technical documents, for example:

Pull out the left-hand ignition element retaining screw pin.

It is not readily apparent what is being referred to, the left-hand ignition element or the retaining screw on the left of the element. A more easily understood version is

Pull out the left-hand pin on the screw that retains the ignition element.

The design of easily understood language can be undertaken using the language comprehension model outlined above to generate a checklist for the systematic evaluation of the material. The contents of such a checklist are brought together for convenience in Table 14-5 and described in more detail below and in other parts of this book.

Pure and applied research on language comprehension has been reviewed by Wright (1978), to whom the reader is referred for more details. Recommendations for the design of visible language have been reviewed by Wright and Barnard (1975) and Broadbent (1977). From the findings of this research, it can be concluded that easily understood sentences have three characteristics:

1 They contain few clauses.
2 They utilize the active rather than the passive voice.
3 They are affirmative rather than negative.

Few Clauses   A complex sentence, such as "The man who met the clerk that made the loan that saved the old firm is in prison" (Johnson-Laird, 1988), may cause comprehension difficulties by placing an **excessive short-term memory load** on the reader.

TABLE 14-5 SOME CHECKPOINTS FOR EVALUATING THE DESIGN OF USER-FRIENDLY LANGUAGE

Level of Analysis	Checkpoint
Sound, spelling, and appearance of words	Signal-to-noise ratio of spoken message and background noise. Digitized speech better quality than synthesized. Typeface of letters (e.g., uppercase or lowercase). Print size (minimum 8-point, preferably 10-point). Color contrast of letters and background. Luminance of active displays. Illuminance in work environment. Visual angle.
Normal meanings of words	Frequency of usage in everyday English. Frequency in particular industry. Avoid technical jargon in nontechnical context (e.g., substitute "paid work" for "gainful employment").
Grammatical rules for utterances	Complexity of syntax (embedded clauses, long sentences). Number of transformations to convert sentence into SVO (subject-verb-object) format. Avoid overly officious style (e.g., unnecessary use of complex grammar and words such as "hereby," "heretofore").
Metaphorical-idiomatic usages	Has the designer sacrificed clarity and directness for high verbal impact? Avoid the use of "exotic" words likely to be understood by a small group of specialists. Replace "purge files" with "delete files," for example.
Contextual knowledge	Is the context made clear in opening remarks, instructions, or screen design (or the change in context with each new screen in a menu-driven system)? Does text have sufficient headings?
General knowledge	Has language designer made unwarranted assumptions about general knowledge of the reader (e.g., use of words such as "debentures" assumes some knowledge of investments, business, and the workings of the financial services sector)? Can general knowledge intrude in an inappropriate way?

The subject of the sentence (the man) is in the first noun phrase of the sentence, but the intervening clauses separate this noun phrase from its corresponding verb phrase which lies at the other end. The danger is that short-term memory will overflow before the final verb phrase is reached—people will forget what the sentence is about before they get to the end. Wright (1978) remarks that excessively long sentences are often found in official documents and forms directed at members of the public, citing a 143-word sentence as an example.

The problem of short-term memory overload arises when clauses are **embedded** within each other in a sentence, for example, "The man that the girl saw stood up." The phrase "that the girl saw" merely specifies something about the man who stood up, but because of its position in the sentence, nothing about the man can be understood until the end of the sentence. The alternative version, "The girl saw the man. The man stood up," involves less memory load and communicates information about its subject more directly. People do seem to be able to cope when two ideas are brought together in a single sentence, but despite this coping ability, performance breaks down rapidly with increasing complexity or when the individual is under stress or suffering from information overload. Wright gives an even more complicated version: "The man that the girl that the dog chased saw stood up."

Active vs. Passive Voice   Passive sentences rarely occur in everyday spoken or written English and, apart from lacking a certain degree of idiomatic force, can also cause comprehension difficulties. Compare the alternative versions of the following instructions:

> The emergency button should be pressed in cases of power failure.
> Press the emergency button if the power fails.

and

> The notes should be read.
> Read the notes.

Passive constructions, nevertheless, *do* have special place in language—when the speaker wishes to avoid mentioning or does not know the agent of a verb. They are commonly used by politicians and scientists, as in the example, "Taxes were increased" as opposed to "The government increased taxes." The passive voice usually places a greater comprehension load on the reader but can sometimes be of use in **maintaining continuity** in a body of text. Suppose that the topic of a body of text is income tax; then it can be advantageous to use the passive voice to begin a new sentence with the word "tax," as in the example above, because this helps maintain context. However, although **passive sentences** are often shorter than their active equivalents, Wright suggests that there is **little justification for using them for everyday purposes.** According to Broadbent, people experience difficulty in recalling both passive and negative sentences in their original form and this difficulty can lead to errors.

Negative vs. Affirmative Sentences   Research has shown that sentences with negative elements are more difficult to understand than affirmative sentences. Consider the question "Is the book unavailable?" To answer this question, many people have to convert it into a positive question ("Is the book available?"), answer it, and then reconvert the answer. Negative questions can increase the mental load of language comprehension for this reason. **Under stress or high primary-task workload, comprehension may break down.** Another problem with negative sentences is that the negative may be missed altogether through hurried reading or noisy speech (Wickens, personal communication).

Double negatives can be even more problematic, for example, "Do not attempt to adjust this equipment if a technician is not present." The alternative "Adjust this equipment only in the presence of a technician" is more easily understood. The **value of negative statements is to correct incorrect assumptions held by readers or to make strong statements,** such as the following:

Do NOT prescribe to children under 5 years of age.

The alternative "Prescribe only to children over 5 years of age" conveys the same information but lacks emphasis. To guard against the reader's mistaking the negative sentence for a positive one, the word "not" can be highlighted—printed in bold or uppercase (as above) or in a different color.

Generally speaking, however, the use of negatives in semantically condensed phrases or complex sentences is to be avoided because of the tortuous mental operations which negatives can introduce. This guideline is aptly illustrated by Fillenbaum's "perverse" sentence:

Don't print that or I won't sue you

which the majority of people are found to comprehend incorrectly.

A final consideration is the level of **abstractness or concreteness** of words. It appears to be the case that **people can cope more successfully with complex syntax when the topic is concrete** or when concrete words are used to describe the underlying concepts (which suggests that syntactic and semantic processing are not completely independent of each other). This suggests that **unnecessarily abstract words should be avoided in the design of forms or instructions,** particularly if complex syntax has to be used. A more concrete, imaginable, and perhaps less perverse version of the previous example serves to illustrate this point:

Don't wear that, or I won't take you to dinner.

Instructions and Warnings   Many of these recommendations also apply to the wording of instructions. Writers of instructions and questions are often concerned with objects and their attributes. Broadbent (1977) points out that the **active, affirmative approach** is best for wording instructions, such as "The large lever controls the depth of cut" as opposed to "Depth of cut is controlled by the large lever."

Research has shown that it is easier to search memory for the attributes of a particular object (to decide whether a particular animal has stripes) than to talk about an attribute and instantiate objects which possess that attribute (name striped animals). As a general rule then, objects should be followed by attributes. However, this guideline does depend to a certain extent on the context for which the instructions or documents are being prepared. When one is writing a tutorial in a user manual, the object-attribute approach might be best—telling users what the various controls or commands enable them to do. However, when one is preparing documentation to be used in fault finding or recovery from abnormal operating conditions, it may be better to emphasize system functions rather than physical characteristics. Documents suitable for an operator (who is concerned with the system functions and attributes rather than technical details about its components) may not be suitable for the maintenance or design engineer.

An important step in the design of instructions and manuals is to characterize the likely users and the context in which they will be using the material.

There is also evidence that people are inclined to carry out tasks in the order in which they read about them. For example, instructions such as "Before completing this section, read note 3" would be more compatible if rephrased as "Read note 3 before completing this section."

Another consideration in the design of instructions is the likelihood of their being read at all. Wright (1982) investigated the factors which determined whether instructions would be read using a sample of consumer products. The **perceived simplicity** of the product was one of the major determinants of willingness to read instructions. Users seem unwilling to read instructions on products, such as packets of ice cream or candy, which appear to be simple. People also seem unwilling to read instructions for frequently used products. In these circumstances, it may be best to do away with instructions altogether and design products which are more "error-proof."

Similar considerations apply to the design of warning labels on products. Friedmann (1988) found that there was a steady decline in the percentage of people noticing (88 percent), reading (46 percent), and complying with (27 percent) warnings on a variety of products. The perceived hazardousness of the product was positively associated with the probability of the warning's being read. Friedmann concluded that labels should be designed to increase the perceived hazardousness of the product. Critical warning information can be placed first, together with a well-designed symbol. In addition, warning labels should be placed on a prominent part of the product where they are likely to be seen and should be made of materials which will last the life of the product under fair conditions of use (e.g., paper labels glued onto products that get wet or are regularly washed are unacceptable). Young (1990) has shown that warnings embedded in instruction manuals are better understood when prominent print is used (e.g., bold print against a colored background) and when accompanied by a warning symbol.

Chapanis (1994) has evaluated some of the perceived hazards associated with warnings. "Danger" is associated with the highest level of hazards, followed by "Warning," then "Caution," which are perceived to be much closer together than "Danger." Perceived hazard increased through the colors white, yellow, orange, and red, with "Danger" on a red background associated with the greatest perceived hazard. Chapanis suggests that this research needs to be repeated on different groups so as to establish international standards for warnings. In view of the perceived proximity of "Caution" and "Warning," he also suggests that many products currently labeled with "Warning" should be relabeled "Danger."

Finally, Frantz and Rhoades (1993) have shown the importance of task analysis in the design of warnings. They compared the effectiveness of four types of warning information while subjects set up an office. The specific task was unpacking and loading a filing cabinet, and the warning was to be sure to load the bottom drawer first and the top drawer last and never to open more than one drawer at a time. The researchers hypothesized that placing the warning information so that it **interfered with or obstructed the potentially hazardous subtask** would lead to greater compliance. Subject behavior confirmed this view—warnings placed on front of the top drawer or in the form of an obstruction inside the drawer were more effective than less interfering warnings. Warning designers must therefore take care to characterize product use and attempt to design warnings which intrude upon potentially hazardous behaviors.

Further information concerning the design of forms, instructions, and warning labels can be found in Wright (1980, 1981), Riley et al. (1982), and Baggett (1983).

HUMAN-COMPUTER DIALOGUES

An ergonomic approach to dialogue design should begin by characterizing all the possible users of the computer system to determine their needs. Several user typologies have been developed over the years. One of the simplest typologies distinguishes between naive users (those with little knowledge of programming or computer science concepts) and expert users (who normally have professional computing skills). Both types of users may be either casual or regular users of a particular system. Some characteristics of naive users are a lack of insight into the workings of the underlying technology, expectations about computers which differ markedly from those of designers and experts, and an inappropriate model or concept of how the system actually works. Regular naive users' knowledge of a system is likely to be **procedural** rather than technical (knowing what to do but with limited understanding of how and why it works). Casual naive users can be provided with procedural information in the form of instructions by means of on-line help facilities. They should not need to have technical knowledge in order to operate the system. Regular expert users will have both technical and procedural knowledge and will require little assistance. Casual expert users may require on-line help to assist with detailed aspects of executing specific procedures.

Language Comprehension and Dialogue Design

In the previous section, the knowledge required to understand natural language was described in a hierarchical manner. At the lowest level, the sound or spelling of words has to be known, and at the highest level, general knowledge about how things are done in the world may be needed in order to understand what is being said. In the case of interactive systems, **general knowledge may be a hindrance rather than a help** in understanding the dialogue, because there is a fundamental difference between the analog mode of operation of the real world and the symbolic world of interactive systems.

Black and Sebrechts (1981) describe some of the ways that prior knowledge and beliefs can affect the usability of systems. A classic example is the use of the word "file" in computer systems. A file in a computer system has properties different from those of a file in a filing cabinet. Commands such as "Get File X," when interpreted using general knowledge, imply that the file is being transferred from one place to another. In fact, the file is in permanent storage, from where it is read and displayed on the VDU. In some systems, when a file is "saved," an active copy of it remains on the screen. Files in systems such as these can be in more than one place at a time, unlike their real-world counterparts. Files in a database resemble files in a filing cabinet in only a superficial way, making the file analogy a poor one in the context of information technology. Filing cabinets in the real world have only one lock and do not have to be plugged in and switched on before any files can be extracted.

The intrusion of inappropriate general knowledge can be exacerbated by the choice of command names. In the real world, the word "print" has a very specific meaning and conjures up images of printing presses, books, etc. Some software packages use the word "print" to mean "display on an output device" and present users with options such as these:

1 Print to screen
2 Print to file
3 Print to printer

Only the third option is compatible with everyday expectations, and the first two could arguably be replaced with "Display on screen" and "Save in a file."

Implementation Modes for Human-Computer Interaction

Shneiderman (1988, 1991) identified five main interaction styles for dialogue design:

1 Menu selection
2 Form fill-in
3 Command languages
4 Natural language
5 Direct manipulation

Table 14-6 summarizes the relative advantages and disadvantages of these interaction styles.

Menu Selection   The main advantage of menus is that the user does not have to remember the options offered—they are explicitly displayed on the screen. The approach requires that the structure of the task to be carried out be made explicit and be implemented in the form of a menu in which decisions or actions can be executed in a fixed sequence of smaller steps. In complex tasks, menus are often organized in a hierarchical manner. The main menu represents the entire task domain in general terms. Each item in the main menu leads to, and provides the context for, a series of more specific choices. The items in a particular screen of the menu can be used as the title line of the screen which they invoke when selected.

The design of menus such as these may depend on the ease with which the task domain can be represented hierarchically. There may be categorization problems of the type described by Lansdale (1988) which complicate menu design. Problems of this type

TABLE 14-6   HUMAN-COMPUTER INTERACTION STYLES*

	Advantage	Disadvantage
Menu selection	Little training required. Reduces keystrokes. Structures decision making. System can "drive" naive users. Can be used with minimal keyboard skills.	Rapid display rate needed. Danger of too many menus. Skilled users may be slowed down.
Form fill-in	Simplifies data entry. Little training needed. Shows context for activity.	Consumes screen space. Typing skills needed.
Command languages	Flexible; user "drives" the system. Appeals to sophisticated users. Macro commands increase power.	Substantial training needed. Language has to be learned, retained, and recalled. Invites errors due to forgetting.
Natural language	No new computer syntax needs to be learned. Clarification may be needed.	Natural language can be ambiguous. Typing skills needed.
Direct Manipulation	What you see is what you get. Encourages use of an explicit model of the task. Good for disabled users who can point but can't type.	Additional pointing devices needed. Visually impaired users may have problems.

*Adapted from Shneiderman (1988) and reproduced with permission of author and publisher.

can occur in semantically rich knowledge domains such as medicine, economics, and many of the newer interdisciplinary subjects such as ergonomics. For example, in a medical information system, the designer may have difficulty deciding whether to categorize a disease such as kwashiorkor under "diseases of children" or "diseases of malnutrition" (assuming these categories have already been created), since the disease in question is both caused by malnutrition and affects mainly children. Depending on the nature of the categories chosen to provide the top-level structure of the menu, the disease might also be placed under "developing countries." It can be seen from this hypothetical example that menu design can present difficult problems if a purely "top-down" approach is taken in which an attempt is made to derive lower-level categories from higher-level ones, particularly if the designer's knowledge of the domain is incomplete. The designer may become locked-in to a way of partitioning the knowledge that leads, unexpectedly, to categorization problems at deeper levels due to the ambiguous nature of particular items with respect to the distinctions implicit between superordinate items.

The alternative "bottom-up" approach requires that higher-level categories be derived from lower-level ones, but this requires extensive knowledge of these lower levels, forcing the designer to confront the complexity of the entire domain. In practice, a combination of both approaches with the assistance of domain experts may be needed.

Schwartz and Norman (1986) use as an example of categorization problems in menu design, a main menu which presents users with the following options:

1 Services for Professionals
2 Home Services
3 Business and Financial
4 Personal Computing

They argue that the menu is unsatisfactory because the **distinction between the first two options is not clear**—first because both options offer services, and second, because "professional" and "home" are not mutually exclusive categories (many professionals work at home). Options 1 and 3 could overlap completely. In options 2 and 4 the distinction between "home" and "personal" is also questionable.

Schwartz and Norman compared the above menu with an alternative which had been modified to be more distinctive:

1 General Home Computing Information
2 Specific Financial Investing Information
3 General Leisure Information
4 Specific Professional Service Information

Subjects using the modified menu took less time to find items, they found them using a more direct path through the hierarchical menu structure, and they gave up on fewer occasions than those using the original menu. Thus, categorization for item **discriminability** seems to be an important consideration in menu design. Shneiderman has suggested that it may be better to design a menu structure as a shallow rather than a deep tree. At each level of the menu, many items would be presented on the screen at once, rather than having a deep hierarchy in which each screen (or level) has a small number of items. This type of menu structure would cut down on the number of screens and the time taken for users to orient to new screens and would facilitate backtracking to previously used screens. Short-term memory would presumably limit the maximum number of items per screen to about seven.

Short-term memory limitations are also potentially problematic in auditory menus used for telephone access to databases because:

1 The user cannot control the rate of information presentation.
2 The user does not know the number of items that will be presented.
3 Presentation mode is sequential, so the user cannot scan the list of items.
4 There are no visual cues to indicate the menu structure.

Another problem with rigid menus is that experienced users cannot interrogate the system much faster or more efficiently than novices. The rigidity of such menu systems is often caused by the constraints involved in traversing the hierarchy. These may cause frustration and annoyance among experienced users who, having learned the menu structure, are still forced to take several small steps to get to a well-known level. For this reason, many modern menu interfaces are provided with extra facilities for experienced users. For example, Laverson et al. (1987) investigated **jump-ahead techniques** which enable experienced users to take shortcuts through the menu structure. A direct-access method was found to reduce learning time and lower error rates. The method worked by assigning a unique name to each menu frame. **A particular frame could then be accessed directly by entering its name.** If the name could not be remembered, the user could fall back on the conventional stepwise method of using the menu.

Form Fill-In   According to Shneiderman (1988), there has been little research on this method of interaction. It requires keyboard skills and therefore may not be appropriate for certain types of users. An advantage of form fill-in is the increased density of items which can be displayed on a single screen compared with item density with menus. Many of the design issues would seem to center on the basic requirement for screen legibility (similar to optimizing the layout of any other visual display or any pencil-and-paper form).

Greene et al. (1992) describe an interesting variation on this theme, called direct entry with autocompletion. Subjects performed an airline reservation task in which they had to specify variables in three fields—departure city, destination city, and airline. The two main methods of specifying these variables were entry and selection. In the entry conditions, subjects had to type in each variable correctly and received either immediate or delayed feedback as to whether the entry was correct. In the entry with the autocompletion condition, the system completed the spelling of the entry once the subject had typed enough letters for it to be recognized. A novel technique, called the sponge, absorbed letters typed after autocompletion had been initiated to prevent surplus letters from spilling over into the next field. In the selection condition, menus were placed in the three fields and subjects used the arrow keys to scroll through these menus and selected the desired variable.

Autocompletion with a sponge to absorb surplus characters proved to be the fastest and most preferred style of interaction. In situations where the user does not need a prompt about what to type in to the system, this variation may be a viable mode of interaction and may be superior to menu selection.

Command Languages   There has been considerable research on the design of command languages (e.g., Carroll, 1982; Barnard et al., 1982). Command languages are appropriate for experienced and/or regular users. Greater effort is required to learn how to use systems driven by command languages, and learning time may be longer than with other forms of interaction. However, a well-designed command language is potentially more powerful than alternative modes of interaction.

Command languages usually consist of a command word, known as a **predicate term,** and various **argument terms.** In the command "Fetch Doc X," for example, the command word "Fetch" is the predicate term and "Doc" and "X" are arguments which specify how the command is to be applied. Two important issues in the design of command languages are:

1 The choice of words to use as predicate terms
2 The specification of appropriate ways of ordering the arguments

Some of the issues concerning the choice of command names have already been dealt with indirectly in previous sections. For example, the abstractness of words is a factor influencing the ease with which they can be learned. Another consideration is the **generality of a command name with respect to the function of its corresponding predicate.** Barnard et al. (1982) compared the effects of a general versus a specific set of command names on performance in a text editing task (e.g., "Transfer" versus "Fetch" and "Edit" versus "Rubout"). Users who had been furnished with general words to describe the commands tended to request Help (in the form of command descriptions) from the text editing system more frequently than those provided with specific command names, presumably because the general words provided fewer clues about the specific nature of the operation being performed. However, when users of specific commands *did* decide to access Help, they spent more time considering the options presented to them. The ability to recall the command operations, however, was greater among users provided with specific command names.

It appears then that specific words are preferable to general words but should be chosen to be semantically compatible with the operation and be distinctive (of low frequency of occurrence).

A second issue in the design of command languages concerns the ordering of arguments. At first glance it may appear that the command string

Save File

is more appropriate than the alternative

File Save

because the former is compatible with natural English. However, a command language should be evaluated in its entirety and not according to individual commands. In an example from Black and Sebrechts (1981), the text editor command string

Delete the 10

has as its predicate "Delete" and two argument terms, "the" (the word to be deleted) and "10" (the number of the line of text from which it is to be deleted). It can be expanded to natural English by saying "delete 'the' from line 10." An alternative argument order might be "Delete 10 the," which can also be expanded as "delete from line 10, 'the.' " The first way of ordering the arguments in the command string appears more compatible with natural language than the second. Evidence suggests, however, that designers of command languages should apply a **consistent set of rules** for the argument order of command strings even if, on occasion, a string is incompatible with natural language (Bernard et al., 1981). For example, in the command string

Select document file X

which means "select a named document from file number X," the document (which is the first argument in the command string) is also the direct object of the verb "select." In this case, the argument order of the command string is **compatible with natural language.** However, in the command string

Search Document File X

which means "search file X for a named document," it is "file X" that is the direct object of the verb "search," and the argument order is **incompatible with natural language.** In order to be compatible with natural language, the command string would have to be changed to

Search File X Document

but this raises problems of consistency in the design of the command language as a whole (argument order would have to vary according to the command used). Barnard et al. (1981) have demonstrated that a consistent command language was most easily learned. **It appears that consistency in a command language is important even if it occasionally generates commands which are incompatible with natural language** (consistency between displays at the expense of occasional incompatibility seems to be recommended in other areas of ergonomics as well; see Andre and Wickens, 1992, for example).

Natural Language   It is sometimes said that natural language will ultimately be the best mode of interaction between a user and a computer, but this view is debatable. It was challenged early on by Hill (1972) who pointed out that, because of its ambiguity, language is often the problem rather than the solution. It is a poor medium for expressing complex or formal ideas, which is why mathematicians replaced it with a symbolic notation thousands of years ago. Hill also suggested that natural language is unsuitable for many of the purposes for which it is currently used, which is why legal language is so difficult for nonexperts to comprehend. Natural language is also unsuitable for dealing with numbers—"twelve and a half" usually means 12.5, whereas "a million and a half" means 1,500,000, not 1,000,000.5.

Many written forms of documentation would be easier to understand if they were rewritten using a limited-vocabulary program style and flow diagrams. There is certainly evidence to this effect—people perform better at problem-solving tasks when the tasks are presented in flowchart form rather than as paragraphs of fully connected prose (Kammann, 1975). Flowcharts themselves are best designed to be compatible with people's reading patterns and should be oriented in a left-to-right, top-to-bottom pattern (Krohn, 1983). Presumably, this conclusion applies only to societies which utilize an alphabet notation which is written in this way (in other societies, people write from top to bottom or right to left).

Table 14-7 summarizes some of the advantages and disadvantages of natural language as a means of interacting with computers (from Shackel, 1980).

According to Shneiderman (1988), one of the main challenges of natural language as an interaction mode is posed by the fact that the permissible syntax and system capabilities (the context) may not be as visible as in other, more structured, systems. Thus, the

TABLE 14-7   ADVANTAGES AND DISADVANTAGES OF NATURAL LANGUAGE*

Advantage	Disadvantage
1 National language provides a familiar way of forming questions.	1 Natural language encourages users to have unrealistic expectations about the intelligence of the system.
2 Complicated queries can be formulated more easily with natural language than with formal languages or menus.	2 The system's linguistic limitations are not as explicit as when more formal languages are used.
3 Users can choose their own wording.	3 Natural language is ambiguous.
	4 Users' choice of vocabulary may be context-dependent. One vocabulary may not be adequate for several different applications.

*From Shackel (1980).

designer has to find some way of providing **contextual** information which makes clear the scope of the system and the types of verbal inputs it is appropriate for users to make. With this in mind, it is noteworthy that several systems have been developed which do enable users to interact with them using natural language. LUNAR was devised for the analysis of geological samples from the moon. Because the context is highly specific, the system's semantic and syntactic boundaries are visible to its expert users. The system's required vocabulary could also be specified fairly clearly. Another example is SHRDLU, which was developed by T. Winograd and consisted of a simplified "world" of building blocks which the user could instruct the system to move around in different ways and to answer questions about. Again, the context is very narrow and clearly defined. Natural language has also been used in other contexts (such as database query) using a limited vocabulary set.

The development of technology for machine-based recognition of natural language and speech is continuing apace. Ballantine (1980) concluded that there are good reasons for not using natural language for human-computer interaction even though it may offer distinct advantages under certain circumstances. Shneiderman (1988) suggested that there is a great need to determine for which applications natural language interaction would be most suitable if it is not to become a solution looking for a problem.

Direct Manipulation   Much of the software now developed for personal computers utilizes concepts of direct manipulation. An attempt is made to provide an explicit model of the system using some physical metaphor, such as a landscape of data or the desktop metaphor of the Apple Macintosh computer. The approach attempts to minimize the requirement for the user to learn computer syntax and to make use of already existing analog reasoning—to make the system less like a manipulator of abstract symbols and more like a concrete object. Despite potential problems of increased development time and the difficulties presented to visually impaired users, this approach seems to be gaining in popularity.

A recent experimental study by Rauterberg (1992) suggests that direct manipulation (using an interface based on the desktop metaphor) may be a more efficient mode of interaction than menu selection. Novice and expert subjects were given 10 database interrogation tasks to carry out on both types of interface. It was found that both groups needed less time to carry out the task in desktop-interface mode than in menu mode (regardless of differences in the number of keystrokes required to operate the two interfaces). There was no support for the view that direct manipulation using the desktop metaphor is suitable only for beginners, and it was concluded that the desktop interface promotes learning. The rather elitist view is sometimes expressed that direct-manipulation systems are for beginners or are themselves simple systems in some way inferior to more conventional systems. This view is incorrect because it confuses the simplicity of the interface with the power and sophistication of the applications software. Some modern systems with these interfaces are highly sophisticated—probably containing more on-screen buttons than a control room in a nuclear power station. The cognitive complexity of learning these systems is reduced by careful use of consistent graphic design rules to increase redundancy.

Further information on the design of the user interface can be found in Smith and Mosier (1986) and Chapanis and Budurka (1990).

CURRENT AND FUTURE RESEARCH DIRECTIONS

The ergonomics literature concerning the application of memory theory to the design and evaluation of systems is not extensive, a fact that seems surprising when the fundamental importance of memory in our everyday lives is considered. Several investigators have carried out applied research in this area which points the way for others. A particular area of research which has important practical and commercial implications is how to design systems which are easy to learn to use as well as easy to use when learned. Some of the issues in the design of learnable interactive systems have been discussed in this chapter. Apart from the choice and design of codes, symbols, and language, there are more general issues such as the selection of an appropriate level of system complexity, the degree to which certain functions are executed automatically or by the user, and the degree to which previous knowledge can be transferred from one system to another. For example, on some word processors, very little of the knowledge learned can be transferred to other word processors. If information technology is to be more learnable, greater knowledge transfer between systems may be needed.

It is not as easy to establish general principles for the design of human-computer interaction as, for example, for the design of manual handling tasks or the auditory environment, because the technology around which interaction takes place is continually advancing. Today's graphic user interface may soon be supplemented with speech (Agou et al., 1993) or give way to tomorrow's pen-and-speech interface, and the detailed design issues will change. However, attempts are being made to establish more general guidelines to guide system designers. Recent European regulations about work with display screen equipment (Health and Safety Commission, 1992) usefully collate informa-

TABLE 14-8   GENERAL PRINCIPLES OF SOFTWARE ERGONOMICS

Appropriate software
1 Is suitable to the task
  a Presents users with no unnecessary obstacles or problems.
  b Is not unnecessarily complex or elaborate.
2 Is easy to use and adaptable
  a Is easy to master.
  b Uses dialogue appropriate to user's ability.
  c Has an interface adaptable to different skill levels.
  d Minimizes consequences of error
    i Lost data are recoverable.
    ii "Undo button" is provided.
3 Provides feedback on system performance
  a Timely error messages.
  b Appropriate level of information.
  c Help on request.
4 Works at the user's own pace
  a User "drives" the system.
  b System displays all keystrokes when they are made.
  c System response time and response time variability are minimized.
5 Doesn't "spy" on the user
  a No covert monitoring of user performance.
  b User is informed of any recording of input.
  c Emphasis is on quality rather than quantity.

---

tion from several sources under the heading "principles of software ergonomics" (Table 14-8 presents this information and other general principles).

The "learnability" of many consumer products also depends on their design. For example, of the 17 buttons on his video camera, the author notes that only 2 have anything directly to do with photography. Most of the photographic functions have been automated, and the remaining buttons control system functions which have to be learned. There is little transfer of knowledge between traditional photography and these camcorder devices.

Finally, research on automatic translation from one language to another has been going on for many years, and real progress has been made. It will be appreciated from the discussion of language in this chapter that system developers are faced with a problem which is rather more complex than it might first appear.

SUMMARY

The concepts and theories developed by cognitive psychologists have direct application to the study of the human-machine interaction at a symbolic level. Memory theory can assist in characterizing the differences between human and computer memory, differences which have implications for interface design. It can also assist in optimizing the design of codes and command names and in designing databases which take into account both the strengths of the human memory system (its multidimensional, associative nature) as well as its weaknesses (the limited capacity of short-term memory and the semantic nature of long-term memory where only the gist of things rather than the specific details is remembered).

The need to develop more usable, error-resistant systems introduces the study of communication into ergonomics. In particular, the study of language provides a framework for approaching the design of artificial means of communication—between users and interactive systems—in a systematic way. It can be seen that for successful human-machine communication to take place, the symbols being exchanged have to be filtered through several layers of knowledge. At a physical level, the letters and words have to be recognized. Rules then have to be applied to disclose the syntactic structure according to which individual words are ordered. Knowledge about the meaning of words and pragmatic (commonsense) knowledge have also to be applied. Communication can fail if either of the parties involved in the symbolic exchange lacks knowledge at any particular level or if the contents of this layered knowledge structure differ between individuals.

This framework can be used to investigate human-computer interaction; it enables design issues (about the choice of words, structure of syntax, etc.) to be disentangled and dealt with at their appropriate levels.

ESSAYS AND EXERCISES

1 Evaluate the operating instructions of some common consumer products. Use the language comprehension framework described in the chapter as a guide. Describe the knowledge required to understand the instructions.
2 Evaluate a menu-based user interface of your choice. Comment on the design of the language, the distinctiveness of menu options, the number of options per level, the provision of context and orienting information, and the depth and navigability of the system. Suggest improvements or an alternative menu structure.
3 Do humans understand language from the top down or from the bottom up? Discuss with reference to the design of verbal material.

FOOTNOTE: Abnormalities of Memory and Language

Abnormally good and bad memory and linguistic ability have been documented in certain exceptional individuals on the one hand and in patients with brain damage on the other (Guyton, 1991). The posterior superior temporal lobe of the cerebral cortex is known as the general interpretative area of the brain. It is thought that the simple sensory outputs of the sensory association areas of the cortex are brought together here. Patients with severe damage to the general interpretative area can recognize different spoken or written words but cannot recognize the idea conveyed by the word (almost as if the link between the mental dictionary and the semantic network had been severed). Electrical stimulation of the general interpretative area can invoke a complex thought (whereas stimulation of the other sensory areas evokes only crude sensations). One part of the general interpretative area is especially important for the analysis of visual stimuli. If it is destroyed while the rest of the interpretative area remains intact, the person's experience of nonvisual stimuli remains normal, but the ability to analyze the meaning of written words is blocked (even though the person recognizes that they are words). This condition is known as dyslexia. The general interpretative area is more developed in the dominant hemisphere of the brain (in which the processing of language takes place) and is closely associated with the auditory areas of the cortex. If this area is destroyed, almost all intellectual functions associated with language and symbolism are lost, including the abilities to read, to solve logic problems, and to speak. This condition is known as aphasia—the loss of verbal cognitive function. Motor aphasia, a separate condition, is caused by damage to an area of the cortex responsible for motor control of the vocal apparatus.

There are several theories of the neurological basis of memory, none of which are generally accepted and all of which are beyond the present scope. Certain parts of the brain do appear to be intimately involved in memory, however. Damage to the hippocampus of the limbic system results in anterograde amnesia—the inability to consolidate new memories (to transfer new information from short-term to long-term memory). Recall of previously stored information remains normal. Retrograde amnesia can occur after a blow to the head or loss of consciousness and entails the inability to recall information stored prior to the event (particularly for recently stored information). Typically, memory for events leading up to the injury returns in the weeks following the injury.

Some aspects of memory decline with age, whereas others (such as vocabulary) can continue to improve for many years. Short-term memory deficits can become noticeable around the sixth decade of life, as can problems in consolidating new information. Severe loss of memory is one of the characteristics of presenile dementias such as **Alzheimer's disease.** Patients with Alzheimer's are often unable to consolidate new information. This inability causes severe problems of daily living (being unable to remember where objects were left, getting lost when going shopping, etc.). Frequently, stored information is lost progressively on a last-in–first-out basis (more recently stored items are lost first). Such patients may have difficulties recognizing their relatives, although social skills, learned early in childhood, remain.

One of the most famous cases of exceptional memory was described by the Russian psychologist A. R. Luria. Luria's subject had the ability to memorize extremely large amounts of information for long periods. In one test, Luria gave his subject an extremely long and complex (meaningless) formula to memorize. Not only could he recall the formula immediately after being presented with it but also on retesting 15 years later. The secret to this exceptional memory appears to have been imagery. Luria's subject had the ability to create vivid images of items which he wished to remember. Furthermore, items in one sensory modality evoked images in other sensory modalities—a 2000-Hz tone or a person's voice evoked visual images, and numbers had colors and shapes. Thus, to-be-remembered items of information could be stored in relation to a rich set of retrieval cues. This exceptional ability to code items multidimensionally using images in different sensory modalities did present problems—first, in being unable to forget unwanted information, and second, because mental images evoked by irrelevant information could cause distraction, even obscuring the mental images associated with desired items of information.

Further discussion of abnormal and exceptional memory can be found in Baddeley (1983).

15

COGNITIVE ERGONOMICS: PROBLEM SOLVING AND DECISION MAKING

In many activities of daily living and at work, people act as **problem solvers and decision makers** rather than as elements in a feedback control system. We can contrast the types of manual control tasks which take place in car driving with problem-solving and decision-making activities which have a large cognitive component. Strictly speaking, there is a very "fuzzy" boundary between manual control and operation of systems and cognitive control because many manual control activities have a large cognitive component. Car drivers and aircraft pilots process feedback at a skill-based, perceptual-motor level but do more than just steer their vehicles from one place to another—they plan, interpret incoming information, and carry out many higher-level cognitive activities. Decision makers and problem solvers also use feedback, as described in Chapter 12, but the feedback is processed at a rule- and knowledge-based level after being acted on by lower-level processes (Rasmussen, 1983). Indeed, many activities which, for convenience, may best be described as cognitive "sit on top of" and depend on the operation of more basic psychological processes. As such, the factors which influence these processes may also influence cognition. Discussion of cognitive processes in ergonomics must occur in the context of the other aspects of human information processing described in previous chapters.

In this chapter, a cognitive approach to human problem solving and decision making is taken and some design guidelines and techniques for assisting the human in carrying out these activities are described.

PROBLEM SOLVING

In problem solving, there is an interaction between the operation of programs and the movement of data between different memory stores (Newell and Simon, 1972). The **memory load** may vary depending on the amount of feedback from the system. Excessive load can constrain the operator's choice of control strategies (since the application of rules also takes up space in memory). Memory limitations can affect problem solving by causing the problem solver to return to previous knowledge states, a process known as **backtracking.**

Models can help specify the conditions under which an operator's ability to control a system will break down. For example, long- and short-term memory limitations can, under certain circumstances, lead to a breakdown in control behavior. Long-term memory contains symbolic structures built up through learning, in which new data can be embedded and represented as a single symbol. Because long-term memory is associative, it enables new data to be represented in the context of past experience. Access to items in long-term memory is fast and does not require search but is blocked in the absence of retrieval cues. Short-term memory contains symbols related to current information processing. Errors associated with short-term memory often involve forgetting important data before it can be consolidated or used.

The well-known "missionaries and cannibals" problem can be used to illustrate some of these points (Table 15-1).

Readers unfamiliar with this problem should attempt to solve it before continuing.

The problem first has to be approached by finding a way of **representing** it, which will permit rules to be generated and applied. Figure 15-1 depicts such a representation. According to Bundy (1978), a first step in solving the problem is the realization that certain types of information can be ignored (the type of boat and the width of the river). The individual missionaries and cannibals can be regarded as interchangeable, since it is only the number of each on either side of the bank that is important. A second step is the realization that the problem is very difficult to solve without some form of external memory aid. The load on memory is very large when exploring the consequences of moves from a current state if the current state itself has to be kept in short-term memory rather than being externally represented (by using a pencil and paper to keep track of the sequence of crossings). For similar reasons, people normally experience difficulties in anticipating the consequences of a particular crossing (since the alternatives cannot all be held in short-term memory at the same time).

Problems such as "missionaries and cannibals" present the solver with **choices** which have to be made at particular **decision points:**

TABLE 15-1 THE "MISSIONARIES AND CANNIBALS" PROBLEM

Three missionaries and three cannibals seek to cross a river from the left bank to the right bank. A boat is available which holds only two people and which can be operated by any combination of missionaries and cannibals involving not more than two people and not less than one person. If the missionaries on either side of the river are outnumbered at any time by cannibals, they will be eaten. When the boat is moored at either bank, the passengers are regarded as being on that bank.

Find the least number of crossings which will permit all the missionaries and cannibals to cross the river safely.

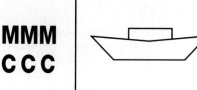

FIGURE 15-1   A way of representing the starting position of the "missionaries and cannibals" problem.

1 To select the most appropriate path for exploring the problem

2 To decide whether or not to continue when a new decision point is reached

3 To decide whether the present knowledge state should be remembered for future reference

4 To abandon a knowledge state and return to a previous one from which further exploration can take place in a different direction

In the "missionaries and cannibals" problem, a state can be reached where there is one missionary and one cannibal on the left bank and two missionaries and two cannibals on the right bank, together with the boat. Since it is impossible to proceed from this point by moving either one cannibal or one missionary back to the left bank (since cannibals will then outnumber missionaries on one of the sides), many people abandon this state and backtrack to the previous one.

Breakdown of Problem-Solving Behavior

Problem-solving behavior can break down for a number of reasons (Table 15-2). Wason has shown that one of the main weaknesses of human problem solvers is the **preference for confirmatory evidence over evidence which will refute their ideas** (Wason and Johnson-Laird, 1972). This preference can lead to rigid, stereotyped behavior which is very difficult for the problem solver to break out of. For example, a novice computer programmer attempting to debug a program may continually search for syntax errors as the cause of a fault, unable to shift attention to search for faults in the structure of the program itself. Preference for confirmatory evidence can have serious consequences for fault diagnosis and correction in process control tasks. One solution is to train operators to recognize this type of behavior and provide them with better specific problem-solving skills, including how to test their hypotheses by searching for evidence which will refute rather than confirm them (Wickens, 1987).

People's ability to solve problems can be improved by providing them with information about the underlying structure of the problem. There are several ways to facilitate the disclosure of a problem's structure:

TABLE 15-2   BREAKDOWN OF PROBLEM-SOLVING BEHAVIOR

1   The one-solution fixation. After failing to solve the problem, the problem solver returns to the same starting point, and the same premises or approaches are used again. The problem solver is unable to think of alternative approaches. This fixation is sometimes accompanied by the irrational belief that it is the problem, rather than the solution, which is incorrect.

2   Getting stuck in a loop. The problem solver repeats a set of moves that lead nowhere except back to the starting point. When this situation occurs, it can be particularly difficult to break out of.

3   Inability to think ahead. This type of breakdown is often due to memory limitations. In the "missionaries and cannibals" problem and in games such as chess and checkers, the number of alternative configurations soon becomes unmanageable when the problem solver tries to consider all the possibilities even two or three steps ahead.

4   Unwillingness to consider counterintuitive actions. In many problems, there are certain essential subgoals which have to be executed in order for the main goal to be attained. Sometimes these may appear to be in direct conflict with the main goal. For example, in the "missionaries and cannibals" problem, the main goal is to transfer the missionaries and cannibals from the left bank to the right bank in as few moves as possible. Intuitively, the best way to do this would appear to be to transfer two people at a time from the left to the right bank and send only one person back each time. However, the problem cannot be solved if this approach is rigidly adhered to. The minimum number of one-way crossings to move everyone across the river is 11. The solution to the problem, for those who still require it, can be found in Bundy (1978).

1 Remove unnecessary and irrelevant information.

2 Choose a representation of the problem which operators are familiar with.

3 Identify the critical information operators need and ensure that it is displayed appropriately.

Shepherd (1993) gives an interesting example of a problem in the design of a computer-based control system for a process control industry which affected operators' ability to handle plant disturbances efficiently. The three control panels in the control room were organized on the basis of geographical areas of the plant. Two VDTs displayed information about the plant which could be retrieved via a hierarchical menu also structured according to geographical area. When a plant disturbance occurred, operators had to decide whether to close the plant down or to set it at some intermediate state while the disturbance was rectified. Although it was the case that disturbances in plant functioning did tend to be confined to one geographical area, information needed to restore functioning had to be collected from all geographical areas of the plant. The result was that, with the control system configured geographically, operators were placed under extreme memory load as they moved from panel to panel and interrogated different areas of the geographically configured menu, while having to retain in STM the information gained to date and the intermediate decisions they had made. Shepherd describes this class of design problem as one of **information fragmentation.** He emphasizes that it can be avoided by carrying out, at an early stage, a particular type of task analysis—an information-requirements analysis—which emphasizes the information operators need to solve the problems they are confronted with at work.

Woods and Roth (1988) give a similar example of a menu-driven database which was designed to computerize paper-based procedures for nuclear power plant emergency operation. Although the user interface had been designed according to ergonomic recommendations, operators were unable to complete recovery from emergencies by following the procedures interactively. They were unable to keep procedure steps in line with plant behavior and got "lost" in the system. The problem occurred because the real cognitive demands of emergency or recovery had not been considered in the design. In real emergencies, operators carry out several activities in parallel and shift from one part of a procedure to a different procedure and back again depending on plant conditions. Paper-based documents support this type of activity well because they permit reading ahead, scanning several documents at once, and leaving open documents in one place while doing something else. They provide a physical trace of the sequences of actions in concurrent activities. However, standard menu systems provide fewer cues as to the sequence of activities. The authors redesigned the interface so as to better support the cognitive demands of the task. The interface was made more "booklike"—two onscreen windows were provided to permit more parallel display of information, steps across procedure boundaries were made easier, and incomplete steps were tracked by the system and flagged with electronic bookmarks to ease resumption of previously interrupted activities.

Problem Representation and Information Design

The representation of a problem (the form in which it is **physically** communicated to the problem solver) influences how it is represented at a **cognitive** level. An appropriate representation of a problem can make its structure more explicit and thus facilitate the selection of appropriate problem-solving procedures. The now classic experiments of Wason can be used to illustrate this point. Note that Wason did not carry out these experiments in order to generate guidelines for ergonomists. They are nonetheless very relevant to problems of information design.

Wason and Shapiro (1971) for example, investigated subjects' problem-solving ability in a falsification task. The subjects were presented with the following rule:

Every card which has an E on one side has a 4 on the other.

Four cards were placed in front of the subjects, who were told that each card had a letter on one side and a number on the other. They were then asked which cards they would have to turn over in order to find out, definitely, if the rule was true or false. Figure 15-2 shows the cards which were presented to the subjects.

Readers unfamiliar with this problem should attempt to solve it before continuing.

A second, and logically equivalent, version of the rule was also tested. It stated:

Every time I go to Manchester I travel by train.

The corresponding cards in Figure 15-2 were presented to subjects.

Abstract
representation

Concrete
representation

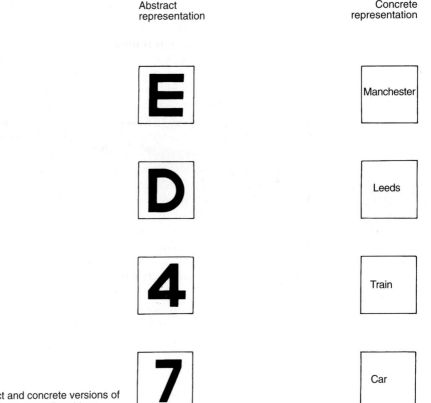

FIGURE 15-2   Abstract and concrete versions of
Wason's "four card" problem.

The first version of the rule can be tested by turning over the cards showing E and 7. If the E card does not have a 4 on the other side, the rule is clearly false; if it does, the rule can be provisionally accepted. The card with 7 on it can also disprove the rule, and for this reason it *must* be turned over as well. If there is an E on the reverse of the 7, the rule is clearly false. Because of the way the problem has been structured (we are told in advance that each card has a letter on one side and a number on the other), the D card can be dismissed as irrelevant as to whether or not the rule is true or false (we *know* that there will be a number on the reverse, not the letter E). Finally, the 4 card is also irrelevant to the testing of the truth of the rule—just because cards with an E on one side are supposed to have a 4 on the other does not mean that cards with a 4 on one side must have an E on the other. There is therefore no point in knowing what is on the reverse of the 4 card.

Wason reports that many subjects experience extreme difficulties in solving the problem and in understanding the solution (i.e., why it is essential to turn over the E and 7 cards and not the others) in the alphanumeric format presented above. However, when the problem is recast in a more familiar form, subjects have less difficulty in selecting the "Manchester" and "car" cards to test the truth of the rule "Every time I go to Manchester I travel by train." If "car" is written on the reverse of the Manchester card, the rule is clearly false, as it is if "Manchester" is written on the reverse of the car card (i.e., the two cards are the same). It is irrelevant what form of transport is used to travel to Leeds or whether any other cities are visited by train (for a review of these and similar experiments, see Wason and Johnson-Laird, 1972). Wason's experiments illustrate that the way a problem is presented to people can have an influence on the ease with which they can solve it. Generally, the less abstract the form of presentation, the easier a problem is to solve (although one advantage of abstraction is that it removes ambiguous or unnecessary information).

Abstract vs. Concrete Problems

People with formal training in mathematics or science are usually fond of using abstract notations or jargon to convey information. This tendency to prefer abstract-general concepts to concrete-particular ones, now firmly entrenched in scientific tradition, is the fault of Plato, who believed in an abstract realm of universal truth accessible only by thought and that empirical observation could provide only a sort of second-rate knowledge because the world was full of imperfect and constantly changing objects.

As Wason's experiment demonstrates, the use of unnecessarily abstract or formal terminology is best avoided on products and machines where practical ends are intended to be achieved. A major trend in software design has been to make more use of graphic user interfaces which provide the user with a concrete metaphor to describe the system.

Problem Solving and Cognitive Style

Individuals have preferred cognitive styles for conceptualizing problems. These depend, at least partly, on prior educational and occupational experiences. The "two liquids" problem (Table 15-3) illustrates some of the different ways in which a problem can be represented in the mind of the problem solver and how the particular representation can influence the types of cognitive operations that are brought to bear on its solution.

Readers unfamiliar with this problem should attempt to solve it before continuing.

TABLE 15-3   THE "TWO LIQUIDS" PROBLEM

Imagine two jars, A and B. Jar A contains a liquid called liquid A and jar B contains exactly the same amount of a different liquid, liquid B. Suppose a teaspoonful of liquid A is taken from jar A and mixed thoroughly with the contents of jar B. Once the contents of jar B are thoroughly mixed, a teaspoonful of the liquid in jar B is removed and mixed thoroughly with the contents of jar A.

If, at the beginning, both liquids were completely pure, which is now the most contaminated by the other?

Intuition leads many people to conclude that jar B is the most contaminated because it was mixed with a teaspoonful of pure liquid A, whereas jar A was subsequently contaminated with a teaspoonful of the B-and-A mixture (less of liquid B was put into jar A than A was put into jar B). This conclusion is incorrect because it does not account for the reduction in the volume of liquid in jar A after the first teaspoon of A was removed.

Engineers and scientists often approach this problem in a straightforward, although rather stereotyped, way. They make some assumptions about the volume of liquid in the two jars (say, 100 mL) and of the teaspoon (5 mL) and from then on, all that is needed is a series of simple arithmetical steps to arrive at the solution. The **arithmetical representation** lends itself to a **solution mode** which requires a knowledge of addition and subtraction, how to handle percentages or fractions, and the use of an external memory aid (such as a notepad) to keep track of the calculations. Mathematicians and statisticians often use their knowledge of algebra to construct a more universal proof using formal notation. This method is a more sophisticated version of the arithmetical approach, but it entails an essentially similar way of thinking.

An alternative way of representing the problem and of solving it is the elegant nonverbal method of Figure 15-3, which is based on the use of **visual imagery.**

A third approach, which depends more on **verbal logic,** begins with an analysis of the problem's structure. Whichever liquid turns out to be the most contaminated after mixing clearly does not depend on the actual volume of liquid in the two jars or on the volume of the teaspoon. What is important is the statement that there is an equal volume of liquid in the two jars initially and that the same volume is transferred from one jar to the other and back again. The **structure of the problem** remains the same regardless of the actual volumes involved. This realization enables the problem to be simplified considerably by taking it to its extremes. If the volume of the teaspoon is the same as the volume of liquid in the jars, then all of A is first mixed with B. The volume of liquid in jar B (the mixture) is doubled and half of it is returned to jar A. Thus, the two liquids are equally contaminated.

The "two liquids" problem is a further illustration of problem representation. It demonstrates that the way a problem is represented determines the types of cognitive operations (verbal, visual, arithmetical, etc.) that can be used to reach the solution. It is interesting to note that there are many other ways of solving the "two liquids" problem, and the reader might find it instructive to explore some of the possibilities.

COGNITIVE CONTROL OF SYSTEMS

As has been described in previous chapters, information technology can free designers from the constraints of electromechanical display technology (such as dials and gauges) because of the increased flexibility which it offers for displaying system variables. This flexibility means that designers have more control over the way a system's behavior is represented to operators.

Kvalseth (1978) investigated subjects' performance on a tracking task in which control was exerted indirectly using digital, as opposed to analog, input and feedback. Kvalseth demonstrated that the control skills required to perform the digital tracking task were different from those required to perform conventional tracking tasks involving continuous manual input. In conventional tracking tasks, considerable differences

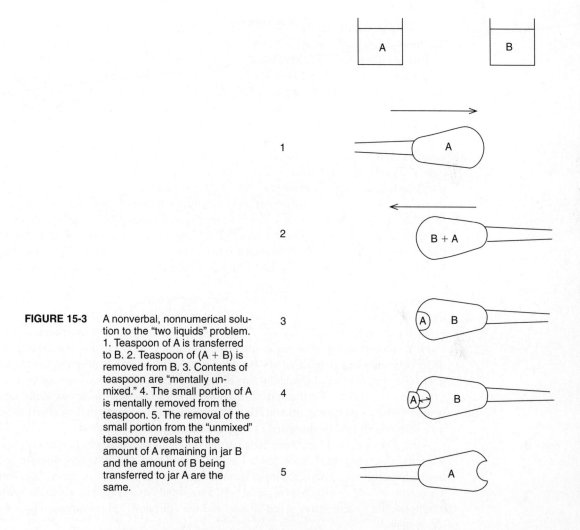

FIGURE 15-3   A nonverbal, nonnumerical solution to the "two liquids" problem. 1. Teaspoon of A is transferred to B. 2. Teaspoon of (A + B) is removed from B. 3. Contents of teaspoon are "mentally unmixed." 4. The small portion of A is mentally removed from the teaspoon. 5. The removal of the small portion from the "unmixed" teaspoon reveals that the amount of A remaining in jar B and the amount of B being transferred to jar A are the same.

are observed between pursuit and compensatory tracking behavior. Under indirect, digital control the difference between these two types of tracking disappeared. Thus, there appear to be differences between the control skills required by direct and indirect control tasks. This raises questions about the design of appropriate interfaces for systems under indirect, cognitive control.

Optimization Tasks

In many cognitive control tasks, the operator is required to maintain one or more output variables at an optimum level by manipulating the values of the inputs under his or her control. Tasks such as these are known as **optimization tasks** and are found in a wide range of industries and in the operation of many types of equipment. Tuning a television set or radio to get the best reception is an example of a one-variable optimization task carried out at a skill-based level. Some of the spectrometers used in chemistry present operators with a multivariate optimization task in which the field strengths of a number of different electromagnets have to be covaried in order to produce a composite, homogeneous magnetic field. In Bainbridge's (1974) research, the control behavior of operators in a steel mill was investigated. They had to direct electrical power to each of five furnaces through different stages of the steelmaking process within the constraint that the total amount of electricity that could be used in any half-hour period was limited.

Course-keeping in ships provides another example of an optimization task. Although course-keeping in large vessels takes place under autopilot control, course-keeping efficiency depends on sea conditions and on the way the autopilot is set (the values of proportional and derivative rudder gain, Marshall, 1981). If the gain settings are too high, the ship keeps tightly to its course but makes many small course corrections which waste fuel. Alternatively, if course corrections are made less frequently, less fuel is required to maintain speed but the ship may wander off course more easily, thus traveling farther than necessary. Somewhere between these two extremes lies an **optimum gain setting** for any set of sea conditions. The autopilot input values (gain) can be manipulated to optimize the ship's course-keeping behavior.

Fowler and Turvey (1978) discussed some of the theoretical issues concerning the behavior of subjects when faced with the problem of finding a combination of several variables to optimize a complex system. They suggest that an **iterative approach** is used and that subjects make use of higher-order information in the feedback, particularly the **difference between successive system output values** (the rate of progress toward the goal state) and the rate of change of this difference (the "acceleration" toward or away from the goal state). This theoretical discussion raises practical questions about **how people extract this type of information from displays** and corresponding issues about **how displays should be designed to support this type of behavior.**

A number of experiments have been carried out on human performance and strategy when carrying out optimization tasks. Laughery and Drury (1979) used a discrete task in which combinations of paired input variables gave an output value (system state) with a bivariate normal distribution (Figure 15-4). This simulates a system in which the space of all possible system states is hill-shaped and the optimum region corresponds to the top of the hill.

Subjects were observed to "tune" the system by manipulating both input (control) variables at once, in two stages. In the first stage, an initial exploration of the space took place to eliminate suboptimum regions using gross control movements (the elimination search). In the second stage, a detailed mapping of the optimum region took place using finely tuned control adjustments as the subjects built up a mental map of the optimum region (the construction search). Human performance on the optimization task was found to be superior to that of a "hill-climbing" algorithm, particularly when the precision and reliability of feedback was degraded. This was thought to be due to the human's ability to build up a mental map of the space. Essentially, the human operator is sensitive to redundancy in the feedback which exists because of **symmetries** in the underlying structure, and it is this redundancy which enables the mental map to be acquired. The superior "immunity" of humans to poor-quality feedback was taken, by the authors, to suggest a continuing role for human optimizers in the control of complex systems.

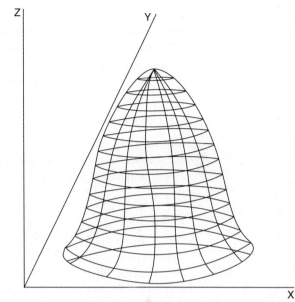

FIGURE 15-4   Hill-shaped space encountered in optimization tasks. The operator has control of X and Y and gets values of Z as feedback. Over time, humans build up a representation of the space structure which enables them to tune the system (get to the top of the hill) with fewer control actions.

Bridger and Long (1984) investigated factors influencing the performance of the two-variable optimization task developed by Marshall (1981). The task was found to have a significant memory load, and unless subjects were provided with an ongoing record of their control actions, performance suffered (in contrast to manual control tasks which impose little, if any, memory load on operators). Analysis of subjects' strategy revealed that performance was characterized by steady progress toward the optimum region, punctuated by exploratory movements about the closest point to the optimum reached up to that point in the search. It was found that the exploratory phase around these "current optima" was the source of memory load and that by redesigning the display to indicate the "current optimum" (the best position reached so far in the search for the true optimum) as well as the present position, the memory load was relieved. In fact, performance was as good as with a complete record of the total sequence of control actions. Essentially, the current optimum is the point to which subjects will backtrack if subsequent moves are unsuccessful. This demonstrates that in many sequential problem-solving tasks (including the "missionaries and cannibals" problem) there are **strategically important points** which are reached on the way to a solution, and it is **these points that must be recorded or displayed to relieve the operator's memory load** and prevent him or her from having to start again from the beginning.

Decortis (1993) has discussed operator strategies in nuclear power station control. It appears to be the case that in this and many other environments requiring complex problem solving, operators use a strategy of breaking down problems into elementary problems and solving them one by one. Problems that are not urgent or have slow dynamics may be left or only partially solved and returned to later (at the risk of increasing memory load unless the interface supports this type of behavior). The effects of remedial action can act as pointers for further action and problem solving if the operator is provided with suitable feedback in space and time.

Display design for cognitive control environments must be based on a model of operator control strategy, critical as well as routine behavior, and memory load.

MODELING OF HUMAN OPERATOR CONTROL STRATEGY

Cognitive control tasks are particularly challenging because cognitive behavior is not directly observable—many of the most important aspects of task performance take place inside the operator's head. The main problems in modeling operators are:

1 Data capture
2 Building a model
3 Validating the model

Data capture involves more than just the recording of observable operator behavior. Data about cognitive behavior has also to be captured and system states have to be characterized with a precision sufficient to enable system behavior to be mapped onto operator behavior (Spencer, 1962). To build a model of a cognitive control skill, the investigator needs to know what operators are thinking and to relate this to what they and the system are doing.

Verbal Methods of Data Capture

There are several verbal methods of capturing data for modeling the human operator. Some basic on-line and off-line methods are noted below.

System state–action state diagrams can be used off-line. The operator is presented with a matrix of possibilities concerning a controlled variable (such as temperature). At each cell in the matrix the operator is asked what action (if any) he or she would take if the system behaved in the way described in the matrix. Although the method provides a structured approach to data capture, it has a number of limitations, such as the validity and reliability of the answers and whether the operator really comprehends or can imagine the occurrence of all the states presented.

Questionnaires can be also be used to gather data off-line about operator control strategy. The general problems of questionnaire design need to be considered—the wording of questions to avoid ambiguity and the avoidance of leading questions which suggest there is a correct answer. With off-line methods there is also the problem of operators' responding by saying what *should* be done under a particular set of circumstances (if they followed the procedures) rather than what they *would* actually do.

Structured interviews can be used to obtain data on the rules and general principles which operators utilize and can be accompanied by the use of static simulation in which the operator is given a number of hypothetical situations which she or he must talk the interviewer through.

Off-line methods have the advantage that the most interesting aspects can be isolated.

The disadvantages include an absence of realism and a lack of time pressure. A wide sample of situations must be investigated to characterize the operator's knowledge.

Verbal protocol collection is an on-line method of capturing data on operator control strategy. The operator is asked to "think out loud" while carrying out the job and what he or she says and does is recorded. The method attempts to "get inside the operator's head" and is intuitively appealing. For example, it may be difficult to determine the difference between a good car driver and a highly skilled car driver using objective measurements alone. Unless exposed to extremely demanding situations, most of what they do will seem the same. However, the running commentary of the two drivers might differ considerably, with the highly skilled driver showing more evidence of higher cognitive activity such as thinking ahead, detecting potentially dangerous situations well in advance, etc.

The verbal protocol technique has a number of advantages over other verbal methods. Objective data about system functioning can be recorded simultaneously with the protocol. The protocol itself can be used to design follow-up questions for use in off-line interview. However, the technique has been questioned, particularly in relation to its validity (e.g., Umbers, 1979):

1 It makes assumptions about the operators' abilities to put into words their control strategies.

2 Operators may neglect to mention things that seem obvious to them but are not obvious to anyone else.

3 Most people can think more quickly than they can talk.

4 The process of thinking aloud may distract operators—it can be seen as a secondary task which, depending on the modality and processing requirements of the primary task, may compete for common processing resources (car drivers often break off in the middle of a conversation when carrying out a difficult maneuver).

Verbal Protocol Analysis

Analysis of verbal protocols can be a complicated and time-consuming procedure, but it can yield important information on how operators make task decisions and on their strategies and information requirements (see Bainbridge, 1979). Such data might otherwise be difficult to obtain and can be used in the evaluation of interfaces, job aids, and training schemes.

Explicit Content Analysis   This type of analysis involves building up categories of statements which are similar in terms of what they refer to. In car driving, for example, there may be categories referring to traffic signals, traffic congestion, road signs, or the condition of the car. This gives information about the predominant cognitive activities and the time spent in different activities.

Implicit Content Analysis   The analyst uses knowledge of the process to try to make inferences about what the operator is referring to in cases where that is not clear.

In the car driving example, the statement "It's getting rather hot" could refer to the weather or the engine.

Grouping and Sequencing of Phrases   Phrases can be grouped together to build up a picture of the operator's strategy. That is, the phrases may be interpreted to suggest how the operator decides on particular actions during different system states. Grouping together these operational tactics provides insights into the nature of the overall strategy.

Representing Operator Strategies   A common method of representing strategy is by means of flow diagrams (e.g., Laughery and Drury, 1979). The flow diagram may correspond to the entire task or just to those parts that are of particular interest from a cognitive point of view. Figure 15-5 shows a flow diagram of the task of an audiologist performing a hearing test on a child (Goemans, 1992). The main cognitive aspects of the task occur at the decision points concerning the reliability of the child's response to an auditory stimulus.

USER MODELS OF INTERACTIVE SYSTEMS

Learning to use products such as word processors, electronic calculators, or videocassette recorders can be seen as a problem-solving exercise. Many of the conclusions reached about problem solving in the above sections can be applied to the analysis and design of consumer products. The way the user interface of a product is designed determines how the operations required to use it are represented in the minds of users and the types of cognitive processing required for successful operation. As a user learns how to operate a product or system, he or she constructs an internal representation of what the system is and how it works. This representation is known as the **user's model of the system.**

Cognitive models enable the user to process data about system behavior in a structured manner. One of the first definitions of these models is from Craik (1943) (quoted in Hollnagel and Woods, 1983):

> If the organism carries a "small scale model" of external reality and of its possible actions within its head it is able to try out various alternatives, conclude which is the best of them, react to future situations before they arise, utilise the knowledge of past events in dealing with present and the future, and in every way to react in a much fuller, safer and more competent manner to the emergencies which face it.

Models of interactive devices guide users' control actions and assist in the interpretation of the device's behavior. Young (1981) investigated the **conceptual models implicit in the design** of a variety of pocket calculators. Of interest here is the difference in control logic of "reverse polish notation" (RPN) calculators and the more familiar "algebraic" (ALG) calculators. In both cases, the model is a metaphor or cover story for thinking about the calculators—how they work and how they can be used to solve problems. This is illustrated in the brief account below. The reader is referred to Young (1981) for a fuller discussion.

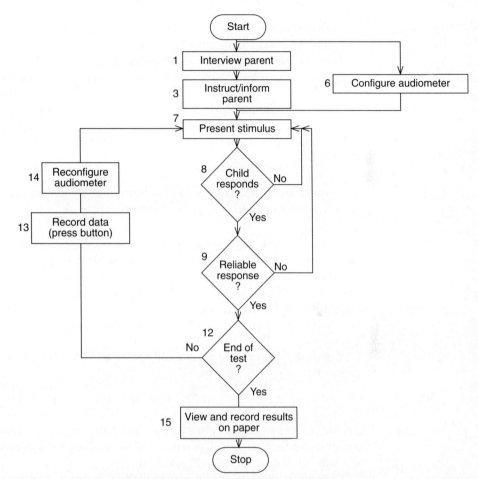

FIGURE 15-5   Flow diagram of the audiologist's task in pediatric audiometry. Flow diagrams can be used to represent control strategies in a wide variety of tasks.

Pocket Calculators

There is a fundamental difference in the operation of RPN and ALG calculators, which arises from the differences in control logic. When one evaluates the sum of 2 and 3 using an ALG calculator, the sequence of key presses is

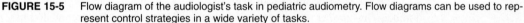

and the number 5 appears on the calculator's display. When the number 2 is first entered into the calculator, it is immediately displayed. However, when $+$ is entered, nothing is displayed. This implies that there is a **memory, or register,** for storing arithmetic functions such as $+$, $-$, $\times$, etc. When the number 3 is entered, it replaces the 2 in the display (the implication being the 2 has not been lost, but is being stored in some kind of memory for later use). Pressing ⊟ causes the ③ to disappear, to be replaced by a ⑤ (the implication being that the ② and ⊞ have been retrieved from somewhere and that the expression has been evaluated). The calculator's behavior implies that it has a memory (or a number of registers) and that if something disappears from the display, it can still be "in" the machine. According to Young, the ALG calculator's behavior when handling more complex expressions leads to a very complex model if the concept of inner registers is adhered to. Most people are unlikely to continue using the implied-register way of thinking and will abandon it in favor of a verbal rule such as the following:

You type in an expression. The machine examines it, analyzes it, and calculates the answer according to the rules of arithmetic.

This is a **verbal model** of the calculator as "an engine for evaluating written formulas" (Young, 1981). It focuses attention on **what** has to be done rather than on **how** the machine works. However, the model doesn't enable the user to infer what to do if a mistake is made in entering expressions for evaluation. If a mistake is made during complex calculations, it is difficult to determine what the error was and how to rectify it. Thus it is advisable to start again from the beginning and concentrate on keying in the formula correctly.

When one evaluates the sum of 2 and 3 using an RPN calculator, the sequence of key presses is

$$\boxed{2}\ \boxed{\text{ENTER}}\ \boxed{3}\ \boxed{+}$$

and the number 5 appears on the display. The RPN calculator's behavior also implies that it has registers for storing numbers, but the registers are organized in a very particular way—as a stack memory (Figure 15-6).

The stack is made up of four registers—X, Y, Z, and T. Each register can store one number. The number in X (the top of the stack) is displayed to the user at all times. Pressing the ENTER key (Figure 15-6B) "pushes" the stack. The number in X goes into Y, that in Y into Z, that in Z into T, and the number that was in T is lost. When one evaluates the sum of 2 and 3, ② is pressed (2 goes into the X register), followed by ENTER, which pushes 2 into the Y register. At this point, a copy of the original 2 remains in X, but it is overwritten when the ③ key is pressed. Pressing ⊞ at the end "pops" the stack. The operation is carried out and the numbers in the Y and X registers are added. The result is written into the X register (the contents of X are always displayed), and the contents of the Z and T registers move up one place (Figure 15-6C).

To evaluate an expression such as

$$2 \times (3 + 4)$$

with an RPN calculator requires the following sequence of key presses:

The ② is pushed into the stack (into register Y) when ENTER is pressed followed by ③.

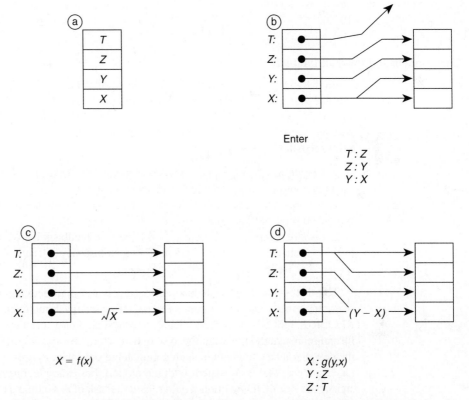

FIGURE 15-6 Stack register model for the RPN calculator. (From, Young, 1981; reproduced with permission of the publisher).

The stack is pushed again; 2 moves into Z and 3 into Y. The copy of 3 in X is overwritten when ⎡4⎤ is entered. When the ⎡+⎤ key is pressed, the contents of Y are added to those of X (3 and 4 are added). The result is written into X and the contents of Z (the number 2) move up to Y. Pressing ⎡×⎤ causes the contents of Y (the number 2) to be multiplied by the contents of X (the number 7) and the result (14) is written into X.

It can be seen from the above brief account that the stack model is **very explicit.** It draws attention to **how** the calculator works rather than **what** has to be done in order to use it. In fact, much of the machine's behavior may seem inexplicable without reference to the stack model. The **RPN calculator** is a good example of **a device whose efficient usage depends on the user's possessing an explicit conceptual model of its operation.** Once a user understands the model, the calculator's behavior for any sequence of key presses (whether or not they are arithmetically meaningful) can be **predicted.**

It is for this reason—predictability of system behavior—that design according to a clear conceptual model is thought to be important.

Empirical studies of the efficiency of different calculator designs indicate that the RPN approach is consistently, if only modestly, superior, but that RPN users are better able to exploit the full power of their machines than are ALG users (Kasprzyk et al., 1979).

User Models and Product Usability

Thimbleby (1991), in a review of the concept of **usability,** argues that user interfaces must be designed to support **algorithms** for their use. An algorithm is a set of procedures for solving a problem or achieving a result or, less formally, a step-by-step method for solving a problem. Thimbleby describes some of the design deficiencies of **videocassette recorders (VCRs):**

1 *Complexity.* A VCR and television have over 100 buttons between them—more than much more complex devices such as computers.

2 *Inconsistency.* There are differences in the logic used to specify the functions of the remote controls and the controls on the VCR front panel.

3 *Lack of meaningful feedback.* Sometimes no feedback is given; other times feedback is given even if nothing happens. Frequent tasks are difficult to perform and infrequent tasks are too easy to perform accidentally.

Thimbleby's main point is that well-designed user interfaces are not just a collection of ergonomically clustered buttons for activating the machine functions (such as STOP, RECORD, TIMER, etc. on a VCR). A well-designed user interface is based on a more formal representation of what the system is (such as an explicit conceptual model). It is the designer's task to develop such a model and only then to implement the user interface (in a way that is consistent with the model). For example, one approach to the design of the REWIND function would regard the task of rewinding a tape as a **subgoal** of a more complex task (e.g., ejecting the tape to end a session, replaying the tape) and not as an end in itself. In this case, it should be permissible to develop simple programs by combining REWIND with EJECT or with PLAY or RECORD. Similar programs could be developed for other functions as well. An alternative model would represent the VCR as a system without a memory—all tasks would have to be broken down into subgoals (directly mappable onto the control keys) and executed sequentially with appropriate feedback after the execution of each subgoal. Thimbleby's argument is that the design of user interfaces should be based on more formal models of the set of tasks involved in using products.

That VCRs are difficult to use or have too many unnecessary functions is supported by the results of a manufacturer's survey of 1000 families—30 percent of people never used the timer on their VCRs. Norman (1988) has made similar observations about the use of other products such as word processors—many of the facilities they offer are never used because of the extra learning required to master them. It is often easy for manufacturers to add more functions to their products (in the hope of adding value and gaining competitive edge), but a point may soon be reached where the perceived usefulness of the new functions is outweighed by the increased cognitive complexity of the system.

Although it is important for products to be designed according to a well-conceived model, conceptual factors are not the only ones which influence usability. Dixon (1991) investigated how conceptual and operational similarity of different devices influenced the transfer of learning from one device to the other. Although conceptual similarity of devices had a small positive effect, **operational similarity** had a substantial positive effect, even if the two devices were conceptually very different (e.g., landing a simulated plane and shutting down a simulated nuclear power station). Knowing a similar procedure when learning a new task helps users structure and organize the task, even if it doesn't help users understand the device. In the design of new products, then, it may not be necessary to make them conceptually similar to existing products, but it is necessary to make them operationally similar. An analysis of the production (if-then) and other rules used in the operation of existing products would appear to be a valuable information source for the design of new ones. This reinforces a point made elsewhere in this text—**designers must strive for consistency in design.**

THE HUMAN OPERATOR AS A DECISION MAKER

People frequently have to make decisions under conditions of uncertainty. For example, whether to play golf or squash depending on how likely it is to rain that day, whether to buy a car now or in 6 months' time depending on how likely it is that prices will go up or come down. Experts such as medical doctors, financial managers, etc. are continually faced with difficult decisions and make them using specialized knowledge and techniques.

There are two main (and complementary) approaches to the study of decision making. **Normative models** attempt to specify the optimum way of making a decision under a particular set of conditions. They specify **how** a decision should best be made and are often couched in mathematical terms. **Descriptive models** attempt to characterize the behavior of human decision makers. They attempt to describe **what** people actually do when making a decision. The first approach is a branch of applied mathematics known as decision theory. The second approach is a branch of applied psychology known as behavioral decision theory. Clearly, both approaches are essential if decision making in complex human-machine systems is to be optimized. If the behavior of human decision makers departs from the optimum, and situations in which suboptimum decisions might occur can be predicted, preventive steps can be taken (such as the provision of decision support systems, as is described in a later section). For a more advanced treatment of decision making, see Edwards (1987).

Cognitive Aspects of Decision Making

General theories of how humans make decisions have an important place in ergonomics, as do specific models of the decision-making behavior of experts in a particular knowledge domain. Jacob et al. (1986) **define a decision as a choice between alternative courses of action.** In reviewing strategies and biases in human decision making, they divide the process into four stages.

Information Acquisition   This stage involves information search and storage to specify the alternatives to be decided between. Selective perception and memory can have an impact on this process; for example, people may choose to ignore certain information because it is unpleasant or they may be unable to recall information which is crucial for making distinctions between possible courses of action.

Evaluation of Alternative Courses of Action   Theorists have developed many models of how people choose between various courses of action. For example, in the "elimination by aspects" strategy, each alternative is selected and alternatives that do not include a key aspect or dimension are eliminated. The process continues until only one alternative remains. A number of models have also been developed in which decision making is based on the utility of the various outcomes and the probability that a particular outcome will occur. If the decision maker does not know the probabilities of the various outcomes, he or she may use subjective estimates of probability, as in subjective expected utility theory in which the choice of a particular alternative depends on its utility and the subjective probability of its occurrence. According to Jacob et al., there is as yet no theory which predicts the circumstances under which any particular decision strategy will be selected.

Execution   Once an alternative has been chosen, a course of action can be taken. The decision maker may choose not to implement an action if inaction involves less responsibility or is emotionally less demanding than action.

Feedback   Feedback can be essential to the learning and improvement of decision-making skills, enabling incorrect decisions to be identified and making the consequences of the chosen course of action explicit. However, the provision of feedback does not, in itself, lead to learning and improvement of decision-making skill if the decision maker does not know that there is something to be learned or if what is to be learned is unclear. The value of feedback can be degraded if it is ambiguous or cannot be mapped onto the behavior which produced it (because of excessive delays, for example).

Heuristics and Biases in Human Decision Making

Tversky and Kahneman (1974, 1981) have attempted to characterize some of the ways humans make decisions. Of particular interest to this discussion is the use of **heuristics** (rules of thumb) to simplify the mental operations required to make a decision. Three main classes of heuristic technique are:

1 Judgment by representativeness
2 Judgment by availability
3 Judgment by adjustment and anchoring

Representativeness   People often have to make judgments which really depend on the probability that an event or object belongs to a particular class or is the result of a certain process. The representativeness heuristic replaces judgments about probability with judgments about whether the event is representative of the class or process. For example, in the attempt to decide whether an individual is a farmer or a librarian, the **sim-**

ilarity of the individual to the **stereotype** of a farmer or librarian is used. Although judgments about similarity and probability are often the same, use of the rule leads to systematic and predictable errors. If one is to come to a decision about someone's occupation, as in the above example, it would be useful to know about the relative proportions of librarians and farmers in the population (the prior probability of a person's being a librarian or a farmer), but judgments about similarity do not take data of this type into account. In the countryside, farmers probably outnumber librarians, so it is more likely that someone living there will be a farmer irrespective of any other considerations, such as whether the person looks like a farmer. In the middle of a large city, the opposite is likely to be the case.

In an example more relevant to ergonomics, if a manual worker complains of arm pain at work, it is often concluded that the pain is caused by the work on the basis of its similarity to occupational arm pain. However, if a correct decision is to be made, it is important to know the probability of someone's experiencing the pain both at work *and* outside of work. Failure to determine that probability causes the role of occupational factors to be overestimated.

Representativeness also leads to **misconceptions of chance.** People expect sequences of random events to be similar to the process which gave rise to them even if the sequence is short. In tossing a coin, for example, people will say that a sequence such as

H T H T T H H T

is more likely than

H H H T T T T

or

H H H H H H T

Although it is true that when a fair coin is tossed, there will be an equal number of heads and tails over the long term (because the probability of either is 0.5), short sequences of tosses need not resemble long sequences. This misconception underpins the well-known **gambler's fallacy** or **law of averages** in which random events are seen as self-correcting processes. The misconception states, for example, that after a long run of heads, a tail is now "due," to restore the appearance of the process. Unfortunately, a fair coin does not "know" that the sequence of prior tosses "looks wrong," and the probability of a head or tail remains the same each time the coin is tossed.

The representativeness heuristic can lead to systematic biases when people accept that a small sample is representative of the process which gave rise to it. Even engineers and scientists are prone to this overreliance on small samples and make unwarranted extrapolations from them. Many complex processes, in economics or meteorology, are affected by overlapping trends with varying periodicities. For example, a person may decide after experiencing five increasingly long and gloomy winters that the earth's climate is getting colder. However, this conclusion is an artifact of extrapolation from a short time series which may not be representative of long-term trends (in the case of Britain, current weather may be slightly cooler than 15 years ago, but is definitely warmer than 250 years ago, definitely cooler than 1000 years ago, and definitely much warmer than 10,000 years ago at the end of the last ice age). In process industries (and other complex situations) where historical plant data can be presented using computer graphics, designers must use a knowledge of system dynamics to ensure that variables are sampled, presented, and displayed so that real trends, rather than random fluctuations or noise, are presented to operators.

Availability   When people use the availability heuristic, they judge the likelihood of events according to the ease with which instances or occurrences of the event can be brought to mind. Things that are difficult to imagine are deemed to be unlikely. Sometimes this heuristic is a useful cue for making probabilistic judgments because instances of large classes of objects can indeed be imagined faster and more frequently than instances of small classes. However, availability, the ease with which events can be imagined or recalled, is influenced by variables which do not influence probability. For example, people judge one activity to be more dangerous than another because it conjures up more vivid images (e.g., working in an explosives factory as opposed to working in your own kitchen) or because more potential causes of failure can be imagined.

Judgments based on availability may be biased by the effectiveness of a search strategy. If people are asked whether more English words have the letter "r" as their first or third letter, many will choose the former even though it is incorrect. It is easier to retrieve from memory words beginning with "r" than words with "r" as the third letter.

Adjustment and Anchoring   When people are required to make quantitative estimates or evaluations, they often begin at an initial value and adjust it to arrive at a final value. As Tversky and Kahneman (1974) point out, the amount of adjustment and the final value can be influenced by the size of the initial value (the anchor). Salespeople and advertisers often use this heuristic as a technique to influence peoples' perception of the fair price of goods (e.g., "How much would you expect to pay for this packet of washing powder—$5? $4.50? Why not take advantage of our special offer of $2.50?").

In one experiment, subjects were first asked whether the percentage of African countries in the United Nations was greater or less than a particular value (the anchor). They were then asked to estimate the actual percentage. With an anchor of 10 percent, an estimate of 25 percent was obtained. With an anchor of 65 percent, an estimate of 45 percent was obtained. In another demonstration, subjects were asked to estimate 8! written in two ways:

$$8 \times 7 \times 6 \times 5 \times 4 \times 3 \times 2 \times 1 \quad \text{or} \quad 1 \times 2 \times 3 \times 4 \times 5 \times 6 \times 7 \times 8$$

and gave estimates of 2250 in the first case and 520 in the second (the correct answer is 40,320).

Adjustment and anchoring effects also occur in the choice of bets. Subjects were given the following types of bets:

1 Drawing a red marble from a bag containing 50 red and 50 white marbles
2 Drawing a red marble 7 times in row from a bag containing 90 red and 10 white marbles

3 Drawing at least 1 red marble from a bag containing 90 white and 10 red marbles in 7 tries

Subjects preferred the second bet followed by the first and preferred the first to the third. The reason is that the probability of the elementary event provides an anchor (0.9 for bet number 2, 0.5 for bet number 1, and 0.1 for bet number 3). However, the combined probabilities for the repeated trials in bets 2 and 3 are 0.48 and 0.52, respectively.

The section on anchoring and adjustment highlights the need for careful design of training programs for operators of systems, since it is at these early stages of learning that anchor and adjustment biases can be inadvertently set.

Misconceptions of Probability

Kurosaka (1986) gives several examples of situations where people's **intuitive judgments about probability** can lead them to make erroneous decisions. For example, given two red cards and two black cards lying facedown on a table, many people can be persuaded that the probability of turning over two cards of the same color is 0.5—that there is a 50 percent chance of picking one red card and one black card. In fact, the probability is 0.66. Because the first card to be turned over stays turned over, two of the remaining three cards are of a different color from the first card; therefore, the probability of the next card's being of a different color from the first is 2 in 3 (0.66). People are susceptible to mistakes such as this because they fail to take into account that the probabilities change as soon as the first card is turned over (they remain anchored to the initial situation in which there were equal numbers of both colors of cards). If we take the second stage as the anchor (the probability that a card of a certain color will match with a card from a pack of three cards, two of which are different) the paradox disappears. It can be appreciated why professional statisticians always specify whether they are sampling with or without replacement when dealing with problems of probability.

Many other extraneous factors can cause biases in human decision making (Table 15-4). Findings such as these demonstrate the ways in which people's intuitive judgments deviate from normative decision models. For much of the time, though, heuristic principles can produce essentially correct decisions and have considerable power in enabling complex problems to be dealt with quickly by simplification.

This discussion of human decision making introduces the notion of human-machine symbiosis—the idea that human and artificial information processors can behave in complementary, rather than contradictory, ways in practical decision environments. Humans are subject to the type of biases documented by Tversky and Kahneman and might benefit from decision aids which can handle complex probabilistic calculations quickly and in a formal way. However, such decision aids often require complete data before the necessary calculations can be performed, whereas, as described above, the human decision maker is capable of arriving at a decision utilizing alternative types of information in the data.

IMPROVING HUMAN DECISION MAKING AND PROBLEM SOLVING

For convenience, problem solving and decision making have been dealt with separately up to this point in the discussion. However, decision making often involves problem

TABLE 15-4   FACTORS WHICH CAN INFLUENCE HUMAN DECISION MAKING*

---

1 Preference of concrete over abstract information
2 Mode of data presentation
3 Ignoring nonoccurrences of an event
4 Illusory correlation (events which occur together are assumed to be related)
5 Selective perception (ignoring contradictory evidence, having the view that things that worked
    in the past will work in the future)
6 Conservatism (unwillingness to use new information)
7 Inconsistent application of criteria
8 Inability to extrapolate growth processes
9 Justification (a rule is deemed usable any time justification can be found for using it)
10 Complexity, time pressure, and overload
11 Emotional stress (can cause cognitive tunnel vision)
12 Social pressure (conformity is rewarded)
13 Illusion of control
14 Wishful thinking (personal preferences may influence the choice of a course of action)
15 Hindsight bias (anything can be explained after the event)
16 Distortion due to problems of retrieval from memory (reconstruction of events is distorted by
    the 'effort after meaning' which distorts retrieval from long-term memory)
17 Tendency to attribute successful decisions to skill and unsuccessful decisions to bad luck

---

* From Jacob et al. (1986).

solving and vice versa. The development of artificial aids to improve decision making and problem solving has a long history—the abacus is one of the earliest examples. Some examples of modern decision aids are summarized below.

Computer Aiding and Decision Support

In military and other settings operators are required to monitor radar or sonar signals displayed on screens. Typically, the information displayed on the screen is complex. The operator's problem is to decide whether something on the screen is a signal (reflected off a real object), to determine what type of signal it is, and to keep a mental picture of the situation. Such tasks can be supported by computerized systems which give a probabilistic response regarding the presence of a signal. The computer-generated probability aids the operator under uncertain situations and provides a useful benchmark to guard against distortions in human judgment caused by fatigue or overload.

In nuclear power stations, the development of one fault can cause numerous subsidiary faults to occur. The operators are presented with a large amount of information from various parts of the system and may experience difficulties in tracing the pattern of faults back to the primary cause. Decision-support systems have been developed to assist operators. These systems have a model of the power station which can process large amounts of fault data rapidly—their success depends on the fidelity of the underlying model.

Systems such as SAMMIE (system for aiding man-machine interaction evaluation) work by extending human visualization. Design specifications can be entered into the system and presented in the form of three-dimensional graphics. The system can vary the anthropometry of its "man-model" to enable decisions about product dimensions to be finalized. Additionally, three-dimensional representations of products such as vehicles, tools, etc. can be rotated in all planes to give different views. Proposed changes to the design can be evaluated quickly using these systems.

As described in a previous chapter, predictor displays have been developed for use in the aerospace industry to enable what-if? questions to be rapidly evaluated, thus assisting the operator in choosing between alternative courses of action.

While human decision makers are good at making judgments under uncertainty, they are not good at integrating them. In military systems, medicine, and several other areas, attempts have been made to develop decision aids which complement the decision-making prowess of humans (the system handles those aspects of the problem that humans are known to be poor at). The examples given below are from the field of medicine (see Shortliffe et al., 1979; Shortliffe, 1984; and Kleinmuntz, 1984, for further information), but the techniques exemplified are applicable to any other semantically rich knowledge domain.

Diagnostic Problem-Solving Aids

Even in moderately complex systems such as automobiles, fault diagnosis is not always easy, as any amateur mechanic knows. Multiple faults can have a single cause (e.g., poor roadholding and fuel consumption due to too low tire pressures) and single faults can have multiple causes (e.g., uneven engine performance due to worn-out spark plugs, incorrect ignition timing, a dirty air filter, and leaky gaskets on the inlet and outlet manifolds). This latter diagnostic scenario may baffle the problem solver because until all faults are fixed, the problem remains—each remedial action, although correct in itself, is not followed by the expected feedback.

In diagnostic problem solving, signs and symptoms are detected and their weighted contributions to a disorder or malfunction are aggregated to determine the most likely disorder. Decision making can be hampered by a lack of knowledge about the problem domain (if there is a shortage of properly trained experts or an expert is not available) and by the use of suboptimal strategies such as the following:

1 Generating a set of competing alternatives and testing them. Humans tend to be mislead by too many alternatives because they focus on irrelevant features of the problem.

2 Being unable to think probabilistically or weighting data in an inappropriate way.

3 Collecting too much data in the belief that redundancy improves decision making. In fact, it can overload the decision maker's capacity to interpret data correctly.

Decision Trees   A decision tree consists of a series of questions with a limited number of responses (sometimes only yes or no). Decision trees have been developed for the diagnosis of biochemical disorders and the screening of patients at rural clinics in developing countries. The user is prompted by a series of questions and enters the data accordingly (either from laboratory results or by asking the patient directly).

The session normally begins with some general questions which define the problem domain or context. In the screening of patients, for example, some general questions about the patient's problem, whether there is pain, fever, injury, etc., would be asked first. The answers to the general questions lead the system into decision pathways in which the questions become more and more specific. Eventually, an end point is reached where either the disorder is diagnosed or it is decided that the ailment is outside the system's knowledge domain and a course of action, such as "refer to gastrointestinal clinic," is suggested. Decision trees are inflexible and cannot easily update themselves in the light of new findings. Each patient is seen as a new case even if it is a follow-up visit. They have an educational impact—with repetition, the user learns the structure of the system's decision pathways and the system makes itself redundant (although the fact that the system makes itself redundant is not necessarily a weakness).

Decision trees are not a new invention, nor do they need to be implemented on computers. They work particularly well in highly structured domains where clearcut distinctions can be made using yes-no responses. Biologists have long used pencil-and-paper decision trees in taxonomy—to classify unknown animals or plants using questions, for example, about the size of the animal, the number of legs, the number of body parts, whether it has an exoskeleton or an endoskeleton, etc.

Statistical Pattern-Recognition Techniques   Statistical pattern-recognition techniques are based on a knowledge of the disease patterns in a population. Surveys of the incidence of a disease and the variables associated with its manifestation (the signs and symptoms in medicine) are carried out and subjected to statistical analysis using regression and other multivariate techniques. This produces a mathematical model of the disease pattern:

$$P\,(D_i\,/\,x) = a_1\,x_1 + a_2\,x_2 + \cdots + a_n\,x_n$$

where a_i = coefficients obtained from sample data
       x_i = value of a feature vector

The model gives the probability that a patient has a particular disease, given that certain signs or symptoms are present.

For example, if a patient is complaining of chest pains, several different diagnoses may be possible (e.g., acute myocardial infarction, coronary insufficiency, or noncardiac pain). Clinicians often misclassify noncardiac cases (because the consequences of misdiagnosing noncardiac pain as cardiac pain are less severe than misdiagnosing cardiac pain as noncardiac). In one study, a model of chest pain was developed using data on 17 variables (signs and symptoms) from a sample of 247 patients whose diagnoses had subsequently been firmly established (Patrick, 1977). The mathematical model was shown to perform better than clinicians in placing patients into one of the three diagnostic categories using data on the 17 variables.

The processes by which a statistical model comes to a decision are very different from those used by a human decision maker and can be regarded as complementary. Statistical models are unlikely to replace human decision makers because of practical problems which limit their validity, including the following:

1 Choosing a set of reliable diagnostic features (predictors)
2 Obtaining a large and representative sample
3 Measuring the features in the sample
4 Ensuring that the methods of measuring the features in a real case are carried out in the same way as in the cases used to build the model

Bayesian Methods   Bayes' theorem was developed in statistics to handle conditional probabilities. It deals with situations which are not easy for human decision makers to deal with intuitively because a judgment about the likelihood of an event depends on some other set of conditions which also have an associated likelihood, and there is an absence of other types of information to help the decision maker come to a conclusion. In a control room, for example, an operator might have to decide the probability that a relief valve is malfunctioning given that pressure in some other part of the system is abnormal. In these type of situations, where there may be many other possibilities, knowledge of the probability of the valve's malfunctioning and the probability of other events causing abnormal pressure may be difficult to process intuitively. An example of the application of Bayes' theorem is given in Table 15-5 (see Edwards, 1987, for more details).

Bayesian models would appear to be of value in situations where, for example, a patient has particular symptoms and the doctor must decide whether they are caused by a particular disease. However, large amounts of data are needed to determine conditional probabilities. The model assumes that the patient has only one disease at a time, which may be incorrect in practice. If the pattern of disease in the population changes over time, the probability estimates can become out of date. There are statistical problems with this type of decision aid, since the Bayesian model assumes conditional independence of symptoms, which may not be the case. However, the model is free of many of the biases which can affect a human decision maker.

Symbolic Reasoning Approaches   In contrast to decision trees and numerical computation, symbolic reasoning approaches attempt to mimic the decision-making behavior of the expert. In diagnostic tasks, judgment is made on the basis of gross chunks of knowledge rather than detailed facts. Heuristic rules are applied to these knowledge chunks to arrive at a decision. The rule-based approach directs attention to pertinent areas and obviates the need for detailed analysis of the problem. Systems based on symbolic reasoning can be quite sophisticated and can be programmed to explain their decisions to the user in terms that the user can understand.

MYCIN is an example of such a system. It was developed for aiding the selection of antibiotics for patients with infections. The designers had the following goals for MYCIN (Shortliffe et al., 1979):

1 It should be able to explain its decisions in terms that physicians can understand.
2 It should be able to respond to questions posed in English.
3 It should be able to learn new information by interacting with experts.
4 Its knowledge should be easily modifiable.
5 The interaction should be engineered with the user in mind.

TABLE 15-5   EXAMPLE OF BAYES' THEOREM*

Suppose you wake up in the middle of the night and go to the bathroom to take some aspirin for a headache. In the dark, you take a tablet from one of three bottles. Later you become ill and remember that two of the bottles contained aspirin and one contained poison.

Given that 80 percent of the people have your symptoms after taking the poison and 5 percent have them after taking aspirin, what is the probability you took the poison?

Let B = symptoms, A = poison, and A' = aspirin. The probability of having the symptoms after taking the poison is given as

$$P(B/A) = 0.8$$

The probability of having the symptoms after taking aspirin is given as

$$P\ B/A = = 0.05$$

If each of the three bottles has an equal chance of being selected in the dark, the probability of taking the poison and aspirin bottles, respectively, is

$$P(A) = 0.33$$
$$P(A') = 0.67$$

Given that you have the symptoms, what is the probability you took the poison? Bayes' theorem can be used to solve the problem as follows:

$$P(A/B) = \frac{P(B/A)\,P(A)}{P(B/A)\,P(A) + P(B/A')\,P(A')}$$

$$= 0.89$$

* From Hays (1973).

Knowledge of infectious diseases is represented in MYCIN in the form of **production rules**—simple conditional statements which relate observations to inferences (such as "**if** a bacterium is a gram-positive coccus growing in chains, **then** it is apt to be streptococcus"). A part of the program, sometimes called the **inference engine,** decides which rules to use and how they should be chained together to make decisions. The rules are stored in a machine-readable format but can be translated into English and displayed to the physician. The system's knowledge structure can be modified by changing the rules or adding new ones, and the rules themselves can be used to explain how the system came to a particular decision.

Systems such as MYCIN are sometimes referred to as **expert systems.** However, a real expert differs from one of these computer programs in a number of ways. Real experts know when to stop asking questions, and they possess **metaknowledge** (they know what they know—and what they don't). Real experts also seem to possess **tacit knowledge**—knowledge that is not explicitly taught or verbalized but is needed in a particular domain. Expert systems, on the other hand, are limited by the information in their databases, the processes by which it got there, and the mechanisms they have for processing it (de Greene, 1986). They cannot give answers that are not already implicit in their databases and lack insight and the capacity for creativity. **However, as decision aids rather than substitutes for real experts, these systems can serve a very useful function in**

providing an external semantic memory for the user. They may also have value in training, since the system can explain **how and why** it came to a particular decision. A review of expert systems and other aspects of **artificial intelligence** can be found in Hillman (1985).

Training

In diagnostic and other problem-solving tasks, training in fundamental principles alone is usually inadequate (Morris and Rouse, 1985). According to Schaafstal (1993), theoretical instruction has to be combined with training in the use of the knowledge to solve problems in the context of the actual equipment which will be used. Morrison and Duncan (1988) have described some of the characteristics of easy and difficult faults. Easily diagnosed faults can be predicted and are associated with an explicit algorithm which will elucidate them. Difficult faults are intermittent or novel, they cannot be predicted from presently available information, and there is no predetermined algorithm for their elucidation. It is with these types of problems that knowledge of fundamental principles is useful and is sometimes the only means of solving the problem. The opportunity for practice is an important component of training and industries such as aerospace and nuclear power spend considerable sums on simulators to enable off-line practice to be carried out. From Schaafstal (1993), a final consideration in the design of training programs for problem solving and decision making seems to be how to optimize the combination of training in the application of procedures with training in the use of fundamental knowledge to decide exactly what to do at any particular point in the application of the procedure.

CURRENT AND FUTURE RESEARCH DIRECTIONS

These cognitive aspects of ergonomics may on first encounter seem rather removed from the practical problems of worksystem design. Adopting this view is to be mistaken—any apparent abstractness is a consequence of the class of problems being investigated. Some of the most important issues in the design of information and communications technology are abstract and concern the conceptual fit between the design of systems and the users' expectations and knowledge. There is a great deal of research currently under way to improve human decision making and problem solving in the process control industries and in military and aerospace systems.

The emerging importance of this field is also indicated by the establishment of "usability" laboratories in universities, research institutes, and private companies. It is in these laboratories that many of the conceptual issues about the design of new products are investigated in relation to human behavior. Usability laboratories also provide an environment for implementing and testing new models and metaphors for the design of the user interface.

SUMMARY

Cognitive ergonomics has been defined by Long (1987) as the attempt to increase compatibility between the users' representations and those of the machine (i.e., of the designers). Accomplishing this goal requires a way of describing users' (and machine) representations so that areas of incompatibility can be disclosed. The concepts and methods of cognitive psychology, artificial intelligence, and linguistics can be used to arrive at descriptions of these representations.

In automated systems the operator's role is largely supervisory. The design of interfaces for such systems must take place at a cognitive level, with the emphasis on the operator's requirements for information to support his or her problem-solving and decision-making skills. Obtaining information about these skills is difficult because cognitive behavior cannot be directly observed. A battery of techniques, including verbal reports, verbal protocol analysis, and objective measures, is needed. Careful collection and analysis of data concerning operator errors and difficulties can provide insights into how operators conceive and misconceive aspects of system behavior.

Many types of decision-support systems have been developed and validated by comparing them with human decision makers. Some have been shown to perform better than human experts, but all are limited by the assumptions under which they operate. Computerized decision aids can improve the decision-making performance of human-machine systems because they are not susceptible to the types of biases that can bedevil humans.

ESSAYS AND EXERCISES

1 Ask a friend or colleague to solve the problems given earlier in this chapter. In addition, ask that person to think out loud. Record the verbal protocol. Analyze it and attempt to construct a representation of the person's problem-solving behavior.
2 Construct an algorithm which will solve the "missionaries and cannibals" problem for any equal number of missionaries and cannibals. Give the algorithm to a friend to use in solving the problem, and ask the friend to think out loud while using it (or implement the algorithm in a programming language of your choice).
3 "Cognitive ergonomics is no different from any other area of ergonomics." Discuss critically with examples from several different application areas.

FOOTNOTE: Cognitive Aspects of Numeracy

Boswell: Sir Alexander Dick tells me that he remembers having a thousand people a year to dine at his house.
Johnson: That, Sir, is about three a day.
Boswell: How your statement lessens the idea.
Johnson: That, Sir, is the good of counting. It brings everything to a certainty which before floated in the mind indefinitely.

From James Boswell, *Life of Samuel Johnson* (1791) (source Flinn, 1966).

In previous sections, the design of verbal material was discussed in relation to ease of comprehension. Similar considerations apply to the design and use of numbers. The human cognitive system has strengths and limitations, as we have seen. The invention of number systems occurred more than 5000 years ago and played a major role in freeing early civilizations from some of these cognitive constraints. However, there are interesting cognitive ergonomic issues concerning the way numbers are represented visually and how easy a number system is to learn and to use. Nickerson (1988) provides an interesting review of these issues and compares our present Arabic system with the Egyptian, Babylonian, Greek, and Roman systems which preceded it. Nickerson argues that the power of the Arabic system lies in its compactness and extensible notation, which is achieved at the cost of greater abstractness. Although the Arabic system may be more efficient for reprinting and manipulating large numbers, it may be more difficult to learn than other systems.

Numbers enable quantities to be represented abstractly and are intellectual tools indispensable for any large-scale activity, such as building a pyramid, which requires record keeping, planning, and coordination. The abstraction of quantity through numbering makes arithmetic possible. Numbers as abstract symbols are amenable to manipulation according to general rules in a way that concrete objects are not. No matter how big or small a number, we can manipulate it using the same set of rules and procedures and get answers to the same questions. Arithmetic enables questions of the what-if variety to be asked and frees the mind from both physical limitations and limitations on human imagination and memory. Hofstadter (1985) has discussed the comprehension of large numbers and concluded that most people are unable to grasp their meaning. For example, many people do not know the difference between 1 million and 1 billion. This lack of comprehension is particularly important in the modern world where members of the public are frequently presented with large numbers in newspaper reports and the media (e.g., numbers in the billions are often quoted in relation to the foreign earnings of countries and the budgetary policies of governments).

Following Hofstadter, it appears that large numbers are neither user-friendly nor usable, which may be due to problems of imaginability and visualization. The difference between 10 and 100 people can be visualized (the number of people in a living room versus the number in a busy restaurant), as can the difference between 100 and 1000 (the number of people in a small cinema versus the number in a large concert hall), but 1 million and 1 billion cannot be visualized. Another problem appears to be lack of practice in the basic numeracy skills needed to handle large quantities. In particular, most people lack practice in making rough estimates. Hofstadter gives some examples of real numbers which would normally cause "number numbness" (i.e., that most people would have difficulty comprehending):

- 10,000: the number of books in a bookshop
- 90,000: the number of cattle slaughtered every day in the United States
- 1,000,000: the number of people in a medium-sized city
- 10,000,000: the number of people in large city
- 100,000,000: the number of cigarette smokers in the United States or the number of cells in the retina of the eye
- 1,000,000,000: the population of China some years ago
- 5,000,000,000: the number of letters in all the books in a bookshop
- 90,000,000,000: the number of hamburgers sold by McDonald's hamburger chain
- 100,000,000,000: the number of neurons in the brain

Large numbers seem to be infiltrating daily life to an increasing extent. A potentially useful and challenging task for ergonomics is to find more user-friendly ways of representing large numbers. One solution proposed by Hofstadter is to popularize the use of existing mathematical notation. Instead of using verbal labels such as "hundreds," "thousands," "millions," and "billions" (when it is doubtful that many people know how many zeros these numbers really contain), a notation such as 10^2, 10^3, 10^6, and 10^{12} should be used, which makes explicit the absolute size of a number and its size in relation to other numbers. It would also remove any confusion about the use of the word "billion" which means 10^9 in the United States and 10^{12} in Europe. Abstract notation may have disadvantages in many applications. However, when the concepts involved cannot be easily imagined in a concrete way, they may as well be represented abstractly and a simple, formal notation may be the answer.

16

ERGONOMICS, WORK ORGANIZATION, AND WORKSYSTEM DESIGN

In addition to engineering and technical expertise, worksystem design requires knowledge about humans from the field of ergonomics. A minimum requirement for ergonomics is that it be based on a scientific model of the human. Its input at this level will consist of more than just empirical statements about good design practice, rules of thumb, or common sense. Ergonomic design guidelines are, if nothing else, implications of existing theoretical knowledge. Conversely, those who claim that their products or systems are "ergonomically designed" must be able to explain the scientific principle underlying the ergonomic design.

Few would deny that scientific knowledge is necessary for the successful design of the human component of modern worksystems—whether it is sufficient is more debatable.

THE DESIGN OF HUMAN-MACHINE SYSTEMS

A system can be defined as a bounded set of related objects, brought together for a specific purpose which transcends any of the constituent parts in isolation. Systems have a hierarchical structure, and systems design and analysis have to take this structure into account. Because systems are more than the sum of their parts, systems design cannot be optimized solely from the bottom up. Real systems are dynamic and interact with their environments at different levels of complexity. Seemingly separate issues may actually overlap, sometimes in complex ways.

The Body as a System

The human body is a system which can be analyzed at different levels—individual cells make up organs which in turn are part of regulatory subsystems. These subsystems constitute the complete organism. Interactions can take place horizontally—between components at the same level of analysis—or vertically—in a top-down or bottom-up fashion. For example, if a particular organ malfunctions, the functioning of other organs, or of the whole body, may be degraded. At the same time, the health of any organ will be influenced by the overall functioning of the rest of the body. It can be seen from this example that complex systems are characterized by both horizontal and vertical interactions between their components.

Analytic and Synthetic Thinking

Systems thinking requires a combination of **analytic** and **synthetic** processes. Analytic processes are used to investigate the individual parts of the system. Synthetic processes are used to understand the behavior of the system as a whole and to account for the ways in which the subsystems are interrelated and can affect each other. For example, a system that is malfunctioning is analogous to a patient who complains of being unwell. Diagnosis of the problem involves the application of analytic processes and the gathering of data about the functioning of specific parts of the system. Remedial intervention can be taken using theories about the functioning of a particular process or part (such as the structure of the organs of the body or the dynamics of a physiological process). This analytic procedure is followed by a shift to synthetic processes—trying to view the system as a whole and evaluate its overall level of functioning. For example, after treating a patient, the doctor may ask, "How do you feel now?" rather than, "How does your temperature feel now?"

Systems thinkers must be able to shift between analytic and synthetic modes and describe system behavior at different levels of analysis. Similarly, after applying ergonomics to the design or redesign of particular aspects of a system, the ergonomist should take a step back and try to assess how the measures which were taken have affected the overall functioning of the system. This involves a shift from analysis and measurement at the level of subsystems and human-machine interaction to the analysis of the functioning of the total system, and requires qualitatively different data such as measures of output, wastage, production costs and efficiency, system reliability, and absenteeism. **Despite many years of debate, ergonomists are often reluctant to take this step back and evaluate the benefits of their intervention in business terms.** According to Chapanis (1991) there are few convincing studies of the cost-effectiveness of ergonomics.

Diagnosing System Malfunction

Problems in the functioning of systems can be misdiagnosed if a systems approach is not taken. For example, there has been much discussion in the literature about the health problems arising out of computerization and the use of VDTs. Although conventional strategies based on detailed analysis of the physical workspace are of value, they do not enable predictions to be made with certainty about the circumstances under which VDT-related health complaints will arise. Smith et al. (1980) compared the incidence of health complaints and indexes of job stress in clerical and professional VDT users and in control subjects. Health complaints were highest in the clerical group, as were indexes of job stress. Factors such as lack of control over the job, boredom, and concerns about career development interacted with VDT use, leading to a higher incidence of health complaints. Findings such as these indicate that the way employees respond to **subsystem** problems such as workplace stressors (e.g., undue noise, badly designed workspaces, or excessive task load) depends on **system** design variables such as work organization. In the VDT example, it can be seen that reduction of health complaints not only would depend on redesign of the physical workspaces but also would require a shift to a higher level of analysis and include redesign of the organization of work and the methods of personnel management.

Industrial Accidents

Further evidence of the need for a systems approach comes from a study of industrial accidents by Dwyer and Raftery (1991). Traditionally, accidents have been characterized at the level of the human-machine interface and concepts such as the "accident-proneness" of workers or unsafe working conditions have been used to explain why an accident happened. However, Dwyer and Raftery point out that many other factors, at higher aspects of system functioning, have been shown to be related to accident frequency at one time or another. Some examples are:

1 Working on a piecework basis
2 Employee undernutrition
3 Extended work hours
4 Absence of integration of different work groups

In their own investigation of industrial accidents, these authors concluded that accidents could be prevented by a system of work in which workers exercised greater **autocontrol** of their activities and by management if, in the absence of conditions favorable to autocontrol, a proper safety management program was initiated.

A Multilayered Approach Is Needed

An ergonomic approach to systems design and evaluation must therefore be multilayered and able to characterize system and subsystem functioning at appropriate levels of analysis. Some aspects of systems design thinking applied to human-machine systems are summarized in Table 16-1 (Singleton, 1972).

Singleton (1989) has reviewed some of the contemporary issues in the application of ergonomics in systems design. In addition to basic knowledge, ergonomics needs **techniques** for applying this knowledge to design. These techniques cannot just be extensions of the methods of anatomy, psychology, or physiology. Bottom-up approaches to the development of ergonomics may lead to a fragmented and piecemeal technology in which the solution to a problem depends more on the personal perspective of the ergonomist (a preference for anatomy over psychology, for example) than on the problem itself.

TABLE 16-1   FEATURES OF SYSTEMS THINKING*

Cause	Feature	Consequence
Development of varieties of ways of carrying out the same activity	Consideration of functions independently of mechanisms	Improved systematic choice between alternative solutions
Increased complexity of new systems	Detailed attention to objectives	Fewer expensive design errors
Hardware performance becoming comparable to human performance	Consideration of human as an integral part of the system	Improvement of system performance by improving human performance

* From Singleton (1972).

Limitations of Reductionism

De Greene (1977, 1980) has discussed some of the shortcomings of the application of ergonomics if it is based solely on the methods and theories of basic sciences:

1 Laboratory findings are not always directly applicable to real-world problems.

2 Apparently relevant findings may lack generality.

3 There is no theory to integrate all the different findings of the basic sciences.

4 The unit of analysis appropriate for basic science may not be appropriate for solving ergonomic problems.

5 The basic sciences do not provide a paradigm for describing and analyzing human-machine systems.

De Greene has argued strongly that the laboratory for ergonomics should be the real world and that ergonomic problems should be analyzed and investigated at the system level at which they occur (rather than simplifying them and simulating them in the laboratory). The power of modern computer systems to handle large data sets using complex modeling techniques should be harnessed for this purpose. Furthermore, as de Greene and others have pointed out, a knowledge of the factors which **motivate** people to interact with machines and enable them to overcome design limitations currently lies beyond the scope of much traditional ergonomics but is crucial to the successful operation of real systems. Display and control design may be limiting factors to human-machine interaction, but they can hardly be compelling ones. At present, the study of job satisfaction and motivation falls within the ambit of occupational psychology and few attempts have been made to integrate the theories of occupational psychology into ergonomics.

THE SYSTEMS APPROACH

Successful systems design requires that the expertise of different specialists be carefully managed and utilized at appropriate stages in the design process. In particular, ergonomics has to be integrated with the engineering and personnel functions. The systems approach is an empirically derived algorithm for managing the design of human-machine systems and of partitioning the design process in time and across groups of specialists. It consists of a number of stages, summarized in Figure 16-1 (from Huchingson, 1981). Some of the critical stages in the systems design process are described below.

FIGURE 16-1   The systems approach. (Adapted from Huchingson, 1981.)

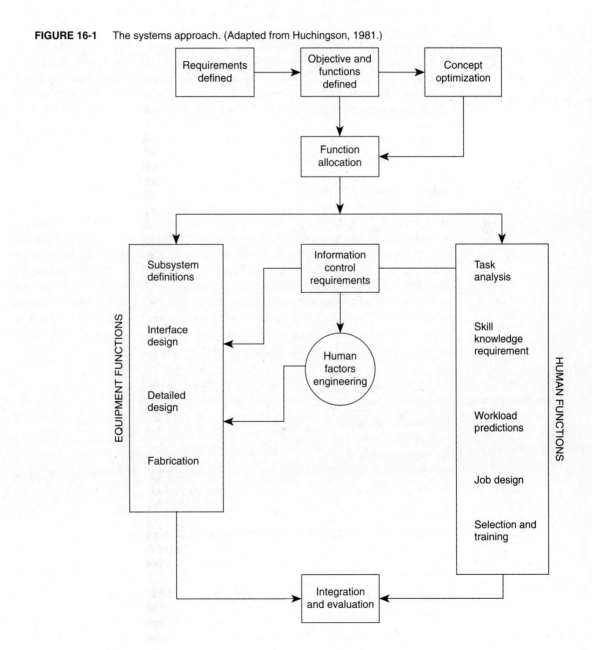

Specification of Requirements

A clear specification of the requirements is necessary to define what the proposed system is supposed to achieve—its scope and objectives. Requirements specification is perhaps one of the most important and easily glossed-over stages in systems design. Two central issues at this stage concern **the choice of techniques used to specify the requirements and the choice of people to sit on the design team.** A common mistake is to leave requirements specification to technical managers, engineers, or business experts and ignore the potential contributions of operators or users. A more **participatory approach** is recommended nowadays, which involves consulting a wider group of people in the organization. The involvement of people closer to the actual work situation can provide more information about the considerations for requirements specification because it takes into account the informal structure and actual working arrangements of the organization (how work is really carried out as opposed to how it is supposed to be carried out).

Systems design is an open-ended process and open-ended methods can be used to gather information for requirements specification. Some examples are given in Table 16-2.

The goal of the requirements specification stage is not to describe how the system is to work so much as it is to describe its purpose—what it must do. In the design of a new airplane or car, for example, the requirements specification is not that the airplane flies or that the car's wheels stay on. The requirements specification describes the characteristics of the product with respect to a defined section of the market (e.g., the number of passengers the airplane should carry, the routes it would operate, its performance characteristics and fuel consumption). Arriving at a requirements specification for products such as these requires that an analysis be made at a higher level of system functioning than that of the product itself. Instead of being concentrated on the technical problems of aircraft or cars, the design process is focused at the level of the overall transport system, consumer demand, and the available market niches.

Identification and Description of Functions

Functions are particular activities which enable a system to attain its objectives (i.e., objectives are reached as a consequence of executing functions). Information on functions can be obtained in several ways. Panel discussions, interviews with users and experts, and observation can provide function-related information—information about current

TABLE 16-2   DATE-CAPTURE METHODS FOR REQUIREMENTS SPECIFICATION

1 Unstructured interviews
2 Structured interviews
3 Unobtrusive observation of the current system
4 Brainstorming sessions
5 Questionnaire surveys
6 Market research

methods of achieving system objectives. System design methods often stress that, initially, functions be identified and described in abstract terms. Some advantages of abstract functional thinking (according to Singleton, 1972) are the following:

1 It encourages the search for new methods.
2 It provides a common language across disciplines.
3 The range of solutions is not limited by particular components or methods.

For example, ignition timing is an essential function in the operation of a gasoline engine. It can be implemented electronically or mechanically. Contact breakers are not a function; they are an example of a particular type of component which implements the function mechanically.

The distinction between a system's objectives and its functions is an important one. Objectives must be specified before functions—the execution of a system's functions is the means by which the system's objectives are achieved to satisfy the requirements. Huchingson (1981) suggests that functions be stated in terms of a verb, a noun, and modifiers, e.g., "make a midcourse correction." The use of noun-verb pairs to describe functions is thought to assist the design team in clearly identifying the important functions around which creative design can take place. For example, the purpose of a chair is to support the mass of the sitter. This purpose is achieved by designing something which implements the following work functions:

- Stabilize body links
- Maintain posture
- Allow movement
- Cushion body tissues
- Fit users
- Fit workspace

Note that the analysis decomposes the purpose (prime function) into a number of secondary functions. These are expressed at a sufficiently abstract level so as not to constrain the design team's creativity. There are many ways of implementing the above chair functions because they say nothing about physical design—whether the chair should have legs, wheels, a backrest, etc.

Concept Optimization

Several activities can constitute the concept optimization stage. Economic analysis of cost-effectiveness and cost benefit can be carried out. Users can be interviewed to obtain more information about the current system—what sorts of problems they experience, which aspects of the present system are viewed favorably and should not be changed, which tasks are the most time-consuming for little gain, and what skills are required to operate the current system.

Allocation of Functions

At this stage, decisions have to be made about whether particular functions can best be performed by humans or by machines or both. In practice, functions usually fall into one of several categories:

1 Those that *must* be carried out by machines (because it is impossible or unacceptable for humans to do them)

2 Those that *must* be carried out by humans (because no adequate machines are available or machine execution of function is not appropriate)

3 Those that *might* be carried out by either humans or machines or both

The third category is of the most interest to ergonomics. Early attempts to formalize decision making about function allocation used comparisons of the relative strengths and weaknesses of humans and machines (Fitts, 1951). For example, machines can generate larger forces and do not require rest periods, but they are less flexible than humans and cannot cope with unforeseen (unprogrammed) events. This approach has now gone out of fashion for several reasons. Apart from its sterility, it is questionable that allocation-of-function issues can be decided **independently of the system context** in which they are made. In an extreme example, automatic systems can be used to fly airplanes, but most passengers would prefer it if there were a pilot in the cockpit as well! Computer programs have been written which can make preliminary diagnoses of diseases and decide on an approach to treatment. Many patients, however, would prefer to interact with a sympathetic clinician.

System functions are normally interdependent and should be allocated in a top-down rather than a bottom-up fashion—according to some higher-level design goals, for example, the desire for greater human-machine symbiosis (based on the view that the relative abilities of humans and machines are complementary) or the desire to utilize the existing skills of the workforce to the fullest extent possible. Function allocation decisions may change over time, depending on the economic climate and availability of labor, and across cultures. In the transfer of technology from one society to another, different ways of utilizing the workforce may be needed, lest "anthropotechnical islands" are created which do not fit in with the surrounding cultural milieu.

Beyond Allocation of Function

Allocation of function on the basis of simple comparisons between humans and machines has been superseded by a more integrated approach in which the goal is to select a logically, physically, and cognitively consistent set of tasks for both machines and people (Singleton, 1972). This approach includes functions' being carried out by both people and machines. Flexible allocation of function is now possible. The human is still very much part of the loop but can be off-loaded by the system at times of high workload and retake the initiative as workload decreases. Adaptive aiding (Rouse, 1988) is an example of this.

Piecemeal allocation of several human functions to one operator can create potentially stressful jobs if there is a conflict between the requirements and responsibilities of the different functions. **Role ambiguity** is an example of a type of job stress in which the worker is unclear about where her or his real responsibilities lie and what she or he is really rewarded for by the organization (Sell, 1980).

The debate about allocation of function has shifted away from the decision of whether to allocate a function to a human or a machine or both. Some of the most critical decisions today concern what type of technology or human resources to employ and extend beyond ergonomics to include, for example, environmental and economic considerations about the choice of different energy sources and the social implications of using skilled, less skilled, or older workers.

Detailed Design

After the function allocation decisions have been made on one basis or another, detailed design can take place. The engineering and human components are designed in parallel to save development time. Traditional ergonomics acts as a link between these activities. The parallel design of the human and engineering activities is a characteristic of the systems approach. It is not necessary for the system to be complete before training and job design can take place. The Apollo space program is an example where all training and development took place off-line using simulators and prototypes.

Many ergonomic aids can be used during design and development. Simple cardboard mock-ups can be used to solve problems of anthropometric fit and to evaluate control panels and user interfaces. A number of more sophisticated development tools have also been developed, such as JACK and MANNEQUIN. In the design of interactive systems, the user interface can be developed early in the design and independently of the applications software and implemented to simulate the tasks which will be carried out on the finished system.

It is during the detailed design phase that jobs are designed, either implicitly or explicitly. Singleton has described how workers have sometimes been thought of as the glue which holds the system together. The operator's function is simply to do all those things that the machines can't do for themselves. Occupational psychologists, however, have proposed alternative ways of designing jobs. **Job-centered** approaches attempted to optimize the content of jobs so that they would be perceived as satisfying to do. **Person-centered** approaches were based on the idea that people would be motivated to work well if the work satisfied their needs. Thus, it was thought that successful job design required some consideration of people's needs. The sociotechnical systems approach attempted to optimize the design of a social system of work to support the technical system. These approaches are described in more detail below.

System Integration and Evaluation

Evaluation of subsystems can take place as soon as they are developed, or prototypes and mock-ups can be used to simulate the real system. At the implementation phase, system characteristics can be evaluated according to previously determined specifications. Specifications may be "off the shelf" (e.g., specifications for lighting and temperature, gleaned from the appropriate literature or from published standards), or they may have been developed in-house. The integration and the implementation phase is often characterized by "teething troubles," which are temporary phenomena not characteristic of the final design. It is for this reason that in activities such as facilities design, a delay is introduced between implementation and final evaluation. This delay enables teething troubles to be overcome so that the true pros and cons of the facility can be more easily identified. For example, postoccupancy evaluations of new buildings are usually carried out after 1 year of occupation.

WORK ORGANIZATION, MOTIVATION, AND JOB SATISFACTION

Much of the emphasis of ergonomics is on human capabilities and limitations and the need to design systems which take these into account. Although this is clearly **necessary,** it can be debated whether it is **sufficient** to optimize the design and performance of human-machine systems. Moores (1972) commented that one of the similarities between ergonomics and work study was the tendency to ignore the behavior patterns of workers either as individuals or as groups. Definitions of ergonomics of the **fitting-the-job-to-the-worker** variety were condemned by Moores for being too mechanistic, conjuring up images of matching at the level of an animal rather than a human being. Sell (1980) has taken a similar line and suggests that a major problem with ergonomics is that it neglects to take account of a person's social and psychological needs. In a paraphrase of de Greene (1980), ergonomic solutions can remove barriers to effective productivity but cannot **compel** people to be productive (or to find work a satisfying experience). This is because productivity depends, to some extent at least, on the level of motivation and motivation depends on higher-level factors such as work organization.

Motivation

Maslow (1956) produced one of the first theories of motivation to be applied to the investigation of work situations. According to Maslow, motivation is the driving force which directs behavior—**the mechanism for the reduction of needs.** If one is to understand what motivates people to work, **a consideration of their needs is essential.** Maslow postulated five classes of needs which he arranged hierarchically. Low-level needs included the basic physiological requirements—food, oxygen, etc. Next came the need for security and avoidance of danger, followed by social needs—companionship, social interaction, etc. A fourth need was for self-respect and the esteem of others, including social feedback about performance. At the top of the hierarchy was the need for self-actualization, which can be seen as an ideal—how we would like to see ourselves in some ideal world. A key point of Maslow's theory is that higher-level needs do not come into play until lower-level needs are satisfied (e.g., there is no point in giving a hungry person a bunch of flowers). Although the theory can be questioned, it provides the ergonomist with a useful framework for understanding employees' conative responses to job design.

Job Enlargement

Several early investigators made the observation that many industrial jobs consisted solely of simple, repetitive tasks and considered this situation to be unsatisfactory for a variety of reasons—these jobs wasted human potential, led to boredom, and caused physical and mental fatigue. Attempts were made to make industrial jobs more interesting. **Job enlargement** is an example of such an approach. According to Barnes (1963):

The job should be redesigned so that it consists of a complete piece of work with at least one unit or component being completed by the operator. Thus, in placing a latch on a vacuum cleaner, one operator could assemble the latch, assemble the plate over the two bolts, start the two nuts and tighten them with a power wrench. If one person were to assemble the latch, another assemble the plate and start the nuts and the third person tighten them, each would be doing but a fragment of the operation. It is not the length of the cycle that is important but the fact that the task must be a complete unit, having a beginning, a duration and an ending.

Job enlargement entails a horizontal extension of the operator's tasks—it gives the operator more tasks at the same level of system operation. Although it increases the physical variety of the job (and may reduce the risk of fatigue or injury), it is only superficially more meaningful and varied. Herzberg (1966) (whose theory of job satisfaction is described below) criticized job enlargement on the grounds that it only increases the number of meaningless tasks given to the operator.

Job Enrichment

Job enrichment programs attempt to extend a worker's duties vertically as well as horizontally. One approach is to make workers responsible for quality control of their own work. This approach does increase the meaningfulness and responsibility of a job since this function is normally carried out by a more senior person—the supervisor. In this approach, a part of the management function is delegated to the worker on the factory floor. However, enrichment of an unskilled worker's job can result in a corresponding impoverishment of the supervisor's job and introduces other complex issues—for example, whether the increased responsibility should be rewarded with a corresponding increase in pay.

Job-centered approaches to improving labor effectiveness have been criticized for a number of reasons—for example, that they are piecemeal and that it is not always clear whether they will fulfill any of an employee's higher-order needs.

Job Satisfaction

According to Maslow's theory, employment in industrialized countries caters well for people's lower-order needs. However, if one wants to motivate people at work, higher-level needs such as self-esteem and the opportunity for self-actualization should be considered. Herzberg (1966) developed a theory of job satisfaction which stressed the importance of those aspects of a job which determine whether or not it is intrinsically satisfying. Herzberg regarded satisfaction and dissatisfaction as separate dimensions rather than opposite ends of a continuum. Job satisfaction was determined by a set of factors known as motivators which were intrinsic to the work itself. Dissatisfaction, on the other hand, was said to be influenced by hygiene factors extrinsic to the work (Table 16-3).

The term "hygiene" was chosen using an analogy with medicine. Unhygienic conditions can cause ill health, but increasing the level of hygiene beyond a certain point cannot lead to positive increases in health. Because satisfaction and dissatisfaction are re-

TABLE 16-3   HYGIENE FACTORS AND MOTIVATORS IN JOB DESIGN

Hygiene Factor	Motivator
Company policy and administration	Achievement
Supervision	The work itself
Relationship with supervisor	Responsibility
Work conditions	Advancement
Salary	Personal growth
Security	

garded as separate factors in Herzberg's theory, the absence of job satisfaction is no satisfaction and the absence of dissatisfaction is no dissatisfaction. Poor working conditions (an absence of hygiene factors) will cause dissatisfaction, but a good working environment cannot make people like the work that they do. According to Herzberg, liking a job depends on factors which are intrinsic to the job itself.

Herzberg proposed a number of principles for job design based on his theory of job satisfaction. Improvements in labor effectiveness and employee motivation can be achieved, it was argued, by changes in the organization of work through better design of the jobs that people do (Table 16-4).

Criticisms of Herzberg's Theory

Herzberg's theory can be criticized in several ways—for example, because of the lack of evidence that satisfaction and dissatisfaction really are separate and because of its job-centered nature. The job itself is seen as the source of satisfaction or lack of satisfaction, without considering what the job means to the worker, his or her expectations,

TABLE 16-4   HOW TO IMPROVE JOB DESIGN*

Do	Don't
1 Encourage responsibility by removing low-level supervision while retaining accountability for performance.	1 Add additional tasks to an already fragmented one in an attempt to make a job more meaningful.
2 Encourage achievement and coherence in the job by giving the worker a natural unit of work (e.g., testing a complete car rather than just the brakes).	2 Rotate low-level tasks in the hope of enriching a job.
3 Provide recognition by giving direct feedback to the worker about his or her performance.	3 Remove the most difficult parts of a job since these are often the most interesting.
4 Introduce new and more difficult tasks to encourage learning.	
5 Assign individuals to specialized tasks to enable them to build up unique skills.	

* From Herzberg (1966).

or other social and cultural factors. It can be argued, for example, that some people do not *expect* to have interesting jobs and will not be dissatisfied if they don't get them. Others may be more interested in their families or in hobbies or other outside interests and see work as merely a means of achieving other, more important, goals. Satisfaction would then be determined by the extent to which the job enables them to achieve these external goals, rather than by the design of the job itself.

Both Maslow's and Herzberg's theories were developed in a **western industrial milieu** where much emphasis is placed on the individual and his or her achievements. Whether these theories are applicable in societies where people customarily view themselves and their work as part of a community endeavor, rather than an exercise in personal growth, is debatable (see Meshkati and Robertson, 1986, for further discussion of this point). **Finally, for many people all over the world, having any job at all is a major goal in life and a salary that raises them above a basic subsistence level is likely to be a major source of satisfaction.**

Relevance of Herzberg's to Theories to Ergonomics

The brief review of Herzberg's ideas is of relevance because of the challenge they represent to ergonomics. According to his theory, the application of ergonomics would *not* be expected to result in increases in job satisfaction because most ergonomics is concerned almost exclusively with hygiene factors rather than motivators.

SOCIOTECHNICAL SYSTEMS THEORY

Most human-machine systems are sociotechnical systems. There is a **technical subsystem** (which integrates the machines themselves and the interconnections between them, the infrastructure and power supply, etc.) and a **social subsystem** (which integrates the activities of the people involved in operating and maintaining the machines and the relationships between them). **Work organization** is the link between machines and the social organization of the individuals who operate them. The optimal utilization of technology depends on an appropriate system of work organization, which itself determines the social organization of the workforce and the relations and interdependencies between individuals. For example, in a classic production-line system of manufacture, workers form a homogeneous group in terms of status and skill. The relationship between them is one of linear dependency in one direction (the direction of the production line). Control is exercised by supervisors, foremen, and managers. This can be contrasted with a patient care team in a hospital, where the members of the team differ in status, and individuals perform different functions, have different expertise, and are mutually interdependent. Individuals can use their own initiative about matters falling within their own area of expertise and can contribute to decision making at the group level.

Sociotechnical systems theory was developed by members of the Tavistock Institute of Human Relations in the years following the Second World War. Cherns (1976) has summarized the design principles derived from sociotechnical systems theory (Table 16-5). Further discussion of these approaches to the design of human-machine systems is beyond the scope of the present text and the reader is referred to the more advanced literature cited in the bibliography.

TABLE 16-5   PRINCIPLES OF SOCIOTECHNICAL JOB DESIGN*

1 *Compatibility.* The way that sociotechnical design is carried out must be compatible with the changes that are made. For example, if one of the reasons for change is to make the most of the cognitive skills of the workforce, these skills must be utilized in the process of change itself.

2 *Minimal critical specification.* In the design of new systems, no more should be specified than is absolutely essential. This leaves individuals with freedom to determine the precise details of how a task is to be carried out. It also leaves open options for change and improvement and enhances adaptability to unanticipated events. Strictly specified rules and procedures can inhibit an organization's ability to adapt.

3 *The sociotechnical criterion.* This principle states that variance in a system (the occurrence of unspecified and unprogrammed events) should be reduced by controlling it as close to its source as possible. Much of the system of supervision, inspection, and maintenance in industry is an attempt to reduce variance from a distance—to correct the consequences of variance rather than control it. If operators carry out their own inspection, the size of the variance control feedback loop is reduced.

4 *The multifunction principle.* This principle is aimed at increasing the adaptability of the organization by allowing workers to fulfill more than one role, but designing jobs so as to reduce the interchangeability of people.

5 *Boundary location.* In all organizations, boundaries between different departments have to be drawn up. Such divisions are usually made on the basis of function, technology, territory, or time. For example, engineering products often pass through several different departments, such as milling or grinding shops, in the manufacturing process. Of the total time taken to manufacture a product, only a portion is spent with the item in contact with machines. The rest is taken up by transport, storage, etc. An alternative is the group or unit production method where each department makes a complete product. This approach has been tried in car assembly, for example.

6 *Information flow.* This principle states that information should go where it is needed in the first instance rather than via senior management to subordinates. The tendency for information to filter down from above can have disadvantages as well as be inefficient. Managers may become preoccupied with matters for which their subordinates should be responsible. An information system designed according to sociotechnical principles should direct information efficiently to those parts of the organization where it is needed and acted upon. It should also support two-way communication.

7 *Support congruence.* Administrative and management systems should be designed to reinforce those behaviors that the organization wishes to encourage. If an aim is to encourage employees to take a more responsible attitude toward their jobs, then supervision and payment systems should be designed to be congruent with this goal.

8 *Design and human values.* This principle emphasizes that systems must be designed to provide high-quality jobs. This is a difficult principle to apply—individuals may respond differently to changes in job design and have different needs.

9 *Incompletion.* Design never ends.

* From Cherns (1976).

TRENDS IN WORKSYSTEMS DESIGN

As worksystems become increasingly complex and costly, there is a need to specify, in advance, more and more aspects of new systems to reduce uncertainty, and to design out errors and give clear briefs to designers, contractors, and equipment suppliers. Ergonomics makes an important contribution to this activity and can provide many specifications for the design of the human-machine interface. This is particularly the case for the design of the physical workspace and environment but less so for the psychological and social aspects of work.

Technological advances frequently force changes in the organization of work and the design of jobs. Old skills may become obsolete and the demand for new skills may change the composition of the workforce itself. Interest in the human and social effects of technology is not new. As Adam Smith pointed out over 200 years ago in "The Wealth of Nations":

> The understandings of the greatest part of men are necessarily formed by their ordinary employment. The man whose life is spent in performing a few simple operations every day, which will always produce the same result, has no occasion to exert his understanding. Since there are no obstacles occurring, he need not think of how to overcome them. He quite normally then forgets how to use his intellect and becomes as stupid and ignorant as a human creature can ever become.

One of the results of the industrial revolution was a new approach to the design of jobs. The new factories were run on a highly centralized basis to ensure tight controls on production and maximize the returns on investment in technology. One of the results of this approach was the emergence of extremely specialized, repetitive tasks, which had a number of advantages:

1 Management gained flexibility in allocating operators to easily learned tasks.

2 Fewer skilled workers were needed. Skill shortages were avoided, and training costs and wages could be more easily contained.

3 The introduction of paced work enabled production schedules to be more rigorously quantified. Better predictions of output could be made.

4 If everyone worked at the same pace, the result was always a finished product.

Many of the early approaches to job design, described above, were a reaction to the apparent shortcomings of existing industrial jobs which were criticized for being monotonous, for lacking challenge, and for denying employees any control over their work.

Attempts to Humanize Work

In the 1960s, 1970s, and 1980s a number of large-scale programs were initiated in several European countries. These were variously titled—"Job Reform" in Sweden, "The Humanisation of Work" in the then West Germany, and the "Quality of Working Life" movement in the United Kingdom. These programs were motivated by a variety of factors. For example, successive generations of school-leavers in the countries in question had increasingly higher levels of education and higher expectations of work. The programs attempted to provide higher-quality jobs through changes in work organization.

Some general characteristics of a good (psychologically rewarding) job are given in Table 16-6.

In Sweden, the Volvo motor car company (which was suffering from high absenteeism and labor turnover in the 1960s) tried to find new ways of assembling cars in an attempt to have a more stable, motivated, and productive workforce. Conventional production-line methods were replaced by unit production. Teams of workers manned electric assembly wagons which moved around the assembly area, stopping at centralized stores to collect the various components.

It is frequently difficult to disentangle the effects of such programs on either productivity or the psychological rewards of work because many different factors are involved. Unit production changes the social relations between people, but it also eliminates pacing (where the rate of work is set by machines) and lengthens cycle times. Both of these latter factors are known to influence job satisfaction. The British "Quality of Working Life" program (Tynan, 1980) attempted to combine new approaches to job design with technological change under the premise that, since technological change forces job redesign anyway, the opportunities presented by new technologies could best be realized by optimizing the work organization and design of jobs. Management, trade unions, and workers were to be included in a participatory approach toward job design.

Sell (1980) has discussed some of the factors which influence the success of the participatory approach. Some examples are:

1 The need for top-level support
2 The need to involve all affected groups in the design of a new worksystem
3 The need to replace negotiation with a problem-solving approach

Success of Work Humanization Programs

The modern workplace, according to this view, is characterized by flexibility and individual discretion over work elements. Traditional fragmented, repetitive tasks and rigid organizational hierarchies have been replaced with more decentralized systems. The extent to which this thinking has really penetrated organizations and replaced traditional styles of management is a question open to empirical investigation. It has been investi-

---

TABLE 16-6   SOME CHARACTERISTICS OF A "GOOD" JOB*

1 Optimum work loading
2 Minimum role ambiguities (each worker has a clearly defined set of responsibilities which he or she can relate to the objectives of the organization)
3 Minimum role conflict (there is a clear relationship between an individual's role, how the individual is rewarded, and the expectations of the individual and his or her co-workers)
4 Support from colleagues or management
5 Social content—interaction with others is built in to the design of the job and the execution of basic work functions

---

* From Sell (1980).

gated by Boreham (1992), who carried out an international comparative study of the organization of work and the amount of discretion available to employees in a variety of organizations in the United States, Australia, Britain, Canada, Germany, Japan, and Sweden. The following employee groups were sampled: managers, professionals, clerks, skilled workers, and semiskilled and unskilled workers. More than half of the employees sampled reported **negligible freedom to put their ideas into practice or to introduce new tasks. Autonomy was found to be a property of higher-status individuals. Lower-status individuals were almost totally excluded from participation in decisions about production in their organizations.** Citing such findings as evidence for the "myth of post-fordist management," Boreham concluded that there was scant evidence to suggest that truly participative organizational practices had been implemented in the countries studied.

It may be noted that these work humanization programs were conceived of in the 1950s and 1960s—a time of expanding economies and low unemployment. Since that time, unemployment has increased in many of the above countries as they sank into recession. Under these circumstances both organizations and individuals may have had to focus on more pressing, low-level needs. As the world economy expands, these issues may again become prominent.

LEGISLATIVE TRENDS: STANDARDS, GUIDELINES, AND INTERVENTION PROGRAMS

Ever since the nineteenth century, government bodies in the developed nations have attempted, for social as well as economic reasons, to influence the way industry runs itself. The Factory Acts were introduced in nineteenth-century Britain to ameliorate the worst excesses of the Victorian industries—long hours, appalling working conditions, and the exploitation of child labor. This trend has continued and industries now have to comply with regulations which limit worker exposure to the health-threatening aspects of their jobs. Ergonomic factors are currently being included in new safety and health legislation in the United States and in Europe.

OSHA: Ergonomic Safety and Health Management Rules

In the United States, OSHA, which monitors about 6 million workplaces with about 93 million workers, aims to implement standards on workplace ergonomics in response to the rapid increase in claims for workers' compensation for incapacitating musculoskeletal injury. According to Bureau of Labor statistics, the number of disorders associated with repeated trauma has more than tripled since 1984. The increase is apparently due to changes in production processes and technologies, increased overtime and piecework, and a lack of integration of ergonomics into production processes. Many tasks are now more specialized, assembly line speeds are higher, and cycles times are lower (Department of Labor, 1992).

The OSHA ergonomic safety and health management standard will cover general, maritime, construction, and agriculture industries and will aim to "prevent, eliminate and reduce ergonomic hazards in the workplace." The ergonomic hazards and disorders covered by the regulation appear mainly to be of a musculoskeletal nature (other ergonomic hazards such as noise are already regulated; less tangible hazards relating to mental stress, error, information processing, etc. are probably outside the scope or are covered by other agencies).

In addition to general guidelines, OSHA intends to implement specific guidelines for high-risk industries such as construction, where there is a high risk of falls, and meatpacking, where there is a high level of musculoskeletal stress on workers. The general thrust of this program is to reduce the prevalence of avoidable ill health by providing employers with specific ergonomic guidelines with which they must comply when deciding how to run their businesses or design their facilities. Employers failing to comply with the guidelines are liable for prosecution. In order to implement its programs, OSHA intends to increase the number of its inspectors by 15 to 20 percent. In 1992, OSHA carried out over 42,000 inspections covering more than 2 million employees and imposed over $116 million worth of fines (National Safety Council, 1993). These data refer to a wide range of industries and violations of all OSHA health and safety rules. As of 1993, the maximum fine which can be imposed on an employer is $70,000 for wilful and repeat violations. Serious violations can lead to a fine of $7000.

The requirement for good working conditions is not a new one. The Occupational Safety and Health Act of 1970 requires all employers to "provide their employees with a workplace free from recognized serious hazards" irrespective of whether these hazards are covered by specific standards. If poor ergonomics constitutes a hazard, then employers are required to act. What the new rules do is to specify what constitutes an "ergonomic hazard" and what action to take to remove the hazard. In this sense, the rules assist employers in complying with already existing legislation.

OSHA: Ergonomics Program Management Guidelines for Meatpacking Plants

The meatpacking guidelines (OSHA 3123, 1993) illustrate the approach of OHSA to the design and implementation of an ergonomic intervention program for health and safety. Meatpacking has received particular attention because it is a high-risk industry for occupational musculoskeletal disorders and because OSHA was approached in 1988 by a meatpacking company to develop an ergonomic standard (Department of Labor, 1992).

The meatpacking guide takes a top-down approach to the management of ergonomic hazards, beginning with the need to get the support of top management and to involve employees (Table 16-7). Health and safety is given a high priority and key people are made responsible for appropriate parts of the program and given the resources and authority to intervene. All employees know what is expected of them, who they are accountable to, and what resources they have to carry out their responsibilities. Deadlines for the attainment of objectives are set and a written plan is made available to all concerned. Employees are involved in the monitoring of injuries, the investigation of problem areas, and the recommendation of solutions. A committee is established to receive information on problem areas and is empowered to act on this information. The program is reviewed regularly and a report is issued detailing current progress and any revisions of goals.

TABLE 16-7   KEY FEATURES OF OSHA'S PROGRAM GUIDELINES FOR MEATPACKING PLANTS*

Management commitment and employee involvement

Commitment by top management	Written program
Eliminating hazards given top priority	Outlines corporate goals and plans
Health and safety as important as production	Advocated at the highest level
Delegation of accountability and responsibility	Communicated to all employees
Provision of resources and authority	Objectives set out with deadlines

Employee involvement	Review and evaluation
Complaint-suggestion procedure implemented to provide feedback	Analysis of trends in injury and illness
Prompt reporting of disorders encouraged	Employee surveys
Health and safety committee established	Pre- and postchange comparisons
Ergonomic intervention teams established	Record of job improvements implemented

Program elements

Worksite analysis	Hazard prevention and control
Examination of injury and illness records	Engineering controls
Identification of problem areas and trends	Workspaces
Visits to worksites to identify risks	Work methods
Follow-up when operations change	Tools
	Work practices
	Work techniques
	Equipment maintenance
	Employee conditioning
	Monitoring
	Personal protective equipment
	Proper fit
	Protection against cold
	Administrative controls
	Increase cycle time
	Provide rest pauses
	Use more people
	Rotate jobs
	Train standby workers
	Use preventive maintenance

Medical management	Training and education
Injury and illness recording	General training
Early recognition and reporting	Job-specific training
Systematic evaluation and referral	Training of supervisors
Conservative treatment	Training for managers
Rehabilitation	Training for engineers and maintenance personnel
Adequate staffing and facilities	

* From OSHA 3123 (1993).

Worksite analysis begins with an analysis of accident and sickness records, to evaluate the severity of any existing ergonomic problems and prioritize areas for intervention. Worksite evaluations utilizing checklists to characterize tasks in terms of force, posture, and repetition are then carried out. Worksite analysis is carried out on an ongoing basis and repeated at least annually. Hazard prevention utilizes a combination of engineering, work practice, and work organization controls as described in previous chapters. Adequate resources for the medical management of patients must be provided. The emphasis is placed on prevention, early diagnosis, conservative treatment (rather than surgical intervention, if possible), and on-the-job rehabilitation. In the case of back injury, it has been shown (Lusted, 1993) that the probability of an injured employee's returning to work is greater the earlier after the injury the patient is referred for rehabilitation. Early referral for rehabilitation is more likely under a well-integrated system of medical management. Finally, employers must provide workers with education and training by qualified persons. The training must be geared toward the needs of particular groups of workers and must include a general component (about ergonomic disorders, their cause, recognition, and prevention) as well as job-specific training (e.g., workers who use knives would receive training on the correct way to hold and cut with knives and proper knife care to maintain sharpness).

As can be seen, the OHSA approach to ergonomics operates at all key levels of the worksystem and includes technical, medical, cognitive, environmental, and organizational components.

The European Community Directives

There has also been renewed interest in health and safety in Europe. Apart from the obvious desire to reduce the amount of disability in society, the European Community also has the objective of harmonizing conditions of trade between its member states. It is therefore concerned to ensure that each country has similar rules concerning health and safety so that manufacturers in one country do not gain an unfair advantage over their competitors in other community countries (Sell, 1993). Six sets of regulations have recently been introduced in the form of directives—each member state is required to introduce its own legislation covering the topics described under the following headings:

- Management of health and safety
- Manual handling of loads
- Work with display screen equipment (i.e., VDTs)
- Use of personal protective equipment
- Use of work equipment
- Workplace design (health, safety, and welfare)

The regulations concerning management of health and safety are particularly interesting in that the onus of responsibility for health and safety is placed on the employer, who is required to do the following:

- Carry out risk assessments
- Make any arrangements necessary for health and safety

- Provide health surveillance
- Develop proper procedures for dealing with emergencies
- Provide employees with specific information about health and safety at their workplace
- Assess employee capacity for work
- Provide training
- Use competent people

Legislation of this nature has enormous ramifications for the way work is carried out and for employers' and employees' expectations. For example, under the directives, employees using VDTs on a regular basis as part of their jobs become entitled to eye and eyesight tests and corrective appliances. This is a radical departure from the traditional expectation of employers that it is the responsibility of the employee to ensure his or her own capability for work in terms of physical characteristics (HSC consultative document, 1992).

Under the directives, risk assessments will have to be carried out at each VDT workstation by competent people. In Britain in 1991, there were 6.75 million VDT workstations, whereas the membership of the Ergonomics Society is less than 1000. If only society members can carry out these assessments, that would leave about 7000 assessments per member! Not surprisingly, the trend has been for qualified people to act in a training capacity, enabling companies to develop in-house expertise in specific areas.

Professional Issues: Competence and Performance in the Practice of Ergonomics

These legislative trends have introduced many issues about the competence of individuals to act as ergonomists. These issues are currently under debate in many countries. In the United States, for example, Swezey (1993) has argued that "certification defines the field," that potential users of ergonomics services need to have some way of recognizing the professional competence of consultants. Swezey suggests that certification is advantageous because it names people who are judged by their peers as meeting a set of predetermined standards. Bain et al. (1994) have registered doubts about certification on the basis that ergonomics is a very diverse field, so it is impossible to have a standard test to cover all areas of ergonomics. Even if someone is certified, that fact doesn't ensure that the individual is competent, and even if the individual is competent now, he or she may not be competent in several years time when the technology has advanced.

Clearly, some form of certification is needed if ergonomics is to become a mature discipline and its practitioners are to have status similar to that of established professionals such as architects, accountants, or medical doctors. The approach of the Ergonomics Society has been to publish a register of professional ergonomists which society members may apply to join. Applicants have to give evidence, which is assessed by the society, of their professional achievements as well as their academic qualifications. Registered members' names appear on the register together with the areas of ergonomics which the society is satisfied they are competent to consult in.

Litigation

The last few decades have seen an increase in the number of cases in which injured employees seek compensation from their employers in the courts. Frequently, the injuries experienced may have been caused or exacerbated by ergonomic factors in the workplace and an ergonomist is summoned as an expert witness by lawyers representing the plaintiff or the defendant. A medical doctor usually carries out an examination to determine the nature and severity of the injury and to give an expert opinion as to the cause. In those cases where there is a real injury and there are grounds for believing it to be work-related, an ergonomist is called in to visit the workplace, analyze the task requirements, and determine the following:

- Whether there were risk factors in the work environment
- Whether the employer could reasonably have been expected to know about the risk factors at the time
- Whether, if there were risk factors and the employer should have known about them, the employer was negligent in not taking all reasonable steps to safeguard the employee's health

This report then forms part of the evidence given in court for consideration. Further discussion of legal aspects (in the context of carpal tunnel syndrome) can be found in Owen (1994).

TRENDS IN PRODUCT DESIGN

Another consequence of the industrial revolution was the emergence of mass markets and mass-produced products, which was accompanied by a distancing of designers, manufacturers, and users. Designers could no longer wait for feedback from users about the adequacy of their products because of the delay and the expense—any design faults would be incorporated into many thousands of items before feedback was received from the market. **Mass production requires a much more explicit and formal model of the user. Building this model is a task which continues to challenge ergonomics to this day.**

Consumer Products

Rutkowski (1982) has discussed some of the factors which determine whether new technologies will develop into successful consumer products. Cars, radios, televisions, etc., all started out as ideas in the minds of experts, and technical knowledge was required to operate the first models. Enthusiastic amateurs became interested in them and a specialist market was established. Then, market penetration occurred on a large scale but only after a significant change had occurred in the knowledge required to use the product. For example, early cars were designed in different ways depending on the inclination of the designer. Some had steering wheels, others had tillers, some had a roof and a windshield, others didn't, and the number of wheels varied from three upward. Many subsidiary tasks had to be mastered and carried out by the user in order to use the car (for example, priming the oil pump, cranking the engine with a cranking handle, and manually advancing the ignition as the engine gained speed).

The enthusiastic amateur stage of market penetration was superseded only after a period of **architectural stabilization** in the design of the product—when designers ceased experimenting with different design formulas and standardized on a number of fundamental characteristics which gave the car a clear product identity. **Design effort shifted from experimenting with different architectures in favor of optimizing the basic features of a standard design.** This is still going on today. The basic architecture of a modern car has changed little in the last 50 years (although the design of individual components has improved). Today's driver would have few problems driving a car made 50 years ago.

Architectural stabilization of the car was followed by improvements in reliability and a corresponding increase in its intrinsic usefulness over and above the curiosity value of being a "horseless carriage." The potential usefulness of the car became real when the car itself became more easily usable. This happened when the knowledge needed to use a car became **procedural** rather than **technical** and when the number of task-irrelevant procedures was reduced. A major trend in product design occurs in the knowledge required to use the product. First, the learning of technical facts is substituted by the learning of **procedures.** Second, the number of learned procedures required to use the product is **minimized.**

For example, the objective of using a car is to reach a destination with little effort and in a reasonably short space of time. The tasks required to achieve this goal are navigation, steering, changing gears, braking, etc. In the evolution of the car, a major trend has been to minimize all task-irrelevant procedures and knowledge (for example, by introducing automatic ignition control, automatic chokes and gearboxes, power steering, and power-assisted braking). Another trend has been to minimize the number of parts which require regular servicing and to lengthen the service intervals (for example, by using sealed or self-lubricating bearings).

The trend in the design of the car has been to make the interface between the user and the task transparent by making the technical aspects invisible. Architectural stabilization has consequences which have an impact on a product and enlarge its potential market. The enlarged market brings about increased production which drives down unit production costs, thus enlarging the market still further. Generalized from the above, some of the principal stages in the development of successful products can be identified as follows:

1 *Standardization.* Products in the same class become interchangeable. After learning how to use one product, one can use all similar products with minimal additional learning—the same cognitive model applies to all products in the same class.

2 *Increased reliability.* The intrinsic usefulness of the product is enhanced and the user is shielded from technical considerations.

3 *Increased usability.* Task-irrelevant procedures and the need for technical knowledge are minimized.

This trend can be seen to have occurred in a wide variety of products (cars, television sets, radios, and washing machines, for example). It may be occurring today in the design of software for personal computers where there is increasing usage of graphic interfaces. Items of software such as spreadsheets or word processors are beginning to be seen as products rather than as computer programs.

Product development proceeds in a stepwise, iterative manner in which a technological advance leads to a new product concept which is then developed (in the manner described above). The company which first develops the product has a **competitive window** in time, during which it can dominate the market before competitive products emerge. James Watt, when he was selling his first steam engines, experienced the same problems as the Apple personal computer company when it introduced its new "desktop" metaphor for its personal computers. These problems include foreshortening of the competitive window as a result of industrial espionage, infringement of patent rights, etc., and affect all innovators of new products. After the original burst of innovation, rival products enter the market and the competitive window closes. Manufacturers are then left to compete for market share on the basis of styling, comfort, ease of use, and similar features—all of which lie at the heart of ergonomics.

SUMMARY AND PROSPECTS FOR THE FUTURE

The field of ergonomics contains much scientific information which can be used to characterize the design issues which affect the health and productive capacity of workers. However, the classic ergonomic approach is really concerned with the integration of human subsystems with machine subsystems—with respect to efficiency and occupational health. Little work has been done to integrate the biological and psychological parts of ergonomics in such a way that they can be related directly to higher-level aspects of system functioning. This has made it difficult to predict the outcomes and benefits of ergonomics to organizations. One of the major challenges for ergonomics is to develop more integrated theories of human-machine systems, to explicate the relationships between lower-level aspects of system functioning and higher-level system outcomes, and to develop more formal procedures for designing the human parts of work systems.

Every manufactured product has, either implicitly or explicitly, a model of the user built into it. The design of a tennis racket is based on assumptions about the anthropometry of the user; the design of some interactive software makes assumptions about the knowledge, skills, and preferences of the user. One clear challenge is to carry out fundamental research to develop improved models of the user to be incorporated into the design of products.

More advanced literature in which these issues are addressed can be found in the Bibliography.

ESSAYS AND EXERCISES

1 "Ergonomics is just common sense." Discuss.

2 What factors determine the implementation of ergonomics in practice?

3 Discuss possible future technological developments of your choice and the ergonomic problems which will have to be solved before they can be successfully implemented as systems or products.

FOOTNOTE: Future of Human-Machine Systems

It will alter the whole system of our internal communications . . . substituting an agency whose ultimate effects can scarcely be anticipated.

Report in the *Liverpool Courier* of 1829 which described the success of George and Robert Stephenson's "Rocket" in a race against its rival locomotives

The world which we have made as a result of the level of thinking that we have done thus far creates problems that we cannot solve at the same level as they were created.

Albert Einstein

Many scientists have pointed out the inadequacy of purely analytical approaches which attempt to reduce complex sociotechnical problems to simple problems of basic science. New technologies frequently have unanticipated effects on the environments into which they are implemented.

The steam engine had many effects on society over and above the increases in production of mines and factories. One of the most profound was on the size of cities. Since ancient times, the size of cities was limited by the time taken to transport fresh produce into the city by wagon or boat. In the seventeenth century, the population of London numbered no more than a few hundred thousand. Suburbs, such as Hampstead, which are now close to the center were then market gardens. An increase in the linear dimensions of a city squares its area and increases the population and need for food accordingly. More food must be brought in from farther and farther afield as size increases. The speed and power of the steam locomotive removed this limiting factor and made possible the growth of cities to modern dimensions. It also brought with it an increased need for control and coordination.

The intellectual climate of the industrial revolution was characterized by confidence in the ability of technology to solve problems and by highly analytic or compartmentalized thinking. Human and material resources were regarded as inexhaustible. This enabled designers and decision makers to judge the appropriateness of their actions by their immediate effects and thus learn rapidly from experience. However, one of the major drawbacks of tightly compartmentalized thinking is that it is insensitive to secondary or indirect effects, as witness the abuses of labor and of the environment which took place in western Europe in the nineteenth century and which are still taking place in other parts of the world today.

Compartmentalized problem solvers confine their efforts to the boundaries of the local system under study and check their opinions against the simplified view of reality so obtained. In the postindustrial era it has become clear that, since the world and its resources constitute a finite, but complex, system, technology cannot be conceptualized in a static way (like a problem in geometry or algebra). Worksystems are interconnected with a wider industrial and environmental system—although compartmentalized thinking simplifies reality, it doesn't make it go away.

Systems thinking is both compartmentalized and holistic—it sees the world as a complex network of interconnected systems. It draws attention to possible secondary or side effects of an action and fosters a systematic attempt to see things in context.

At the level of simple worksystems, ergonomics comes to the fore—inefficiency, errors, and accidents are seen as system problems rather than human or technological ones. The way a control is designed may, deliberately or not, make it possible for an operator to make certain types of errors and not others. For this reason many modern components are often designed using the principle that they can be assembled in only one way, the correct way, to design out errors in assembly. Alternatively, systems can be designed to be immune to certain types of errors that humans are known to make (such as reversal errors). At the level of complex worksystems, a more thorough approach to the analysis of system inputs and outputs is carried out. The environmental and social impacts of new developments are often assessed together with the financial and technical viability. Narrow views of productivity and profit may be included under wider corporate goals such as "organizational effectiveness" where the organization is likened to an organism in an environment and the emphasis is on establishing a mutually satisfactory relationship between the two.

Modern economies face many new challenges. Manufacturing productivity has increased rapidly over the last 20 years and many previously industrial countries have developed to the point that the majority of their citizens today work in offices. This trend explains the upsurge in interest in office ergonomics over the last 10 years. Newly emergent east Asian nations now manufacture many of the products previously made in the United States and Europe, as well as many entirely new ones. Together with better management and order-of-magnitude increases in quality control, these trends have resulted in structural blue-collar unemployment in many countries. To date, similar increases in white-collar productivity have not been realized on a large scale and white-collar unemployment is less of a problem than blue-collar unemployment. The trend toward industrialization and postindustrialization has brought with it changes in expectations and values. Modern families often have more materialistic parents who desire fewer, but better-educated, children. In industrially nondeveloping countries, populations continue to increase and struggle to survive at low levels of subsistence. The gap between the most and least technologically and economically advanced nations has never been wider. It is hoped that this introductory text has illustrated some of the ways ergonomics can contribute to the design of worksystems at both technological and social extremes.

BIBLIOGRAPHY

The following books are recommended for readers in need of more advanced study and for those who require more information on specific topics or areas. The list is intended to be representative rather than exhaustive.

Textbooks

2 Spaces

General

Huchingson RD. 1981. *New Horizons for Human Factors in Design.* McGraw-Hill, New York. Similar in content to Sanders and McCormick, but the material has been arranged in terms of application areas, such as aerospace systems, transportation systems, etc.

Salvendy G (ed.). 1987. *Handbook of Human Factors.* Wiley, New York. A comprehensive tome of over 1000 pages. A recommended reference on ergonomics and human factors.

Sanders MS, McCormick EJ. 1993. *Human Factors in Engineering and Design,* 7th edition. McGraw-Hill, New York. The seventh edition of this standard reference work in human factors and ergonomics. Recommended source material which will complement the present volume.

Singleton WT. 1982. *The Body at Work: Biological Ergonomics.* Cambridge University Press, Cambridge, England. Contains excellent reviews of physiology, anthropometry, noise, vibration, climate, and lighting in ergonomics. Will prepare the reader for advanced work.

Wilson JR, Corlett EN (eds.). 1990. *Evaluation of Human Work: A Practical Ergonomics Methodology.* Taylor and Francis, London. An edited volume which covers a wide area. The emphasis is on methods, but many of the basic principles are also presented. Can be used as a companion to the present text.

Anatomy and Biomechanics

Chaffin D, Andersson GBJ. 1984. *Occupational Biomechanics.* Wiley, New York. The standard reference on occupational biomechanics. Industrial and mechanical engineers with a specialized interest in ergonomics should read this book.

Kapandji IA. 1974. *The Physiology of the Joints.* Churchill Livingstone, New York. A beautifully illustrated primer on the anatomy of the musculoskeketal system. All the main joints are illustrated in detail. Many otherwise obscure anatomical facts are related to everyday life. Recommended sourcebook for the ergonomist with a background in social sciences or engineering who needs a better understanding of anatomy.

Anthropometry, Posture, and Design

Clark TS, Corlett EN. 1984. *The Ergonomics of Workspaces and Machines: A Design Manual.* Taylor and Francis, London. A great deal of information in a compact design manual. Will be of use to industrial designers and engineers.

Panero J, Zelnik M. 1979. *Human Dimension and Interior Space.* Whitney Library of Design, New York. Recommended for the designer, architect, or industrial engineer. The book is extremely well illustrated and the design applications are easy to understand.

Pheasant ST. 1986. *Bodyspace.* Taylor and Francis, London. A standard reference on anthropometry and its applications in ergonomics. Suitable for a wide audience.

Zacharkow D. 1986. *Posture: Sitting, Standing, Chair Design and Exercise.* Charles C Thomas, Springfield, IL. A well-illustrated and very interesting book on posture. Contains much fundamental information otherwise difficult to find. The ergonomics and clinical aspects will interest a wide readership.

Ergonomics and Occupational Health

Grandjean E. 1984. *Ergonomics and Health in Modern Offices.* Taylor and Francis, London. A varied collection of conference papers compiled by Grandjean. Includes papers on ophthalmic and toxicological aspects not normally dealt with in the ergonomics literature.

Pearce B. 1986. *Health Hazards of VDU's.* Wiley, New York. A useful collection of papers on a topic which remains controversial. Particularly useful for industrial medical personnel, although it will appeal to a wide audience.

Pheasant ST. 1991. *Ergonomics, Work and Health.* Aspen, Gaithersburg, MD. Recommended for the student or professional with a particular interest in occupational health. It is written in a style that will appeal to the nonspecialist.

Putz-Anderson V. 1988. *Cumulative Trauma Disorder: A Manual for Musculoskeletal Diseases of the Upper Limbs.* Taylor and Francis, London. A practical guide to work-related musculoskeletal disorders of the upper body, their causes, cures, and how to recognize them.

Physiology

Astrand PO, Rodahl K. 1977. *Textbook of Work Physiology.* McGraw-Hill, New York. A standard reference on work physiology. Still relevant although somewhat dated.

Guyton AC. 1991. *Textbook of Medical Physiology,* 8th edition. Saunders, Philadelphia. A very advanced reference book of nearly 1000 pages. Recommended for the specialist.

Noakes T. 1992. *The Lore of Running.* Oxford University Press, London. A useful reference on exercise physiology. Explains many of the key physiological concepts used in ergonomics but in the context of running and personal fitness. Easy to read, with many interesting anecdotes. User-friendly physiology for the nonphysiologist.

Human Performance and Systems Psychology

Bailey RW. 1982. *Human Performance Engineering: A Guide for System Designers*. Prentice-Hall, Englewood Cliffs, NJ. A comprehensive introduction to modern engineering psychology. A useful reference but computer section needs to be supplemented with a more recent text.

Shackel B, Richardson S. 1991. *Human Factors for Informatics Usability*. Cambridge University Press, Cambridge, England. The emerging concept of usability is put into an organizational-technological perspective. An important collection of papers by experts in the field. Useful for practitioners and academics alike.

Singleton WT. 1989. *The Mind at Work*. Cambridge University Press, Cambridge, England. The long-awaited psychological version of *The Body at Work*. Also serves to update his earlier *Man-Machine Systems*. Good introduction to systems psychology.

Wickens CD. 1992. *Engineering Psychology and Human Performance*. Charles E Merrill, Columbus, OH. Theories and models of human information processing are described in depth with examples of their practical application. Advanced reading for the specialist or graduate student.

Cognitive Ergonomics

Carroll JM. 1991. *Designing Interaction: Psychology at the Human-Computer Interface*. Cambridge University Press, Cambridge, England. A useful collection of papers on a rapidly growing topic. It will appeal to HCI practitioners and researchers.

Howard D. 1983. *Cognitive Psychology: Memory, Language and Thought*. Macmillan, New York. An extremely readable and well-illustrated textbook on cognitive psychology. It is one of the few such books which is readily digestible by the nonpsychologist. Recommended background reading for nonpsychologists interested in cognitive ergonomics who need a better understanding of cognitive psychology.

Long JB, Whitfield A (eds.). 1989. *Cognitive Ergonomics and Human-Computer Interaction*. Cambridge University Press, Cambridge, England. A collection of papers on user-interface design projects written from a cognitive perspective. The book attempts to illustrate some of the ways of, in the editors' words, "achieving compatibility between the users' representations and the computer's representations." One of the few books explicitly concerned with cognitive ergonomics.

Information Technology: Hardware Design

Cakir A, Hart DJ, Stewart TFM. 1980. *Visual Display Terminals*. Wiley, New York. This text is rather dated but contains a great deal of useful technical information which is still relevant.

Grandjean E. 1986. *Ergonomics in Computerized Offices*. Taylor and Francis, London. A very useful summary of the literature by one of the world's experts in this field. Recommended, in particular, for office designers, architects, facilities managers, suppliers of office equipment, and students.

Occupational and Environmental Psychology

Sundstrom ED. 1986. *Workplaces*. Cambridge University Press, Cambridge, England. Essentially this is a book about the psychology of the workplace. It contains a great deal of useful information about people's psychological responses to the physical environment at work. Recommended, in particular, for facilities managers, designers, architects, students, and ergonomics practitioners.

Warr P (ed.). 1987. *Psychology at Work*, 3rd edition. Penguin Books, New York. A useful collection of essays on a wide range of topics relevant to ergonomics.

Journals

Ergonomics. Published by Taylor and Francis Ltd. The journal of the Ergonomics Society and the first academic ergonomics journal to be published. Articles on all aspects of ergonomics are published, but the journal tends to contain more papers on physical than on psychological ergonomics.

Human Factors. The journal of the Human Factors and Ergonomics Society. Articles on all aspects of ergonomics and human factors are published, but the journal usually contains more papers on psychological and human performance issues than on physical ergonomics.

Ergonomics in Design. The magazine of human factors applications published by the Human Factors and Ergonomics Society. A very readable journal written and produced in magazine format.

International Journal of Industrial Ergonomics. Published by Elsevier in collaboration with the International Ergonomics Association. Publishes high-quality papers dealing with practical industrial ergonomics issues.

Applied Ergonomics. Published by Butterworths. Established as an alternative publication outlet for ergonomics practitioners rather than academics. Publishes practical papers on a wide variety of topics.

Behaviour and Information Technology. Published by Taylor and Francis. Offered as an alternative journal by the Ergonomics Society to members more interested in interface design and interactive systems than in physical ergonomics. Published papers are written in the academic style, however.

REFERENCES

Abeysekera JDA, Shahnavaz H. 1989. Body size variability between people in developed and developing countries and its impact on the use of imported goods. *International Journal of Industrial Ergonomics,* 4:139–149.

Adams MA, Hutton WC. 1980. The effect of posture on the role of the apophyseal joints in resisting intervertebral compressive forces. *Journal of Bone and Joint Surgery,* 2B:358–362.

Adams MA, Hutton WC. 1983. The effect of posture on the fluid content of the intervertebral discs. *Spine,* 8:665–671.

Adams MA, Hutton WC. 1985. The effect of posture on the lumbar spine. *Journal of Bone and Joint Surgery,* 67B:625–629.

Aghazedeh F, Mital A. 1987. Injuries due to handtools. *Applied Ergonomics,* 18:273–278.

Agnew J, Suruda AJ. 1993. Age and fatal work-related falls. *Human Factors,* 35:731–736.

Agou S, Raskin V, Salvendy G. 1993. Combining natural language with direct manipulation: The conceptual framework for a hybrid human-computer interface. *Behaviour and Information Technology,* 12:48–53.

Akerstedt T, Torsvall L. 1981. Shift work: Shift-dependent well-being and individual differences. Ergonomics, 24:265–273.

Allen AA, Fischer GJ. 1978. Ambient temperature effects on paired associate learning. *Ergonomics,* 21:95–101.

Allport DA, Antonis B, Reynolds P. 1972. On the division of attention: A disproof of the single channel hypothesis. *Quarterly Journal of Experimental Psychology,* 24:225–235.

Alnutt MF, Allen JR. 1973. The effects of core temperature elevation and thermal sensation on performance. *Ergonomics,* 16:189–196.

Altman I. 1975. *The Environment and Personal Space.* Wadsworth, Belmont, CA.

American Industrial Hygiene Association. 1975. *Industrial Noise Manual.*

Anderson JR. 1983. *The Architecture of Cognition.* Harvard University Press, Cambridge, MA.

Anderson LE. 1993. Biological effects of extremely low-frequency electromagnetic fields: In vivo studies. *Journal of the American Industrial Hygiene Association,* 54:186–196.

Anderson R. 1992. The back pain of bus drivers. *Spine,* 17:1481–1489.

Andersson GBJ. 1986. Loads on the spine during sitting. In *The Ergonomics of Working Postures,* edited by EN Corlett, J Wilson, and I Mannenica. Taylor and Francis, London.

Andre DA, Wickens CD. 1992. Compatibility and consistency in display-control systems: Implications for aircraft decision aid design. *Human Factors,* 34:639–653.

ANSI/HFS 100-1988. *American National Standard for Human Factors Engineering of Visual Display Terminal Workstations.* Human Factors Society Inc.

Argyle M. 1975. *Bodily Communication.* Methuen, London.

Armstrong TJ, Buckle PD, Fine LJ, Hagberg M, Jonsson B, Kilbom A, Kuorinka I, Silverstein BA, Sjogaard G, Viikari-Juntura ERA. 1993. A conceptual model for work-related neck and upper limb musculoskeletal disorders. *Scandinavian Journal of Work Environment and Health,* 19:73–84.

Aschoff J. 1965. Circadian rhythms in man. *Science,* 148:1427–1432.

Aschoff J. 1978. Features of circadian rhythms relevant for the design of shift schedules. *Ergonomics,* 21:739–754.

Ashby P. 1979. *Ergonomics Handbook 1: Body Size and Strength.* SA Design Institute, Pretoria.

Aspden RM. 1987. Intra-abdominal pressure and its role in spinal mechanics. *Clinical Biomechanics,* 2:168–174.

Aspden RM. 1989. The spine as an arch: A new mathematical model. *Spine,* 14:266–274.

Astrand PO, Rodahl K. 1977. *Textbook of Work Physiology.* McGraw-Hill, New York.

Ayoub MM. 1973. Work place design and posture. *Human Factors,* 15:265–268.

Ayoub MM. 1982. Control of manual lifting hazards: II. Job redesign. *Journal of Occupational Medicine,* 24:668–676.

Azer NZ, McNall PE, Leung HC. 1972. Effects of heat stress on performance. *Ergonomics,* 15:681–691.

Baddeley A. 1982. *Your Memory: A User's Guide.* Macmillan, New York.

Baggett. 1983. Four principles of designing instructions. *IEEE Transactions on Professional Communication,* PC-26:99–105.

Bailey RW. 1982. *Human Performance Engineering: A Guide for System Designers,* Prentice-Hall, Englewood Cliffs, NJ.

Bailey RW. 1983. *Human Error in Computer Systems.* Prentice-Hall, Englewood Cliffs, NJ.

Bain L, Burns D, Davies S, Keahy B, King M, Marics M, Messamer P, Rudman C, Somers P, Tsoi KC, Webb J. 1994. Still talking about certification. *Ergonomics in Design,* January: 38.

Bainbridge L. 1974. Analysis of verbal protocols from a process control task. In *The Human Operator in Process Control,* edited by E Edwards and FP Lees. Taylor and Francis, London.

Bainbridge L. 1979. Verbal reports as evidence of the process operator's knowledge. *International Journal of Man-Machine Studies,* 11:411–436.

Bainbridge L. 1982. Ironies of automation: Analysis, design and evaluation of man-machine systems. *Proceedings of IFAC/IFIP/IFORS/IEA Conference.* Baden-Baden, Germany, 151–157.

Ballal MA, Fentem PH, MacDonald IA, Sukkar MY, Patrick JM. 1982. Physical condition in young adult Sudanese: A field-study using a self-paced walking test. *Ergonomics,* 25:1185–1196.

Ballantine M. 1980. Conversing with computers—The dream and the controversy. *Ergonomics,* 23:935–945.

Ballantine M. 1983. Well, how do children learn population stereotypes? In *Proceedings of the 1983 Ergonomics Conference,* edited by K Coombs. Taylor and Francis, London.

Bandyopadhyay B, Chattopadhyay H. 1981. Assessment of physical fitness of sedentary and physically active male college students using a modified Harvard step test. *Ergonomics,* 24:15–20.

Banerji D. 1988. The knowledge of human nutrition and the peoples of the world. *World Reviews of Nutrition and Dietetics,* 57:1–20.

Barber P. 1988. *Applied Cognitive Psychology.* Methuen, London.

Barnard PJ, Hammond NV, McLean A, Morton J. 1982. Learning and remembering interactive commands in a text-editing task. *Behaviour and Information Technology,* 1:347–358.

Barnard PJ, Hammond NV, Morton J, Long JB. 1981. Consistency and compatibility in human-computer dialogue. *International Journal of Man-Machine Studies,* 15:87–134.

Barnes RM. 1963. *Motion and Time Study: Design and Measurement of Work,* 5th edition. Wiley, New York.

Barnett BJ, Wickens CD. 1988. Display proximity in multicue information integration: The benefits of boxes. *Human Factors,* 30:15–24.

Bartholomae RC, Kovac JG. 1979. *USBM Develops a Low-noise Percussion Drill.* Coal Age Conference and Expo V, Louisville, KY, 23–25 October.

Barlett FC. 1932. *Remembering.* Cambridge University Press, Cambridge, England.

Barton NJ, Hooper G, Noble J, Steel WM. 1992. Occupational causes of disorders in the upper limb. *British Medical Journal,* 304:309–311.

Battie MC, Bigos SJ, Fisher LD, Spengler DM, Hansson TH, Nachemson AL, Wortley MD. 1990. The role of spinal flexibility in back pain complaints within industry. *Spine,* 15:768–773.

Bauk DA. No date. *Saude, Seguranca E Terminais de Video.* Departmento de Saude Ocupacional, IBM, Rio de Janeiro.

Beard DV, Walker II JQ. 1990. Navigational techniques to improve the display of large two-dimensional spaces. *Behaviour and Information Technology,* 9:451–466.

Becker WGE, Ellis H, Goldsmith R, Kaye AM. 1983. Heart rates of surgeons in theatre. *Ergonomics,* 26:803–807.

Begault DR. 1993. Head-up auditory displays for traffic collision avoidance system advisories: A preliminary investigation. *Human Factors,* 4:707–717.

Beiers JL, 1966. A study of noise sources in pneumatic rockdrills. *Journal of Sound and Vibration,* 3:166–194.

Beiring-Sorensen F. 1984. Physical measurements as risk indicators for low back trouble over a 1-year period. *Spine,* 9:106–119.

Bejjani FJ, Gross CM, Pugh JW. 1984. Model for static lifting: Relationship of loads on the spine and the knee. *Journal of Biomechanics,* 17:281–286.

Bell CR, Crowder MJ, Walters JD. 1971. Durations of safe exposure for men at work in high temperature environments. *Ergonomics,* 14:733–757.

Bemis SV, Leeds JL, Winner EA. 1988. Operator performance as a function of type of display: Conventional versus perspective. *Human Factors,* 30:163–169.

Bendix T, Beiring-Sorensen F. 1983. Posture of the trunk when sitting on forward inclining seats. *Scandinavian Journal of Rehabilitation Medicine,* 15:197–203.

Bendix T, Hagberg M. 1984. Trunk posture and load on the trapezius muscle whilst sitting at sloping desks. *Ergonomics,* 27:873–882.

Biedermann HJ, Shanks GL, Forrest WJ, Inglis J. 1991. Power spectrum analysis of electromyographic activity. *Spine,* 16:1179–1184.

Black JB, Sebrechts MM. 1981. Facilitating human-computer interaction. *Applied Psycholinguistics,* 2:146–175.

Blaker JW. 1980. Towards an adaptive model of the human eye. *Journal of the Optical Society of America,* 70:220–223.

Blattner MM, Sumikawa DA, Greenberg RM. 1989. Earcons and icons: Their structure and common design principles. *Human Computer Interaction,* 4:11–44.

Blumberg BS, Sokoloff L. 1961. Coalescence of caudal vertebrae in the giant dinosaur Diplodocus. *Arthritis and Rheumatism,* 4:592–601.

Bone J. 1993. Chisel out a hand-tool ergonomics plan. *Safety and Health,* May:64–68.

Borg GAV. 1982. Psychophysical bases of perceived exertion. *Medicine and Science in Sports and Exercise,* 14:377–381.

Boreham P. 1992. The myth of post-fordist management: Work organisation and employee discretion in seven countries. *Employee Relations,* 14:13–24.

Boshuizen HC, Bongers PM, Hulshof CT. 1992. Self-reported back pain in fork lift truck and freight container tractor drivers exposed to whole body vibration. *Spine,* 17:59–65.

Bough B, Thakore J, Davies M, Dowling F. 1990. Degeneration of the lumbar facet joints. *Journal of Bone and Joint Surgery,* 72-B:275–276.

Bower GH. 1972. Mental imagery and associative learning. In *Cognition in Learning and Memory,* edited by LW Gregg. Wiley, New York.

Bowler JN. 1982. Welbeck colliery noise level survey. *The Mining Engineer,* September:159–162.

Boyce PR. 1982. Vision, light and colour. In *The Body at Work,* edited by WT Singleton. Cambridge University Press, Cambridge, England.

Bracken TD. 1993. Exposure assessment for power frequency electric and magnetic fields. *Journal of the American Industrial Hygiene Association,* 54:165–177.

Brand JL, Judd KW. 1993. Angle of hard copy and text editing performance. *Human Factors,* 35:57–69.

Brandt LPA, Nielsen CV. 1990. Congenital malformations among children of women working with visual display terminals. *Scandinavian Journal of Work Environment and Health,* 16:329–333.

Bransford JD, Johnson MK. 1973. Consideration of some problems of comprehension. In *Visual Information Processing,* edited by W Chase. Academic Press, New York.

Branton P. 1969. Behaviour, body mechanics and discomfort. *Ergonomics,* 12:316–327.

Bridger RS. 1988. Postural adaptations to a sloping chair and worksurface. *Human Factors,* 30:237–247.

Bridger RS, Jaros GG. 1986. Ergonomics: Ergosystems and occupational health. *Continuing Medical Education,* 4:39–51.

Bridger RS, Long JB. 1984. Some cognitive aspects of interface design in a two variable optimisation task. *International Journal of Man-Machine Studies,* 21:521–539.

Bridger RS, Orkin D. 1992. Effect of a footrest on standing posture. *Ergonomics SA,* 4:42–48.

Bridger RS, Orkin D, Henneberg M. 1992. A quantitative investigation of lumbar and pelvic postures in standing and sitting: Interrelationships with body position and hip muscle length. *International Journal of Industrial Ergonomics,* 9:235–244.

Bridger RS, Ossey S, Fourie G. 1990. The effect of lumbar traction on stature. *Spine,* 15:522–525.

Broadbent DE. 1958. *Perception and Communication.* Pergamon, New York.

Broadbent DE. 1977. Language and ergonomics. *Applied Ergonomics,* 8:15–18.

Brooks LR. 1968. Spatial and verbal components in the act of recall. *Canadian Journal of Psychology,* 22:349–368.

Brouwer WH, Waternik W, Van Wolfelaar-Rothengatter T. 1991. Divided attention in experienced young and older drivers: Lane tracking and visual analysis in a dynamic driving simulator. *Human Factors,* 33:573–582.

Brown CD, Nolan BM, Faithful DK. 1984. Occupational repetition strain injuries: Guidelines for diagnosis and management. *The Medical Journal of Australia,* March:329–332.

Brown ID. 1966. Subjective and objective comparisons of successful and unsuccessful trainee drivers. *Ergonomics,* 9:50–56.

Brown ID, Poulton EC. 1961. Measuring the spare mental capacity of car drivers by a subsidiary task. *Ergonomics,* 4:35–40.

Brunswic M. 1984. Ergonomics of seat design. *Physiotherapy,* 70:39–43.

BS 7179. 1990. *Ergonomics of Design and Use of Visual Display Terminals (VDT's) in Offices. Part 6: Code of Practice for the Design of VDT Work Environments.* British Standards Institute.

Bull NL. 1988. Studies of dietary habits, food consumption and nutrient intakes of adolescents and young adults. *World Review of Nutrition and Dietetics,* 57:24–67.

Bundy A. 1978. *Computational Models for Problem Solving.* Open University Press, Milton Keynes, U.K.

Burgess R, Neal RJ. 1989. Document holder usage when reading and writing. *Clinical Biomechanics,* 4:151–154.

Burton K. 1991. Measuring flexibility. *Applied Ergonomics,* 22:303–307.

Butler DS. 1991. *Mobilisation of the Nervous System.* Churchill Livingstone, Edinburgh.

Butler DL, Acquino AL, Hissong AA, Scott PA. 1993. Wayfinding by newcomers in a complex building. *Human Factors,* 35:159–173.

Cakir A, Hart DJ, Stewart T. 1980. *Visual Display Terminals.* Wiley, New York.

Carroll JM. 1982. Learning, using and designing filenames and command paradigms. *Behaviour and Information Technology,* 1:327–346.

Cartas O, Nordin M, Frankel VH, Malgady R, Sheikhzadeh A. 1993. Quantification of trunk muscle performance in standing, semi-standing and sitting postures in healthy men. *Spine,* 18:603–609.

Casali JG, Park MY. 1990. Attenuation performance of four hearing protectors under dynamic movement and different user fitting conditions. *Human Factors,* 32:9–25.

Casali SP, Williges BH, Dryden RD. 1990. Effects of recognition accuracy and vocabulary size of a speech recognition system on task performance and user acceptance. *Human Factors,* 32:183–196.

Ceesay SM, Prentice AM, Day KC, Murgatroyd PR, Goldberg GR, Spurr GB. 1989. The use of heart rate in the estimation of energy expenditure: A validation study using indirect whole-body calorimetry. *British Journal of Nutrition,* 61:175–186.

Chaffin DB. 1987. Biomechanical aspects of workplace design. In *Handbook of Human Factors,* edited by G. Salvendy. Wiley, New York.

Chaffin DB, Andersson GBJ. 1984. *Occupational Biomechanics.* Wiley, New York.

Chapanis A. 1990. Short-term memory for numbers. *Human Factors,* 32:123–137.

Chapanis A. 1991. The business case for human factors in informatics. In *Human Factors for Informatics Usability,* edited by B Shackel and S Richardson. Cambridge University Press, Cambridge, England.

Chapanis A. 1994. Hazards associated with three signal words and four colours on warning signs. *Ergonomics,* 37:265–275.

Chapanis A, Budurka WJ. 1990. Specifying human-computer interface requirements. *Behaviour and Information Technology,* 9:479–492.

Chapanis A, Lindenbaum LE. 1959. A reaction time study of four control display linkages. *Human Factors,* 1:1–7.

Chaplin B. 1983. Anti-noise: The Essex breakthrough. *CIM Bulletin,* January:41–47.

Charteris J, Scott PA, Nottrodt JW. 1989. Metabolic and kinematic responses of African women headload carriers under controlled conditions of load and speed. *Ergonomics,* 32:1539–1550.

Charteris J, Wall JC, Nottrodt JW. 1982. Pliocene hominid gait: New interpretations based on available footprint data from Leatoli. *American Journal of Physical Anthropology,* 58:133–144.

Chase WG, Simon HA. 1973. Cognitive psychology. In *Visual Information Processing,* edited by WG Chase. Academic Press, New York.

Chatterjee DS. 1992. Workplace upper limb disorders: A prospective study with intervention. *Occupational Medicine,* 42:129–136.

Chavalitsakulchai P, Shahnavaz H. 1990. *Woman Workers and Technological Change in Industrially Developing Countries from an Ergonomic Perspective.* Center for Ergonomics of Developing Countries, CEDC, Dept. of Human Work Sciences, Lulea University, Sweden.

Cherns A. 1976. The principles of sociotechnical design. *Human Relations,* 29:783–792.

Chomsky N. 1957. *Syntactic Structures.* Mouton, The Hague.

Chong I, McDonough A. 1984. A new angle on sports equipment. *Brief,* 1:66–68.

Christ RE. 1975. Review and analysis of colour coding research for visual displays. *Human Factors,* 17:542–570.

Ciriello VM, Snook SH, Hughes GH. 1993. Further studies of psychophysically determined maximum acceptable weights and forces. *Human Factors,* 35:175–186.

Citron N. 1985. Femoral neck fractures: Are some preventable? *Ergonomics,* 28:993–997.

Clarke TS, Corlett EN. 1984. *The Ergonomics of Workspaces and Machines: A Design Manual.* Taylor and Francis, London.

Cleary. 1993. A review of in vitro studies: Low frequency electromagnetic fields. *Journal of the American Industrial Hygiene Association,* 54:178–185.

Cole S. 1982. Vibration and linear acceleration. In *The Body at Work,* edited by WT Singleton. Cambridge University Press, Cambridge, England.

Colligan MJ, Frock IJ, Tasto DL. 1979. Frequency of sickness absence and worksite clinic visits among nurses as a function of shift. *Applied Ergonomics,* 10:79–85.

Collins AM, Quillian MR. 1972. Experiments on semantic memory and language comprehension. In *Cognition in Learning and Memory,* edited by LW Gregg. Wiley, New York.

Colquhoun WP, Goldman RF. 1972. Vigilance under induced hyperthermia. *Ergonomics,* 15:621–632.

Constable SH, Bishop PA, Nunneley SA, Chen T. 1994. Intermittent microclimate cooling during rest increases work capacity and reduces heat stress. *Ergonomics,* 37:277–285.

Cook TM, Neumann DA. 1987. The effects of load placement on the activity of the low back muscles during load carrying by men and women. *Ergonomics,* 30:1413–1423.

Corlett EN, Eklund JAE. 1986. How does a backrest work? *Applied Ergonomics,* 15:111–114.

Costall A, Still A. 1987. In place of cognitivism. In *Cognitive Psychology in Question,* edited by A Costall and A Still. St. Martin's, New York.

Courtney AJ. 1994. Hong Kong Chinese direction-of-motion stereotypes. *Ergonomics,* 37:417–426.

Cowley KC, Jones DM. 1992. Synthesised or digitised? A guide to the use of computer speech. *Applied Ergonomics,* 23:172–176.

Cox R, Shephard J, Corey P. 1981. Influence of an employee fitness programme upon fitness, productivity and absenteeism. *Ergonomics,* 24:795–806.

Craik K. 1943. *The Nature of Explanation.* Cambridge University, Cambridge.

Craik FIM, Lockhart RS. 1972. Levels of processing: A framework for memory research. *Journal of Verbal Learning and Verbal Behaviour,* 11:71–84.

Croney J. 1980. *Anthropometry for Designers.* Batsford, London.

Crossman ERFW. 1974. Manual control of slow response systems. In *The Human Operator in Process Control,* edited by E Edwards and FP Lees. Taylor and Francis, London.

Culver CC, Viano DC. 1990. Anthropometry of seated women during pregnancy: Defining a fetal region for crash protection research. *Human Factors,* 32:625–636.

Cushman WH, Crist B. 1987. Illumination. In *Handbook of Human Factors,* edited by G. Salvendy. Wiley, New York.

Datta SR, Chatterjee BB, Roy BN. 1983. The energy cost of pulling handcarts ("thela"). *Ergonomics,* 26:461–464.

Datta SR, Ramanathan NL. 1971. Ergonomic comparison of seven modes of carrying loads on the horizontal plane. *Ergonomics,* 14:269–278.

Davies DR, Jones DM. 1982. Hearing and noise. In *The Body at Work,* edited by WT Singleton. Cambridge University Press, Cambridge, England.

Decortis F. 1993. Operator strategies in a dynamic environment in relation to an operator model. *Ergonomics,* 36:1291–1304.

de Greene KB. 1973. *Sociotechnical Systems: Factors in Analysis, Design and Management.* Prentice-Hall, Englewood Cliffs, NJ.

de Greene KB. 1977. Has human factors come of age? *Proceedings of the Human Factors Society 21st Annual Meeting,* 457–461.

de Greene KB. 1980. Major conceptual problems in the systems management of human factors/ergonomics research. *Ergonomics,* 23:3–11.

de Greene KB. 1986. Systems theory, macroergonomics and the design of adaptive organisations. In *Human Factors in Organisational Design and Management: II,* edited by O Brown Jr and HW Hendrick. North Holland, New York.

Department of Labor. 1992. Ergonomic safety and health management: Proposed Rule, 20 CFR Part 1910. *Federal Register,* 57(149):34192–34200.

de Puky P. 1935. Physiological oscillation of the length of the body. *Acta Orthopaedica Scandinavica,* 6:338–347.

de Wall M, van Riel MPJM, Aghina JCFM, Burdorf A, Snijders CJ. 1992. Improving the sitting posture of CAD/CAM workers by increasing VDU monitor height. *Ergonomics,* 35:427–436.

de Wall M, van Riel MPJM, Snijders CJ. 1991. The effect on sitting posture of a desk with a 10-degree inclination for reading and writing. *Ergonomics,* 34:575–584.

Dempster WT. 1955. Anthropometry of body action. *Annals of the New York Academy of Sciences,* 63:559–585.

Diamond J. 1991. Pearl Harbor and the Emperor's physiologists. *Natural History,* 12:5–7.

Diaz E, Goldberg GR, Taylor M, Savage JM, Sellen D, Coward WA, Prentice A. 1989. Effects of dietary supplementation on work performance in Gambian labourers. *Proceedings of the Nutrition Society,* 49(44-A):45.

Diehl GM. 1973. *Machinery Acoustics.* Wiley, New York.

Dimberg L. 1987. The prevalence and causation of tennis elbow (lateral humeral epicondylitis) in a population of workers in an engineering industry. *Ergonomics,* 30:573–580.

Dixon P. 1991. Learning to operate complex devices: Effects of conceptual and operational similarity. *Human Factors,* 33:103–120.

DonTigny RL. 1985. Function and pathomechanics of the sacro-iliac joint. *Physical Therapy,* 65:35–44.

Dowell J, Long J. 1989. Towards a conception for an engineering discipline of human factors. *Ergonomics,* 32:1513–1535.

Dray SM. 1988. From tier to pier: Organisational adaptation to new computing architectures. In *Designing a Better World, Proceedings of the 10th International Congress of the IEA.*

Dreyfus HL, Dreyfus SE. 1987. The mistaken psychological assumptions underlying belief in expert systems. In *Cognitive Psychology in Question,* edited by A Costall and A Still. St. Martin's, New York.

Drury CG, Addison JL. 1973. An industrial study of the effects of feedback and fault density on inspection performance. *Ergonomics,* 16:159–169.

Drury CG, Francher M. 1985. Evaluation of a forward sloping chair. *Applied Ergonomics,* 16:41–47.

Ducharme RE. 1977. Women workers rate "male tools" inadequate. *Human Factors Society Bulletin,* 20:1–2.

Duncan PW, Studenski S, Chandler J, Bloomfield R, LaPointe LK. 1990. Electromyographic analysis of postural adjustments in two methods of balance testing. *Physical Therapy,* 70:36–44.

Durrett J, Trezona J. 1982. How to use colour displays effectively. *Byte,* April:50–53.

Dwyer T, Raftery AE. 1991. Industrial accidents are caused by the social relations of work: A sociological theory of industrial accidents. *Applied Ergonomics,* 22:17–178.

Eason KD. 1986. Job design and VDU operation. In *Health Hazards of VDT's?* edited by B Pearce. Wiley, New York.

Easterby RS. 1970. The perception of symbols for machine displays. *Ergonomics,* 13:149–158.

Edholm OG. 1967. *The Biology of Work.* World University Library, London.

Edwards W. 1987. Decision making. In *Handbook of Human Factors,* edited by G. Salvendy. Wiley, New York.

Ekdahl C, Jarnlo GB, Andersson SI. 1989. Standing balance in healthy subjects. *Scandinavian Journal of Rehabilitation Medicine,* 21:187–195.

Ekholm J, Schuldt K, Harms-Ringdahl K, Arborelius UP, Nemeth G. 1985. Effects of different sitting postures on the level of neck and shoulder muscular activity during work movements. *Biomechanics XA,* 29–33.

Eklund JAE, Corlett EN. 1984. Shrinkage as a measure of the effect of load on the spine. *Spine,* 9:189–194.

Eklund JAE, Frievalds A. 1993. Hand tools for the 1990s. *Applied Ergonomics,* 24:146–147.

Eklund JAE, Liew M. 1991. Evaluation of seating: The influence of hip and knee angles on spinal posture. *International Journal of Industrial Ergonomics,* 8:67–73.

English W. 1994. The validation of slipmeters. In *Contemporary Ergonomics,* edited by SA Robertson. Taylor and Francis, London.

Fahrni WH, Trueman GE. 1965. Comparative radiological study of the spines of a primitive population with North Americans and northern Europeans. *Journal of Bone and Joint Surgery*, 47B:552–555.

Farfan HF. 1978. The biomechanical advantage of lordosis and hip extension for upright activity. *Spine*, 3:336–342.

Fernstrom E, Ericson MO, Malker H. 1994. Electromyographic activity during typewriter and keyboard use. *Ergonomics,* 37:477–484.

Fillenbaum S. 1973. *Syntactic Factors in Memory?* Mouton, The Hague.

Filley RD. 1982. Opening the door to communication through graphics. *IEEE Transactions on Professional Communication*, PC-25:91–94.

Fisher GH. 1973. Current levels of noise in an urban environment. *Applied Ergonomics*, 4:211–218.

Fisher J, Levin N. 1989. Display control compatibility in the design of domestic cookers for the South African population. *Ergonomics SA,* 1:29–41.

Fisher J, Olivier JA. 1992. Display-control stereotypes amongst South African children in urban and rural settings. *Ergonomics SA*, 4:20–29.

Fisk JW, Baigent ML. 1981. Hamstring tightness and Scheuermann's disease. *American Journal of Physical Medicine*, 60:122–125.

Fitts PM. 1951. Engineering psychology and equipment design. In *Handbook of Experimental Psychology*, edited by SS Stevens. Wiley, New York.

Flinn MW. 1966. *The Origins of the Industrial Revolution*. Longmans, London.

Flowers KA. 1978. The predictive control of behaviour: Appropriate and inappropriate actions beyond the input in a tracking task. *Ergonomics*, 21:109–122.

Floyd WF, Roberts MA. 1958. *Anatomical, Physiological and Anthropometric Considerations in the Design of Office Chairs and Tables*. British Standards Institution, BS 3044.

Floyd WF, Silver PHS. 1955. The function of the erectores spinae muscles in certain movements and postures in man. *Journal of Physiology*, 129:184–203.

Folkard S. 1987. Circadian rhythms and hours of work. In *Psychology at Work*, Penguin Books, New York.

Folkard S. 1992. Is there a 'best' compromise shift system? *Ergonomics*, 35:1453–1463.

Forrester-Brown MF. 1930. Improvement of posture. *The Lancet*, July.

Foss CL. 1989. Tools for reading and improving hypertext. *Inf Processing Manage*, 25:407–418.

Fowler CA, Turvey MT. 1978. *Skill Acquisition: An Event Approach with Special Reference to Searching for the Optimum of a Function of Several Variables*. Haskins Laboratories Status. Report on Speech Research, SR-53, Vol. 1.

Fox JG. 1971. Background music and industrial efficiency: A review. *Applied Ergonomics*, 2:70–73.

Fox RH, Bradbury PA, Hampton IF. 1967. Time judgement and body temperature. *Journal of Experimental Psychology*, 75:88–98.

Frantz JP, Rhoades TP. 1993. A task-analytic approach to the temporal and spatial placement of product warnings. *Human Factors*, 35:719–730.

Frey JK, Tecklin JS. 1986. Comparison of lumbar curves when sitting on the Westnofa Balans Multichair, sitting on a conventional chair, and standing. *Physical Therapy*, 66:1365–1369.

Friedmann K. 1988. The effect of adding symbols to written warning labels on user behaviour and recall. *Human Factors*, 30:507–515.

Frievalds A, Kim YJ. 1990. Blade size and weight effects in shovel design. *Applied Ergonomics*, 21:39–42.

Fulton JP, Cobb S, Lorrena P, Leone L, Forman E. 1980. Electrical wiring configurations and childhood leukemia in Rhode Island. *American Journal of Epidemiology*, 111:292–296.

Gallagher S, Hamrick CA. 1991. The kyphotic lumbar spine: Issues in the analysis of the stresses in stooped lifting. *International Journal of Industrial Ergonomics*, 8:33–47.

Garg A, Owen B. 1992. Reducing back stress to nursing personnel: An ergonomic intervention in a nursing home. *Ergonomics*, 35:1353–1375.

Genaidy AM, Karwowski W. 1993. The effects of neutral posture deviations on perceived joint discomfort ratings in sitting and standing postures. *Ergonomics,* 36:785–792.

Genaidy AM, Waly SM, Khalil TM, Hidalgo J. 1993. Spinal compression tolerance limits for the design of manual material handling operations in the workplace. *Ergonomics*, 36:415–434.

Gilad I, Kirschenbaum A. 1988. Rates of back pain incidence associated wth job attitudes and worker characteristics. *International Journal of Industrial Ergonomics,* 5:267–272.

Goldhaber MK, Polen MR, Hiatt RA. 1988. The risk of miscarriage and birth defects among women who use visual display terminals during pregnancy. *American Journal of Industrial Medicine,* 13:695–706.

Goemans B. 1992. *Design and Evaluation of a Remote Control System for Use in Paedo-audiometry.* MSc thesis, University of Cape Town, Cape Town.

Gomez LM, Lochbaum CC. 1984. People can retrieve more objects with enriched keyword vocabularies. But is there a performance cost? In *Interact '84, First IFIP Conference on Human Computer Interaction,* Vol. 1:1429–1433.

Goodman TJ, Spence R. 1981. The effect of computer system response time variability on interactive graphical problem solving. *IEEE Transactions on Systems, Man and Cybernetics,* SMC-11:207–217.

Gordon AM, Julian FJ, Huxley AF. 1966. The variation in isometric tension with sarcomere length in vertebrate muscle fibres. *Journal of Physiology,* 184:170–192.

Gracovetsky O, Farfan HF. 1986. The optimum spine. *Spine,* 11:543–572.

Grahame R, Jenkins JM. 1972. Joint hypermobility: Asset or liability? *Annals of Rheumatic Diseases,* 31:109–111.

Grandjean E. 1973. *Ergonomics of the Home.* Taylor and Francis, London.

Grandjean E. 1980. *Fitting the Task to the Man: An Ergonomic Approach.* Taylor and Francis, London.

Grandjean E. 1987. *Ergonomics in Computerised Offices.* Taylor and Francis, London.

Green RA, Briggs CA. 1989. Effect of overuse injury and the importance of training on the use of adjustable workstations by keyboard operators. *Journal of Occupational Medicine,* 31:557–562.

Greene J. 1975. *Thinking and Language.* Methuen, New York.

Greene J, Cromer R. 1978. *Language as a Cognitive Process.* Open University Press, Bristol, PA.

Greene SL, Gould JD, Boies SJ, Rasamny M, Meluson A. 1992. Entry and selection-based methods of human-computer interaction. *Human Factors,* 34:97–113.

Gregg HD, Corlett EN. 1988. Developments in the design and evaluation of industrial seating. In *Designing a Better World: Proceedings of the 10th Triennial Conference of the International Ergonomics Association,* edited by AS Adams, RR Hall, BJ McPhee, and MS Oxenburg. Ergonomics Society of Australia Inc.

Grieco A. 1986. Sitting posture: An old problem and a new one. *Ergonomics,* 29:345–372.

Grieve D, Pheasant S. 1982. Biomechanics. In *The Body at Work,* edited by WT Singleton. Cambridge University Press, Cambridge, England.

Griffin MJ. 1992. Causes of motion sickness. In *Contemporary Ergonomics 1992,* edited by EJ Lovesy. Taylor and Francis, London.

Gunnarson E, Soderberg I. 1983. Eye strain resulting from VDT work at the Swedish Telecommunications Administration. *Applied Ergonomics,* 14:61–69.

Guo L, Genaidy A, Warm J, Karwowski W, Hidalgo J. 1992. Effects of job-simulated flexibility and strength/flexibility training protocols on maintenance employees engaged in manual handling operations. *Ergonomics,* 35:1103–1117.

Guyton AC. 1991. *Textbook of Medical Physiology,* 8th edition. Saunders, Philadelphia.

Hagberg M. 1987. *Occupational Shoulder and Neck Disorders.* The Swedish Work Environment Fund, Stockholm.

Handbook of Noise and Vibration Control. 1979. Trade and Technical Press, Morden, Surrey, U.K.

Hanna JM, Brown DA. 1979. Human heat tolerance: Biological and cultural adaptations. *Yearbook of Physical Anthropology*, 22:163–186.

Hansson JE, Eklund L, Kihlberg S, Ostergren CE. 1987. Vibration in car repair work. *Applied Ergonomics,* 18:57–63.

Hardman AE, Hudson A, Jones PRM, Norgan NG. 1989. Brisk walking and plasma high cholesterol concentration in previously sedentary women. *British Medical Journal*, 299:1204–1205.

Harrison J. 1993. Hand transmitted vibration. *British Medical Journal*, South African Edition, 2:55.

Harrison MH, Brown GA, Belyavin AJ. 1982. The "Oxylog": An evaluation. *Ergonomics*, 25:809–820.

Harvey SM. 1984. Electric-field exposure of persons using video display units. *Bioelectromagnetics,* 5:1–12.

Haslegrave CM. 1990. Auditory environment and noise assessment. In *Evaluation of Human Work: A Practical Ergonomics Methodology*, edited by JR Wilson and EN Corlett, Taylor and Francis, London.

Hayne CR. 1981. Manual transport of loads by women. *Physiotherapy,* 67:226–231.

Hays WL. 1973. *Statistics for the Social Sciences*, 2d edition. Holt, Rinehart and Winston, New York.

Hawkins LH. 1984. The possible benefits of negative ion generators. In *Health Hazards of VDU's*, edited by B Pierce. Wiley, New York.

Health and Safety Commission. 1991. *Manual Handling of Loads: Proposals for Regulation and Guidance*. Health and Safety Executive, London.

Health and Safety Commission. 1992. *Work with Display Screen Equipment: Proposals for Regulation and Guidance*. Health and Safety Executive, 1 London.

Hellebrandt FA. 1938. Standing as a geotropic reflex. *American Journal of Physiology*, 121:471–474.

Henneberg M. 1992. Continuing human evolution: Bodies, brains and the role of variability. *Transactions of the Royal South African Society, Part 1*:159–182.

Herbert R. 1988. The passive mechanical properties of muscle and their adaptations to altered patterns of use. *The Australian Journal of Physiotherapy*, 34:141–149.

Herzberg F. 1966. *Work and the Nature of Man*. World Publishing, Celina, TN.

Hewes GW. 1957. The anthropology of posture. *Scientific American*, 196:122–132.

Hill ID. 1972. Wouldn't it be nice if we could write computer programs in ordinary English—or would it? *The Computer Bulletin*, July.

Hillman DJ. 1985. Artificial intelligence. *Human Factors*, 27:21–31.

Hitchcock D. 1992. A convert's primer to socio-tech. *Journal for Quality and Participation*, June.

Hofstadter DR. 1985. *Metamagical Themas: Questing for the Essence of Mind and Pattern*. Viking, New York.

Holdo J. 1958. Energy consumed by rockdrill noise. *The Mining Magazine*, August.

Hollands JG, Spence I. 1992. Judgments of change and proportion in graphical perception. *Human Factors*, 34:313–334.

Hollnagel E, Woods DD. 1983. Cognitive systems engineering: New wine in new bottles. *International Journal of Man-Machine Studies*, 18:583–600.

Howard D. 1983. *Cognitive Psychology: Memory, Language and Thought*. Macmillan, New York.

Howarth HVC, Griffin MJ. 1990. The relative importance of noise and vibration from railways. *Applied Ergonomics*, 21:129–134.

Howarth PA. 1990. Assessment of the visual environment. In *Evaluation of Human Work: a*

Practical Ergonomics Methodology, edited by JR Wilson and EN Corlett. Taylor and Francis, London.

Hsu SH, Peng Y. 1993. Control/display relationship of the four-burner stove: A reexamination. *Human Factors*, 35:745–749.

Hsu SH, Wu SP. 1991. An investigation for determining the optimum length of chopsticks. *Applied Ergonomics*, 22:395–400.

Huchingson RD. 1981. *New Horizons for Human Factors in Design*. McGraw-Hill, New York.

Hughes PC, Meer RM. 1981. Lighting for the elderly: A psychobiological approach to lighting. *Human Factors*, 23:65–85.

Hunting W, Laubli T, Grandjean E. 1981. Postural loads at VDT workplaces: Constrained postures. *Ergonomics*, 24: 917–931.

Irvine CH, Snook SH, Sparshatt JH. 1990. Stairway risers and treads: Acceptable and preferred dimensions. *Applied Ergonomics*, 21:215–225.

Jacob V, Gaultney LD, Slavendy G. 1986. Strategies and biases in human decision-making and their implications for expert systems. *Behaviour and Information Technology*, 5:119–140.

Jacobsen FM, Wehr TA, Sack DA, James SP, Rosenthal NE. 1987. Seasonal affective disorder: A review of the syndrome and its public health implications. *American Journal of Public Health*, 77:57–60.

Jaschinski W. 1982. Conditions of emergency lighting. *Ergonomics*, 25:363–372.

Jaschinski-Kruza W. 1991. Eyestrain in VDU users: Viewing distance and the resting position of ocular muscles. *Human Factors*, 33:69–83.

Johnson SL. 1988. Evaluation of powered screwdriver design characteristics. *Human Factors,* 30:61–69.

Johnson-Laird PN. 1988. *The Computer and the Mind: An Introduction to Cognitive Science*. Harvard University Press, Cambridge, MA.

Jones DM, Frankish CR, Hapeshi K. 1992. Automatic speech recognition in practice. *Behaviour and Information Technology*, 11, 2:109–122.

Jonsson BG, Persson J, Kilbom A. 1988. Disorders of the cervicobrachial region among female workers in the electronics industry: A two-year follow-up. *International Journal of Industrial Ergonomics*, 5:1–12.

Jorna PGA. 1993. Heart rate and heart rate variations in actual and simulated flight. *Ergonomics* 36:1043–1054.

Kammann R. 1975. The comprehensibility of printed instructions and the flowchart alternative. *Human Factors*, 17:183–191.

Kapandji IA. 1974. *The Physiology of the Joints*, vols. 1, 2, and 3. Churchill and Livingstone, Edinburgh.

Kasprzyk DM, Drury CG, Bialas WF. 1979. Human behaviour and performance in calculator use with Algebraic and Reverse Polish Notation. *Ergonomics*, 22:1011–1019.

Kasra M, Shirazi-Adl A, Drouin G. 1992. Dynamics of human lumbar intervertebral joints. *Spine*, 17:93–102.

Keegan JJ. 1953. Alternations of the lumbar curve related to posture and seating. *Journal of Bone and Joint Surgery*, 35A:589–603.

Kelsey JL. 1975. An epidemiological study of acute herniated lumbar intervertebral discs. *Rheumatology and Rehabilitation*, 14:144–159.

Kemp M. 1985. Repetition strain injuries: A need for proper management. *Work and People*, 10:25–28.

Kendall HO, Kendall FP, Wadsworth GE. 1971. *Muscles: Testing and Function*. Williams and Wilkins, Baltimore.

Kerslake DMcK. 1982. Effects of climate. In *The Body at Work*, edited by WT Singleton. Cambridge University Press, Cambridge, England.

Keyserling WM, Punnett L, Fine L. 1988. Trunk posture and back pain: Identification and control of occupational risk factors. *Applied Industrial Hygiene*, 3:87–92.

Keyserling WM, Stetson DS, Silverstein BA, Brouwer ML. 1993. A checklist for evaluating ergonomic risk factors associated with upper extremity cumulative trauma disorders. *Ergonomics*, 36:807–831.

Khaleque A. 1981. Job satisfaction, perceived effort and heart rate in light industrial work. *Ergonomics*, 24:735–742.

Kilbom A. 1988. Intervention programmes for work-related neck and upper limb disorders: Strategies and evaluation. In *Designing a Better World, Proceedings of the 10th International Congress of the IEA*.

Kim JY, Stuart-Buttle C, Marras WS. 1994. The effects of mats on back and leg fatigue. *Applied Ergonomics*, 25:29–34.

Klausen K. 1965. The form and function of the loaded human spine. *Acta Physiologica Scandinavica*, 65:176–190.

Klausen K, Rasmussen B. 1968. On the location of the line of gravity in relation to L5 in standing. *Acta Physiologica Scandinavica*, 75:45–52.

Klein AB, Snyder-Mackler L, Roy SH, DeLuca CJ. 1991. Comparison of spinal mobility and isometric trunk extensor forces with EMG spectral analysis in identifying LBP. *Physical Therapy*, 71:445–454.

Kleinmuntz. 1984. Diagnostic problem solving by computer: A historical review and the current state of the art. *Computers in Biology and Medicine*, 14:255–270.

Kline DW, Fuchs P. 1993. The visibility of symbolic highway signs can be increased among drivers of all ages. *Human Factors*, 35:25–34.

Kline TJ, Ghali LM, Kline DW, Brown S. 1990. Visibility distance of highway signs among young, middle-aged and older observers: Icons are better than text. *Human Factors*, 30:609–619.

Knave B. 1984. Ergonomics and lighting. *Applied Ergonomics*, 15:15–20.

Knowlton RG, Gilbert JC. 1983. Ulnar deviation and short-term strength reductions as affected by a curve-handled ripping hammer and a conventional claw hammer. *Ergonomics*, 26:173–179.

Konig HL, Kreuger AP, Lang S, Sonning W. 1980. *Biologic Effects of Environmental Electromagnetism*. Springer Verlag, New York.

Konz S. 1986. Bent hammer handles. Human Factors, 28:317–323.

Konz S, Bandla V, Rys M, Sambasivan J. 1990. Standing on concrete versus floor mats. In *Advances in Industrial Ergonomics and Safety II*, edited by B Das. Taylor and Francis, London.

Kroemer KHE. 1970. Human strength, terminology, measurement and interpretation of data. *Human Factors*, 20:481–497.

Kroemer KHE. 1989. Cumulative trauma disorders: Their recognition and ergonomics measures to avoid them. *Applied Ergonomics*, 20:274–280.

Kroemer KHE. 1991. Working strenuously in heat, cold, polluted air and at altitude. *Applied Ergonomics*, 22:385–389.

Krohn GS. 1983. Flowcharts used for procedural instructions. *Human Factors*, 25:573–581.

Kukkonen-Harjula K, Rauramaa R. 1984. Oxygen consumption of lumberjacks in logging with a power saw. *Ergonomics*, 27:59–65.

Kumar S. 1990. Cumulative load as a risk factor for back pain. *Spine*, 15:1311–1316.

Kurosaka RT. 1986. Paradoxes of probability. *Byte*, November:373–375.

Kvalseth TO. 1978. Human performance comparisons between digital pursuit and digital compensatory control. *Ergonomics*, 21:419–425.

Labadarios D. 1992. Nutrition in the 1990's in South Africa. *Hospital and Specialist Medicine*, March:28–30.

Lachman R. 1989. Comprehension aids for on-line reading of expository text. *Human Factors*, 31:1–15.

Lansdale M. 1988. The psychology of personal information management. *Applied Ergonomics*, 19:55–66.

Lansdale M, Simpson M, Stroud TRM. 1990. A comparison of words and icons as external memory aids in an information retrieval task. *Behaviour and Information Technology*, 9:111–131.

Laughery KR, Drury CG. 1979. Human performance and strategy in a two variable optimisation task. *Ergonomics*, 22:1325–1336.

Laverson A, Norman K, Shneiderman B. 1987. An evaluation of jump-ahead techniques in menu selection. *Behaviour and Information Technology*, 6:97–108.

Lawrence JS. 1969. Disc degeneration: Its frequency and relation to symptoms. *Annals of Rheumatic Diseases*, 28:121–138.

Lazarus H, Hoge H. 1986. Industrial safety: Accoustic signals for danger situations in factories. *Applied Ergonomics*, 17:41–46.

Leamon TB. 1980. The organisation of industrial ergonomics: A human:machine model. *Applied Ergonomics*, 11:223–226.

Lee KS, Chaffin DB, Herrin GD, Walker AM. 1991. Effect of handle height and lower-back loading in cart pushing and pulling. *Applied Ergonomics*, 22:117–123.

Lee K, Swanson N, Sauter S, Wickstrom R, Waikar A, Mangum M. 1992. A review of physical exercises recommended for VDT workers. *Applied Ergonomics:* 23:387–408.

Leithead C, Lind A. 1964. *Heat Stress and Heat Disorder*. Davis, Philadelphia.

Lhuede EP. 1980. Ear muff acceptance among saw-mill workers. *Ergonomics*, 23:1161–1172.

Levine F, de Simone L. 1991. *Journal of the International Association for the Study of Pain*, 44:69–72.

Lewy AJ, Wehr TA, Goodwin FK. 1980. Light supresses melatonin secretion in humans. *Science*, 210:1257–1269.

Lie I, Watten R. 1987. Oculomotor factors in the aetiology of occupational cervicobrachial diseases (OCD). *European Journal of Applied Physiology and Occupational Physiology*, 56:151–156.

Lie I, Watten R. 1993. VDT work, oculomotor strain and subjective complaints: An experimental study. *Ergonomics,* 36.

Liublinskaya AA. 1957. The development of children's speech and thought. In *Psychology in the Soviet Union*, edited by B Simon. Routledge Kegan Paul, New York.

Long GM, Garvey PM. 1988. The effects of target wavelength on dynamic visual acuity under photopic and scotopic viewing. *Human Factors*, 30:3–13.

Long JB. 1987. Cognitive ergonomics and human-computer interaction. In *Psychology at Work*, edited by PB Warr. Penguin Books, New York.

Loslever P, Ranaivosoa A. 1993. Biomechanical and epidemiological investigation of carpal tunnel syndrome at workplaces with high risk factors. *Ergonomics*, 36:537–554.

Lovejoy CO. 1988. Evolution of human walking. *Scientific American*, November:82–89.

Lusted M. 1993. Predicting return to work after rehabilitation for low back injury. *Australian Journal of Physiotherapy*, 39:203–210.

McClelland L, Cook SW. 1979. Energy conservation effects of continuous in-home feedback in all-electric houses. *Journal of Environmental Systems*, 9:169–173.

McCormick EJ, Sanders MS. 1982. *Human Factors in Engineering and Design*. McGraw-Hill, New York.

McHenry HM, Temerin LA. 1979. The evolution of hominid bipedalism: Evidence from the fossil record. *Yearbook of Physical Anthropology*, 22:105–131.

Magnusson M, Granquist M, Jonson R, Lindell V, Lundberg U, Wallin L, Hanson T. 1990. The loads on the lumbar spine during work at an assembly line: The risks for fatigue injuries of vertebral bodies. *Spine*, 15:774–779.

Magora A. 1972. Investigation of the relation between low back pain and occupation. *Industrial Medicine*, 39:504–510.

Maida JC. 1993. Review of Jack 5.4. *Ergonomics in Design*, July:35–36.

Malone TB, Kirkpatrick M, Mallory K, Eike D, Johnson JH, Walker RW. 1980. *Human Factors Evaluation of Control Room Design and Operator Performance at Three Mile Island: Part 2*. National Technical Information Service, U.S. Department of Commerce, Springfield, VA.

Mandal AC. 1981. The seated man (Homo sedens): The seated work position, theory and practice. *Applied Ergonomics*, 12:19–26.

Mandal AC. 1991. Investigation of the lumbar flexion of the seated man. *International Journal of Industrial Ergonomics*, 8:75–87.

Mannenica I, Corlett EN. 1977. A study of a light repetitive task. *Applied Ergonomics*, 8:103–109.

Marha K, Charron D 1985. The distribution of pulsed, very low frequency electric field around video display terminals. *Health Physics*, 49:517–521.

Marics MA, Williges BH. 1988. The intelligibility of synthesised speech in data inquiry systems. *Human Factors,* 30:719–732.

Marras WS, Kim JY. 1993. Anthropometry of industrial populations. *Ergonomics*, 36:371–378.

Marras WS, Lavender SA, Leurgans SE, Rajulu SL, Allread WG, Fathallas FA, Ferguson SA. 1993. The effects of dynamic three-dimensional trunk position and trunk motion characteristics on risk of injury. *Spine*, 18:617–628.

Marras WS, Mirka GA. 1992. A comprehensive evaluation of asymmetric trunk motions. *Spine*, 17:318–326.

Marras WS, Schoenmarklin RW. 1993. Wrist motions in industry. *Ergonomics*, 4:341–351.

Marriot IA, Stuchley MA. 1986. Health aspects of work with visual display terminals. *Journal of Occupational Medicine*, 28:833–848.

Marshall L. 1981. *A Manually Tuned Autopilot for Optimal Course-Keeping in Ships*. PhD thesis, University of London, London.

Martin JP. 1967. *The Basal Ganglia and Posture*. Pitman Medical Publishing, London.

Martin J, Long J. 1984. The division of attention between a primary tracking task and secondary tasks of pointing with a stylus or speaking in a simulated ship's gunfire control task. *Ergonomics*, 27:397–408.

Matthews ML. 1987. The influence of colour on CRT reading performance and subjective comfort under operational conditions. *Applied Ergonomics*, 18:323–328.

Meese GB, Kok R, Lewis MI, Schiefer RE, Kustner PM. 1989. Performance of a five choice serial reaction time task and an associated vigilance task in the cold. *Ergonomics SA*, 1:56–67.

Merrifield HH. 1971. Female gait patterns in shoes with different heel heights. *Ergonomics*, 14:411–417.

Meshkati N, Robertson MM. 1986. The effects of human factors on the success of technology transfer projects to industrially developing countries: A review of representative case studies. In *Human Factors in Organisational Design and Management: II*, edited by O Brown Jr and HW Hendrick. North Holland, New York.

Milerad E, Ekenvall L. 1990. Symptoms of the neck and upper extremities in dentists. *Scandinavian Journal of Work Environment and Health*, 16:129–134.

Miller GA. 1956. The magical number seven, plus or minus two: Some limits on our capacity for processing information. *Psychological Review*, 63:81–97.

Miller ME, Beaton RJ. 1994. The alarming sounds of silence. *Ergonomics in Design*, February:21–23.

Miller RJ. 1990. Pitfalls in the conception, manipulation and measurement of visual accommodation. *Human Factors*, 30:27–44.

Milne RA, Mireau DR. 1979. Hamstring distensibility in the general population: Relationship to pelvic and low back stress. *Journal of Manipulative and Physiological Therapeutics*, 2:146–150.

Ministry of Science and Technology. 1988. Pesquisa Antropométrica e Biomecânica dos Operários da Indústria de Transfermaçáo RJ. Instituto Nacional de Tecnologia, Rio de Janeiro, Brazil.

Mital A, Founooni-Fard H, Brown M. 1993. Fatigue in high and very-high frequency manual lifting, lowering and carrying and turning. *Proceedings of the Human Factors and Ergonomics Society 37th Annual Meeting*. Human Factors and Ergonomics Society Inc, Santa Monica, CA, 674–678.

Moores B. 1972. Ergonomics or work study? *Applied Ergonomics*, 3:147–154.

Morris JN, Chave SPW, Adam C, Sirey L, Epstein D, Sheehan DJ. 1973. Vigorous exercise in leisure time and the incidence of coronary heart disease. *The Lancet*.

Morris JN, Heady JA, Raffle PA, Roberts CG, Parks JW. 1953. Coronary heart disease and physical activity of work. *The Lancet*, 2:1053–1057, 1111–1120.

Morris JN, Lucas DB, Bresler MS. 1961. Role of the trunk in the stability of the spine. *Journal of Bone and Joint Surgery*, 43A:327–351.

Morris NM, Rouse WB. 1985. Review and evaluation of empirical research in troubleshooting. *Human Factors*, 27:503–530.

Morrison DL, Duncan KD. 1988. Strategies and tactics in fault diagnosis. *Ergonomics*, 31:761–784.

Motohashi Y. 1992. Alteration of carcadian rhythm on shift-working ambulance personnel: Monitoring of salivary cortisol rhythm. *Ergonomics*, 11:1331–1340.

Mullaly J, Grigg L. 1988. RSI: Integrating the major theories. *Australian Journal of Psychology*, 40:19–33.

Murdock BB. 1962. The serial position effect in free recall. *Journal of Experimental Psychology*, 64:482–488.

Murrell KFH. 1965. *Ergonomics: Man and His Working Environment*. Chapman and Hall, New York.

Murrell KFH. 1971. *Ergonomics: Man in His Working Environment*. Chapman and Hall, London.

Nachemson A. 1966. The load on the lumbar discs in different positions of the body. *Clinical Orthopaedics*, 45:107–122.

Nachemson A. 1968. The possible importance of the psoas muscle for stabilisation of the lumbar spine. *Acta Orthopaedica Scandinavica*, 39:47–57.

National Research Council. 1983. *Video Displays, Work and Vision*. Panel on Impact of Video Viewing on Vision of Workers, Committee on Vision, National Academy Press, Washington, DC.

National Safety Council. 1993. OSHA update. *Safety and Health Magazine*, May 10.

Neisser U. 1976. *Cognition and Reality*. Freeman, San Francisco.

Nemecek J, Grandjean E. 1973. Noise in landscaped offices. *Applied Ergonomics*, 4:19–22.

Neumann DA, Cook TM. 1985. Effect of load and carrying position on the electromyographic activity of the gluteus medius muscle during walking. *Physical Therapy*, 65:305–311.

Newell A, Simon HA. 1972. *Human Problem Solving*. Prentice-Hall, New York.

Nickerson RS. 1988. Counting, computing and the representation of numbers. *Human Factors*, 30:181–191.

Nielsen J. 1990. Miniatures versus icons as a visual cache for videotex browsing. *Behaviour and Information Technology*, 9:441–449.

Nielsen J, Schaefer L. 1993. Sound effects as an interface element for older users. *Behaviour and Information Technology,* 12:208–215.

NIOSH. 1981. *Work Practices Guide for the Design of Manual Handling Tasks.* NIOSH.

Noakes T. 1988. Implications of exercise testing for prediction of athletic performance: A contemporary perspective. *Medicine and Science in Sports and Exercise*, 20:319–329.

Noakes T. 1992. *The Lore of Running*. Oxford University Press, Cape Town.

Noe DA, Mostardi RA, Jackson ME, Porterfield JA, Askew MJ. 1992. Myoelectric activity and sequencing of selected trunk muscles during isokinetic lifting. *Spine,* 17:225–229.

Norman DA, 1981. Categorisation of action slips. *Psychological Review*, 88:1–15.

Norman DA. 1984. Stages and levels in human-machine interaction. *International Journal of Man-Machine Studies*, 21:365–375.

Norman DA. 1988. *The Psychology of Everyday Things*. Basic Books, New York.

Norman DA. 1993. Toward human-centred design. *Technology Review*, July:47–53.

Noyes JM, Frankish CR. 1989. A review of speech recognition technology in the office. *Behaviour and Information Technology*, 6:475–486.

Nurminen T, Kurppa K. 1988. Office employment, work with video display terminals and course of pregnancy. *Scandinavian Journal of Work Environment and Health*, 14:293–298.

Oberoi K, Dhillon MJ, Miglani SS. 1983. A study of energy expenditure during manual and machine washing of clothes. *Ergonomics*, 26:375–378.

Oborne, 1982. *Ergonomics at Work.* Wiley.

Oborne DJ. 1983. Cognitive effects of passive smoking. *Ergonomics*, 26:1163–1171.

Oborne DJ, Boarer PA. 1982. Subjective responses to whole-body vibration: The effects of posture. *Ergonomics*, 25:673–681.

Ogden GD, Levine JM, Eisner EJ. 1979. Measurement of workload by secondary tasks. *Human Factors*, 21:529–548.

Ohnaka T, Tochihara Y, Murumatsu T. 1993. Physiological strains in hot-humid conditions while wearing disposable protective clothing commonly used by the asbestos removal industry. *Ergonomics*, 36:1241–1250.

Olsen PL. 1989. Motorcycle conspicuity revisited. *Human Factors*, 31:141–146.

OSHA 3123. 1993. *Ergonomics Program Management Guidelines for Meatpacking Plants.* U.S. Dept of Labor, OSHA.

Ostberg ON, Reddan WG, Swanson NG, Kleman JE, Miezio KR. 1988. Assessment of a cold air breathing aid. *Applied Ergonomics*, 19:325–328.

Ostberg ON, Warell B, Nordell L. 1984. ComforTable: A generic desk for the automated office. *Behaviour and Information Technology*, 3:411–416.

Owen RD. 1994. Carpal tunnel syndrome: A products liability perspective. *Ergonomics*, 37:449–476.

Paffenbarger RS, Hyde RT, Jung DL, Wing Al. 1984. Epidemiology fo exercise and coronary heart disease. *Clinics in Sports Medicine*, 3:297–318.

Panero J, Zelnik M. 1979. *Human Dimension and Interior Space.* Whitney Library of Design, Architectural Press, London.

Parsons KC. 1993. *Human Thermal Environments,* Taylor and Francis, London.

Parsons KC. 1993. Safe surface temperatures of domestic products. In *Contemporary Ergonomics,* edited by EJ Lovesey. Taylor and Francis, London.

Patkin M. 1989. Hand and arm pain in office workers. *Modern Medicine of South Africa,* March:53–67.

Patrick EA. 1977. Pattern recognition in medicine. *IEEE Transactions on Systems, Man and Cybernetics,* 6:4.

Patterson R, Martin LM. 1992. Human stereopsis. *Human Factors,* 34:669–692.

PB91-226274. *Scientific Support Documentation for the Revised 1991 Lifting Equation.* National Technical Information Service, Springfield, VA.

Pearcy MJ. 1993. Twisting mobility of the human back in flexed postures. *Spine,* 18:114–119.

Perrot DR, Sadralodabai T, Saberi K, Strybel TZ. 1991. Aurally aided visual search in the central visual field: Effects of visual load and visual enhancement of the target. *Human Factors,* 33:389–400.

Peterson LR, Peterson MJ. 1959. Short-term retention of individual verbal items. *Journal of Experimental Psychology,* 58:193–218.

Pheasant ST. 1982. A technique for estimating anthropometric data from the parameters of the distribution of stature. *Ergonomics,* 25:981–982.

Pheasant ST. 1986. *Bodyspace.* Taylor and Francis, London.

Pheasant ST. 1992. Does RSI exist? Balance of opinion. *Occupational Medicine,* 42:167–168.

Pheasant ST. 1992. *Ergonomics Work and Health.* Taylor and Francis, London.

Pheasant ST, O'Neill D. 1975. Performance in gripping and turning: A study in hand/handle effectiveness. *Applied Ergonomics,* 6:205–208.

Pheasant ST, Scriven JG. 1983. Sex differences in strength: Some implications for the design of hand tools. In *Proceedings of the Ergonomics Society's Conference,* edited by Coombes. Taylor and Francis, London.

Pope MH, Bevins T, David RPT, Wilder G, Frymower JW. 1984. The relationship between anthropometric, postural, muscular and mobility characteristics of males ages 18–55. *Spine,* 10:644–648.

Porritt, The Rt Hon Lord. 1985. Slipping, tripping and falling, familiarity breeds contempt. *Ergonomics,* 28:947–948.

Porter JM, Gyi DE, Robertson J. 1992. Evaluation of a tilting computer desk. In *Contemporary Ergonomics,* edited by EJ Lovesy. Taylor and Francis, London.

Posner M. 1980. Orienting of attention. *Quarterly Journal of Experimental Psychology,* 32:3–25.

Posner MI, Nissen MJ, Klein RM. 1976. Visual dominance: An information processing account of its origins and significance. *Psychological Review,* 83:157–171.

Poulton EC. 1970. *Environment and Human Efficiency.* Thomas, Springfield, IL.

Poulton EC. 1976a. Arousing environmental stresses can improve performance, whatever people say. *Aviation, Space and Environmental Medicine,* 47:1193–1204.

Poulton EC. 1976b. Continuous noise interferes with work by masking auditory feedback and inner speech. *Applied Ergonomics,* 7:79–84.

Poulton EC. 1977. Continuous intense noise masks auditory feedback and inner speech. *Psychological Bulletin,* 84:977–1001.

Prumper J, Zapf D, Brodbeck FC, Frese M. 1992. Some surprising differences between novice and expert errors in computerised office work. *Behaviour and Information Technology*, 11:319–328.

Putz-Anderson V. 1988. *Cumulative Trauma Disorders: A Manual for Musculoskeletal Disease of the Upper Limbs*. Taylor and Francis, London.

Pyykko I, Jantti P, Aalto H. 1990. Postural control in elderly subjects. *Age and Ageing*, 19:215–221.

Quillian MR. 1969. The teachable language comprehender: A simulation program and theory of language. *Communications of the Association for Computing Machinery*, 12:459–476.

Quinter JL, Elvey RL. 1993. Understanding "RSI": A review of the role of peripheral neural pain and hyperalgesia. *The Journal of Manual and Manipulative Therapy*, 1:99–105.

Radwin RG, Lin ML. 1993. An analytical method for characterising repetitive motion and postural stress using spectral analysis. *Ergonomics*, 36:379–389.

Ralston JV, Pisoni DB, Lively SE, Greene BG, Mullenix JW. 1991. Comprehension of synthetic speech produced by rule: Word monitoring and sentence-by-sentence listening times. *Human Factors*, 33:471–491.

Ramazzini B. 1964. *Diseases of Workers*. Hafner Publishing, New York.

Randall G. 1980. Interviewing at work. In *Psychology at Work*, edited by PB Warr. Penguin Books, New York.

Rasmussen J. 1983. Skills, rules and knowledge: Signals, signs and symbols and other distinctions in human performance models. *IEEE Transactions on Systems, Man and Cybernetics*. SMC-13:257–266.

Rasmussen J, Rouse WB. 1981. *Human Detection and Diagnosis of System Failures*. Plenum, New York.

Rauterberg M. 1992. An empirical comparison of menu-selection (CUI) and desktop (GUI) computer programmes carried out by beginners and experts. *Behaviour and Information Technology*, 11:227–236.

Reason J. 1990. *Human Error*. Cambridge University Press, Cambridge, England.

Reddell CR, Congleton JJ, Huchingson RD, Montgomery JF. 1992. An evaluation of a weightlifting belt and back injury prevention training class for airline baggage handlers. *Applied Ergonomics*, 23:319–329.

Reilly T. 1991. Assessment of some aspects of physical fitness. *Applied Ergonomics*, 22:291–294.

Riley MW, Cochran DJ, Ballard JL. 1982. Designing and evaluating warning labels. *PC*, 25:127–130.

Richardson C, Toppenberg R, Jull G. 1990. An initial evaluation of eight abdominal exercises for their ability to provide stabilisation for the lumbar spine. *The Australian Journal of Physiotherapy*, 36:6–11.

Ring L. 1981. *Facts on Backs*. Institute Press, Loganville, GA.

Rivas FJ, Diaz JA, Santos R. 1984. Valores maximos de esfuerzo admisibles en los puestos de trabajo. *Prevencion*, 87:20–25.

Robinson JT. 1972. *Early Hominid Posture and Locomotion*. University of Chicago Press, Chicago.

Rogers Y, Oborne DJ. 1985. Some psychological attributes of potential computer command names. *Behaviour and Information Technology*, 4:349–365.

Rohmert W. 1987. Physiological and psychological work load measurement and analysis. In *Handbook of Human Factors*, edited by G Salvendy. Wiley, New York.

Rolt LTC. 1963. *George and Robert Stephenson: The Railway Revolution*. Longmans, London.

Rosa RR, Bonnet MH. 1993. Performance and alertness on 8h and 12h rotating shifts at a natural gas utility. *Ergonomics*, 36: 1177–1193.

Roscoe AH. 1993. Heart rate as a psychophysiological measure for in-flight workload assessment. *Ergonomics*, 36:1055–1062.

Rouse WB. 1988. Adaptive aiding for human-computer control. *Human Factors*, 30:431–443.

Roy SH, De Luca CJ, Casavant DA. 1989. Lumbar muscle fatigue and chronic lower-back pain. *Spine*, 14:992–1001.

Roy SH, De Luca CJ, Snyder-Mackler L, Emley MS, Crenshaw RL, Lyons JP. 1990. Fatigue, recovery and low-back pain in varsity rowers. *Medical Science in Sports and Exercise,* 22:515–523.

Rutkowski C. 1982. The human applications standard computer interface. Part 1: Theory and practice. *Byte*, October.

Rys M, Konz S. 1994. Standing. *Ergonomics*, 37:677–687.

Sanders AF. 1970. Some aspects of the selective process in the functional visual field. *Ergonomics*, 13:101–117.

Sanders MS, McCormick EJ. 1993. *Human Factors in Engineering and Design*, 7th edition. McGraw-Hill, New York.

Sauter SL, Schleifer LM, Knutson SJ. 1991. Work posture, workstation design and musculoskeletal discomfort in a VDT task. *Human Factors*, 33:151–17.

Savitz DA. 1993. Overview of epidemiological research on electric and magnetic fields and cancer. *Journal of the American Industrial Hygiene Association*. 54:197–204.

Savitz DA, Wachtel H, Barnes FA, John EM, Tvrdik JG. 1988. Case-control study of childhood cancer and exposure to 60-Hz magnetic fields. *American Journal of Epidemiology*, 128:21–38.

Scansetti G. 1984. Toxic agents emitted from office machines and materials. In *Ergonomics and Health in the Modern Office*, edited by E. Grandjean. Taylor and Francis, London.

Schaafstal A. 1993. Knowledge and strategies in diagnostic skill. *Ergonomics*, 11:1305–1316.

Schaffer LH. 1975. Multiple attention in continuous verbal tasks. In *Attention and Performance V*, edited by PMA Rabbitt and S Dornic. Academic Press, London.

Schierhout G, Myers JE, Brider RS. 1992. Musculoskeletal pain, postural stress and ergonomic design factors in factory floor occupations in sectors of manufacturing in South Africa. *Proceedings of the PREMUS '92 Conference*. Stockholm, Sweden, 7-10 May.

Schleifer LM, Ley R. 1994. End-tidal PCO2 as an index of psychophysiological activity during VDT data entry work and relaxation. *Ergonomics*, 37:245–254.

Schnorr TM, Grajewski BA, Hornung RW, Thun MJ, Egeland GM, Murray WE, Conover DL, Halpern WE. 1991. Video display terminals and the risk of spontaneous abortion. *The New England Journal of Medicine*, 324:727–733.

Schuldt K, Ekholm J, Harms-Ringdahl K, Nemeth G, Arborelius P. 1986. Effects of changes in sitting work posture on static neck and shoulder muscle activity. *Ergonomics*, 29:1525–1537.

Schutte PC, Kielblock AJ, Van der Walt WH, Celliers CP, Strydom NB. 1982. *An Analysis of the Viability of Microclimate Acclimatisation*. Chamber of Mines Research Report No. 28/82, Chamber of Mines, Johannesburg.

Schwartz JP, Norman KL. 1986. The importance of item distinctiveness on performance using a menu selection system. *Behaviour and Information Technology*, 5:173–182.

Sell RG. 1980. Success and failure in implementing changes in job design. *Ergonomics*, 23:809–816.

Sell RG. 1986. The politics of workplace participation. *Personnel Management*, June.

Sell RG. 1993. *The New European Community Directives and Other International Developments in Ergonomics*. Paper presented at the Graduate School of Business, University of Cape Town, Cape Town, 17 March.

Seminara JL, Smith DL. 1983. Remedial human factors engineering: Part 1. *Applied Ergonomics*, 14:253–264.

Seminara JL, Smith DL. 1984. Remedial human factors engineering: Part 2. *Applied Ergonomics*, 15:31–44.

Sen RN. 1984. Application of ergonomics to industrially developing countries. *Ergonomics*, 27:1021–1032.

Sen RN, Ganguli AK, Ray GG, Chakrabarti D. 1983. Tea-leaf plucking: Workloads and environmental studies. *Ergonomics*, 26:887–893.

Shackel B. 1980. Dialogues and language: Can computer ergonomics help? *Ergonomics*, 23:857–880.

Shah RK. 1993. A pliot survey of the traditional use of the patuka round the waist for the prevention of back pain in Nepal. *Applied Ergonomics*, 24:337–344.

Shepard RN, Cooper LA. 1982. *Mental Images and Their Transformations*. Bradford Books, MIT Press, Cambridge, MA.

Shepherd A. 1993. An approach to information requirements specification for process control tasks. *Ergonomics*, 36:1425–1437.

Shilling R, Anderson N. 1986. Occupational epidemiology in developing countries. *Occupational Health and Safety* (Australia and New Zealand), 2:468–471.

Shirazi-Adl A. 1992. Finite-element simulation of changes in the fluid content of human lumbar discs: Mechanical and clinical implications. *Spine*, 17:206–212.

Shneiderman B. 1988. We can design better user interfaces: A review of human-computer interaction styles. In *Designing a Better World, Proceedings of the 10th Congress of the International Ergonomics Association*, edited by AS Adams, RR Hall, BJ McPhee, and MS Oxenburgh, Ergonomics Society of Australia, Sydney.

Shneiderman B. 1991. A taxonomy and rule base for the selection of interaction styles. In *Human Factors for Informatics Usability*, edited by B Shackel and S Richardson. Cambridge University Press, Cambridge, England.

Shorland FB. 1988. Is our knowledge of human nutrition soundly based? *World Reviews of Nutrition and Dietetics*, 57:126–205.

Shortliffe EH. 1984. Reasoning methods in medical consultation systems: Artificial intelligence approaches. *Computer Programs in Biomedicine*, 18:5–14.

Shortliffe EH, Buchanan BG, Feigenbaum EA. 1979. Knowledge engineering for medical decision making: A review of computer-based decision aids. *Proceedings of the IEEE*, 67:1209–1224.

Shute SJ, Starr SJ. 1984. Effects of adjustable furnitue on VDT users. *Human Factors*, 2:157–70.

Siekmann H. 1990. Recommended maximum temperatures for touchable surfaces. *Applied Ergonomics*, 21:69–73.

Simon HA. 1979. Information processing models of cognition. *Annual Reviews of Psychology*, 30:33–396.

Sims M, Gillies G, Drury R. 1977. Using a cool spot to improve the thermal comfort of glassmakers. *Applied Ergonomics*, 8:2–6.

Singleton WT. 1972. *Man-Machine Systems*. Penguin, London.

Singleton WT. 1989. *The Mind at Work*. Cambridge University Press, Cambridge, England.

Sliney DH. 1983. Biohazards of ultraviolet, visible and infrared radiation. *Journal of Occupational Medicine*, 25:203–206.

Smith CS. 1981. *A Search for Structure: Selected Essays on Science, Art and History*. MIT Press, Cambridge, MA.

Smith MJ, Colligan MJ, Tasto DL. 1982. Health and safety consequences of shift work in the food processing industry. *Ergonomics*, 25:133–144.

Smith MJ, Stammerjohn LW, Cohen GF, Lalich NR. 1980. Job stress in video display operations. In *Ergonomics Aspects of Visual Display Terminals*, edited by E Grandjean and E Vigliari. Taylor and Francis, London.

Smith PJ, Wilson JR. 1993. Navigation in hypertext through virtual environments. *Applied Ergonomics*, 24:271–278.

Smith SL, Mosier JN. 1986. *Guidelines for Designing User-Interface Software*. Mitre Corporation,

Smith W, Cronin DT. 1993. Ergonomic test of the Kinesis keyboard. *Proceedings of the Human Factors and Ergonomics Society 37th Annual Meeting*, 318–322.

Smythe H. 1988. The "repetitive strain injury syndrome" is referred pain from the neck. *The Journal of Rheumatology*, 15:1604–1608.

Snook SH. 1978. The design of manual handling tasks. *Ergonomics*, 21:963–985.

Snook SH, Ciriello VM. 1991. The design of manual handling tasks: Revised tables of maximum acceptable weights and forces. *Ergonomics*, 34:1197–1213.

Soames RW, Evans AA. 1987. Female gait patterns: The influence of footwear. *Ergonomics*, 30:893–900.

Soderberg GL, Cook TM. 1984. Electromyography in biomechanics. *Physical Therapy*, 64:1813–1820.

Soderberg I, Calissendorff B, Elofsson S, Knave B, Nyman KG. 1983. Investigation of the visual strain experienced by microscope operators at an electronics plant. *Applied Ergonomics*, 14:297–305.

Solomonow M, D'Ambrosia R. 1987. Biomechanics of muscle overuse injuries: A theoretical approach. *Clinics in Sports Medicine*, 6:241–257.

Sommerich CM, McGlothlin JD, Marras WS. 1993. Occupational risk factors associated with soft tissue disorders of the shoulder: A review of recent literature. *Ergonomics*, 36:697–717.

Sorkin RD. 1987. Design of auditory and tactile displays. In *Handbook of Human Factors*, edited by G Salvendy. Wiley, New York.

Soule RG, Goldman RF. 1969. Energy cost of loads carried on the head, hands and feet. *Journal of Applied Physiology*, 27:687–690.

Spence R, Apperly M. 1982. Data base navigation: An office environment for the professional. *Behaviour and Information Technology*, 1:43–54.

Spence R. 1993. *A Taxonomy of Graphical Presentation*. Information Engineering Section Report #93/3, July. Dept. of Electrical and Electronic Engineering, Imperial College of Science, Technology and Medicine, London.

Spender J. 1962. An investigation of process control skill. *Occupational Psychology*, 36:30.

Sperling G. 1960. The information available in brief visual presentations. *Psychological Monographs*, 74.

Spurr GB, Prentice AM, Murgatroyd PR, Goldberg GR, Reina JC, Christman NT. 1988. Energy expenditure from minute-by-minute heart-rate recording: A comparison with indirect calorimetry. *American Journal of Clinical Nutrition*, 48:552–559.

Stamper MT. 1987. Good health is not for sale. *Ergonomics*, 30:199–206.

Stanton N, Edworthy J. 1994. Towards a methodology for constructing and evaluating representational auditory alarm displays. In *Contemporary Ergonomics*, edited by SA Robertson. Taylor and Francis, London.

Starr S. 1983. *Video Display Terminals: Preliminary Guidelines for Selection, Installation and Use.* Bell Telephone Laboratories, Short Hills, NJ.

Sterling TD, Sterling E, Dimich-Ward. 1983. Building illness in the white-collar workplace. *International Journal of Health Services*, 13:277–287.

Stewart-Buttle C, Marras WS, Kim JY. 1993. The influence of anti-fatigue mats on back and leg fatigue. In *Proceedings of the Human Factors and Ergonomics Society 37th Annual Meeting.* Human Factors and Ergonomics Society, 769–773.

Stock SR. 1991. Workplace ergonomic factors and the development of musculoskeletal disorders of the neck and upper limbs: A meta analysis. *American Journal of Industrial Medicine*, 19:87–107.

Stone PT. 1986. Issues in vision and lighting for users of VDU's. In *Health Hazards of VDTs?* edited by BG Pearce. Wiley, New York.

Stroop JR. 1935. Studies of interference in serial verbal reactions. *Journal of Experimental Psychology*, 18:643–662.

Stubbs DA, Buckle PW, Hudson MP, Rivers PM. 1983a. Back pain in the nursing profession: I. Epidemiology and pilot methodology. *Ergonomics*, 26:755–75.

Stubbs DA, Buckle PW, Hudson MP, Rivers PM. 1983b. Back pain in the nursing profession: II. The effectiveness of training. *Ergonomics*, 26:767–779.

Stuhlen FB, De Luca CJ. 1982. Muscle fatigue monitor: A non-invasive device for observing localised muscular fatigue. *IEEE Transactions on Biomedical Engineering*, 29:760–768.

Sulotto F, Romano C, Dori S, Piolatto G, Chiesa A, Chiacco C, Scansetti G. 1994. The prediction of recommended energy expenditure for an 8-hour work-day using an airpurifying respirator. *Ergonomics*, 36:1479–1487.

Sundstrom ED. 1986. *Workplaces.* Cambridge University Press, Cambridge, England.

Swenson EE, Purswell JL, Schlegel RE, Stanevich RL. 1992. Coefficient of friction and subjective assessment of worksurfaces. *Human Factors*, 34:67–77.

Swezey RW. 1993. Certification defines the field. *Ergonomics in Design*, July: 4.

Sykes JM. 1988. *Sick Building Syndrome: A Review.* Specialist Inspector Reports No. 10. Health and Safety Executive, London.

Thimbleby H. 1991. Can humans think? The Ergonomics Society Lecture 1991. *Ergonomics*, 34:1269–1287.

Tichauer ER. 1978. *The Biomechanical Basis of Ergonomics.* Wiley, New York.

Tile M. 1984. *Fractures of the Pelvis and Acetabulum.* Williams and Wilkins, Baltimore.

Tilley AJ, Wilkinson RT, Warren PSG, Watson B, Drud M. 1982. The sleep and performance of shiftworkers. *Human Factors*, 24:629–641.

Tola S, Riihimaki H, Videman T. 1988. Neck and shoulder symptoms among men in machine operating, dynamic physical work and sedentary work. *Scandinavian Journal of Work Environment and Health*, 14:299–305.

Tomenius L. 1986. 50-Hz electromagnetic environment and the incidence of childhood tumours in Stockholm county. *Bioelectromagnetics*, 7:191–207.

Tomlinson RW. 1971. Estimation and reduction of risk to hearing: The background and a case study. *Applied Ergonomics*, 2:112–119.

Tomlinson RW, Mannenica I. 1977. A study of physiological and work study indices of forestry work. *Applied Ergonomics*, 8:165–172.

Tower SS, Pratt WB. 1990. Spondylolysis and associated spondylolisthesis in Eskimo and Athabascan populations. *Clinical Orthopaedics and Related Research*, 250:171–175.

Traffimow JH, Schipplein OD, Novak GJ, Andersson GBJ. 1993. The effects of quadriceps fatigue on the technique of lifting. *Spine*, 18:364–367.

Trist EL, Bamforth KW. 1951. Some social and psychological consequences of the longwall method of coal getting. *Human Relations*, 4:3–38.

Troup JDG. 1978. Drivers' back pain and its prevention: A review of the postural, vibratory and muscular factors, together with the problem of transmitted roadshock. *Applied Ergonomics*, 9:207–214.

Tulving E. 1972. Episodic and semantic memory. In *Organisation of Memory*, edited by E Tulving and W Donaldson. Academic Press, New York.

Tversky A, Kahneman D. 1974. Judgement under uncertainty: Heuristics and biases. *Science,* 185:1124–1131.

Tversky A, Kahneman D. 1981. The framing of decisions and the psychology of choice. *Science*, 211:453–458.

Tynan O. 1980. *Improving the Quality of Working Life in the 1980's*. WRU Occasional Paper 16, November. Work Research Unit, London.

Umbers IG. 1979. Models of the process operator. *International Journal of Man-Machine Studies*, 11:263–284.

University of California. 1993. Not seeing red. *University of California at Berkeley Wellness Letter*, 9:2.

US Dept of Health and Human Services. 1981. *Work Practices Guide for Manual Lifting*. Centers for Disease Control. National Institute of Occupational Health and Safety, Cincinnati, OH 45226.

Van Akerveeken PJ. 1985. Teaching aspects. *Ergonomics*, 28:371–377.

Van Cott HP. 1984. From control systems to knowledge systems. *Human Factors,* 26:115–122.

Van Dieen JH, Toussant HM, Thissen C, Van de Ven A. 1993. Spectral analysis of erector spinae EMG during intermittent isometric fatiguing exercise. *Ergonomics*, 36:407–414.

Van Nes FL. 1986. Space, colour and typography on visual display terminals. *Behaviour and Information Technology*, 5:99–118.

Van Wyk RJ. 1992. *Corporate-wide Technological Literacy*. Seminar given at the Graduate School of Business, University of Cape Town, Cape Town, 10 November.

Verbeek J. 1991. The use of adjustable furniture: Evaluation of an instruction programme for office workers. *Applied Ergonomics*, 22:179–184.

Verhagen P, Bervoets R, Debrandere G, Millet F, Santermans G, Stuyck M, Vandermoere D, Willems G. 1975. Direction of movement stereotypes in different cultural groups. In *Ethnic Variables in Human Factors Engineering*, edited by A Chapanis. Johns Hopkins University Press, Baltimore.

Vernon HM. 1924. The influence of rest pauses and changes of posture on the capacity for muscular work. *The Medical Research Council Report*, 29:28–55.

Vernon-Roberts B. ;1989. Pathology of intervertebral discs and apophyseal joints. In *The Lumbar Spine and Back Pain*, 3d edition, edited by I Jayson and IV Malcolm. Churchill Livingstone, Edinburgh.

Vidulich MA. 1988. Speech responses and dual task performance: Better time-sharing asymmetric. *Human Factors*, 30: 517–529.

Vihma T, Nurminen M, Mutanen P. 1982. Sewing machine operators work and musculoskeletal complaints. *Ergonomics*, 25:295–298.

Visick D, Johnson P, Long J. 1984. The use of some simple speech recognisers in industrial applications. In *Interact '84, Proceedings of the First IFIP Conference on Human Computer Interaction*, Vol. 1, 99–103.

Vogt JJ, Libert JP, Candas V, Daull F, Mariaux PH. 1983. Heart rate and spontaneous work-rest cycles during exposure to heat. *Ergonomics,* 2:1173–1185.

Waddell G. 1982. An approach to backache. *British Journal of Hospital Medicine*, 28:187–219.

Wall TD. 1980. Job redesign and employee participation. In *Psychology at Work*, edited by PB Warr. Penguin Books, New York.

Wanner HU. 1984. Indoor air quality in offices, in *Ergonomics and Health in Modern Offices*, edited by E. Grandjean. Taylor and Francis, London.

Wason PC, Shapiro D. 1971. Natural and contrived experience in a reasoning problem. *Quarterly Journal of Experimental Psychology,* 23:63–71.

Wason PC, Johnson-Laird PN. 1972. *Psychology of Reasoning: Structure and Content*. Batsford, London.

Waters TR, Putz-Anderson V, Garg A, Fine LJ. 1993. Revised NIOSH equation for the design and evaluation of manual lifting tasks. *Ergonomics*, 36:749–776.

Webb JM, Kramer AF. 1990. Maps or analogies? A comparison of instructional aids for menu navigation. *Human Factors*, 32:251–266.

Weber BH. 1970. Silencing of hand-held percussive rockdrills for underground operations. *CIM Bulletin*, February.

Wedderburn AAI. 1992. How fast should the shift rotate? A rejoinder. *Ergonomics*, 35:1447–1451.

Welch RB, Langley EO, Lamaev O. 1971. The measurement of fatigue in hot working conditions. *Ergonomics*, 14:85–90.

Wertheimer N, Leeper E. 1979. Electrical wiring configurations and childhood cancer. *American Journal of Epidemiology*, 109:273–284.

Wertheimer N, Leeper E. 1986. Possible effects of electric blankets and heated waterbeds in fetal development. *Bioelectromagnetics*, 7:13–22.

Wetherall A. 1981. The efficacy of some auditory-vocal subsidiary tasks as measures of the mental load on male and female drivers. *Ergonomics,* 24:197–214.

White AA, Panjabi MM. 1978. *Clinical Biomechanics of the Spine*. Lippincott, Philadelphia.

Wickens CD. 1980. The structure of attentional resources. In *Attention and Performance VIII*, edited by R Nickerson. Erlbaum, Hillsdale, NJ.

Wickens CD. 1984. *Engineering Psychology and Human Performance*. Merrill, Columbus, OH.

Wickens CD. 1987. *Decision Making*. In *Human Factors Design Handbook*, edited by G Salvendy. Wiley, New York.

Wieder DL. 1992. Impingement syndrome: A question of mechanics. *Rehab Management*, February/March:87–92.

Wierwille WW. 1979. Physiological measures of aircrew mental workload. *Human Factors*, 21:575–593.

Wiker SF, Chaffin DB, Langolf GD. 1989. Shoulder posture and localised muscle fatigue and discomfort. *Ergonomics*, 32:211–237.

Wilkinson RT. 1992. How fast should the shift rotate? *Ergonomics*, 35:1425–1446.

Williamson AM, Gower CGI, Clarke BC. 1994. Changing the hours of shiftwork: A comparison of 8 and 12 hour shift rosters in a group of computer operators. *Ergonomics*, 37:287–298.

Williges RC, Wierwille WW. 1979. Behavioural measures of aircrew mental workload. *Human Factors*, 21:549–574.

Withey WR. 1982. The provision of energy. In *The Body at Work*, edited by WT Singleton. Cambridge University Press, Cambridge, England.

Wolgater MS, Young SL. 1991. Behavioural compliance to voice and print warnings. *Ergonomics*, 34:79–89.

Woods DD, Roth EM. 1988. Cognitive engineering: Human problem solving with tools. *Human Factors*, 30:415–430.

Woodson W. 1981. *Human Factors Design Handbook.* McGraw-Hill, New York.

Wright P. 1978. Feeding the information eaters: Suggestions for integrating pure and applied research on language comprehension. *Instructional Science*, 7:249–312.

Wright P. 1980. Strategy and tactics in the design of forms. *Visible Language*, XIV:151–193.

Wright P. 1981. Five skills technical writers need. *IEEE Transactions on Professional Communication*, PC-24:10–16.

Wright P. 1982. Some factors determining when instructions will be read. *Ergonomics*, 25:225–237.

Wright P, Barnard P. 1975. Just fill in this form: A review for designers. *Applied Ergonomics*, 6:213–220.

Wyon DP. 1974. The effects of moderate heat stress on typewriting performance. *Ergonomics*, 17:309–318.

Yamada H. 1970. *Strength of Biomaterials.* Williams and Watkins, Baltimore.

Yang KH, King AI. 1984. Mechanism of facet load transmission as a hypothesis for low-back pain. *Spine*, 9:557–565.

Young LR. 1973. Human control capabilities. In *Bioastronautics Data Book,* 2d edition, edited by Parker and West. NASA SP-3006, 751–805.

Young RM. 1981. The machine inside the machine: Users' models of pocket calculators. *International Journal of Man-Machine Studies*, 15:51–85.

Young SL. 1990. Comprehension and memory of instruction manual warnings: Conspicuous print and pictorial icons. *Human Factors*, 32:637–649.

Zacharkow D. 1988. *Posture: Sitting, Standing, Chair Design and Exercise.* Thomas, Springfield, IL.

Zihlman A, Brunker L. 1979. Hominid bipedalism: Then and now. *Yearbook of Physical Anthropology*, 22:132–162.

Zipp P, Haider E, Halpern N, Rhomert W. 1983. Keyboard design through physiological strain measurements. *Applied Ergonomics*, 14:117–122.

NAME INDEX

Adams, MA, 44, 58, 61, 62
Addison, JL, 333
Aghazadeh, F, 130
Agnew, J, 177
Agou, S, 146
Allen, AA, 241
Allen, JR, 240
Allport, DA, 329, 331
Alnutt, MF, 24
Altman, I, 92
Anderson, GBJ, 58, 61, 65, 105
Anderson, JR, 339, 341
Anderson, LE, 286
Anderson, N, 137
Anderson, R, 126
Appelby, M, 389
Argyle, M, 91, 92
Armstrong, TJ, 128, 134–135, 137, 154
Ashby, P, 71, 73
Aspden, RM, 36, 160
Astrand, PO, 186, 189, 196, 206, 245
Ayoub, MA, 166
Ayoub, MM, 110, 112
Azer, NZ, 241

Baddeley, A, 337, 420
Baggett, D, 408

Baigent, ML, 126
Bailey, RW, 4, 312, 315, 336, 387, 392
Bain, L, 473
Bainbridge, L, 323, 383, 390, 430, 433
Ballal, MA, 214
Ballantine, M, 372, 380, 415
Bandyopadhyay, B, 214
Banerji, D, 221–222
Banforth, KW, 14
Barber, P, 380, 398
Barnard, P, 403, 412–414
Barnes, RM, 204, 462
Barnett, BJ, 351
Bartholomae, RC, 309
Bartlett, FC, 335, 395
Barton, NJ, 49, 138, 152
Battie, MC, 165
Beard, DV, 360
Beaton, RJ, 313, 314
Becker, WGE, 212
Begault, DR, 314
Beiers, JL, 309
Beiring-Sorensen, F, 112, 218
Bejjani, FJ, 162
Bell, CR, 238
Bendix, T, 112, 132
Biedermann, HJ, 57

SUBJECT INDEX